Mauna Loa Revealed

Structure, Composition, History, and Hazards

Geophysical Monograph Series

Including

IUGG Volumes

Maurice Ewing Volumes
Mineral Physics Volumes

GEOPHYSICAL MONOGRAPH SERIES

Geophysical Monograph 92

Mauna Loa Revealed

Structure, Composition, History, and Hazards

J. M. Rhodes
John P. Lockwood

Editors

 American Geophysical Union

A contribution to the Decade Volcano Program of the International Association of Volcanology and Geochemistry of the Earth's Interior (IAVCEI), a part of the International Decade for Natural Disaster Reduction (IDNDR).

Published under the aegis of the AGU Books Board

Library of Congress Cataloging-in-Publication Data

Mauna Loa revealed : structure, composition, history, and hazards
    / J. M. Rhodes and John P. Lockwood, editors.
        p.   cm. — (Geophysical monograph ; 92)
    Includes bibliographical references.
    ISBN 0-87590-049-6 (alk. paper)
    1. Mauna Loa (Hawaii Island, Hawaii)   I. Rhodes, J. M.
II. Lockwood, John P.   III. Series.
QE523.M38M38   1995
559.69'1—dc20

95-42029
CIP

ISBN 0-87590-049-6
ISSN 0065-8448

Printed in the United States of America.

CONTENTS

CONTENTS

PREFACE

Mauna Loa is a volcano of superlatives: it is the largest active volcano on Earth and among the most productive. This volume serves to place on record the current state of our knowledge concerning Mauna Loa at the beginning of the *Decade Volcano Project*. The scope is broad, encompassing the geologic and exploratory history of the volcano, an overview of its submarine geology, its structure, petrologic and geochemical characteristics, and what Mauna Loa has to tell us about the Hawaiian mantle plume; it covers also remote sensing methods and the use of gravity, seismic and deformational studies for eruption monitoring and forecasting, hazards associated with the volcano, and even the importance of a changing volcanic landscape with a wide spectrum of climate zones as an ecological laboratory.

We have made a deliberate effort to present a comprehensive spectrum of current Mauna Loa research by building on a December 1993 symposium at the AGU Fall Meeting that considered (1) what is currently known about Mauna Loa, (2) critical problems that need to be addressed, and (3) the technical means to solve these problems, and by soliciting contributions that were not part of the symposium. We encouraged authors to consider how their papers relate to others in the volume through cross-referencing. The intent was that this monograph should be a book about Mauna Loa rather than a collection of disparate papers.

Every introductory-level geology student is probably aware that Mauna Loa is the largest active volcano on Earth and possibly also that it is the world's highest mountain. Although rising only 4 km above sea level, the submarine flanks descend to the north-central Pacific sea floor an additional 5 km. The sea floor of the Pacific plate, in turn, is isostatically depressed by Mauna Loa's great mass a further 8 km, resulting in a total height of the volcanic edifice of about 17 km. It covers over half the surface area of the Island of Hawaii ($>5,000$ km^2), and recent estimates of its volume (Garcia et al. and Lipman, this volume) have nearly doubled the previous estimates of about 45,000 km^3 to over 75,000 km^3.

Mauna Loa is not only Earth's largest volcano, it is also among the most active, having erupted more than 30 times since its first well-documented historical eruption in 1843.

The last eruption occurred in 1984, and another is anticipated before the end of this century (Decker et al. and Ryan, this volume). Mauna Loa was exceptionally active in the middle of the 19th century (1843-1887) at a time when its more diminutive sibling, Kilauea Volcano, was attracting visitors to Hawaii. As a consequence, Mauna Loa figured prominently in the formative years of the science of volcanology, along with Kilauea, Vesuvius, Etna, the Puy du Dôme region of France, and the Eifel region of Germany. It was on Mauna Loa that initial concepts of flow morphology, lava tubes, caldera collapse, and even the establishment of volcano observatories were discussed and formulated (Barnard, this volume), invariably finding their way into the *American Journal of Science* through the continuing enthusiasm for Mauna Loa and Hawaiian volcanism of J. D. Dana. Although in recent years Mauna Loa has not erupted with the frequency of Kilauea, the volumetric lava production rates, on average, are very similar (about 0.02 km^3/year). Consequently, Mauna Loa eruptions tend to be much larger, resulting in extensive lava flow fields that cover the volcano's surface at a rate of about 40% per 10^3 years, with flows that travel many kilometers from the vents (five have reached the coast since 1868). It is these frequent, high-volume, far-travelling lava flows that pose the principal hazard from Mauna Loa. They are a threat to people and their property on the lower slopes of the volcano and along the coast. Although the population at risk is relatively small (about 75,000), the recent rapid growth of the tourist industry has placed many expensive developments in harm's way from future eruptions. Additionally, the recent recognition of the extent of enormous prehistoric submarine and subaerial landslides from the south and west flanks of Mauna Loa (see Moore and Chadwick, this volume, for an overview) constitute a long-term hazard. These infrequent giant landslides and their attendant tsunamis pose perhaps the worst-case scenario for volcanic disaster on Mauna Loa and at many other oceanic volcanoes around the world.

These attributes and concerns lay behind the selection of Mauna Loa, along with 14 other volcanoes from the world's 500 or so active volcanoes, for special study as part of the *Decade Volcano Project* by the International Association of Volcanology and Chemistry of the Earth's

Interior (IAVCEI). This status provides increased opportunities for multi-disciplinary and multi-national efforts to improve understanding of volcanic processes and monitoring at this magnificent volcano. Unlike other designated volcanoes, however, the principal importance of Mauna Loa to the *Decade Volcano Project* derives not from the magnitude of the immediate associated risk, but from its potential as a laboratory to investigate processes that must be understood in order to reduce the volcanic risks associated with basaltic shield volcanoes and large volume lava flows. Again unlike some of the other *Decade* volcanoes, Mauna Loa is likely to erupt before the end of the project, providing a test of ideas concerning volcanic processes and procedures for monitoring and forecasting eruptions. The excellent exposures and accessibility of the volcano make past and future lava flows ideal for detailed study. Mauna Loa has the best-documented prehistoric eruptive chronology of any volcano (Lockwood, this volume), making it an ideal testing ground for models of volcano genesis and evolution. The records of instrumental monitoring over the past half century by the Hawaiian Volcano Observatory are an important resource upon which to build, and its present instrumental network makes it one of the better-monitored of all *Decade* volcanoes.

J. M. Rhodes
University of Massachusetts
Amherst

John P. Lockwood
U.S. Geological Survey
Hawaii Volcanoes National Park

Editors

ACKNOWLEDGMENTS

The editors would like to thank the authors and co-authors for their excellent contributions, and for their punctuality and persistent good humor in the face of ever-pressing deadlines. We all owe a debt of gratitude to the reviewers, too numerous to mention, who provided thoughtful reviews in a timely manner, and did their best to keep everyone honest.

Finally, we would like to dedicate this volume to the memory of Maurice and Katia Krafft, who, like so many of us, cherished this splendid volcano. The Kraffts studied and photographed erupting volcanoes all over the world; tragically, they were killed by a hot ash flow while photographing an eruption in Japan in 1991. It is fitting that Katia's evocative photograph of the 1984 lava flow, *Pele Dancing*, graces the cover of the monograph. This image has been reproduced as a 45 x 60 cm poster and is available to contributors to the Maurice and Katia Krafft Memorial Fund at the Center for the Study of Active Volcanoes, c/o Geology Department, University of Hawaii at Hilo, 200 West Kawili Street, Hilo, Hawaii 96720. This fund will provide scholarships so that young volcanology students and technicians may study volcano monitoring and hazards mitigation techniques at the Center.

Mauna Loa Volcano: Historical Eruptions, Exploration, and Observations (1779-1910)

Walther M. Barnard

Department of Geosciences, SUNY College at Fredonia, Fredonia, New York

The period 1779-1910 spans the years from the first Western contact with the Island of Hawaii through the decade immediately preceding the establishment of the Hawaiian Volcano Observatory and the beginning of systematic scientific investigation of Hawaii's volcanoes. During this period Mauna Loa apparently erupted a minimum of 30 times. Many of those eruptions were visited and described by non-scientists, foremost among whom was the Reverend Titus Coan. Some highlights in the exploration of Mauna Loa include the first recorded attempt to ascend the summit, in 1779, by John Ledyard; the first recorded successful ascent, in 1794, by Archibald Menzies; the ascent in 1834 by David Douglas, whose "incredible" description provoked considerable controversy; the topographic mapping of the summit area by the U. S. Exploring Expedition under Charles Wilkes in 1840-41; and the first ascents by women in 1873. Among the many contributions made to the emerging science of volcanology were Coan's recognition of lava tubes as conduits of lava, and his recognition of the processes of formation of tumuli and lateral outbreaks of lava; Dana's precepts on a variety of topics relating to Hawaiian volcanism; W. D. Alexander's and Haskell's descriptions of pahoehoe and aa lava, and early accounts of flow morphology and structure; Dutton's recognition that the calderas of Kilauea and Mauna Loa resulted from collapse; Green's description of formation of pillow lavas; and Guppy's suggestion and justification for the establishment of an observatory that subsequently led to the founding of the Hawaiian Volcano Observatory. Many contributed insights to the continuing debate on the relationship between Mauna Loa and Kilauea.

INTRODUCTION

This chapter is an abridgment of the first volume of a comprehensive three-volume history of the historical eruptions and exploration of Mauna Loa Volcano covering the years 1778 through 1991 [*Barnard*, 1990, 1991a, 1992].

The volcanic nature of Mauna Loa was well-known to the Polynesian settlers of Hawaii, who had recognized basic volcanic relationships and the overall age relationships of the Hawaiian islands. It was also apparent to the first Europeans to visit the Island of Hawaii, sailing under the command of Captain James Cook in 1779 [*Ledyard*, 1963 ed., pp. 122-123; *Cook and King*, 1784, III, p. 104]. The volcano's first eruption in historical times occurred shortly thereafter [*C. H. Hitchcock*, 1911, p. 80]. In the more than two centuries since its

Mauna Loa Revealed: Structure, Composition, History, and Hazards
Geophysical Monograph 92

"discovery" by Europeans, Mauna Loa (both the world's largest active volcano and largest single mountain) has erupted at least 39 times: from 1832 through 1950, on the average, every 3.6 years and about 6.2 percent of the time. Since 1950 it has erupted only twice, in 1975 and in 1984. Its *recorded* historical eruptions are listed in Table 1.

ERUPTIONS, EXPLORATION AND OBSERVATIONS

Earliest Recorded Attempted Ascent

Prior to the landing of Cook, Hawaiians roamed over Mauna Loa and may even have climbed to the summit, but they left no written record or other evidence of having been in the summit area. In January 1779, during Cook's first visit to Kealakekua Bay, the tropical snow-capped "Colossus of the Pacific" captured the attention of the men aboard the *Resolution* and the *Discovery*. With Cook's approval, a party was organized to attempt an ascent to the summit. On the 26th, John Ledyard, an American corporal of marines aboard the *Resolution*, and three companions, Robert Anderson, David Nelson, and Simeon Woodruff, started off with some Hawaiians who served as porters and guides. (In his memoirs, *John Rickman* [1781, pp. 306-307] referred to the attempted ascent as being undertaken by "Mr. Nelson and four other gentlemen"; Ledyard was not identified by name.) Two days later their way was blocked by impenetrable thickets. The group abandoned their goal and arrived back at Kealakekua Bay the evening of the 29th. Ledyard overestimated that they had "penetrated about 24 miles and we suppose within 11 of the peak...According to an attitude of the quadrant, the Peak of Owyhee [Hawaii] is 35 miles distant from the surface of the water, and its perpendicular elevation nearly 2 miles" [*Ledyard*, 1963 ed., p. 123]. The summit is actually only 20 miles east of Kealakekua Bay, and its elevation is approximately 13,677 feet.

Earliest Historical Eruptions

The earliest eruption of Mauna Loa within the time of written history appears to have occurred about 1780, according to testimony by Keaweehu, the old guide in Wilkes' party which occupied the summit area in 1840-1841: Mokuaweoweo, the summit caldera, "emitted fire not long after Cook's visit, and again five years since, on the north side" [*Wilkes*, 1845, IV, p. 150].

The Ascent of Menzies

The English expedition (1791-1795) under the command of Captain George Vancouver made three visits to the Hawaiian Islands (March 1792, February-March 1793, and January-

TABLE 1. Historical Eruptions of Mauna Loa

| Date of commencement | | [a]Duration (days) | | | | | | | |
Year	Month and day	Summit eruption	Flank eruption	Location of principal outflow	[b]Altitude of main vent (m)	Repose period (mo)	[c]Area of lava flow (km^2)	[c]Volume of lava (million m^3)
1832	June 20	~21	(?)	Summit	(?)3960	--	--	--
1843	Jan. 10	5	~90	North flank	2990	126	45	202
1849	May	~15	0	Summit	3960	73	5	25
1851	Aug. 8	4	0	Summit	3960	26	12	35
1852	Feb. 17	1	20	Northeast rift	2560	6	33	182
1855	Aug. 11	<1	~450	Northeast rift	(?)3200	41	66	280
1859	Jan. 23	<1	~300	North flank	2800	26	[d]91	[d]383
1865	Dec. 30	~122	0	Summit	3960	73	5	50
1868	Mar. 27	<1	[e]4	Southwest rift	1010	23	[d]24	[d]123
1871	Aug. (?)	[f]?	0	Summit	3960	18	3	20
1872	Aug. 9	[g]60	0	Summit	3960	--		
1873	Jan. 6	2(?)	0	Summit	3960	[h]__		
1873	Apr. 20	547	0	Summit	3960	[h]__	5	630
1875	Jan. 10	30	0	Summit	3960	[h]__		
1875	Aug. 11	7	0	Summit	3960	[h]__		
1876	Feb. 13	Short	0	Summit	3960	[h]__		
1877	Feb. 14	<1	<1	West flank	-55+	[h]__	[d]1	[d]8
1879	Mar. 9	<1	0	Summit	3960	24	1	1
1880	May 1	6	0	Summit	3960	14	5	10
1880	Nov. 5	0	280	Northeast rift	3170	6	51	130
1887	Jan. 16	<1	~12	Southwest rift	1740	65	[d]29	[d]128
1892	Nov. 30	3	0	Summit	3960	68	3	12
1896	April 21	16	0	Summit	3960	41	5	25
1899	July 1	4	~20	Northeast rift	3260	38	23	81
1903	Sept. 1	<1	0	Summit	3960	49	1	3
1903	Oct. 6	63	0	Summit	3960	1	5	70
1907	Jan. 9	<1	15	Southwest rift	1890	37	28	121
1914	Nov. 25	48	0	Summit	3960	94	5	55
1916	May 19	0	12	Southwest rift	2260	16	17	31

TABLE 1.   Historical Eruptions of Mauna Loa

Date of commencement		Duration (days)[a]						
Year	Month and day	Summit eruption	Flank eruption	Location of principal outflow	Altitude[b] of main vent (m)	Repose period (mo)	Area[c] of lava flow (km²)	Volume[c] of lava (million m³)
1919	Sept. 26	<1	38	Southwest rift	2350	40	[d]28	[d]183
1926	Apr. 10	<1	14	Southwest rift	2320	77	[d]35	[d]121
1933	Dec. 2	17	0	Summit	3960	91	6	100
1935	Nov. 21	6	40	Northeast rift	3690	23	33	87
1940	Apr. 7	134	0	Summit	3960	51	13	110
1942	Apr. 26	2	13	Northeast rift	2800	20	34	176
1949	Jan. 6	145	0	Summit	3960	79	22	116
1950	June 1	1	23	Southwest rift	2440	12	[d]112	[d]376
1975	July 5	1	0	Summit	3960	181	13	30
1984	Mar. 25	<1	22	Northeast rift	2860	108	48	220
TOTAL							806	4124

Modified from *Barnard* [1990, p. 10; after *Lockwood and Lipman*, 1987, p. 518, *Macdonald et al.*, 1983, pp. 64-65, and *Stearns*, 1985, pp. 160-161].

[a]The duration for most of the eruptions prior to 1903 is only approximate. Heavy columns of fume at Mokuaweoweo, apparently representing copious gas release accompanied by little or no lava discharge, were observed in January 1870, December 1887, March 1921, November 1943, and August 1944. They are not indicated in the table.

[b]All eruptions in the caldera are listed at 3960 m altitude, although many of them were a little lower.

[c]Areas and volumes are from *Lockwood and Lipman* [1987, p. 518].

[d]Area above sea level. The volume (in millions of cubic meters) below sea level has been estimated by *Lockwood and Lipman* [1987, p. 518] as follows: 1859--95; 1868--50; 1877--7; 1887--10; 1919--100; 1926--5; 1950--70. These are included in the volumes given in the table.

[e]Flank eruption started April 7.

[f]Eruption was still in progress on Oct. 22, 1871--82 days after commencement [*Coan*, 1871].

[g]Activity in the summit caldera may have been essentially continuous from August 1872 to February 1877. Only the most violent activity was visible from Hilo.

March 1794). Aboard the *Discovery* as naturalist was the distinguished botanist Archibald Menzies.

Prior to his undertaking the first known successful ascent of Mauna Loa, Menzies made two unsuccessful attempts to ascend from the western side. His first attempt began on February 25, 1793. After traversing a distance which he overestimated at 16 miles from Kealakekua Bay, his party's way, like that of Ledyard's, was blocked by impenetrable underwood and ferns. The mission was abandoned. [*Menzies*, 1907; 1920, pp. 73-86.]

Menzies' second unsuccessful attempt was in January 1794. (W. F. Wilson, editor of *Menzies* [1920], erroneously subtitled this event as "First Attempt To Climb Mauna Loa," after having presented the account of the actual first attempt earlier.) From Kealakekua Bay, his party first ascended Hualalai Volcano, descended on the southeast side, and crossed onto the high plateau between it and Mauna Loa. Progress of those selected

to reach the summit was halted on the 22nd by dense growth of shrubby wood and ferns and by rough ground. [*Menzies*, 1920, pp. 163-168.]

Though thwarted twice, Menzies successfully accomplished his goal in February 1794. After consulting with Kamehameha I, the ruling chief, on the means and best route to gain the summit, Menzies and his party set out on February 6 in the King's double canoes for a small village near the island's south point, from where they traveled overland northeast to Kapapala and, on the 13th, began the ascent up the southeast slope of Mauna Loa. En rôute, as the temperature dropped below freezing and snow was encountered, Menzies sent most of his Hawaiian porters down to an encampment at the edge of the forest to await his return. In the forenoon of the 16th, Menzies and his companions reached the southeast rim of Mokuaweoweo. Leaving one of the party, "Padre" Howell, and a few Hawaiians on the south side, Menzies, Third Lieutenant

Joseph Baker, Midshipman George McKenzie, and the unidentified Hawaiian who carried the barometer made their way to the highest point on the western rim of the caldera. Menzies determined its elevation as 13,634 feet, or 13,564 feet when corrected for temperature. The party did not have instruments to ascertain whether Mauna Loa or Mauna Kea was the higher, but Menzies correctly judged that the latter was, inasmuch as its summit was more whitened over with snow. They returned to the south side of the caldera to find that Howell and all but two of the Hawaiians had deserted them, whereafter they "set out on our return to the encampment where we were so fortunate as to arrive safe at ten at night, after the most persevering and hazardous struggle that can possibly be conceived." [*Menzies*, 1920, pp. 175-199; *C. H. Hitchcock*, 1911, pp. 63-79.]

Throughout the next century, Menzies' feat apparently was forgotten, except, perhaps, by the Hawaiians. This may have occurred because *Vancouver* [1798] in his own narrative of the expedition omitted any reference to Menzies' ascent (possibly because Menzies had refused to deliver up his journals upon formal demand at the end of the expedition), or because Menzies' own manuscript account was not readily accessible.

An Eruption of 1803 (?)

Mauna Loa apparently was in eruption again in early 1803. John Turnbull, in his narrative of a voyage around the world from 1800 to 1804, claimed that as he was leaving Kealakekua on January 21, he had a full view of some eruptions from the "volcanic center of the island of Owhyhee." Thus the eruption was either on the west or north side of Mauna Loa. [*C. H. Hitchcock*, 1911, p. 80.]

The Improbable Ascent of Goodrich

Several sources (*e. g.*, *Sparks* [1828, p. 100]; *Anonymous* [1838, p. 98]; Löwenstern [*Barnard*, 1990, p. 56]; *Whitney* [1875, p. 93]; and *Thrum* [1908, p. 134]) erroneously assert that the first ascent was made (as early as 1824) by Joseph Goodrich, an American missionary of the Second Company who arrived on the Big Island in 1823.

As late as 1832 *Goodrich* [1834, p. 201] claimed he had *not* climbed Mauna Loa but contemplated attempting to do so in the following January. Nothing further is heard of an ascent. It would appear reasonable to expect that if he had made the ascent, he would have followed through on his promise to report his observations to the editor of the *American Journal of Science*. Goodrich left with his family for the United States in 1836.

The misconception may have originated from the probability that on one of his climbs to the summit of Mauna *Kea*, Goodrich viewed Mokuaweoweo from there and stated to the effect that he looked into the crater; this was then misinterpreted by others as suggesting that Goodrich viewed Mokuaweoweo directly from the summit of Mauna Loa [*Barnard*, 1991b, pp. 60-61].

The Eruption of 1832

On June 20, 1832, the summit erupted and "continued burning for two or three weeks; the lava was also seen running out of the sides of the mountain, in different places; it discharged the red hot lava from so many vents, that it was seen on every side of the mountain; it was visible as far as Lahaina, upwards of one hundred miles..." [*Goodrich*, 1834, p. 201]. This eruption is commonly listed as Mauna Loa's first *recorded* historical eruption. No one ascended the mountain during the eruption, and to date the flows have not been positively identified.

The Ascent of Douglas

In January 1834, the Scottish botanist David Douglas, fired with a passion to collect plant specimens on the island's volcanoes, arrived at Hilo. There he hired a Hawaiian as his interpreter and guide, and porters to carry baggage. After visiting the summit of Mauna Kea, he traveled to Kilauea. His next goal was the summit of Mauna Loa. After making preparations and hiring guides at Kapapala, he set off on January 28 with his party of 13. The ascent was strenuous and slowed by heavy rains, mud, and rough lava. Near the summit area on the next day Douglas and the four men who accompanied him that far encountered deep snow. Only Douglas and his "trusty man" reached the edge of Mokuaweoweo and walked to the highest point on the eastern side of the caldera, which Douglas apparently (and mistakenly) believed to be the actual summit. He took several barometric and temperature readings. He also made some very inaccurate measurements and estimates of the caldera's dimensions and other faulty observations. (Douglas' "incredible account" generated controversy and doubt as to its authenticity; subsequently it was attributed by *Dana* [1890, pp. 58-59] and *Harvey* [1947, pp. 228-231] to his deteriorated state of health.) Their descent on the 29th and 30th was "even more fatiguing, dangerous, and distressing than the ascent had proved." [*Douglas*, 1838; 1914 ed., pp. 309-317; *Hooker*, 1836, pp. 158-177.]

As with Goodrich, the first ascent of Mauna Loa was erroneously attributed to Douglas by several individuals, including *Dana* [1849, p. 208; 1850, pp. 236-237; 1890, pp. 58-59], Remy [*Summers*, 1988, p. 49], and *Brigham* [1909, p. 63]. (See also *Barnard* [1991b, pp. 62-66].)

The Ascent of Löwenstern

On February 6, 1839, climbing from Kilauea, scientist M. Isidor Löwenstern reached the eastern rim of Mokuaweoweo. He described his excursion in a letter to the Royal Geographical Society which remained unpublished until transcribed by *Barnard* [1990, pp. 55-56].

Löwenstern described the caldera's features and stated that the "circumference of the present nearly circular Crater on the Top of Mouna Roa, is to my opinion no more than 2 miles, having from 1/2 to 3/4 of a mile in Diameter." *Wright and Takahashi* [1989, p. 140] concluded that because Löwenstern described an inactive circular crater no more than two miles in circumference from which the distance down to a black ledge was not more than 635 feet, he apparently did not reach the summit. "This is a confusing description. Although the ascent was of Mauna Loa, the description sounds like Kilauea."

Wilkes' party surveyed Mokuaweoweo in January 1841, and Wilkes specifically stated that there was no black ledge, as there was at Kilauea [*Wilkes*, 1845, IV, p. 159]. It is possible that Löwenstern confused the two calderas. Though undated, his account was written after, not during, the ascent, and a full year lapsed between the time of his ascent and the receipt of his letter by the Royal Geographical Society. Nevertheless, Löwenstern's description of a circular crater is consistent with that of the center of Mokuaweoweo caldera as it existed in 1840, and his "ledge" may have referred to the crescent-shaped benches that stood above the level of the central floor at that time [*Macdonald et al.*, 1983, pp. 60-62]. The dimensions of the caldera are admittedly underestimated, and may be due to the fact that he apparently lacked instruments to measure. He was giving, in his own word, an "opinion" only.

Wilkes and the U.S. Exploring Expedition

Following Löwenstern's ascent, Mauna Loa summit was occupied and mapped from late December 1840 to mid-January 1841 by U.S. Navy Lieutenant Charles Wilkes and the United States Exploring Expedition. The U.S. Exploring Expedition also made atmospheric and geophysical measurements. The undertaking, which required extraordinary exertion and unyielding perseverance, is recorded by *Wilkes* [1845]; a recent narration is by *Sprague* [1989].

On his trek around the caldera, and still in doubt about the relative heights of Mauna Loa and Mauna Kea, Wilkes set up his transit on the highest point on the western edge of Mokuaweoweo. He found the summit of Mauna Kea to be 193 feet higher than the place where he stood.

The topographic map of Mokuaweoweo produced by the U.S. Exploring Expedition portrays relief not by contours, but by hachuring, in which lines of varying width simulate shading and texture to produce an illusion of sloping land. Unfortunately,

the technique doesn't convey the scientific precision upon which the Expedition's maps are based.

Dana [1890, pp. 183-184] noted two problems with Wilkes' map. Seven small cones are portrayed on the floor of the center of Mokuaweoweo, although only one cone was reported by Dr. G. P. Judd and Lieut. Henry Eld, who had actually descended onto the caldera floor. Judd's description [*Wilkes*, 1845, IV, p. 152] is probably just incomplete, and Wilkes' reference to Eld [p. 156] states only "the bottom of the crater he found much as Dr. Judd had described it...." The second problem is that the NNE-SSW axis of the caldera is misoriented, being depicted as running NNW-SSE. The error may have been made during the compilation or engraving processes. Forty years earlier, *Dana* [1850b, p. 235] himself described the long axis of the caldera as "lying in a north-by-west and south-by-east direction."

Mokuaweoweo

In 1840-1841 the center of the caldera was a deep, nearly circular pit, whose floor was nearly 900 feet below the highest point of the western rim. Crescent-shaped benches or platforms, about 200 feet above the central floor, stood at the NNE and SSW ends. Throughout the later part of the century the central basin gradually filled. Lava of the 1914 eruption flooded over the northern bench, which was completely buried by lavas of the 1933 and 1940 eruptions. The southern bench lost its identity in 1949 when lavas completed the burial of its northward-facing scarp, already partly buried by the lavas of 1933 and 1940. [*Macdonald and Abbott*, 1970, p. 55; *Macdonald and Finch*, 1949.]

The summit caldera has changed markedly since it was first mapped by Wilkes. *Jaggar* [1931] gave a history of the changes and depicted modified topographic maps made by Wilkes in 1841, Lydgate in 1874, J. M. Alexander in 1885, and B. Friedlaender in 1896, in addition to the 1926 map by E. G. Wingate.

The Eruption of 1843 and the Reverend Titus Coan

On January 9, 1843, clouds of smoke near the summit of Mauna Loa were followed the next day by a brilliant light: the eruption of 1843 had begun. Shortly thereafter lavas were discharged from the volcano's north flank at approximately 9800 feet elevation and flowed toward Mauna Kea, dividing at the saddle between the volcanoes into two broad streams, one flowing toward Waimea and the other toward Hilo. A third stream flowed toward Hualalai. This flank eruption lasted approximately 90 days.

The outbreak and early phases were described by Dr. *Seth Lathrop Andrews* [1843], the missionary physician stationed at Kailua, who was apparently in Hilo at this time. During the eruption, Kilauea was in its usual active condition, in constant undisturbed ebullition. Because Kilauea caldera was unaffected by the Mauna Loa eruption, Andrews concluded that there was no great central reservoir of fire beneath the island connecting the two volcanoes "which many have supposed to exist under Hawaii" (according to the editor of the *Missionary Herald*). Andrews reasoned that Kilauea, at the lower elevation, would have been filled to overflowing before Mauna Loa erupted.

The principal chronicler of Hawaiian volcanic activity for the four decades from 1843 was the Reverend Titus Coan (1801-1882). Coan had entered the service of the American Board of Commissioners for Foreign Missions and arrived in Hilo in July 1835. The larger part of his communications were published in the *American Journal of Science*, edited by James Dwight Dana, whom Coan had met in 1840. Notes on some of the eruptions were published also in his autobiography, *Life in Hawaii* [Coan, 1882], and in the Honolulu newspapers.

On March 6, 1843, Coan, with the Rev. John Davis Paris of Waiohinu and seven Hawaiians, set out to ascend to the source of the eruption of that year. It was the first (partial) ascent of Mauna Loa by either of these missionaries [Coan, 1882, p. 270]. Their route took them up through the forests above Hilo to the saddle area between Mauna Loa and Mauna Kea via the rocky bed of the "River of Destruction" (Wailuku River) and thence up toward the summit of Mauna Loa.

Coan [1843] briefly described his excursion to the source and witnessed lava through "openings in the superincumbent stratum" being "conveyed down the side of the mountain in a subterranean duct from fifty to one hundred feet below the surface, at the rate of fifteen or twenty miles an hour." Later, in describing the excursion in more detail, he coined the word *pyroduct* for what is now referred to as a "lava tube" [Coan, 1844]. Slightly edited versions of Coan's descriptions are reproduced in the *American Journal of Science* [Dana, 1850b], where Dana commented on his own surprise at the relative quietness of the outbreak.

The Eruption of 1849

A brilliant and lofty column of light on Mauna Loa was seen for two or three weeks in May 1849. *Coan* [1851] reported no outflow of lava or earthquakes. The eruption immediately followed action in Halemaumau at Kilauea and may have been coincident. Coan could not determine whether there was any connection between the two eruptions.

The Eruption of 1851

A brief flow began a few miles west of the summit on August 8, 1851. No earthquake was reported. One source [Anonymous, 1851] reported the eruption as having continued 12 days; *Coan* [1852a, p. 396] stated "the eruption continued

but three or four days". Both *Dana* [1890, p. 186] and *C. H. Hitchcock* [1911, p. 86] gave a three- or four-day duration. Without offering an explanation for a longer period, *Stearns and Macdonald* [1946, p. 79] listed a 21-day duration, repeated in their later works [Stearns, 1985, p. 160; Macdonald et al., 1983, p. 64].

Coan [1852a] wrote on October 1, 1851, that two gentlemen had ascended the mountain from Kaawaloa until they reached the point where the stream terminated but had unfortunately not explored the stream to its source; another two (Mr. Sawkins, an English artist, and a Mr. Grist) had left Hilo 10 days previously for Kilauea and Mauna Loa, and probably were then on the summit to visit and sketch the late eruption. No connection was noticed between Kilauea and the eruption.

Eventually, on June 25, 1855, *Sawkins* [1855] read a paper to the Royal Geographic Society of London in which he described the summit of Mauna Loa and concluded that the pit crater south of South Bay ("Pohakuhanalie") was the source of the 1851 eruption.

In 1864 William T. Brigham visited this flow. He noted that the eruption broke out about 1000 feet below the summit and estimated that the flow, asserted to be mostly pahoehoe, was 10 miles long, but less than a mile in average width [Brigham, 1909, p. 65]. (Lockwood mapped the flow during the summer of 1994 and claims it to be greater than 95 percent aa [Lockwood, personal communication, 1995].)

The Eruption of 1852

A one-day summit eruption on February 17 heralded the eruption of 1852. On the 20th, the eruption renewed from a "curtain of fire" along the northeast rift between 8,200 and 8,600 feet. The escaping lava rose in a lofty fountain and flowed eastward towards Hilo for 20 miles. Letters from both *Coan* [1852b] and his first wife [F. C. Coan, 1852] announced the eruption.

Coan [1852c] also described the outbreak of the eruption and his week-long trip to the source. Leaving Hilo on February 23, he reached the active crater in mid-afternoon on the 27th ahead of his guide, and revelled in "the beauty, the grandeur, the terrible sublimity of the scene." Noting that the eruption commenced at the summit, Coan remarked that "it would seem that the lateral [hydrostatic] pressure of the emboweled lava was so great as to force itself out at a weak point in the side of the mountain; at the same time cracking and rending the mountain all the way down from the summit to the place of ejection." Lava emerged from "an inclined subterranean tube" in fountains that reached heights of 400 or 500 feet.

Traveling from the Waiohinu-Honuapo area near the southern end of the island, J. Fuller and H. Kinney visited the source crater on March 3. *Fuller* [1852a, b] described the cone as 1000 feet in diameter, from 100 to 150 feet high, and sustaining a lava fountain which varied from 200 to 700 feet in height. *Kinney* [1852] also described the source crater and emphasized the accompanying whirlwinds, which he cited as being the most dangerous aspect of the volcano.

When Coan visited the inactive cone in July, accompanied by his son and six other youths, he found it to be "seventy-four chains in circumference at its base, nearly five hundred feet high, and has a crater four hundred or five hundred feet deep." The flow was half a mile wide, but at the base of the mountain it spread to a breadth of three or four miles; its length, "including its windings, may be forty or fifty miles, and it approached within about ten miles of Hilo" [*Coan*, 1853a]. In the following year *Coan* [1853b] described the eruption in verse.

The Rev. E. P. Baker of Hilo visited the scene of this eruption in July 1888. He noted the transition from aa to pahoehoe as the source was approached [*Baker*, 1889].

James Dwight Dana

It was to Dana (1813-1895) in his capacity as editor of the *American Journal of Science* that communications from Coan and others were sent. Dana had served for four years as mineralogist and geologist for the U.S. Exploring Expedition, commanded by Wilkes. (Although he visited Kilauea in 1840, 1841, and in 1887, Dana never ascended Mauna Loa.)

Dana was a principal in establishing modern volcanology. In his classic monograph on the geology of the Pacific [*Dana*, 1849], and in sections on Mauna Loa and Kilauea derived from it for publication in the *American Journal of Science*, *Dana* [1850a, 1850b] made observations and drew conclusions on a variety of topics relating to Hawaiian volcanism, including the origin of the Hawaiian Archipelago, the structure of volcanoes, the relationship of steepness of cones to their method of formation, the relative quietness of eruptions, the fluidity of lavas, the causes of summit and flank eruptions and of summit collapse, and the relationship between Mauna Loa and Kilauea. Many, but not all, of his precepts are consistent with current views.

Dana [1852] again commented on some of these topics in light of the 1852 eruption which he believed confirmed his earlier views. He still entertained notions, however, now considered inaccurate (*e. g.*, that volcanic action is caused by an influx of surface water through channels to the magma source), and he disputed Coan's observations that lavas could flow over long distances (as it actually does in lava tubes) without additional source fissures along their courses.

Remy and Brenchley

In June 1853 French scientist Jules Remy and his English companion Julius Brenchley climbed to the summit area of Mauna Loa. *Remy's* [1892] account, in French, was not published until almost 40 years later; an English translation by *Summers* [1988] finally appeared 135 years after the excursion.

Remy and Brenchley set out from Kilauea with four Hawaiians on June 15 (but released one on the next day). When they reached the edge of Mokuaweoweo at 1 P.M. on the 17th, Remy was disappointed: he had hoped "to be able there to look out over an enchanted archipelago, over an unlimited wave", but ended up "on a narrow, snow-covered, monotonous plateau". In descending into Mokuaweoweo, Remy accidentally slid and found himself, unharmed but with a good fright, at the base of the precipice. He explored the floor of the caldera, and, with Brenchley, the neighborhood of Mokuaweoweo. After four hours at the summit, they descended to Kilauea. Throughout the excursion Remy carefully noted the fauna and flora they encountered.

The Eruption of 1855-1856

On August 11, 1855, began one of the greatest flows seen by modern observers. It lasted 16 months. From its source on Mauna Loa's northeast rift zone along fissures extending from 10,000 to 12,000 feet elevation, the lava flowed directly towards Hilo, causing much anxiety for the safety of the town.

Coan [1856a] left Hilo via the channel of the Wailuku River on October 2 with Lawrence M'Cully and four Hawaiians, and reached the eruption source on the 6th. En rôute, through numerous windows in a lava tube, they saw lava rushing along; in one place, uncovered for a distance of 30 rods, the lava flowed at 40 miles per hour. Coan estimated that the principal stream (there were many lesser and lateral ones), including all its windings, was 60 miles long; average breadth, three miles; and depth, from three to 300 feet, varying with the surface over which it flowed. The party then went to Kilauea.

Both at Kilauea and Mauna Loa, Coan measured angles of slopes down which lava flowed and cited an extreme value of 95° ("flowing down a rock or bank until it came to where said rock *retreated*, it would follow the inward curve in a thin layer like molasses, adhering to the rock"), thus refuting the idea of some, including Dana, that lava flowed only on relatively gentle slopes.

Coan [1856b] also described the flow as he witnessed it on a trip undertaken on October 31-November 1 with a Mr. Ritson and three Hawaiians. They encamped at a point about two miles above the terminus, which Coan estimated was progressing at one mile per week. By the 20th, the flow had advanced another two miles and was estimated to be within eight miles of the shore.

Coan's description of the development of tumuli on top of the flow appears to be the first use of the term *tumuli* in its present meaning in reference to Mauna Loa volcanism: "At thousands of points on the solidified crust of the stream, the accumulating fusion, fed from above, was swelling and raising this superincumbent stratum into tumuli of endless form and size, and then, bursting open the cone or dome thus raised, either laterally or at the apex, when flowing off for several rods over the old substratum, the stiffening flood became solid." (In an earlier communication on this same eruption, *Coan* [1956a] had used the expression "smoking tumuli" to describe a series of cones along a great fissure.)

The Englishman *F. A. Weld* [1857] visited the 1855 flows on November 16 and provided a good description of the vents. In his hike from Kilauea to the eruption site and return, he wore out two pairs of English shooting-boots.

When Sereno Edwards Bishop visited and described the flow in January 1856, the terminus had slowed and advanced to about seven miles from the shore [*Bishop*, 1856]. In a letter dated January 26, *Coan* [1856c] noted the flow was "now about six miles from us, coming at the rate of a half mile a week, probably."

In still another letter, *Coan* [1856d] described the processes of formation of tumuli and lateral outbreaks of lava resulting from obstruction by hardening of lavas at and near its terminus, and the effect of hardening in reducing the rate of advance of lava upon Hilo. Coan's observations of flow dynamics are unparalleled for his time. They, like those of later studies on flow dynamics (*e. g., Lipman and Banks* [1987]), have important implications for assessing volcanic hazards.

Dana [1856] again took up the subject of volcanic action at Mauna Loa. He commented on Coan's observations and speculated on a connection between Mauna Loa and Kilauea. He also challenged Coan's observation that fissures feeding the flow were restricted to the source area, rather than underlying the entire length of flow. In his final remarks on the eruption of 1855-1856, *Coan* [1857] pressed his point that lava flowed in newly formed conduits (lava tubes), a fact that Dana apparently had not earlier understood correctly and had repudiated.

Coan's observations of "pyroducts" appear to be the first in establishing the importance of lava tubes in the emplacement of lava and in the evolution of volcanoes. Later, *Dutton* [1884] wrote, "Probably no great eruption takes place without the formation of several such tunnels...There are literally thousands of these tunnels throughout the mass of Mauna Loa...So numerous are these caverns that it seems as if they must form some appreciable part of the entire volume of the mountain." Recently, *Greeley* [1987] discussed the role of lava tubes in Hawaiian volcanoes in transporting lava between vents, along rift zones, and directly from vents to flow fronts, in addition to their critical role in the submarine extension of many flows; his own analysis found that 30 percent of the surface area of Mauna Loa is covered by flows spread at least partly via lava tubes.

The Eruption of 1859

On January 23, 1859, a brief summit eruption was followed by the 10-month-long north flank eruption that sent lavas north and northwestward to the ocean, destroying the village at Wainanalii about 12 miles south of Kawaihae.

An early account edited by *Dana* [1859] included narratives by Coan, the editorial writer of Honolulu's *Pacific Commercial Advertiser*, and the Rev. Lorenzo Lyons of Waimea, along with a map of the flow.

An Oahu party visited the active craters and flows during the second week of February and provided early accounts of flow morphology and structure. *W. D. Alexander* [1859] of Oahu College described distinctions between pahoehoe and aa lava, and formation of the latter: "The difference between 'pahoehoe' or smooth lava, and 'aa' or clinkers, seems to be due more to a difference in their mode of cooling than to any other cause." Alexander also provided Dana with a map giving the course of the lava that differed slightly from the one Dana had recently published (per note in *Haskell* [1859a, p. 71]). Prof. *R. C. Haskell* [1859a] provided details of movement of "pahoihoi" and aa, and noted construction of levees and melting of the channel bottom. On February 18, *J. H. Sleeper* [1859] visited the active cones; he witnessed pahoehoe flowing on the surface and in lava tubes but did not observe formation of aa.

Although it has been asserted, *e. g.,* by *Macdonald* [1972, p. 71], that the Hawaiian terms for the two common types of lava were first introduced into scientific literature by C. E. Dutton in 1884, *W. D. Alexander* [1859], *Haskell* [1859a], and *Sleeper* [1859] in fact used these very terms 25 years earlier, and while it may be argued that Alexander's and Sleeper's letters were not published in scientific literature (they were published in the *Pacific Commercial Advertiser*, a newspaper), Haskell's letter was published in the *American Journal of Science*. From the context of the articles published in 1859, it is clear that the terms were not being introduced and defined for the first time. Even earlier, *Dana* [1849, p. 167] stated "The party which ascended Mount Loa to its summit from the Vincennes, under the direction of Captain Wilkes, represent the country passed over as abounding in lava streams, *pahoihoi* and *clinkers*". Similarly, *Dana* [1850a] described tracts of lava at Kilauea as "pahoihoi" and "clinkers".

In June, on his second trip to the flow, Haskell made it to the summit and found no activity in Mokuaweoweo. He described a crack which extended four miles above the cones of the 1859 eruption and cited it as being the real source of the eruption [*Haskell*, 1859b]. On another trip, apparently in late October or early November, *Haskell* [1859c] noted that lava was still flowing into the sea at a velocity of two or three miles per hour at a point 40 miles from the source and at least 25 miles from the lowest point to which the fissure extended. *Coan* [1860] reported the destruction of the small village of "Kibele" (Kiholo) in November.

Geologist William Lowthian Green described the entry of lava into the sea at Wainanalii and the formation of spheroids (pillow lavas). He also offered an explanation for the detonations so common during the flow of lava from Hawaiian volcanoes as being the explosions of highly heated compressed air [*Green*, 1887].

Brigham and Others

In early August 1864, William Tufts Brigham, with Hilea *luna* (overseer) Horace Mann, made his first ascent of Mauna Loa, via a western route from Kaawaloa, and reported that Mokuaweoweo had varied little from the conditions described by Wilkes. He noted that when the U.S. Exploring Expedition was on Mauna Loa in 1841, the floor of Mokuaweoweo was rough, and contained "eight or ten cones", some of considerable height; now there were only two cones, each about 200 feet high, near the eastern wall; the whole bottom had been overflowed by fresh black lava, and appeared no rougher than an ordinary lava stream [*Brigham*, 1909, pp. 15-19].

J. L. Wisley, Charles Hall, and M. Worman ascended Mauna Loa in 1865 by the north side. There was a line of openings or gashes up the mountain along the line of the 1859 flow, as well as pumice and sand at the point of outburst. They described the shape of Mokuaweoweo as being like the figure 8. Descending to the floor, they found two steam holes on the west side [*C. H. Hitchcock*, 1911, p. 104].

The Eruption of 1865-1866

In late December 1865, light, which continued for four months, was seen at the summit. There are no records of an ascent during this eruption, nor was there any report of flowing lava on the mountain side [*Coan*, 1866, 1867].

The Eruption of 1868

In 1868 Mauna Loa erupted in Mokuaweoweo and on its southwest rift, generating an intense earthquake swarm that culminated in a shock of magnitude 8, the largest earthquake recorded in Hawaii within historical time [T. L. Wright, personal communication, 1994]. Kilauea also erupted, apparently in response to the earthquake.

A column of smoke over Mokuaweoweo was observed at Kawaihae and Kealakekua on March 27 and a day later at Hilo. A summit outbreak of lava was seen from Kau on the 30th. Earthquakes became pronounced at Kona, Kau, and Kilauea. Between March 28 and April 11, 2000 distinct shocks were felt in Kau. The culminating shock came on April 2. At Kilauea caldera, the great shock of April 2 was followed by another on April 4; lava receded from the caldera, followed by an eruption in the adjacent pit crater Kilauea-iki; large portions of the walls tumbled down; and more than half of the caldera floor subsided. Coincident with the April 2 earthquake were a landslide ("Mud Flow") which occurred between Kapapala and Keaiwa in Kau,

about 26 miles southwest of Kilauea, and a tsunami which washed the shore from Hilo to the South Cape, being most destructive at Keauhou, Punaluu, and Honuapo. Finally, came the flow of lava from Mauna Loa to Kahuku: on April 7 lava began pouring down from an opening at 5600 feet elevation, about ten miles distant from the shore, gushing out chiefly in a gash one mile long. People fled their homes and escaped, but 37 buildings were destroyed. The flow ceased after four days. [*C. H. Hitchcock*, 1911, supplement pp. 1-5, and revision by T. L. Wright, personal communication, 1994.]

Contributors to the *American Journal of Science* article edited by *Dana* [1868] include *Coan* [1868], *Lyman* [1868], *Whitney* [1868], and *Hillebrand* [1868]. Additional comments by Hillebrand on the Kahuku flow are given in *Brigham* [1909, pp. 111-112].

In August, following the eruptions of Kilauea and Mauna Loa earlier that year, Coan made a tour through Puna and Kau; his letters include descriptions of the eruption-related phenomena and their effects [*Coan*, 1869; *Brigham*, 1909, pp. 112-117].

The Eruptions in the 1870s

During the first two weeks of January 1870, smoke and steam rose from the summit caldera. On January 10, Judge David H. Hitchcock, in company with Dr. Hans Berag (or Beraz) and Lord Charles Hervey, rode on mules to the summit via Kapapala, the first time this feat had been accomplished. Steam issued from the banks and floor of Mokuaweoweo, but there were no indications of recent flows [*C. H. Hitchcock*, 1911, p. 111]. *Coan* [1870] provided a few details on the new route. On June 22, 1870, L. Severance, J. D. Brown and S. L. Austin reported conditions at the summit as being similar to those found by Hitchcock and his colleagues in January [*C. H. Hitchcock*, 1911, p. 111].

The *Pacific Commercial Advertiser* of September 9, 1871 [*Anonymous*, 1871a], reported the occurrence of smoke in Mokuaweoweo, and its issue of September 30 [*Anonymous*, 1871b] published a brief eyewitness report of an eruption lasting a few days in a crater about five miles southwest of Mokuaweoweo. According to both *Stearns* [1985, p. 160] and *Macdonald et al.* [1983, p. 64], this eruption broke out on or about August 1 and lasted 30 days. *Coan* [1871], however, stated that it was still active on October 22, and "From week to week and month to month it burns in solitude...." Inexplicably, this eruption is not mentioned in the historical summaries of eruptions by *Dana* [1890], *Brigham* [1909], and *C. H. Hitchcock* [1909, 1911].

Mokuaweoweo erupted again on August 10, 1872, and 17 days later *Coan* [1872] wrote that "still the great furnace is in full blast." Throughout the eruption, lava was confined to the caldera.

John M. Lydgate was at the summit during the latter part of August and reported a fountain of fire in its crater. An account [*Anonymous*, 1872, attributed to J. H. Black by *Bevens et al.*, 1988, I, p. 500] described both an ascent to the place of eruption and Mokuaweoweo caldera. The original article was headed "By our special correspondent" and ended with the name "Quirk", who may have been the special correspondent and author (Ruth Horie, Bishop Museum Library, personal communication, 1989). Other observers of the eruption noted in *C. H. Hitchcock* [1911, p. 112] are W. T. Conway, H. C. Dimond, G. M. Curtis and H. N. Palmer, and on September 8, a party of 13 men and a guide, who confirmed the existence of a 150-foot high fountain.

The first week of January 1873 ended with vigorous action in Mokuaweoweo, with light vapors visible at night rising thousands of feet above the summit. *Coan* [1873] observed the event from Hilo and noted that parties were planning a visit to the source, but the eruption suddenly ceased.

Again on April 20 Mauna Loa burst into eruption and continued in action for 18 months. *Coan* [1874] remarked that the "great marvel of this eruption is its duration." The Rev. A. F. White climbed to the summit in late May and saw the lava fountaining from 150 to 300 feet [*C. H. Hitchcock*, 1911, p. 112].

Riding mules and leading pack-horses from Ainapo, Isabella L. Bird, who represented herself as the second woman to ascend Mauna Loa, and William Lowthian Green reached Mokuaweoweo on June 6, 1873, and camped overnight. Bird's descriptions of the environs and the eruption are vivid--and her imagination occasionally ran wild (*e. g.*, "It is probable that the whole interior of this huge dome is fluid...") [*Bird*, 1974 ed., pp. 262-274]. Apparently the first woman to ascend Mauna Loa had done so within the past few months, based on a statement in the *Hawaiian Gazette* of September 3, 1873 [*Anonymous*, 1873], that Mrs. O. B. Adams was the third lady who had made the ascent (in early August) during the past year. *Green* [1887, pp. 165-168] described the same lava-fountaining activity in Mokuaweoweo as that described by Bird.

Early in August O. B. Adams and his wife ascended to the summit area and found a column of molten lava rising from 200 to 500 feet in height [*Anonymous*, 1873]. W. W. Hall, with his guide John B. Kitu, explored the floor of Mokuaweoweo on September 19; his account in the Volcano House record book includes a sketch map of the caldera [*Brigham*, 1909, pp. 122-123]. In October, E. G. and H. R. Hitchcock spent one night at the summit near the site of Wilkes' camp. They found a fountain of lava playing to a height of 600 feet in the southwestern end of the crater. The descending lava of the fountain, falling into the basin, flowed off northward nearly the whole length of the western side of the pit [*Merritt*, 1889, p. 52].

On June 24, 1874, John M. Lydgate drew a plan of Mokuaweoweo from actual survey by triangulation (see also modified map in *Jaggar* [1931]) and left notes in the Volcano House record. He reported that its greatest length including the basin at the north end was 17,000 feet, or about 3.2 miles; excluding the basin it was 15,000 feet; its greatest breadth was 8600 feet or about 1.7 miles; its greatest depth 1050 feet. The floor, however, was continually rising owing to repeated overflows, and the lake was about 500 feet in diameter [*Brigham*, 1909, p. 501]. In his chronology, *C. H. Hitchcock* [1911, p. 114] inserted this comment about Lydgate between that of the January 6, "1873" (should be 1874), letter of Coan and the August 27, 1873, ascent of the Adamses, thereby implying that Lydgate's map was drawn in 1873; later, however, in Hitchcock's account of the ascent of Dutton in 1882, he refers to the "Lydgate map of 1874" [*C. H. Hitchcock*, 1911, p. 120].

In another entry in the Volcano House record, dated September 9, 1874, B. F. Dillingham noted that volcanic activity in Mokuaweoweo had diminished [*Brigham*, 1909, pp. 124-125]. The activity probably ceased in mid-October.

In his summary of eruptions from 1873 through 1877, *Coan* [1877] referred to an eruption commencing in January 1874 and lasting for 15 months. Both the year and the duration present puzzles. If the eruption actually began in January 1874, there would have been two simultaneous summit eruptions, as it would have occurred while the eruption of April 20, 1873 to mid-October 1874 was in progress, and apparently continued through March 1875, after the earlier eruption had ceased. The beginnings and endings of these summit eruptions cannot possibly be distinguished from each other. More likely, Coan must have meant to state that the eruption began in January 1875, which would harmonize with the report of Green that Mokuaweoweo was in action in this month, and that the activity lasted several weeks. Yet Coan followed his comment on the January "1874" eruption with a statement that another summit eruption occurred on August 11, 1875 (with subsequent eruptions in February 1876 and February 1877). If of 15 months duration, a January 1875 summit eruption would have overlapped with that beginning in August of that year, again requiring distinction between two summit eruptions. No mention of Coan's reference to a January 1874 eruption, or the problems it presents, is made by *Dana* [1890], *Brigham* [1909], or *C. H. Hitchcock* [1911].

Of the August 1875 eruption, *Coan* [1877] wrote that he thought it commenced on the 11th, and continued for one week.

After a repose period of six months, Mauna Loa again erupted at the summit on February 13, 1876 [*Coan*, 1877]. The eruptions from 1873 through 1876 were all confined to the summit. The summit eruption that began on February 14, 1877, however, was followed by a submarine outpouring on

February 24. On land, new fissures opened in the mountain in a westward course toward the place of submarine disturbance [*Coan*, 1877]. *Whitney* [1877] witnessed this submarine eruption and described the associated events.

Mauna Loa's next eruption, at the summit on March 9, 1879, has long been overlooked or ignored. J. P. Lockwood (personal communication, 1989) re-discovered the following references to it: (1) In the *Volcano House Register*, under date of March 9, 1879, Walter Foote, of St. Ives, Huntingdonshire, England, and I. H. Postlethwaite and G. F. Postlethwaite, of Liverpool, England, entered the note, "Fire Seen on The Summit of Mauna Loa on the Evening of March 9th by the above", and Mrs. Geo. P. Gordon and H. DuBois Van Wycke (or Wyck), of Norwalk, Virginia, wrote on the same date, "Large and very bright light distinctly visible for over two hours on the summit of Mauna Loa, which seemed to increase in size and volume before being shut out by immense black clouds, which covered the whole top of the mountain"; (2) News of this eruption was embedded in a short article [*Anonymous*, 1879] noting a lava flow at Kilauea and earthquake shock at Kau: "There was some fire on the top of Maunaloa about two weeks ago, and all the people in Kau expect a lava-flow down there sooner or later." Apparently *Coan* [1879] was unaware of this eruption; only 12 weeks later he wrote, "Mokuaweoweo...has been quiet for a long time...."

A party was on the summit of Mauna Loa on April 6 and 7, 1880. "[A]s far as we could see, deep snow; walked across [the caldera?] more than 1 hour, got into the snow to our shoulders, saw the place of the crater [vent of recent eruption?] but couldn't look down into it..." [Bevens, D., transcriber, Entry for April 8, 1880, author unknown, *Volcano House Register*, 3 vol.: Hawaii National Park, Hawaii Natural History Association, unpublished transcription, II (1873-1885), unpaginated, per T. J. Takahashi, personal communication, 1995]. The caldera was inactive at the time.

The Eruptions of 1880-1881

A summit eruption broke out on May 1, 1880, and probably lasted six days [*Coan*, 1880]. *Brigham* [1888, p. 35] stated that his guide Ahuai was at the summit at the time of the outbreak, and estimated the height of the jets at 1000 feet, a figure that *Dana* [1890, p. 203] maintained was probably too high.

In late July, Brigham ascended Mauna Loa from Ainapo and compared the appearance of Mokuaweoweo then with its appearance during his earlier 1864 visit and with descriptions by Luther Severence in 1870 and John Lydgate in 1874 [*Brigham*, 1888; 1909, pp. 143-146].

On November 5, 1880, began a nine-month-long eruption which produced three principal flows. Initially, from a point about 11,100 feet elevation, at the extremity of a long fissure

where there is now the pit crater Puka Uahi, west of Puu Ulaula (Red Hill), a flow extended down the slope toward Mauna Kea. This source fissure followed a topographic divide and ran by the north side of Puu Ulaula. At nearly the same time, a second stream from the same source fissure, but at a lower elevation and diverted by the obstacle of Puu Ulaula, flowed southeastward into Kau. A third stream from the same source but at a still lower elevation began toward Hilo, flowing between the lavas of the 1852 and 1855 eruptions. The Kea stream stopped on the intermontain plateau east of Kalaieha, having a length of 10 to 12 miles; the Kau stream reached nearly the same length; and the Hilo stream came nearly giving Hilo a burial, stopping within three-quarters of a mile of Hilo.

The Rev. *E. P. Baker* [1889] described the source of these flows, which he examined in July 1888. A vivid description of the eruption appeared in *Harper's Weekly* [*Anonymous*, 1881b] and gave a general overview of the activity. Though undated, its contents reveal that it was composed sometime between May and the end of the eruption. Judge *D. H. Hitchcock* [1880a] of Hilo gave graphic descriptions of the outbreak and an account of an early visit to the flow. *Whitney* [1880] also provided details of the outbreak and progress in the first week. Coan [1881a] communicated the earliest report of this eruption to appear in the *American Journal of Science* and in July provided an update [*Coan*, 1881b]. *W. B. Oleson* [1880] made observations of the lava flow at the base of Mauna Loa in mid-November. Lady *C. F. Gordon Cumming* [1883] concluded her two-volume *Fire Fountains* with a chapter devoted to "The River of Fire of 1881--The Story of a Great Deliverance", in which she included abridged quotations from numerous eyewitnesses. The December 8, 1880, issue of the *Hawaiian Gazette* [*Anonymous*, 1880] carried the latest news of the eruption and a second letter of *D. H. Hitchcock* [1880b], who made another trip to the flow; Hitchcock's letter also contained some final comments which corrected observations by Oleson. The July 6, 1881, issue published a letter of *D. H. Hitchcock* [1881] and extracts of letters from *Coan* [1881c] and his second wife, *Lydia Bingham Coan* [1881]. The August 31, 1881, issue contained two descriptive letters written at or near the conclusion of the flow, one by an individual identified only as "N. B. E." [*Anonymous*, 1881a] and the other by *W. R. Castle* [1881]. In one of his final communications to the editor of the *American Journal of Science*, Coan [1881d] wrote of the cessation of the flow. (Coan died the following year at the age of 81.) Publication of notes by *E. D. Baldwin* [1953] on the 1880-1881 lava flow was delayed for seven decades.

In her book, *Hawaii's Story*, Liliuokalani [1964 ed.], Queen of Hawaii from 1891 to 1893, recalled this eruption. Her visit to the Big Island in 1880 during the reign of her brother, King David Kalakaua, coincided with the outbreak of that eruption.

She made a second trip in August 1881, during her regency while Kalakaua was touring Europe.

Finally, *G. H. Barton* [1884] described formation of tree molds, lava tubes, and stalactites and stalagmites.

Dutton, C. H. Hitchcock, E. P. Baker and J. M. Alexander

In the summer of 1882, Capt. Clarence Edward Dutton, Ordnance Corps, U.S.A., examined the source of the recent great eruption of 1880-1881 and twice visited Mokuaweoweo. Dutton gave both a brief description of his excursions in Hawaii [*Dutton*, 1883] and a more extensive report [*Dutton*, 1884]. He apparently was the first to use the term caldera for the summit depressions on Kilauea and Mauna Loa, whose formation he attributed, not to explosions as had others, but to collapse, probably in the middle and later stages of growth of the volcanoes [*Dutton*, 1884, pp. 142-143]. He also remarked on summit eruptions preceding flank eruptions.

In January 1883 Prof. Charles Henry Hitchcock, with F. J. Perryman of the Government Survey, followed the trails taken by Dutton to the sources of the 1880 flow near Puu Ulaula and to Mokuaweoweo, which at the time was blanketed in snow. He was later to author his last extended geologic work, *Hawaii and its Volcanoes* [*C. H. Hitchcock*, 1909, 1911]. The Rev. *E. P. Baker* [1885], successor to Coan in pastoral duties at Hilo, reached Mokuaweoweo from Ainapo on April 20, 1885, and spent the evening next to steam cracks on the caldera's floor. He too found the walls and floor of the caldera bedecked with snow.

In 1886, a recent survey of Mokuaweoweo by "Mr. Alexander, Surveyor General of the Sandwich Islands" was announced [*Anonymous*, 1886]. *Dana* [1887] referred to W. D. Alexander as the Surveyor General, and *C. H. Hitchcock* [1911, p. 57] identified W. D. Alexander as a Government surveyor. It appears, however, that the survey referred to is the one conducted in 1885 by the Rev. *J. M. Alexander* [1886a, b; 1888], notwithstanding the fact that figures for the length and breadth of Mokuaweoweo are not exactly the same in the two communications. Alexander's modified map is depicted in *Jaggar* [1931, p. 4].

Some suggestions made by *J. M. Alexander* [1886a] are no longer tenable: (1) that Haleakala on Maui may have originated by coalescence of several craters; (2) that lava rises higher in Mauna Loa than in Kilauea without overflow in the lower because of Mauna Loa's smaller conduit diameter; and (3) that volcanism is caused by the production of heat generated by Earth's contracting crust.

The Eruption of 1887 and an Ascent in 1888

Following the cessation of the 1880-1881 flank eruption, Mauna Loa remained in a non-eruptive state until January 16, 1887, when lava burst forth briefly near the small pit crater south of the summit caldera, and later, on the 18th, through fissures at about 6500 feet elevation on the southwest flank north of Kahuku in Kau. The lavas reached the sea at noon on the 19th and extended the shore outward up to 500 feet. By noon on the 24th the flow had nearly stopped, but some activity continued through the end of the month. The outbreak was accompanied by frequent and sometimes strong earthquakes, which began in the preceding month. Summary accounts of the eruption are provided by *Dana* [1887] and *Bishop* [1887b], and several issues of the *Hawaiian Gazette* contained communications from *Maby* [1887], *Paris* [1887], *Jones* [1887], *Martin* [1887], *Spencer* [1887], *D. H. Hitchcock* [1887], *McKay* [1887], and *Bishop* [1887a].

W. C. Merritt, president of Oahu College, and the Rev. Baker together ascended Mauna Loa in July 1888. *E. P. Baker's* [1889] notes and an abstract of *Merritt's* [1889] letter to Dana described the caldera. It is with this entry that *Dana* [1890; see also 1888] ended his summary account of Mauna Loa and Kilauea, upon which those of *Brigham* [1909] and *C. H. Hitchcock* [1909, 1911] drew heavily.

An Eruption (?) and Ascents in 1890

In July 1890 W. B. Clark, Julian Monsarrat, W. Gates, and L. A. Thurston, guided by Kanae of Ainapo, made an expedition to Mauna Loa summit. Thurston noted "There was evidence of recent eruption from a blow-hole about the centre of the crater which was still uncomfortably warm. The eruption was mostly of a dark pumice stone and a very thin black pahoehoe." [*Brigham*, 1909, pp. 174-175] *This eruption appears to have gone unnoticed by compilers of listings of historical eruptions of Mauna Loa.*

Ernest E. Lyman made an ascent in August 1890 and left a description of the summit area in a five-page letter which, although unaddressed, apparently was written to Dana, but was not published. The manuscript, now the property of the Lyman House Memorial Museum in Hilo, was brought to the attention of the writer by J. P. Lockwood and is reproduced in *Barnard* [1990, pp. 263-264].

The Eruption of 1892

An eruption lasting three days began on November 30, 1892 [*Brigham*, 1909, p. 185]. In late June 1893 Julian Monsarrat, E. P. Baker, and others noted that a small crater, 200 feet in diameter and 200 feet deep, on the floor of Mokuaweoweo had disappeared. Baker had seen it in 1888, as had Monsarrat in 1889, but now it had been obliterated by the infilling of lava [*Anonymous*, 1893].

The Eruption of 1896

Following a repose of 41 months, a summit eruption broke out on April 21, 1896, and was confined there for its entire duration of 16 days. Among the earliest visitors to this eruption were volcanologist Benedict Friedlaender of Naples and his party of four others, who ascended Mauna Loa via the western route taken by Brigham and Mann in 1864. *Friedlaender* [1896] described activity in Mokuaweoweo and closed his account with remarks on differences in activity between Mokuaweoweo and Kilauea, erroneously hypothesizing that the differences were due to a larger amount of gasses in the former volcano.

Immediately following the ascent of Friedlaender, a party of 13 made the ascent and spent April 29-30 at the summit. *F. S. Dodge* [1896] described the excursion and conditions at Mokuaweoweo, including the active lake at the southern end of the caldera. Another of the party, Daniel Logan, described the fire fountains [*C. H. Hitchcock*, 1911, pp. 129-130].

H. B. Guppy

In August 1897 the English naturalist H. B. Guppy camped at Mokuaweoweo, which at the time was essentially inactive. *Guppy* [1897] suggested that "By a regular study of this volcano the forecasting of its eruptions would come within the domain of the possible...The establishment of an observatory would not be beneath the notice of an enlightened government." (This statement by Guppy presaged T. A. Jaggar's similar plea and justification that led to the founding of the Hawaiian Volcano Observatory in 1912 [*Wright and Takahashi*, 1989, p. 101].) Later, *Guppy and Salcombe* [1906] described Mokuaweoweo as it appeared in August 1897, and concluded that the caldera formed by coalescence of several pit craters and was subsequently enlarged by collapse of its sides. They argued (probably erroneously) that the depth of the caldera had not changed significantly since the time of Wilkes' expedition in 1840-1841, although they had not measured it and several intervening summit eruptions had undoubtedly caused some infilling.

The Eruption of 1899

Although contemporary newspapers announced the commencement of this eruption as July 4 [*Anonymous*, 1899a; *Fennell*, 1899], the eruption was witnessed by at least one observer as early as July 1 [*C. H. Hitchcock*, 1911, p. 132]. Lockwood (personal communication, 1989) suggested that the initial activity on July 1 may have diminished and then intensified on 4 July, and that for patriotic reasons in the wake of American successes at Cuba and at Manila Bay during the Spanish-American War, the date of the outbreak was identified with Independence Day.

Following what was probably a four-day summit eruption, a flank eruption on the northeast rift broke out with two fountains at about 11,000 and 10,800 feet, and nearly one mile apart; lava flowed north-northeastward for about 20 days. Dewey Cone, named in honor of the distinguished admiral, was the principal cone formed.

C. H. Hitchcock [1900; 1911, pp. 132-137] gave a retrospective of the eruption, with abridged accounts by several individuals, in addition to his own. Visits in mid-July were made by *Edgar Wood* [1899], *C. W. Baldwin* [1900], *A. B. Ingalls* [1900] and his guide John Gaspar, and *W. R. Castle* [1899].

(In 1899 an interesting account [*Lyons*, 1899; *Anonymous*, 1899b] correlated Hawaiian eruptions with minimum sunspot activity, but the relationship fell apart in the first quarter of the 20th century.)

The Eruptions of 1903

An eruption in Mokuaweoweo was witnessed on the evening of September 1, 1903, from the summit of Haleakala on Maui by C. J. Austin and his companion, Hansted [*Anonymous*, 1903a]. References to this overlooked eruption were brought to the attention of the writer by J. P. Lockwood.

The main eruption broke out in the summit crater on October 6; activity ceased on December 8. The eruption was confined to the caldera, except for a small flow on the southwest side of the mountain. An account, with details of visits by several parties, is provided by *Thrum* [1903]; *Brigham* [1909, pp. 202-204] gives an abridged version.

Another account [*Anonymous*, 1903b] not included in the foregoing summary is that of a party of six (including two women) and their guide who reached the summit on October 20. They witnessed fountaining, construction and destruction of cones, and spreading of lava on the main caldera floor. That trip was described nine years later by *Margaret (Mrs. W. L.) Howard* [1912], one of the women in the party; her prose is reminiscent of that of Isabella Bird, whom Howard erroneously inferred was the only woman to have ascended Mauna Loa prior to her own ascent. Prof. *Edgar Wood* [1904] also described a display from the summit in October which included reports of flows outside the summit caldera.

Honolulu Prof. Willis T. Pope ascended Mauna Loa with R. O. Reiner and guide Joseph Gaspar in 1905 via the western route; he described Mokuaweoweo as it appeared on July 17 [*C. H. Hitchcock*, 1911, pp. 140-142].

The Eruption of 1907

On January 9, 1907, Mauna Loa erupted along the southwest rift. The flow ceased after 15 days, but activity continued at the source for some time. News of the eruption and accounts of those who visited it were given in the *Hilo Tribune* [*Anonymous*, 1907a]. The *Hawaiian Annual* for 1908 [*Anonymous*, 1907b] provided a summary. *Bishop* [1907] left notes on this eruption, and *A. S. Baker* [1907] described his trip to the flow. *Brigham* [1909] and *C. H. Hitchcock* [1909, 1911] concluded their summaries of the history of Hawaii's volcanoes with the foregoing descriptions and comments on the eruption of 1907.

The second decade of the 20th century would see the establishment of the Hawaiian Volcano Observatory and the commencement of systematic scientific investigation of these volcanoes.

RELATIONSHIP OF MAUNA LOA AND KILAUEA

One of the longest debated (and still unresolved) matters in Hawaiian volcanism is the relationship of Mauna Loa and Kilauea. *Dana* [1849, p. 171; 1850a; 1850b] regarded Kilauea as "the great pit-crater on the flanks of Mauna Loa"; forty years later, he [*Dana*, 1890, pp. 260-263] considered Kilauea to be an independent volcano, the southernmost on the easterly of two crustal fissures believed to mark the location of volcanoes of the Hawaiian Archipelago. *Dutton* [1884] pointed out that Kilauea was an independent topographic form, a fact later repeated by *Stone* [1926], who also demonstrated that the flows in the walls of Kilauea caldera originated from Kilauea, not Mauna Loa.

The possibility of an interconnection between their magma reservoirs was believed unlikely by *Andrews* [1843], because Kilauea caldera was unaffected by the 1843 Mauna Loa eruption; he believed that Kilauea, at the lower elevation, would have filled to overflowing before Mauna Loa erupted. *Dana* [1849, pp. 218-221] also noted an apparent independency of volcanic action, but believed the two volcanoes were connected at depth; he attributed the difference in altitude of the molten magma in the individual conduits to a difference in specific gravity resulting from different gas contents. Independence of activity and the great difference in altitude of the vents led both *Scrope* [1862, p. 262] and *Judd* [1881, pp. 138 and 326] to conclude that the two volcanoes were independent. *Dutton* [1884] argued that the two volcanoes had separate magma reservoirs: any connection between simultaneously active vents differing more than 9000 feet in altitude was "so thoroughly opposed to all hydrostatic laws as to be incredible." Although *Green* [1887, II, pp. 155-165, 179-181] believed an interconnection could exist in a liquid substratum at a depth of 20 miles, with lava rising to different altitudes because of differences in density cause by temperature differences, he maintained that Mauna Loa and Kilauea are independent volcanoes. *J. M. Alexander* [1886a] insisted that the density of Mauna Loa's lava is no lighter than that of Kilauea's lava, and suggested that lava rises higher in

Mauna Loa without an overflow at Kilauea because Mauna Loa's conduit is smaller and more constricted; it "is therefore by no means certain that there is no subterranean connection between the two volcanoes." Others who believed that Mauna Loa and Kilauea are interconnected at depth are *Friedlaender* [1895], *Cartwright* [1913], and *Jaggar* [1917]. More recently, *Klein* [1982a] noted that a "deep source of tremor beneath Kilauea's lower SW rift happens to be nearly equidistant between Kilauea, [the submarine volcano] Loihi, and Mauna Loa, and this source may represent a common zone through which magma is supplied to the three active volcanoes." Further, his study [*Klein*, 1982b] of patterns of historical eruptions of Mauna Loa and Kilauea appears to show that when one is most active, the other is in repose and vice versa, which may suggest that the two volcanoes compete for the same magma supply. On the other hand, *Rhodes et al.* [1989] noted that petrological, geochemical and isotope data obtained by several researchers provide evidence that the two volcanoes are supplied by parental magmas from physically and geochemically distinct mantle sources. *Rhodes et al.* [1989] showed that in the past 2000 years Kilauea has erupted a spectrum of lava compositions resembling historical Kilauea lavas at one end and Mauna Loa lavas at the other, and speculated that magma from Mauna Loa may have invaded Kilauea's "high-level" magmatic plumbing system. Geochemical differences and their implications for source compositions, melting process and magma ascent paths were recently considered by *Frey and Rhodes* [1993], who also presented plume models consistent with those differences. The controversy over possible relationships between the two volcanoes continues to the present.

SYNOPSIS OF SCIENTIFIC CONTRIBUTIONS

In the early exploration and observations of Mauna Loa (and of neighboring Kilauea), many ideas emerged to contribute to the science of volcanology. A few are singled out here. Coan's recognition of "pyroducts" as conduits of lava was a first step in establishing the importance of lava tubes in the emplacement of lava and in the evolution of volcanoes. His recognition of the processes of formation of tumuli and lateral outbreaks of pahoehoe resulting from obstruction by hardening of lavas and the effect of hardening in reducing the rate of advance of lava upon Hilo were significant contributions to the understanding of flow dynamics and have important implications for assessing volcanic hazards. Many of Dana's precepts on a variety of topics relating to Hawaiian volcanism are consistent with current views. W. D. Alexander and Haskell described distinctions between pahoehoe and aa lava, and provided early accounts of flow morphology and structure. Green described the formation of pillow lavas. Dutton appears to be the first to use the term "caldera" for the summit depressions on Kilauea and Mauna Loa, and to attribute their formation to collapse rather than to explosions. Guppy suggested the establishment of an observatory and provided justification that subsequently led to the founding of the Hawaiian Volcano Observatory in 1912 and the beginning of systematic scientific study of Hawaiian volcanoes. Finally, many contributed insights to the continuing debate on the relationship between Mauna Loa and Kilauea.

Acknowledgments. I would like to thank R. A. Apple, J. B. Halbig, J. P. Lockwood, J. M. Rhodes, T. J. Takahashi, and T. L. Wright for their contributions to this work.

REFERENCES

Alexander, J. M., The craters of Mokuaweoweo, on Mauna Loa, *Nature*, *34*, 232-234, 1886a (reprinted from *Pacific Commercial Advertiser*, Oct. 1885).

Alexander, J. M., Summit crater of Mt. Loa, Mokuaweoweo, *Am. J. Sci.*, *132* [*3rd ser., 32*], 235-236, 1886b.

Alexander, J. M., On the summit crater in October, 1885, and its survey, *Am. J. Sci.*, *136* [*3rd ser., 36*], 35-39, 1888.

Alexander, W. D., Later details from the volcano on Hawaii, *Pacific Commercial Advertiser*, *3*, 2/24/1859, 2, 1859.

Andrews, S. L., [On the 1843 eruption of Mauna Loa], *Missionary Herald*, *39*, Oct., 381-382, 1843.

Anonymous, Great crater on the summit of Mauna Loa, Hawaii, *Hawaiian Spectator*, *1*, 98, 1838. (Initial comments are anonymous; the principal entry is by Douglas, 1838.)

Anonymous, Eruption of Mauna Loa, *Polynesian*, 8/23/1851 (reprinted in *Am. J. Sci.*, *63* [*2nd ser., 13*], 299, 1852).

Anonymous, Volcanic signs, *Pacific Commercial Advertiser*, 9/9/1871, 3, 1871a.

Anonymous, The outbreak of Mauna Loa, *Pacific Commercial Advertiser*, 9/30/1871, 3, 1871b.

Anonymous, Volcanic eruption at Hawaii, *Am. J. Sci.*, *104* [*3rd ser., 4*], 331, 1872a.

Anonymous, Ascent of Mauna Loa to the scene of eruption, *Pacific Commercial Advertiser*, 17 supp., 9/21/1872 (reprinted in *Am. J. Sci.*, *104* [*3rd ser., 4*], 407-409, 1872b).

Anonymous, The volcano of Mauna Loa [Ascent of O. B. Adams and wife], *Hawaiian Gazette*, *9*, 9/3/1873, 2, 1873.

Anonymous, [Eruption of March 9, 1879], *Hawaiian Gazette*, 4/16/1879, 3, 1879.

Anonymous, Latest from Mauna Loa, *Hawaiian Gazette*, *16*, 12/8/80, 3, 1880.

Anonymous, ["N.B.E."], [On the 1881 lava flow of Mauna Loa], *Hawaiian Gazette*, *17*, 8/31/1881, 3, 1881a.

Anonymous, A river of fire, *Harper's Weekly*, *25*, 9/3/1881, 598, 1881b.

Anonymous, [On the vent of the 1880 eruption], *Hawaiian Annual* for 1883, 35, 1882.

Anonymous, Summit crater of Mt. Loa, Mokuaweoweo, *Am. J. Sci.*, *132* [*3rd ser., 32*], 235-236, 1886.

Anonymous, Mokuaweoweo. Last ascent of the mountain Mauna Loa, *Pacific Commercial Advertiser*, 7/11/1893, 5, 1893.

Anonymous, [On the 1899 eruption of Mauna Loa], *Hawaiian Gazette*, *34*, 7/11/1899, 1, 1899a.

Anonymous, Eruption of Mauna Loa, *Am. J. Sci.*, *158* [*4th ser., 8*], 237-238, 1899b.

Anonymous, [Eruption of Sept. 1, 1903], *Pacific Commercial Advertiser*, 9/7/1903, 1; *Hawaiian Gazette*, 9/8/1903, 1, 1903a.

Anonymous (editor of Hilo Tribune), [On a trip to the summit of Mauna Loa during eruption in October 1903], *Hilo Tribune*, *8*, 10/23/1903, 1, 4, 1903b.

Anonymous, [On the eruption of 1907], *Hilo Tribune*, *12*, 1/15/1907, 1; 12, 1/22/1907, 1, 6, 1907a.

Anonymous, Mauna Loa's outbreak and lava flow of 1907, *Hawaiian Annual* for 1908, 131-135, 1907b.

Baker, A. S., Nature's pyrotechnics, *The Friend*, *64*, Feb., 4-6, 1907.

Baker, E. P., The ascent of Mauna Loa, *Hawaiian Gazette*, 5/27/1885, 4, 1885.

Baker, E. P., Notes on Mauna Loa, *Am. J. Sci.*, *137* [*3rd ser., 37*], 52-53, 1889.

Baldwin, C. W., To the lava flow of 1899, *Hawaii's Young People*, *4*, Mar., 139-144, Apr., 178-181, May, 203-207, 1900.

Baldwin, E. D., Notes on the 1880-81 lava flow from Mauna Loa, *The Volcano Letter*, No. 520, April-June, 3 pp., 1953.

Barnard, W. M. (Ed.), *Mauna Loa--A Source Book, Historical Eruptions and Exploration, Volume One: From 1778 Through 1907*, xv + 353 pp., privately published, SUNY College at Fredonia, Fredonia, New York, 1990.

Barnard, W. M. (Ed.), *Mauna Loa--A Source Book, Historical Eruptions and Exploration, Volume Two: The Early HVO and Jaggar Years (1912-1940)*, xvii + 452 pp., privately published, SUNY College at Fredonia, Fredonia, New York, 1991a.

Barnard, W. M., Earliest ascents of Mauna Loa Volcano, Hawaii, *The Hawaiian J. of History*, *25*, 53-70, 1991b.

Barnard, W. M. (Ed.), *Mauna Loa--A Source Book, Historical Eruptions and Exploration, Volume Three: The Post-Jaggar Years (1940-1991)*, xvii + 373 pp., privately published, SUNY College at Fredonia, Fredonia, New York, 1992.

Barton, G. H., Notes on the lava flow of 1880-81 from Mauna Loa, *Science*, *3*, 410-413, 1884.

Bevens, D., T. J. Takahashi, and T. L. Wright (Eds.), *The Early Serial Publications of the Hawaiian Volcano Observatory*, 3 vols., Hawaii Natural History Assoc., Hawaii National Park, 1988.

Bird, I. L., *Six Months among the Palm Groves, Coral Reefs and Volcanoes of the Sandwich Islands*, xxvi + 318 pp., Charles E. Tuttle Co., Rutland, Vermont and Tokyo, Japan, 1974 ed.

Bishop, S. E., The new volcano--Hilo threatened, &c., *The Friend*, *13*, Mar., 23, 1856.

Bishop, S. E., [On the 1887 eruption of Mauna Loa], *Hawaiian Gazette*, *22*, 2/8/1887, 1, 5, 1887a.

Bishop, S. E., The recent eruption of Mauna Loa, *Science*, *9*, 205-207, 1887b.

Bishop, S. E., Hawaii's volcanic eruption, *The Friend*, *64*, Feb., 6-7, 1907.

Brigham, W. T., Notes on an ascent in 1880, about three months before the great eruption of that year, *Am. J. Sci.*, *136 [3rd ser., 36]*, 33-35, 1888.

Brigham, W. T., *The Volcanoes of Kilauea and Mauna Loa on the Island of Hawaii*, Mem. 2, no. 4, 222 pp. + plates, Bernice Pauahi Bishop Mus., Honolulu, Hawaii, 1909.

Cartwright, B., Halemaumau and Mokuaweoweo, *Paradise of the Pacific*, *26*, no. 8, 10-11, 1913.

Castle, W. R., [On the 1880-81 lava flow of Mauna Loa], *Hawaiian Gazette*, *17*, 8/31/1881, 3, 1881.

Castle, W. R., [On the 1899 eruption of Mauna Loa], *Hawaiian Gazette*, *34*, 7/25/1899, 1, 1899.

Coan, F. C., [On the 1852 eruption of Mauna Loa], *Polynesian*, 3/8/1852 (reprinted in *Am. J. Sci.*, *64 [2nd ser., 14]*, 106-107, 1852).

Coan, L. B., [On the 1881 lava flow of Mauna Loa], *Hawaiian Gazette*, *17*, 7/6/1881, 3, 1881.

Coan, T., Volcano of Mauna Loa, *Missionary Herald*, *39*, Dec., 463-464, 1843.

Coan, T., Journey to Mauna Loa [On the 1843 eruption of Mauna Loa], *Missionary Herald*, *40*, Feb., 44-47, 1844.

Coan, T., [On the recent condition of Kilauea and 1849 eruption of Mauna Loa], *Am. J. Sci.*, *62 [2nd ser., 12]*, 80-82, 1851.

Coan, T., On the eruption of Mauna Loa in 1851, *Am. J. Sci.*, *63 [2nd ser., 13]*, 395-397, 1852a.

Coan, T., [On the 1852 eruption of Mauna Loa], *Am. J. Sci.*, *64 [2nd ser., 14]*, 105-106, 1852b (from *Polynesian*, *8*, 3/13/1852, 2).

Coan, T., On the eruption of Mauna Loa, Hawaii, February, 1852, *Am. J. Sci.*, *64 [2nd ser., 14]*, 219-224, 1852c.

Coan, T., Notes on Kilauea and the recent eruption of Mauna Loa, *Am. J. Sci.*, *65 [2nd ser., 15]*, 63-65, 1853a.

Coan, T., Eruption of Mauna Loa, February 20, 1852 [in verse], *The Friend*, June, 45, 1853b (revision in C. H. Hitchcock, 1911).

Coan, T., Eruption of Mauna Loa, *Am. J. Sci.*, *71 [2nd ser., 21]*, 139-144, 1856a.

Coan, T., On the recent eruption of Mauna Loa, *Am. J. Sci.*, *71 [2nd ser., 21]*, 237-241, 1856b.

Coan, T., [On the 1855 eruption of Mauna Loa], *The Friend*, March, 23, 1856c.

Coan, T., On the [1855] eruption at Hawaii, *Am. J. Sci.*, *72 [2nd ser., 22]*, 240-243, 1856d.

Coan, T., Volcanic action on Hawaii, *Am. J. Sci.*, *73 [2nd ser., 23]*, 435-437, 1857.

Coan, T., [On Kilauea caldera and the 1859 eruption of Mauna Loa], *Am. J. Sci.*, *79 [2nd ser., 29]*, 302, 1860.

Coan, T., New eruption of Mauna Loa, *Am. J. Sci.*, *91 [2nd ser., 41]*, 424-425, 1866.

Coan, T., Volcanic eruptions in Hawaii, *Am. J. Sci.*, *93 [2nd ser., 43]*, 264-265, 1867.

Coan, T., [On the earthquakes and eruption of 1868], *Am. J. Sci.*, *96 [2nd ser., 46]*, 106-109, 1868.

Coan, T., Notes on the recent volcanic disturbances of Hawaii, *Am. J. Sci.*, *97 [2nd ser., 47]*, 89-98, 1869.

Coan, T., Volcanic action on Hawaii, *Am. J. Sci.*, *99 [2nd ser., 49]*, 393-394, 1870.

Coan, T., On Kilauea and Mauna Loa, *Am. J. Sci.*, *102 [3rd ser., 2]*, 454-456, 1871.

Coan, T., Recent eruption of Mauna Loa, *Am. J. Sci.*, *104 [3rd ser., 4]*, 406-407, 1872.

Coan, T., Volcanoes of Hawaii, *Am. J. Sci.*, *105 [3rd ser., 5]*, 476-477, 1873.

Coan, T., Note on the recent volcanic action in Hawaii, *Am. J. Sci.*, *107 [3rd ser., 7]*, 516-517, 1874.

Coan, T., Volcanic eruptions on Hawaii, *Am. J. Sci.*, *114 [3rd ser., 14]*, 68-69, 1877.

Coan, T., On a recent silent discharge of Kilauea, *Am. J. Sci.*, *118 [3rd ser., 18]*, 227-228, 1879.

Coan, T., Recent action of Mauna Loa and Kilauea, *Am. J. Sci.*, *120 [3rd ser., 20]*, 71-72, 1880.

Coan, T., Volcanic eruptions of Mauna Loa, Hawaii, *Am. J. Sci.*, *121 [3rd ser., 21]*, 79, 1881a.

Coan, T., Progress of the volcanic eruption on Hawaii, *Pacific Commercial Advertiser*, 7/30/1881b (abridged version in *Am. J. Sci.*, *122 [3rd ser., 22]*, 227-229, 1881b).

Coan, T., [On the 1881 lava flow of Mauna Loa], *Hawaiian Gazette*, *17*, 7/6/1881, 3, 1881c.

Coan, T., Volcanic eruption on Hawaii, *Am. J. Sci.*, *122 [3rd ser., 22]*, 322, 1881d.

Coan, T., *Life in Hawaii, An Autobiographic Sketch of Mission Life and Labors (1835-1881)*, 340 pp., Anson D. F. Randolph & Co., New York, 1882.

Cook, J., and J. King, *A Voyage to the Pacific Ocean. Undertaken, by the Command of his Majesty, for making Discoveries in the Northern Hemisphere to determine the Position and Extent of the West Side of North America; its Distance from Asia; and the Practicability of a Northern Passage to Europe. Performed under the direction of Captains Cook, Clerke, and Gore, in his Majesty's Ships the*

Resolution and Discovery. In the Years 1776, 1777, 1778, 1779, and 1780, 3 vols., W. and A. Strahan, London, 1784.

Dana, J. D., *United States Exploring Expedition. During the Years 1838, 1839, 1840, 1841, 1842. Under the Command of Charles Wilkes, U.S.N. Vol. X. Geology*, C. Sherman, Philadelphia, 1849.

Dana, J. D., Historical account of the eruptions in Hawaii, *Am. J. Sci., 59 [2nd ser., 9]*, 347-364, 1850a.

Dana, J. D., On the volcanic eruptions of Hawaii, *Am. J. Sci., 60 [2nd ser., 10]*, 235-244, 1850b.

Dana, J. D., Note on the eruption of Mauna Loa, *Am. J. Sci., 64 [2nd ser., 14]*, 254-257, 1852.

Dana, J. D., On volcanic action at Mauna Loa, *Am. J. Sci., 71 [2nd ser., 21]*, 241-244, 1856.

Dana, J. D., Eruption of Mauna Loa, Hawaii, *Am. J. Sci., 77 [2nd ser., 27]*, 410-415, 1859 (includes letters of Coan, Lyons, and editorial writer for the *Pacific Commercial Advertiser*).

Dana, J. D., Recent eruption of Mauna Loa and Kilauea, Hawaii, *Am. J. Sci., 96 [2nd ser., 46]*, 105-123, 1868 (includes letters of Coan, Lyman, and Whitney).

Dana, J. D., Eruption of Mauna Loa, Hawaii, in January, *Am. J. Sci., 133 [3rd ser., 33]*, 310-311, 1887.

Dana, J. D., History of the changes in the Mt. Loa craters, II. Mokuaweoweo, the summit crater of Mt. Loa, *Am. J. Sci., 136 [3rd ser., 36]*, 14-32, 81-112, 167-175, 1888 (includes maps).

Dana, J. D., *Characteristics of Volcanoes, with Contributions of Facts and Principles from the Hawaiian Islands*, 399 pp., Dodd, Mead, and Co., New York, 1890.

Dodge, F. S., On Mokuaweoweo's brink, *Hawaiian Gazette, 31*, 5/8/1896 1, 6, 1896.

Douglas, D., Great crater on the summit of Mauna Loa, Hawaii, *Hawaiian Spectator, 1*, no. 2, 98-103, 1838.

Douglas, D., *Journal kept by David Douglas during his Travels in North America 1823-1827* (with appendices), 364 pp., William Wesley & Son, London, 1914 ed.

Dutton, C. E., Recent exploration of the volcanic phenomena of the Hawaiian Islands, *Am. J. Sci., 125 [3rd ser., 25]*, 219-226, 1883.

Dutton, C. E., Hawaiian volcanoes, in *Fourth Annual Report of the United States Geological Survey, 1882-83*, pp. 75-219, Government Printing Office, Washington, D. C., 1884.

Fennell, W. P., [On the 1899 eruption of Mauna Loa], *Hawaiian Gazette, 34*, 7/11/1899, 1, 1899.

Frey, F. A., and J. M. Rhodes, Intershield geochemical differences among Hawaiian volcanoes: implications for source compositions, melting process and magma ascent paths, *Phil. Trans. R. Soc. Lond.* A, 342, 121-136, 1993.

Friedlaender, B., Der Vulkan Kilauea auf Hawaii, *Himmel und Erde, 8*, no. 1, pp. 41-51, 105-130, 1895.

Friedlaender, B., Mokuaweoweo in activity, *Hawaiian Annual* for 1897, 71-79, 1896.

Fuller, J., Great volcanic eruption [On the 1852 eruption of Mauna Loa], *The Friend, 9*, May, 3-4, 1852a.

Fuller, J., [On the 1852 eruption of Mauna Loa], *Am. J. Sci., 64 [2nd ser., 14]*, 258-259, 1852b.

Goodrich, J., Notices of some of the volcanos [sic] and volcanic phenomena of Hawaii, (Owyhee,) and other islands in that group, *Am. J. Sci., 25*, 199-203, 1834.

Gordon Cumming, C. F., *Fire Fountains: The Kingdom of Hawaii, its Volcanoes, and the History of its Missions*, 2 vols., Wm. Blackwood and Sons, Edinburgh and London, 1883.

Greeley, R., The role of lava tubes in Hawaiian volcanoes, in *Volcanism in Hawaii*, edited by Decker, R. W., T. L. Wright, and H. Stauffer, pp. 1589-1602, U.S. Geol. Surv. Prof. Pap. 1350, 1987.

Green, W. L., *Vestiges of the Molten Globe*, Parts I and II, 59 + 337 pp., Hawaiian Gazette Publishing Co., Honolulu, Hawaii, 1887.

Guppy, H. B., On the summit of Mauna Loa, *Pacific Commercial Advertiser*, 9/18/1897, 1, 5, 1897.

Guppy, H. B., and M. B. Salcombe, Observations on the Mokuaweoweo Crater, *Pacific Commercial Advertiser*, 9/6/1906, 7; 9/7/1906, 7; 9/8/1906, 6, 1906.

Haskell, R. C., On a visit to the recent eruption of Mauna Loa, Hawaii, Am. J. Sci., 78 [2nd ser., 282], 66-71, 1859a.

Haskell, R. C., Eruption of Mauna Loa, Sandwich Islands, *Am. J. Sci., 78 [2nd ser., 29]*, 284, 1859b.

Haskell, R. C., Eruption of Mauna Loa, Sandwich Islands, *Am. J. Sci., 79 [2nd ser., 29]*, 301-302, 1860.

Hillebrand, W., [On the eruptions of 1868], *Hawaiian Gazette, 4*, 5/6/1868, 2, 1868 (reprinted in *Am. J. Sci., 96 [2nd ser., 46]*, 115-121, 1868).

Hitchcock, C. H., Volcanic phenomena on Hawaii, *Geol. Soc. Am. Bull., 12*, 45-56, 1900.

Hitchcock, C. H., *Hawaii and its Volcanoes*, 1st ed., 314 pp., Hawaiian Gazette Co., Ltd., Honolulu, Hawaii, 1909.

Hitchcock, C. H., *Hawaii and its Volcanoes*, 2nd ed. with suppl., 314+ pp., Hawaiian Gazette Co., Ltd., Honolulu, Hawaii, 1911.

Hitchcock, D. H., [On the Mauna Loa eruption of 1880], *Hawaiian Gazette, 16*, 11/17/1880, 3, 1880a.

Hitchcock, D. H., [On the Mauna Loa eruption of 1880], *Hawaiian Gazette, 16*, 12/8/1880, 3, 1880b.

Hitchcock, D. H., [On the 1881 lava flow of Mauna Loa], *Hawaiian Gazette, 17*, 7/6/1881, 3, 1881.

Hitchcock, D. H., [On the 1887 eruption of Mauna Loa], *Hawaiian Gazette, 22*, 2/1/1887, 8, 1887.

Hooker, W. J., A brief memoir of the life of David Douglas, *Companion of the Botanical Magazine, 2*, 158-177, 1836 (reprinted in *Oregon Historical Society Quarterly*, *6*, 417-441, 1905).

Howard, M., A woman's ascent of Mauna Loa, *Mid-Pacific Magazine*, *3*, Apr., 310-317, 1912.

Ingalls, A. B., Mauna Loa's eruption of 1899, *Hawaiian Annual* for 1900, 51-60, 1900.

Jaggar, T. A., Jr., Lava flow from Mauna Loa, 1916, *Am. J. Sci.*, *193, 4th ser., 43*], p. 255-288, 1917.

Jaggar, T. A., Jr., The Crater of Mauna Loa, *The Volcano Letter*, no. 360, 11/19/1931, 1-4, 1931.

Jones, G., [On the Mauna Loa eruption of 1887], *Hawaiian Gazette*, *22*, 2/1/1887, 1, 1887.

Kinney, H. W., [On the 1852 eruption of Mauna Loa], *Am. J. Sci.*, *64* [*2nd ser., 14*], 257-258, 1852.

Klein, F. W., Earthquakes at Loihi submarine volcano and the Hawaiian hot spot, *J. Geophys. Res.*, *87*, no. B9, 7719-7726, 1982a.

Klein, F. W., Patterns of historical eruptions at Hawaiian volcanoes, *J. Volcanol. Geotherm. Res.*, *12*, 1-35, 1982b.

Ledyard, J., *A Journal of Captain Cook's Last Voyage to the Pacific Ocean, and in Quest of a North-West Passage, Between Asia and America; Performed in the Years 1776, 1777, 1778, and 1779*, 208 pp., Quadrangle Books, Inc., Chicago, 1963 ed.

Liliuokalani, *Hawaii's Story*, 414 pp., Charles E. Tuttle Company, Inc., Rutland, Vermont and Tokyo, Japan, 1964 ed.

Lipman, P. W., and N. G. Banks, Aa flow dynamics, Mauna Loa 1984, in *Volcanism in Hawaii*, edited by Decker, R. W., T. L. Wright, and H. Stauffer, pp. 1527-1567, U.S. Geol. Surv. Prof. Pap. 1350, 1987.

Lockwood, J. P., and P. W. Lipman, Holocene eruptive history of Mauna Loa Volcano, in *Volcanism in Hawaii*, edited by Decker, R. W., T. L. Wright, and H. Stauffer, pp. 509-535, U.S. Geol. Surv. Prof. Pap. 1350, 1987.

Lyman, F. S., [On the 1868 eruption of Mauna Loa], *Am. J. Sci.*, *96* [*2nd ser., 46*], 109-112, 1868.

Lyons, C. J., Sun spots and Hawaiian eruptions, *Monthly Weather Review*, *27*, 144, 1899.

Maby, J. H., [On the Mauna Loa eruption of 1887], *Hawaiian Gazette*, *22*, 1/25/1887, 8, 1887.

Macdonald, G. A., *Volcanoes*, 510 pp., Prentice-Hall, Inc., Englewood Cliffs, New Jersey, 1972.

Macdonald, G. A., and A. T. Abbott, *Volcanoes in the Sea*, 1st ed., 441 pp., University of Hawaii Press, Honolulu, Hawaii, 1970.

Macdonald, G. A., A. T. Abbott, and F. L. Peterson, *Volcanoes in the Sea*, 2nd ed., 517 pp., University of Hawaii Press, Honolulu, Hawaii, 1983.

Macdonald, G. A., and Finch, R. H., Activity of Mauna Loa during April, May, and June, 1949, *The Volcano Letter*, no. 504, 1-4, 1949.

Martin, J. H. S., [On the 1887 lava flow of Mauna Loa], *Hawaiian Gazette*, *22*, 2/1/1887, 1, 1887.

McKay, W. A., [On the 1887 eruption of Mauna Loa], *Hawaiian Gazette*, *22*, 2/15/1887, 8, 1887.

Menzies, A., An early ascent of Mauna Loa, *Hawaiian Annual* for 1908, 99-112, 1907.

Menzies, A., *Hawaii Nei 128 Years Ago by Archibald Menzies*, (Wilson, W. F., Ed.), 199 pp., T. H., Honolulu, Hawaii, 1920.

Merritt, W. C., On an ascent of Mount Loa, *Am. J. Sci.*, *137* [*3rd ser., 37*], 51-52, 1889.

Oleson, W. B., [On the Mauna Loa lava flow of 1880], *Hawaiian Gazette*, *16*, 12/1/1880, 3, 1880.

Paris, J. D., [On the Mauna Loa eruption of 1887], *Hawaiian Gazette*, *22*, 1/25/1887, 8, 1887.

Remy, J., Ascension de Mm. Brenchley et Remy au Mauna Loa [The Ascent by Brenchley and Remy of Mauna Loa, Island of Hawaii], Extrait du Journal de Jules Remy, Chalons-sur-Marne, 45 pp., 1892 (published in English translation by M. C. Summers in *Hawaiian J. of History*, *22*, 33-69, 1988).

Rhodes, J. M., K. P. Wenz, C. A. Neal, J. W. Sparks, and J. P. Lockwood, Geochemical evidence for invasion of Kilauea's plumbing system by Mauna Loa magma, *Nature*, *337*, 257-260, 1989.

Rickman, J., *Journal of Captain Cook's Last Voyage to the Pacific Ocean, on Discovery; Performed in the Years 1776, 1777, 1778, 1779*, E. Newbery, London, 1781 (reprinted by University Microfilms, Inc., Ann Arbor, Michigan, 396 pp., 1966).

Sawkins, J. G., On the volcanic mountains of Hawaii, Sandwich Islands, *R. Geog. Soc. Lond. J.*, *25*, 191-194, 1855.

Sleeper, J. H., [On the 1859 eruption of Mauna Loa], *Pacific Commercial Advertiser*, *3*, 3/10/1859, 2, 1859.

Sparks, J., *The Life of John Ledyard, the American Traveller, Comprising Selections from his Journals and Correspondence*, 325 pp., Hilliard and Brown, Cambridge, Mass., 1828.

Spencer, C. N., [On the 1887 eruption of Mauna Loa], *Hawaiian Gazette*, *22*, 2/1/1887, 1, 8, 1887.

Sprague, R. A., Measuring the mountain: The United States Exploring Expedition on Mauna Loa, 1840-1841, *Hawaiian J. of History*, *25*, 71-91, 1991.

Stearns, H. T., *Geology of the State of Hawaii*, 2nd ed., 335 pp., Pacific Books, Palo Alto, Calif., 1985.

Stearns, H. T., and G. A. Macdonald, *Geology and Ground-Water Resources of the Island of Hawaii*, 363 pp., Hawaii Division of Hydrography, Bull. 9, Honolulu, Hawaii, 1946.

Stone, J. B., *The Products and Structure of Kilauea*, 59 pp., B. P. Bishop Mus. Bull. 33, Honolulu, Hawaii, 1926.

Summers, M. C., 1988 (see Remy, J., 1892).

Thrum, T. G., Activity of Mauna Loa's summit crater, *Hawaiian Annual* for 1904, 163-171, 1903.

Thrum, T. G., Notable Events, *Hawaiian Almanac* for 1909, 134, 1908.

Vancouver, G., *Voyage of Discovery to the North Pacific Ocean and Round the World*, 3 vols., 1798 (reprinted by Da Capo Press, New York, 1967).

Weld, F. A., On the volcanic eruption at Hawaii in 1855-56, *Geol. Soc. London Quarterly J.*, *13*, 163-169, 1857.

Whitney, H. M., [On the 1868 eruption of Mauna Loa], *Am. J. Sci.*, *96* [*2nd ser., 46*], 112-115, 1868.

Whitney, H. M., *The Hawaiian Guide Book, Containing a Brief Description of the Hawaiian Islands, Their Harbors, Agricultural Resources, Plantations, Scenery, Volcanoes, Climate, Population, and Commerce*, Henry M. Whitney, Honolulu, Hawaii, 1875 (republished by Charles E. Tuttle Co., Rutland, Vermont and Tokyo, Japan).

Whitney, H. M., A new and remarkable volcanic outbreak in Kealakakua Bay!, *Hawaiian Gazette*, *13*, 2/28/1877, 3, 1877.

Whitney, H. M., The great eruption of 1880, *Hawaiian Gazette*, *16*, 11/17/1880, 3, 1880.

Wilkes, C., *Narrative of the United States Exploring Expedition. During the years 1838, 1839, 1840, 1841, 1842*, 5 vols., Lea and Blanchard, Philadelphia, Penna., 1845.

Wood, E., Eruption of Mauna Loa, 1899, *American Geologist*, *24*, 300-304, 1899.

Wood, E., Eruption of Mauna Loa, 1903, *American Geologist*, *34*, 62-64, 1904.

Wright, T. L., and T. J. Takahashi, *Observations and Interpretation of Hawaiian Volcanism and Seismicity, 1779-1955, An Annotated Bibliography and Subject Index*, xxiv + 270 pp., University of Hawaii Press, Honolulu, Hawaii, 1989.

W. M. Barnard, Department of Geosciences, State University of New York College at Fredonia, Fredonia, NY 14063-1198.

Offshore Geology of Mauna Loa and Adjacent Areas, Hawaii

James G. Moore

Branch of Volcanic and Geothermal Processes, U.S. Geological Survey, Menlo Park, California

William W. Chadwick, Jr.

Hatfield Marine Science Center, Oregon State University, Newport,Oregon

Recent high-resolution multibeam sonar surveys of the seafloor near the island of Hawaii provide a new detailed view of the morphology and structure of the island's submarine flanks. We have merged this multibeam data with other digital topographic and bathymetric data from several sources to create a series of maps showing the topography, bathymetry, and overall morphology of the south part of the island of Hawaii (including all of Mauna Loa volcano) and an offshore region including the leading end of the Hawaiian Ridge from 20° to 18.5° north latitude. These morphologic maps cover an area of 167 km north-south (1.5° of latitude) by 277 km east-west (2.5° of longitude), with an area of ~46,000 km^2. The morphologic maps include a contoured bathymetric map, a slope map, and a shaded-relief map that show in previously unattained crispness the texture and detail of geologic features both above and below sea level. A geologic map was prepared of this subaerial and submarine area by making use of the morphologic detail of the maps coupled with previous subaerial and submarine geologic studies, including regional side looking sonar (GLORIA) surveys of the Hawaiian Exclusive Economic zone, surface ship dredging, coring, bottom photography, and submersible dives. This map outlines the areas covered by major rock units including terrestrial and subaqueous lava, fragmental quenched lava, submerged terrestrial lava, landslides, Cretaceous seamounts, and abyssal sediment. The map documents the importance of landslides, best exposed in the submarine realm, in the growth and decline of Mauna Loa and adjacent volcanoes.

INTRODUCTION

The island of Hawaii is made up of five coalesced volcanoes: Kohala, Mauna Kea, Hualalai, Mauna Loa, and Kilauea. Mauna Loa covers the largest area of the island (~5000 km^2) and has an offshore area somewhat larger than its area above sea level. This study presents a new series of maps of the island and its submerged flanks out to the flat

Mauna Loa Revealed: Structure,
Composition, History, and Hazards
Geophysical Monograph 92
Copyright 1995 by the American Geophysical Union

sea floor between 18.5° and 20° N latitude,which includes all of Mauna Loa volcano. The bulk of the submarine rocks in this region are volcanic products from Mauna Loa, but a considerable part are on the submarine flanks of other volcanoes, principally Kilauea, but also Hualalai, Mauna Kea, and Loihi. Much of the bathymetry of offshore Mauna Loa and adjacent regions has recently been mapped by high-resolution multibeam methods and these maps, in conjunction with side-scan sonar surveys (GLORIA), provide a new basis for studies of the submarine geology. In addition, our understanding of the offshore part of Mauna Loa and adjacent volcanoes has grown in the last decade or two as a result of other oceanographic investigations including surface ship dredging, photography, coring, and submersible dives.

Geologic investigations offshore are much more limited than those on land because of the shorter time during which they have been undertaken, the larger area, and the difficulty and expense of marine studies. Hence, the offshore geologic map of south Hawaii is a reconnaissance map, and is speculative in many places. Nevertheless, the marine studies have provided ground truth for the remote imaging data leading to a better understanding of the nature of Mauna Loa, both below and above sea level. These new maps provide a more detailed view of the seafloor around Hawaii Island than has previously been available and may stimulate new investigations into volcanic processes and island growth.

BATHYMETRIC-TOPOGRAPHIC DATA

The general study area covers ~46,000 km^2 between 18.5° to 20° north latitude. It includes the southern three-quarters of the island of Hawaii and the offshore flanks of the Hawaiian Ridge out beyond the axis of the Hawaiian Deep. This offshore region is covered by reasonably complete bathymetric mapping, and digital bathymetric and topographic data from several sources with differing resolutions (Figure 1), were mathematically gridded to produce a single grid with a uniform spacing of 200 m for the small scale maps. Three morphologic maps were then produced from this grid, including a contour map (Figure 2), a slope map (Figure 3), and a shaded relief map (Figure 4). Three more detailed slope maps were made from 100-m gridded data to highlight the morphology of selected areas.

The four main data sources (with varying resolutions) include: digital elevation data on land, single-beam hydro-graphic data near shore, recent high resolution multibeam sonar data in deep water, and older low-resolution single-beam soundings in deep water. The seafloor below about 1000 m in the western two thirds of the map area has been surveyed with multibeam sonar (Figure 1). In the eastern third of the map, however, only Kilauea's south flank, and most of the Puna and Hilo Ridges have been surveyed with high-resolution methods (Figure 1).

We have attempted to remove all artifacts on the maps that are created by mismatches between data sets or data resolutions, but a few remain. The main unavoidable artifact is that the areas covered by lower resolution data must be interpolated and generally appear smoother in texture. Also, the quality of the sparser soundings may vary and sometimes create unrealistic seafloor features. For example, on our slope and shaded-relief maps (Figures 3 and 4), the southeast corner of the map area and the area

between the Puna and Hilo Ridges are only covered by low-resolution data (Figure 1). Another artifact, especially apparent on the slope map, is the north-south oriented lineament along 156° west longitude, west of southern Mauna Loa (Figure 3). This line is caused by an artificially abrupt change in slope at the boundary between high-resolution multibeam data to the west and lower-resolution hydrographic data to the east. An additional minor artifact is a north-south fabric on the flat seafloor of the abyssal plain in the northwest corner of the slope map (Figure 3), which is parallel to the ship tracks that collected the multibeam data.

METHODS

A geologic map (Plate 1) was prepared of the area mapped morphologically (Figures 2-4). The subaerial parts of the geologic map were derived from previous sources including the geologic map of the island by *Stearns and Macdonald* [1946] as modified by *Peterson and Moore* [1987]. The position of some subaerial faults and other features have been modified by use of oblique illumination images generated from modern digital elevation data [*Moore and Mark*, 1992].

The generalized breakdown of offshore geologic units is based on seafloor morphology as depicted in our new high-resolution bathymetry (Figs. 2-4), and on seafloor reflectivity as shown on side-looking sonar images (GLORIA) [*Lipman et al.*, 1988, and *Moore et al.*, 1989]. The slope map (Figure 3) is a powerful tool in interpreting the submarine geology around the island because of its effective display of seafloor texture. Ground truth for the interpretation of these subtle differences in morphology is based on marine geophysical surveys, sampling, and bottom photographic data.

Because of the interfingering relationships and commonly obscure boundaries between the lavas that were erupted from Mauna Loa and its neighboring volcanoes, these boundaries have not been extended beneath the sea on the geologic map (Plate 1). Kilauea volcano is generally younger than Mauna Loa and represents a thin construct perhaps about 1 km thick on the flank of Mauna Loa. Hence, lava erupted from Mauna Loa may actually crop out on some of the steeper slopes offshore from Kilauea.

MAPPED GEOLOGIC UNITS

Terrestrial lava

Within the map area the subaerial parts of four volcanoes on the island of Hawaii are covered by terrestrial lava. The

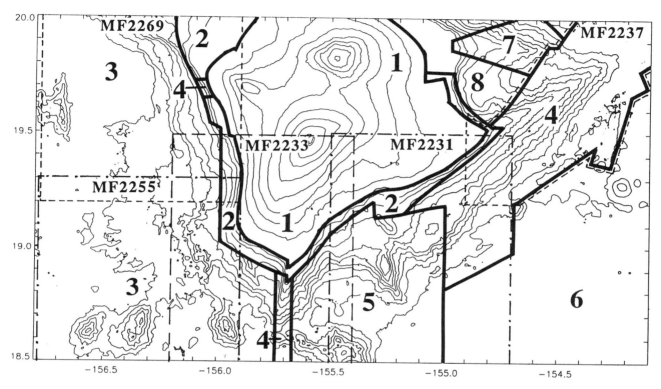

Fig. 1. Index map showing sources of topographic-bathymetric data in numbered areas bounded by heavy lines: 1. Subaerial topography based on recent digital elevation data from 7.5' quadrangles of the U. S. Geological Survey. 2. Nearshore bathymetric mapping, conducted by single-beam sonar surveys, from the NOAA National Ocean Service Hydrographic Database, National Geophysical Data Center (NGDC). 3. Multibeam bathymetry collected from the National Oceanic and Atmospheric Administration (NOAA) ship *Surveyor*, R.V. Forster commanding, during November 1986 and April 1987. 4. Multibeam surveys made from the U.S. Navy vessel *Laney Chouest*, Victor Gisclair commanding, in October 1992 as part of a joint program of NOAA, National Undersea Research Program, and the U.S. Navy, Chief Scientists: D.A. Clague, M. O. Garcia, J. G. Moore, J. R. Smith, and R.E. Young. 5. Multibeam bathymetry collected from the NOAA ship *Discoverer*, R.V. Smart commanding, during August-September 1991. 6. Data from single beam soundings along ship tracks from NOAA's Marine Geophysical Database, NGDC, edited and gridded courtesy of John R. Smith and Theresa Duennebier, University of Hawaii. 7. Multibeam data from R/V Thomas Thompson, Univ. of Washington (C. L.Wilkerson, R. T. Holcomb, J. C. Wiltshire, J. G. Sempere, written communication, 1994). 8. Single beam soundings from NOAA's Marine Geophysical Database, NGDC, and other sources edited, combined, and gridded courtesy of Courtenay Wilkerson, University of Washington. All NOAA multibeam data is available from NGDC. Dashed lines show boundaries of published USGS 1:150,000 scale MF bathymetric maps with numbers at top centers of mapped areas [*Chadwick et al.*, 1993-94].

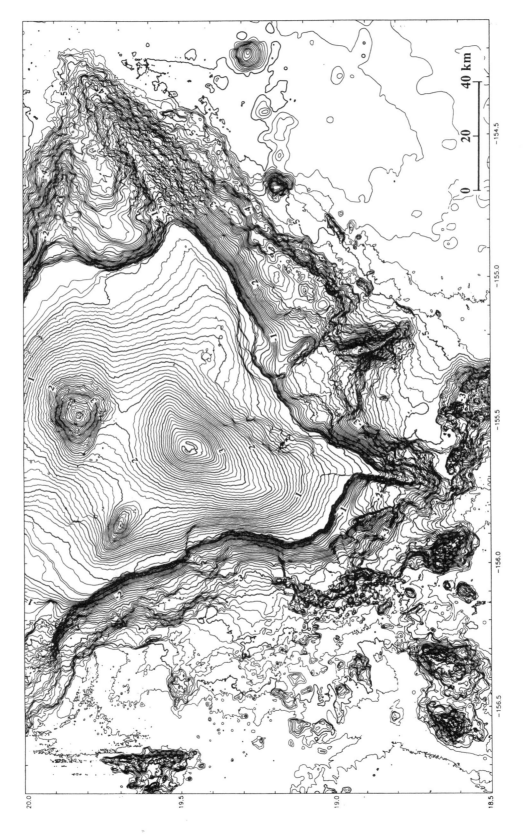

Fig. 2.   Offshore bathymetry and onshore topography of south Hawaii region, depth and elevation contour interval 100 m, bold contours labeled in kilometers.   Data sources in figure 1.

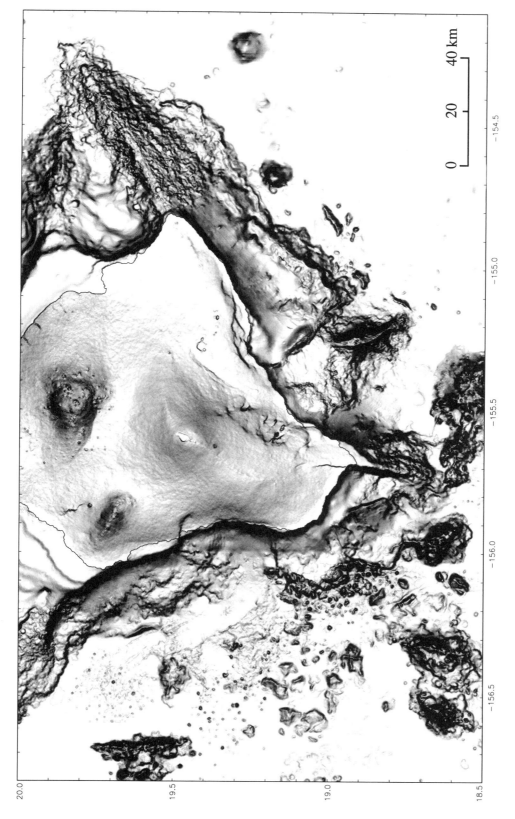

Fig. 3. Slope map of south Hawaii region; dark areas are steepest slopes. Coastline is shown by solid black line. Image derived from 200-m gridded data.

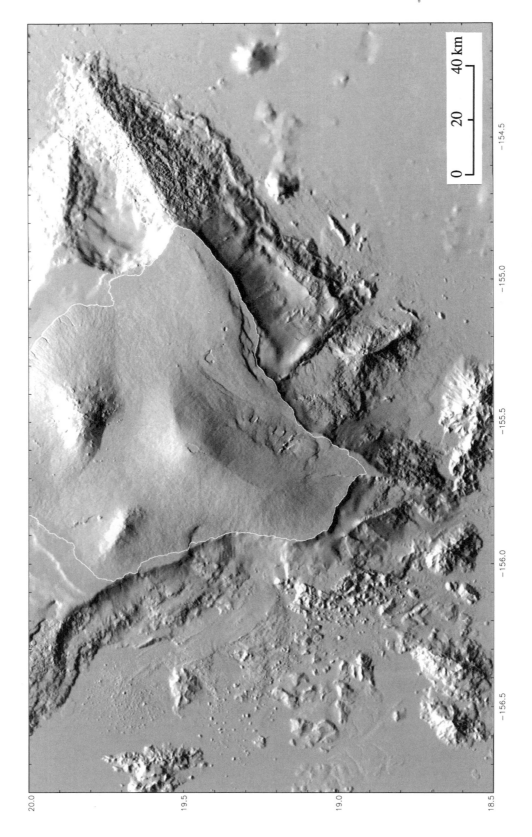

Fig. 4. Shaded relief map of south Hawaii region; apparent illumination from the north. Coastline is shown by solid white line. Image derived from 200-m gridded data.

great bulk is basaltic lava flows, both aa and pahoehoe, which were erupted and cooled on land (except for minor subglacial lavas on Mauna Kea). Cinders, scoria, and spatter make up the cones that are aligned on the rift zones of Mauna Loa, Kilauea, and Hualalai, and are scattered over the shield of Mauna Kea.

The terrestrial lava unit is divided on the geologic map (Plate 1) into prehistoric lava (TbQ) and historic (post-1790) lava (Tbh). The historic lavas of Mauna Loa, Kilauea, and to a lesser extent Hualalai are concentrated at the volcano summits and rift zones.

The terrestrial lava generally constructs much gentler slopes than subaqueous lava so that a prominent slope break occurs at the shoreline of young volcanoes (Figure 3). This produces a characteristic tablemountain, or flat topped, aspect for the entire edifice from its base at abyssal depths to its subaerial summit.

Subaqueous lava

The subaqueous lava unit includes that erupted from submarine vents and quenched below sea level. It is largely basaltic pillow lava, but includes subaqueous aa, pahoehoe, and sheet flows and various types of breccia and talus generated by the downslope breakup and movement of such lava. We distinguish between Quaternary subaqueous lava on the flanks of the Hawaiian Ridge (SbQ) and Cretaceous subaqueous lava (SbK) on seamounts related to the pre-Hawaiian Ridge oceanic crust.

The primary areas of subaqueous lava on the Hawaiian Ridge include the offshore rift zones of Kilauea [*Moore and Fiske*, 1969] (Figure 7), Mauna Loa [*Moore et al.*,1990; *Garcia et al.*, this volume] Mauna Kea, and Hualalai volcanoes [*Moore and Clague*, 1992], and virtually all of Loihi volcano [*Malahoff*, 1987; *Garcia et al.*, 1993,] (Figure 8). The individual hills along the rift zone ridges are apparently mounds of pillowed lava at vents. In addition to hills and cones, many small pit craters and collapse features have been mapped by multibeam sonar along the crest of the submarine east rift zone ridge of Kilauea. About four dozen pit craters averaging 100 m in diameter occur along a 16-km reach of the ridge crest between 1900-3000 m depth [*Lonsdale*, 1989].

The flanks of the submarine rift zones are characterized by a distinctive lobate-hummocky morphology. The lobes are commonly about 0.5-2 km in size with steep sides and gently sloping tops that are well depicted on the slope map (Figures 3, 7). The nature of these flank lobes is uncertain. Some may be vent cones fed from inclined fissures that pond lava on the upslope side, producing flat-topped lobes. Others may be thick flow lobes, or slumps. The more

elongate and larger benches along the lower, southeast flank of the east rift zone ridge of Kilauea seem to be shaped by rift-parallel fault scarps. These features may represent large scale spreading and thrusting of the basal part of the ridge as a result of the weight of overlying lava. They merge to the west with the structures of the Hilina slump and have been grouped with it on the geologic map (Plate 1).

The submarine expression of the southwest rift zone of Kilauea is subdued, and no prominent ridge has been built as compared with Kilauea' east rift zone (Figure 3). The distinctive texture of subaqueously erupted lava is apparent on the east edge, and upper surface, of a 900-1300-m-deep bench that extends about 10 km west of the zone where the southwest rift zone crosses the shoreline (Figures 3, 8). In addition, the west side of the next bench below, at 1600-2000-m depth, also is mantled on its southern margin with lobate apparent subaqueous lava and displays an apparent volcanic cone several hundred meters high on its west side. East of these benches and extending to the west margin of Papa'u seamount the seafloor morphology is dominated by landslide terrain, interpreted to be a part of the Punaluu slump (Plate 1, and landslide section below).

The plateau-like, rather than ridge-like, nature of subaqueous lava offshore of the southwest rift zone as compared to the east rift zone is mirrored by the subaerial rift zone topography with the southwest rift zone ridge being smaller and more subdued than the east rift zone ridge (Figures 2, 5, 8). This view is consistent with the observation that Kilauea's southwest rift zone has shown much less eruptive and seismic activity than the volcano's east rift zone and has a subordinate role in volcano deformation [*Duffield et al.*, 1982]. Moreover, the southwest rift zone has apparently migrated east considerably. It previously erupted lava from vents (near the mapped boundary of Kilauea and Mauna Loa, Plate 1) some 5 km west of the location of its last vents near the coast in 1823. Hence the distribution of subaqueous lava on the offshore benches may reflect offshore eruption from the rift zone as it migrated eastward.

These marked differences between the southwest rift zone and the east rift zone of Kilauea apparently result from the greater age and stability of the east rift zone. The southwest rift zone in its present form has grown on top of the Punaluu slump and consequently has erupted far less volume of material than the east rift zone.

Several areas of large, flat-lying lava flows occur at the base of the steeper flanks of the rift zone ridges, and are mapped separately in Plate 1. They appear as zones of high reflectivity on GLORIA side-looking sonar images [*Holcomb et al.*, 1988]. Such flows are mapped at the distal end of Kilauea's east rift zone ridge, off the southeast

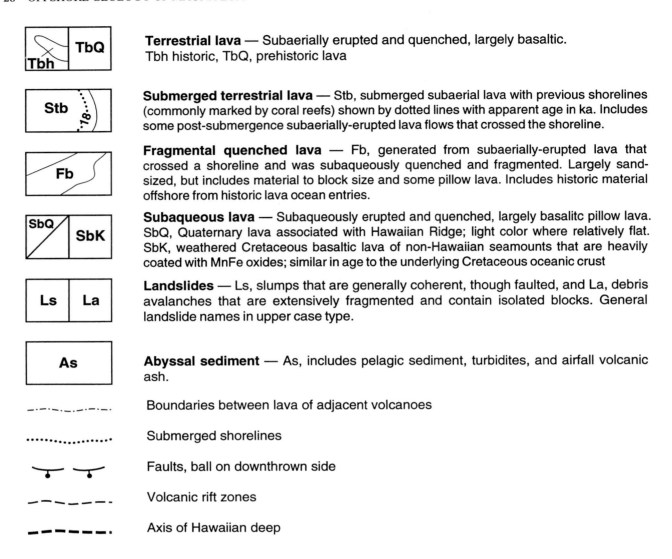

Terrestrial lava — Subaerially erupted and quenched, largely basaltic. Tbh historic, TbQ, prehistoric lava

Submerged terrestrial lava — Stb, submerged subaerial lava with previous shorelines (commonly marked by coral reefs) shown by dotted lines with apparent age in ka. Includes some post-submergence subaerially-erupted lava flows that crossed the shoreline.

Fragmental quenched lava — Fb, generated from subaerially-erupted lava that crossed a shoreline and was subaqueously quenched and fragmented. Largely sand-sized, but includes material to block size and some pillow lava. Includes historic material offshore from historic lava ocean entries.

Subaqueous lava — Subaqueously erupted and quenched, largely basalitc pillow lava. SbQ, Quaternary lava associated with Hawaiian Ridge; light color where relatively flat. SbK, weathered Cretaceous basaltic lava of non-Hawaiian seamounts that are heavily coated with MnFe oxides; similar in age to the underlying Cretaceous oceanic crust

Landslides — Ls, slumps that are generally coherent, though faulted, and La, debris avalanches that are extensively fragmented and contain isolated blocks. General landslide names in upper case type.

Abyssal sediment — As, includes pelagic sediment, turbidites, and airfall volcanic ash.

Boundaries between lava of adjacent volcanoes

Submerged shorelines

Faults, ball on downthrown side

Volcanic rift zones

Axis of Hawaiian deep

Fig. 5. Expanded explanation for geologic map, Plate 1.

flank of the south rift zone ridge of Mauna Loa, and flanking the lower slopes of Loihi volcano (Plate 1). Ocean bottom photographs and dredge hauls of the submarine flow off the east end of Kilauea's east rift zone ridge show a wrinkled and rubbly lava surface near the flow center and pillows near its margin. A sample collected from a flow north of Green Seamount is distinctly alkalic in composition (Clague et al., 1995).

Two small submarine vents with associated lava flows that apparently erupted from Mauna Loa in 1877 have been mapped by submersible offshore from Kealakekua Bay [*Fornari et al.*, 1980; *Moore et al.*, 1985], but are too small to show on the geologic map. A small flat area about 5 km in diameter and rising 25 m above the flat sea floor off

the north end of the Alika 2 debris avalanche is apparently a subaqueous flow related to Hualalai volcano (Figures 5, 9).

Also included in the subaqueous lava unit are the much older seamount volcanoes that predate the Hawaiian Ridge. They are similar in overall morphology to the submarine rift zone subaqueous lavas, but have been extensively modified by mass wasting and sedimentation. They were built somewhat after the time of formation of the Cretaceous oceanic crust about 110 Ma [*Waggoner*, 1993] which underlies the Hawaiian Ridge. Age estimates of Apuupuu and Dana seamounts have been made by comparing paleomagnetic poles of the volcanoes, derived from surface ship magnetic surveys, with the Pacific

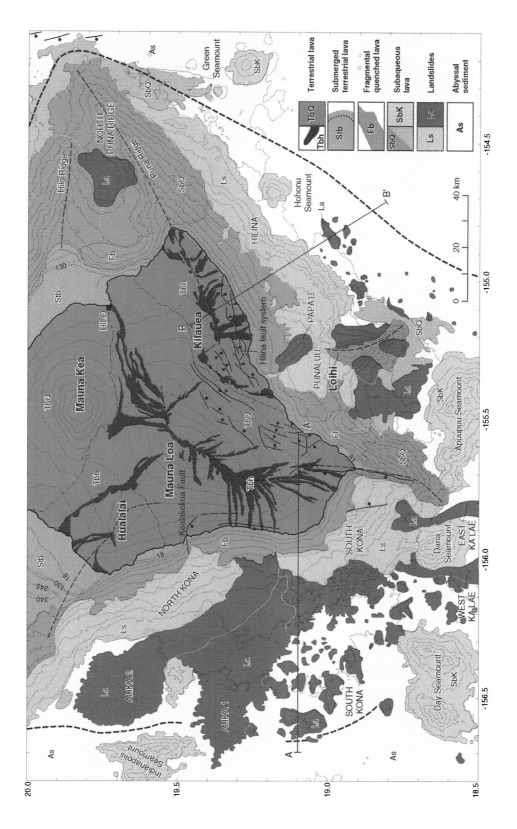

Plate1. Reconnaissance geologic map of south Hawaii region. See figure 6 for expanded explanation.

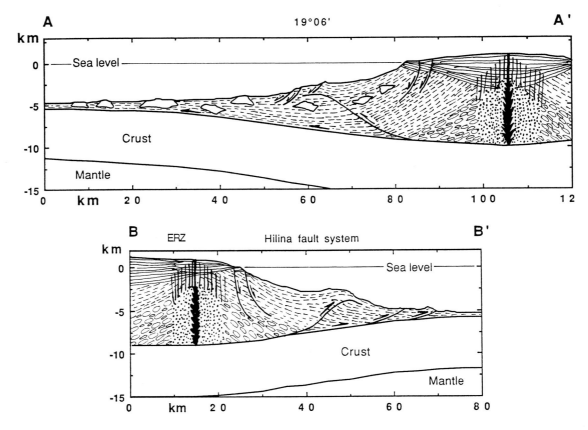

Fig. 6. Geologic cross sections extending from land to sea, as located in figure 5. A-A' West-east section along latitude 19°06' with two-fold vertical exaggeration after Moore et al., 1995. Horizontal layering--subaerial lava; dashed layering--fragmental lava, pillow lava; ellipses--pillow lava; vertical lines--sheeted dikes; dotted pattern--gabbro; unpatterned--giant blocks of first phase of South Kona landslide; solid black--magma or plastic core of southwest rift zone of Mauna Loa. B-B' Cross section of the Hilina slump with east rift zone (ERZ) of Kilauea at its head; two-fold vertical exaggeration. Symbols same as A-A'; after Moore et al., 1994.

apparent polar wander path, and both seamounts fit on the Late Cretaceous part of the wander path [*Sager and Pringle*, 1990]. Dredged basalt samples from the east crest of Day seamount and from Apuupuu seamount are deeply weathered and coated with thick layers of Mn-Fe oxides also indicating their great age relative to the Quaternary rocks of the Hawaiian island flanks [*Moore*, 1965].

Other seamounts in the mapped area, including Indianapolis, Hohonu, and Green, have not been sampled and their designation is based on morphology and position. The small seamount to the east of Indianapolis at 156.5° west longitude was earlier considered a block of the Alika slide complex [*Lipman et al*, 1988]; it is here designated as Cretaceous based on its similarity in morphology to others of that age (Plate 1), but this question can only be answered by detailed sampling, photography, or sediment thickness

measurements. The small cones in deep water at the east end of Kilauea's east rift zone are interpreted as small, sediment-covered Cretaceous volcanic cones [*Clague et al.*, 1994].

Fragmental quenched lava

Steep-sloped banks of fragmental basaltic lava (Fb) mantle the upper subaqueous slopes of the active volcanoes and extend 5-20 km seaward from the shoreline (Plate 1). They are relatively smooth with a uniform texture on the slope map (Figure 3). This material, largely hyaloclastite, is produced when subaerially-erupted lava flows enter the sea and shatter from the combined effects of falling over the sea-cliff, seawater quenching with cooling-induced fracture, and disintegration by waves and surf [*Moore and Fiske*,

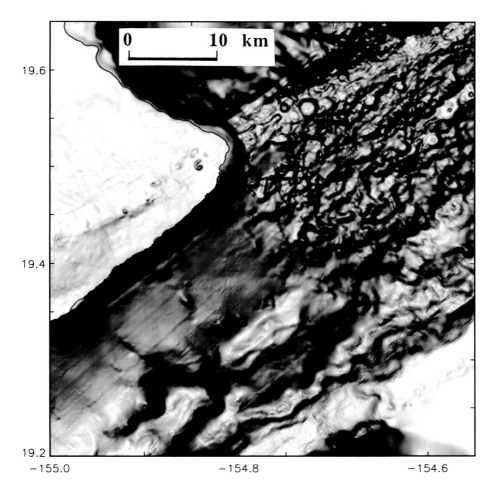

Fig. 7. Slope map of east cape of the Island of Hawaii where east rift zone of Kilauea volcano crosses the shoreline (horseshoe-shaped subaerial cone is Kapoho Cone). Steeper slopes are darker; data gridded at 100 m spacing. Note textural contrast between smooth fragmental quenched basalt draping offshore part of flank downslope from subaerial vents on rift zone with irregular lobate subaqueous lava draping flank downslope from submarine vents on rift zone.

1969; *Moore et al.*, 1973]. In some places, benches of new subaerial lava are eventually built up above sea level seaward of the previous sea cliff. These benches are unstable because of their foundation of steep-bedded offshore fragmental material and they commonly slump into deeper water [*Hon et al.* 1993]. They are probably reduced to rubble during the slumping process.

Despite the fragmentation of most shoreline-crossing flows, some flows maintain their coherence through the surf zone and feed submarine lava tongues as occurred during the 1969 and 1971 Kilauea eruptions [*Moore et al.*, 1973], and is also well displayed by the 710±150-year-old Waha Pele lava flow from Hualalai volcano that flowed about 2 km offshore [*Moore and Clague*, 1987]. Most

ocean-entering lava, however, disintegrates into chiefly sand-size hyaloclastite that may contain blocks exceeding a meter in size. This material commonly extends down to ~1 km in depth and produces the steepest regional slopes on the volcanoes with slopes averaging >10° over the map area and attaining >20° in places [*Mark and Moore*, 1987].

The remarkable difference in the gross texture of subaqueously-deposited basalt that was erupted above and below sea level is graphically shown near where the east rift zone of Kilauea enters the sea at the east cape of the island (Figures 2, 3, 4, 7). Downslope from the subaerial part of the rift, a bank of fine fragmental material mantles the submarine rift zone flank producing a smooth, steep slope apparent in both bathymetry and digital terrain maps.

Downslope from the subaqueous part of the rift, submarine pillowed flows, mounds at vents, and perhaps small slumps produce a bumpy, lobate terrain. The abrupt boundary between these two terrains projects upslope nearly directly to the east cape where the rift zone vents make the transition from subaerial eruption to submarine eruption (Figure 7). With time, as eruptions from the rift zone build the edifice upward and the east cape migrates east, this boundary will also migrate east as the shoreline-generated, subaqueously-erupted fragmental lava buries the subaqueously-erupted lava.

Submerged terrestrial lava

Areas of terrestrial lava on the lower volcano slopes that have subsided below sea level have been mapped as submerged terrestrial lava (Stb, Plate 1). Parts of the subaerial volcanic shields may be submerged during island subsidence when volcanic activity wanes near the end of subaerial shield-building, and lava no longer crosses the shoreline in large amounts. These submerged regions can be identified because they have about the same gentle slope as the adjacent on-land part of the shield, and they are flanked on their offshore side by a steep slope which represents the former shoreline (Figure 2, Plate 1). The present depth of the transition from gentle (previously subaerial) to steep slopes is a measure of the amount of subsidence (and hence time) since lava stopped flowing from the land into the sea in the area in question [*Moore and Clague*, 1992].

The shallow water regions of the submerged terrestrial lava shelves are favorable places for coral to grow. In order for terrestrial lava to be preserved long enough to subside, it must occur in a region where eruptive activity has waned and where there is a lack of cover by rubble generated by shoreline-crossing lava flows. When sea level is falling at a rate comparable to island subsidence, the shoreline will be stable and reef growth is favored. When sea level is rising (as at the end of an ice age), then the submergence rate may exceed the rate of maximum reef coral growth (~10 mm/yr) and the reef drowns. The result of these processes is to produce a stair-step array of reefs on the older volcanoes, each representing the end of an ice age, with each deeper reef being about 100 kyrs older than the one above [*Moore and Campbell*, 1987; *Ludwig et al.*, 1991].

The best mapped and sampled array of reefs occurs on the submerged flanks of Hualalai and Mauna Kea, and on the older Kohala volcano to the north. Here a total of six sequential reefs are preserved, with ages ranging from 18-463 ka [*Moore and Clague*, 1992]. The younger four of them occur in the northwest corner of the maps (Figure 3,

Plate 1), and those that have been dated are indicated (Plate 1). They are useful in dating rates of island subsidence as well as adjacent features such as volcanoes, faults, and landslides.

Photographic evidence indicates that a reef surmounts the edge of the -375 m terrace that flanks the entire east flank of Mauna Kea volcano [*Moore and Fiske*, 1969]. A U-series age of 170 ka on a coral fragment dredged from the east scarp of this terrace [*K. Ludwig*, written comm., 1992] suggests that the terrace is possibly correlative with the -430 m (130 ka) terrace on the west side of the island, but more work is needed on this major terrace to define its history.

A -1100 m terrace occurs on the Hilo Ridge [*C. L.Wilkerson, R. T. Holcomb, J. C. Wiltshire, J. G. Sempere*, written communication, 1994] which is the apparent east rift zone ridge of Mauna Kea (Figs. 3, 5). This terrace may be correlative with the -1,145 m (360-475 ka) reef of west Hawaii [*Moore and Clague*, 1992].

The 18-ka reef at a depth of about 150 m has been sampled and dated offshore of Mauna Loa both on the west central side north and south of the Kealakekua fault and off the south cape of the island (Plate 1) [*Moore et al.*, 1990]. A probable 18-ka reef occurs offshore from Hilo on Mauna Loa where limited bathymetry defines a prominent terrace at about 140 m depth. Other Mauna Loa reefs of this age may occur about 10 km south of the Kealakekua fault on the west Mauna Loa slope, and possibly in small areas east of the south cape. No submerged reefs have been found on Kilauea, though shallow benches beginning about 15 km northwest of the east cape (Figures 2-4, 7) may support reefs.

The distribution of the regions of submerged subaerial lava (Plate 1) indicates that Mauna Kea has been out of the shield-building stage for more than 130 kyrs, Hualalai has sizable coastal regions that have not been covered by lava for 18 ka, and in places much longer, Mauna Loa has only limited such areas, and Kilauea is still actively extending its shoreline (both by lava and slumping) almost everywhere.

Landslides

Most of the submarine flanks of Mauna Loa and Kilauea that are downslope from the island are mantled by landslide deposits [*Moore et al.*, 1994b]. Relatively undisrupted submarine volcano flanks occur mainly on the rift zones. The large landslides mapped here have been grouped by morphology and degree of dislocation into two groups. Slumps (Ls) are movements of largely undeformed masses along discrete faults in which large blocks rotate along curved slip surfaces yet maintain some coherence of the volcano flanks; they are mainly slow-moving. Debris avalanches (La) are long-runout mass movements in which fragmentation has affected original volcanic forms and reduced the landslide mass to individual blocks and hummocks during sudden, catastrophic events. In some places the slumps seem to merge downslope and feed, or overly, debris avalanches.

As displayed by the slope map (Figure 3), the slump terrain resembles subaqueous lava terrain, but differs by the presence of non-uniform slopes with relatively flat benches separated by steep, sinuous scarps or ramps produced by internal deformation during the movement of the slump. In addition, circular constructs in subaqueous lava terrain are absent. The debris avalanches generally display fields of hummocks or isolated blocks up to 10 km in size in their lower reaches and amphitheaters near their heads. The lower reaches of the debris avalanches, especially near and beyond the axis of the Hawaiian deep, are commonly sedimented to the extent that only isolated blocks or hummocks of avalanche material have been mapped. Apparently these blocks represent only high points of the landslide, with the lower-lying matrix covered by sediment. In such cases, the distal extent of the landslide is known only to be seaward of the last mapped block.

The ages of the mapped landslides are poorly known, except for the currently moving Hilina slump. None of the larger, fast-moving debris avalanches are younger than ~10-30 ka, the general age of the oldest lava surfaces on Mauna Loa and Kilauea, because no tsunami deposits overlie these lavas and such landslides would have produced large waves. On the island of Hawaii only one, poorly-preserved tsunami deposit is present--on the ~100-ka west Kohala surface [*Moore and Moore*, 1988]. Well-preserved tsunami deposits on Lanai are ~100 ka and those on Molokai are ~200 ka [*Moore et al.*, 1994a]. The younger may well have been laid down by a giant wave from one of the Kona debris avalanches such as the Alika landslides. The ages of all the landslides in the map area (Plate 1) are less than ~1 Ma, the maximum age of the volcanoes on which they occur, and probably less than ~250-500 ka, the postulated time of emergence above sea level of Mauna Loa and Hualalai respectively [*Moore and Clague*, 1992]. The following review of the morphology and history of the mapped south Hawaii landslides is in approximate order from youngest to oldest.

Hilina slump. The currently active Hilina slump involves much of the south flank of Kilauea volcano (Plate 1, Figure 8). During a magnitude 7.2 earthquake in 1975, a 60-km length of Kilauea's south coast subsided as much as 3.5 m and moved seaward as much as 8 m [*Tilling et al.*,

1976; *Lipman et al.*, 1985]. In addition to such episodes of rapid movement, the south flank of the volcano also is creeping seaward continuously with the east rift zone as the northern limit of the currently slumping terrain. GPS measure-ments indicate that the region north of the rift zone is stable whereas that to the south, extending to the coast and beyond, is moving southeast at rates up to 10 cm/yr [*Owen et al.*, 1995]. The Hilina slump is moving in the same direction and probably at about the same rate, as the leading edge of the Hawaiian Ridge is advancing over the Pacific plate. Hence, the Hilina slump's movement keeps it continuously over the Hawaiian hotspot. This fact indicates that the steady slumping and spreading of volcanoes may alternate with the episodic jumping forward to initiate a new volcanic conduit as important processes that advance the Hawaiian Ridge over the Pacific plate.

The upslope limit of the Hilina slump is the east rift zone of Kilauea. The rift zone is not obvious as a boundary between the stable region to the north and a moving terrain to the south because eruptive lava largely covers this primary suture. The Hilina faults on land to the south of the rift zone are a well-defined concave extensional zone near the head of the slump which are more removed from the lava source and hence better exposed (Plate 1, Figure 8). Below the Hilina fault system, and below sea level, is a broad system of benches about 30 km long at about 3000 m depth (Figures 2-4). These benches are indented by several large closed depressions that should be rapidly filled by fragmental material moving downslope from shoreline lava entries. Apparently, active slump movement in the lower convex-downslope compressional zone near the toe of the slump favors upward thrusting [*Borgia and Treves*, 1992] and development of the ridge bounding the benches on the southeast side. This thrusting and folding is forming the depressions faster than they are being filled.

Several large elongate blocks up to 14 km in length occur on the abyssal plain at about 5000 m depth 10-20 km seaward from the primary base of the volcano slope (Figure 3). They may represent large blocks that calved off the front of the slump, blocks of the Mauna Loa edifice that were transported by an earlier debris avalanche before growth of Kilauea, or perhaps blocky upthrust terrain representing the discontinuous leading edge of a chiefly blind thrust forming the toe of an advancing slump structure.

The western edge of the Hilina slump appears to extend downslope from the western limit of the Hilina faults seaward along the southwest edge of Papa'u seamount and along the western edge of the main mid-slope bench below the Hilina fault system (Plate 1, Figures 3,8). This is consistent with the current pattern of active deformation on

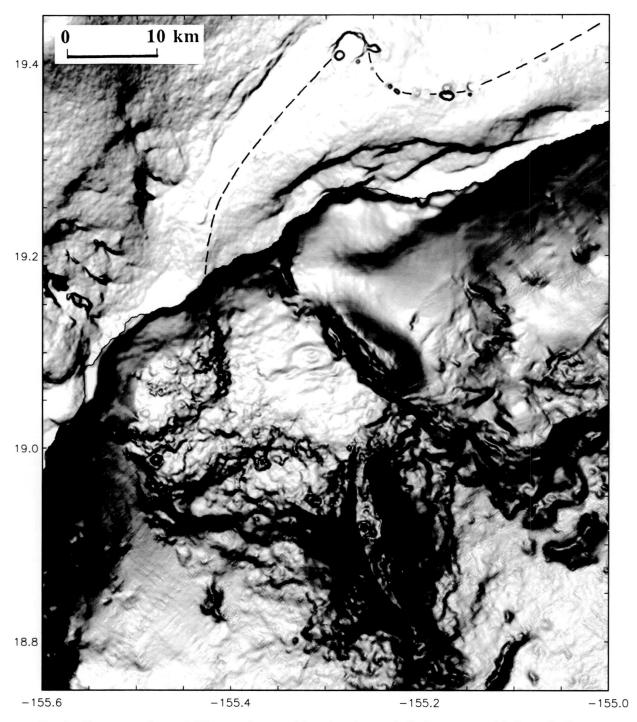

Fig. 8. Slope map of central Kilauea volcano and its submarine south flank; compare with figure 5. Steeper slopes are darker; data gridded at 100 m spacing. Dashed lines delimit southwest rift zone and east rift zone of Kilauea volcano which meet at caldera at volcano summit.

Kilauea's south flank [*Owens et al*, 1995] and the location of the Hilina faults (representing a longer, time-integrated expression of deformation), which both show that the most active part of the slump is to the east of this lineament, and that deformation abruptly decreases to the west. This boundary is also consistent with the pattern of seismicity in Kilauea's south flank which decreases to the west of this lineament [*Klein and Koyanagi,* 1989].

The eastern edge of the most active part of the Hilina slump appears to extend about to the east limit of the subaerial Hilina fault system as based on electronic distance measurements [*Swanson, et al.,* 1976; *Lipman, et al.,* 1985; *Delaney et al,* 1993] and GPS measurements [*Owens et al,* 1995] of horizontal deformation, and nearshore seismicity [*Klein and Koyanagi,* 1989]. An anomalous terraced terrain, however, extends east from the main part of the Hilina slump along the south base of the rift zone ridge to about 154° 18' W longitude. These benches extend down from about 4000 m depth to the base of the ridge where it abuts the Hawaiian Deep at 5500 m depth. They are commonly highest at their outer, southeastern margin and resemble rotational slump blocks [*Clague et al.,* 1994]. The benches may represent the distal margin of a massive slump that was driven seaward by the overlying weight of the rift zone ridge. The apparent paucity of seismicity along the submarine rift zone ridge [*Klein and Koyanagi,* 1989] suggests that current movement is minimal or aseismic.

Papa'u sand rubble flow. The Papa'u landslide is a lobate debris tongue south of central Kilauea that has moved southeast on top of the Hilina slump. It is 19 km long with a volume of about 40 km^3. Bottom photographic surveys coupled with analyzed samples collected by dredging and submersible dives indicate that it is composed of unconsolidated angular glassy basalt sand and blocks up to about 1 m in size that were originally erupted on land [*Fornari et al.,* 1979]. This is consistent with its smooth morphology in the slope maps (Figures 3,8). The general source of the landslide is reflected by the upslope reentrant of the shoreline extending down to the -200 m contour (Figure 2), which occurs directly downslope from the summit of Kilauea volcano. The sand-rubble flow was apparently fed from the oversteepened nearshore bank of basaltic hyaloclastite generated when subaerial lava flows poured into the sea. Estimates based on the present lava production of Kilauea volcano suggest that Papa'u landslide was emplaced several thousand years ago [*Fornari et al.,* 1979].

The northeast margin of the Papa'u debris lobe is smooth and grades into the surrounding fragmental debris slope, whereas the southwest margin is steeper, and bounded by ridges, troughs, and scarps (Figure 8). Perhaps the southwest margin is banked against a pre-existing structural ridge. This interpretation is supported by the observation that the structures on the southwest margin of Papa'u appear to merge southeast into structures bounding the west margin of the Hilina slump to produce a lineament nearly 40 km long. We speculate that this lineament is a system of compressional folds that accommodate convergence between the motion of the Hilina and Punaluu slumps. The west part of the Hilina slump is moving south southeast [*Owen, et al.,* 1995] whereas movement on the Punaluu slump is southeast [*Endo,* 1985; *Wyss,* 1988]. This convergence may have produced a zone of compressional folds that later guided movement of the Papa'u sand rubble flow.

Loihi debris avalanches. Loihi, the active submarine volcano growing on the south flank of Kilauea, will presumably emerge above sea level in a few tens of thousands of years and produce the next Hawaiian island. Loihi has grown up through, and is abutted on the northwest by, the Punaluu slump. It is bordered on the northeast by the active Hilina slump originating from the mobile south flank of Kilauea. Simultaneous growth of Loihi and movement of the Hilina slump has, no doubt, produced a complex interfingering of material related to each.

The Loihi edifice itself is scored by three debris avalanches, on the east, west, and south flanks, which have produced cookie-cutter-style amphitheaters reaching nearly to the crest of the volcano, and masses of hummocky debris extending downslope (Plate 1). These landslides cover about half of the volcano [*Malahoff,* 1987]. The slide headwalls are among the steepest slopes in Hawaii, attaining 40° in the first vertical kilometer at the east landslide [*Moore et al.,* 1989]. The extraordinary extent of mass wasting on Loihi has left its rift zones as narrow blade-like ridges that do not display as much of the lobate and hummocky morphology of the other submarine rift zone ridges within the map area (Figures 3,8).

North Puna Ridge debris avalanches. The presence of a steep amphitheater on the north side of the submarine east rift zone ridge of Kilauea, the Puna Ridge, suggests the presence of a sizable debris avalanche (Figure 2, Plate 1). Unfortunately the area below this 10-km-wide amphitheater lies within the region for which only limited bathymetric data are available (Figure 1). We can only estimate the extent of this landslide from available bathymetry. Other such landslides may also be present on the north side of the ridge as indicated by the generally steeper slopes on this side relative to the south flank [*Moore,* 1971], but confirmation must await more detailed work.

Ka Lae east and west debris avalanches. Two narrow debris avalanches lead from the upper reaches of the South Kona landslide south along both sides of Dana Seamount to the Hawaiian deep [*Moore et al.*, 1989]. They originate just west of Ka Lae, the south cape of the island, from which they derive their names (Plate 1). The West Ka Lae debris avalanche extends 85 km south and has produced a broad field of hummocky terrain at its foot. The east Ka Lae debris avalanche is about 75 km long and originated in a funnel shaped depression flanked on the east by the steep west-facing scarp of the rift zone ridge of Mauna Loa extending south of the south cape. The relatively small size and sharpness of the hummocks in these debris fields resemble those of the Alika 2 debris avalanche and indicate the youth of these landslides (Figure 3).

Alika debris avalanches. The Alika phase 1 and 2 debris avalanches [*Lipman et al.*, 1988; *Moore et al.*, 1989] are among the youngest and largest adjacent to the island of Hawaii. They produce sharp and distinct GLORIA images of their lower hummocky fields as a result of thin sediment cover, and hence relative youth. Ocean floor photography from surface ships and observations from submersibles indicate that they are mantled with only a thin sediment cover, less than 1 m in most places. The Alika 1 debris avalanche moved directly west about 80 km down the steep west flank of Mauna Loa and produced a broad hummocky apron.

The Alika 2 avalanche is considered younger because its upper fine-textured debris flanked by natural levees covers the Alika 1 material, and its distal hummocks are sharper and apparently less sedimented than are the more isolated and larger hummocks of the Alika 1 avalanche (Figures 3,4,9). The younger Alika 2 debris avalanche moved west from near the Mauna Loa shoreline, but turned to a northwesterly course apparently blocked by elements of the Alika 1 deposit and followed the west base of the Hawaiian Ridge into the Hawaiian Deep for a total length of about 80 km. The middle course of the avalanche is about 10 km wide and is flanked by discontinuous levees up to 100 m high [*Moore et al.*, 1992]. The broad lobe at the lower part of the avalanche is 34 km in diameter and is studded by more than 13 hummocks > 1 km in diameter, and 92 hummocks between 0.5-1 km in diameter. Detailed side-looking acoustic surveys, ocean-floor photographs, and one submersible dive demonstrate that fragmental material of all sizes occurs between the larger hummocks shown on the maps [*Moore et al.*, 1992].

Movement of the Alika slide was before ~30 ka, the general upper age limit of most surface lavas on Mauna Loa and Hualalai volcanoes, because if movement had postdated these surface lavas, traces of faulting or tsunami deposits would probably be preserved on the surface [*Moore et al.*, 1995]. Nevertheless, the source area for these slides still forms a broad embayment in Mauna Loa's west coastline and steep submarine slopes near the shore (Figure 3). The Alika 2 debris avalanche is a likely candidate to have produced the giant tsunami that swept Lanai about 105 ka [*Moore and Moore*, 1988].

North Kona slump. The North Kona slump is restricted to the submarine flanks of west Hualalai and northwest Mauna Loa volcanoes (Figure 2, Plate 1). The landslide head reached back in the shield about to the axis of the northwest rift zone of Hualalai and produced a concave scarp more than 40 km wide and up to 4 km high. The slump is marked by benches and scarps and one back-tilted block in its lower central part. The considerable age of the slump is suggest by the absence of rubble at its base. Perhaps post-slide subsidence of the volcano has depressed the toe of the slide permitting burial by the lower reaches of the Alika 2 slide, the age of which is estimated at ~100 ka [*Moore and Moore*, 1988].

The Kealakekua fault [*Stearns and Macdonald*, 1946] trends east where it comes ashore in the southern part of the slump but curves 90° on land to extend south parallel with the coast for about 15 km (Plate 1). Even though covered by younger lava its trace can be followed as a zone of steepening revealed by a detailed slope map [*Moore and Mark*, 1992]. The terrain south and seaward of the fault moved down, producing a scarp 500 m high on land, which is a minimal indicator of fault displacement because of later extensive cover by subaerial lavas particularly of the lowered (southwestern) side of the fault. The fault offsets lava flows with a K-Ar age of 166± 53 ka [*Lipman*, this volume] and is overlain by a 16-ka, 150-m deep offshore coral reef [*Moore et al.*, 1990]. The principle movement of the North Kona slump probably occurred during the period of active shield building on Hualalai prior to 130 ka [*Moore and Clague*, 1992].

Three offshore volcanic vents extending from near the shoreline to 1000 m depth are aligned east-west, parallel with, and two kilometers south of, the offshore extension of the Kealakekua fault. These vents which are too small to show on figure 5 erupted small pads of lava in 1877 that are apparently associated with the Kealakekua fault [*Moore et al.*, 1985]. Apparently the disruption of Mauna Loa's west flank by movement on the fault was profound enough to have tapped magma within the volcano so as to feed these small eruptive vents.

South Kona debris avalanche and slump. The South Kona landslide is a large and complex gravity failure extending west offshore from southwestern Mauna Loa. It appears to be composed of giant distal blocks deposited by

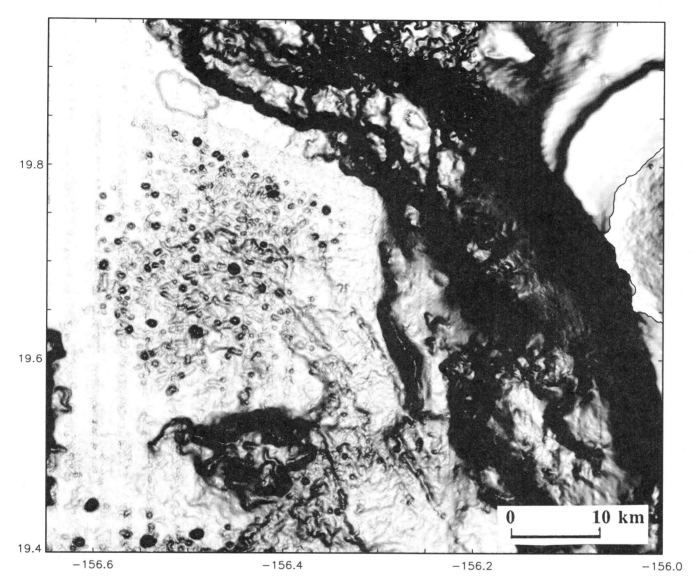

Fig. 9 Slope map of the distal apron of the Alika 2 landslide as shown in the northwest part of figure 3. Steeper slopes are darker; data gridded at 100 m spacing.

an earlier episode of avalanching, that were later overridden by a younger episode of slumping (Figure 2, Plate 1). The later slumping producing two convex-seaward benches closer inshore at mid-slope depths [*Moore et al*, 1995]. A large number of smaller blocks averaging about 2 km in size flank the outer edge of the slump benches (Figures 3,4). These were apparently produced by calving from the steep, seaward facing scarps bounding the mid-slope benches.

The distal blocks commonly have their steepest slope facing land (Figure 3). They are the largest free-standing

landslide blocks identified in the map area. Some are >10 km long and attain volumes of 7 km^3. Submersible diving and dredging on two of these large blocks 50 km from the present shoreline indicate that they contain lava sequences from Mauna Loa, some of which have been derived from vents on land [*Moore et al*, 1995].

Initial failure of the major debris avalanche that moved the blocks to their present position was in the late Pleistocene after ~250 ka, the estimated emergence of Mauna Loa volcano above sea level [*Moore and Clague*, 1992]. The general appearance of the distal blocks of the

South Kona slump when compared with the Alika debris avalanches suggests that the blocks are older. The lack of any small hummocks between the large blocks as well as the incision of gullies in the abyssal sediment between the large distal blocks (Figures 3, 4) suggest a thick sediment cover and hence considerable age, especially when compared with the Alika slides.

Punaluu slump and debris avalanche. The Punaluu slump [*Lipman, et al.*, 1990], on the southeast submarine flank of Mauna Loa, is marked by mid-slope benches and seaward bulge similar to those of the South Kona slump and the Hilina slump (Figures 2-4, Plate 1). The slump is bounded on its south side by a steep slope extending from 3 to 4 km depth below which a debris avalanche can be traced 30-50 km to abyssal depths (Plate 1). The debris avalanche may have been initiated by movement of the Punaluu slump or it may represent an older event that preceded, and is overlain by the slump. The slump is located directly downslope from the Ninole Hills which are the interfluve ridges of southeast-directed giant canyons (now nearly filled by younger lava) on the southeast subaerial flank of Mauna Loa (Figures 4,8). These canyons are believed to have been cut in the amphitheater of an ancestral Punaluu landslide. The headwalls of other large submarine landslides throughout the Hawaiian Islands are commonly cut by major erosional canyons, apparently as a result of landslide-induced coastal oversteepening [*Moore et al.,* 1989[.

The initial Punaluu landsliding occurred before much of the growth of both Kilauea and Loihi volcanoes, both of which then helped to stabilize the southeast flank of Mauna Loa. However, the Punaluu slump may be still active because a zone of active seismicity on Mauna Loa's southeast flank underlies the source area of the landslide in a pattern that is similar to the seismicity beneath the Hilina faults on Kilauea's south flank landward of the Hilina slump [*Klein and Koyanagi*, 1989]. Also, two earthquakes > magnitude 7 in 1868 may have been generated by movement on both the Punaluu and Hilina slumps [*Clague and Denlinger*, 1993].

The upper part of the Punaluu slump appears to be overlain on its west side by lava, erupted subaqueously from the southwest rift zone of Kilauea. This region which forms a plateau from 900-1,200 m depth and 10 km wide may have formed from lava erupted from the southwest rift as the rift migrated ~10 km east from near the Mauna Loa-Kilauea boundary to its present position (Plate 1, Figure 8). The east edge of the Punaluu slump is obscured by the younger Hilina and Papa'u slides which overlie it. About 10 km east of Papa'u seamount is a steep east-facing zone of irregular lobate terrain (Figures 2-4, 8) that may mark the buried eastern edge of the Punaluu slump which has now been incorporated into the younger Hilina slump.

A distinct elliptical shape 5 km in major diameter is visible in digital images west of Papa'u seamount on the Punaluu slump (Figures 3, 4, 8). Detailed bathymetry [*Chadwick et al.*, 1993] suggests that this is a shallow slump headwall and toe.

ABYSSAL SEDIMENT

Most of the flat ocean floor at abyssal depths seaward of the base of the volcano flanks is covered by sediment (As). Isostatic subsidence of the Hawaiian Ridge has produced a giant downwarp beneath it due to the weight of the young volcanoes making up the ridge. The axis of the trough adjacent to the ridge caused by this downwarp, the Hawaiian Deep, is close to the base of the ridge. It is a zone of active sedimentation by turbidity currents, and is a common destination for major debris avalanches (Plate 1). Outboard from the deep the sediment thins progressively to a general thickness of ~100 m where the sedimentation rate is little affected by proximity to the island. The age of the oceanic crust in the Hawaiian region is ~110 million years [*Waggoner*, 1993] and hence the average rate of this pelagic sedimentation is ~1 m/m.y. Closer to the volcanoes, sediment thickness increases rapidly because of the contribution of detritus from the volcanoes carried by windblown dust, volcanic ash, and turbidity currents. The average rate of sedimentation on the distal part of several debris avalanches as determined by sediment thickness measured by sonar and age estimated from the age of the end of shield building of the host volcano is ~ 2.5 m/m.y. [*Moore et al.*, 1994b], and the sedimentation rate at Ocean Drilling Program site 842 (320 km west of the island of Hawaii) is 4-5 m/Ma [*Garcia and Hull,* 1994].

In the southwest part of the map area abyssal sediment is particularly thick and is eroded and scored by a dendritic pattern of gullies a few tens of meters deep (Figures 3, 4). Some of the gullies head between the giant blocks of the distal part of the South Kona debris avalanche and lead north into the axis of the Hawaiian Deep north of the terminus of the Alika 2 landslide. They were apparently carved by turbidity currents. The absence of such gullies in the other landslides flanking Mauna Loa, indicates a thinner sediment cover and more recent movement than the distal part of the South Kona landslide, assumed, therefore, to be the oldest mapped gravity failure in the region.

In the southeast and northeast corners of the mapped area the sediment is thinner and hence contains the largest pelagic component. These regions of thinnest sediment can be identified by the fact that linear, north-trending, fault-controlled ridges have been mapped from GLORIA images

(Plate 1) since they are not yet drowned by sediment. Although several of these fault controlled ridges occur in the southeast corner of the map of Plate 1, they are obscured by the map legend. These nearly north-trending, fault-bounded ridges are part of the sea floor grain, exposed over much of the deep ocean, and represent fault scarps produced at, and parallel with, the Cretaceous spreading ridge at the time that the oceanic crust was formed.

STRUCTURE

Mauna Loa and Kilauea are shaped like table mountains. The volcanoes are much flatter above sea level than below [*Mark and Moore*, 1987] (Figures 2, 3, 6, 10). This shape results from the fact that subaerial lava behaves in a more fluid fashion and forms lower slopes as compared to lava that crosses the shoreline or is erupted beneath the sea. The classic shield shape of Hawaiian volcanoes refers only to that above sea level. The subaqueous lava does not flow as freely as that on land; where lava crosses the shoreline it is rapidly quenched and undergoes cooling-contraction granulation. In addition, the buoyant effect of water reduces the effective specific gravity of lava (or rubble) under water and inhibits its movement downslope [*Mark and Moore*, 1987]. These factors cause subaqueous material to pile up and produce steeper submarine slopes. As the shoreline advances, the width and height of this steep offshore slope increases, until instability is attained and some form of gravitational failure occurs.

A striking feature of the geologic map (Plate 1) is the predominance of landslides below sea level as opposed to little-faulted lava flows above. The realization of the importance of these landslides can be attributed directly to marine surveys, particularly the side-scan sonar (GLORIA) surveys beginning in 1986. The overall gravity-induced spreading of the volcanoes, of which the landslides are a prime indicator, is proving to be a fundamental process of volcanic growth and decline [*Borgia and Treves*, 1992]. This spreading has shaped most of the fault systems and rift zones, and modulates eruptive activity.

The contrast of rock types on- and off shore points to the overall internal structure of the volcanoes that make up the island of Hawaii. Coherent subaerial lava flows are the dominant rock type above sea level, whereas fragmented basalt and landslide debris in all stages of disruption and fragmentation are most common below sea level. Most lava is erupted above sea level except on the submarine extension of the rift zones, and hence as the island grows, the subaerial lava overrides the fragmental subaqueous facies. This major boundary, therefore, must extend beneath the island to its center. The original boundary

between the subaerial and submarine facies, however, has subsided more (and is hence deeper) toward the center of the volcanoes because of the greater age of the inner, now covered, shorelines.

Two cross sections extending from the island to the offshore region (Figure 6) provide some speculative views of the gross structure of the south Hawaii region. Sections A-A' and B-B' of south Mauna Loa and Kilauea respectively are drawn assuming that the volcanoes are spreading seaward away from the molten or plastic cores of their rift zones [*Delaney et al*, 1990] as a result of their height and weight. Volcano spreading may occur on a zone of hot olivine cumulates at the base of the volcano beneath its magma chamber and rift zones [*Clague and Denlinger*, 1994]. This spreading (slumping) toward the unbuttressed volcano flanks will produce a three-fold layering both in the stable part of the volcano and in the upper part of the slump. The layers would consist of, from the surface down, a shallow layer of lava flows and hyaloclastites fed from subaerial vents on the rift zone, an intermediate sheeted dike complex of rift zone feeders, and a deep gabbroic layer (perhaps with cumulates) produced from the solidified walls and floors of the magma chamber and master rift zone feeding system. This three-fold structure can be compared to the generation of oceanic crust by spreading at mid-ocean ridges [*Borgia and Treves*, 1992].

One view of the substructure of the volcanoes near Hilo was afforded by a one-km-deep scientific drill hole [*Stolper et al.*, 1994] completed in 1994. The hole, near sea level, penetrated 280 m of Mauna Loa materials including subaerial lava with minor hyaloclastites and reef-derived sediments, indicating that the Mauna Loa shoreline was generally close to the drill hole site for about the last 100,000 years. Below 280 m the hole entered lava from Mauna Kea volcano which persisted to the bottom of the hole at one kilometer depth. All of the Mauna Kea lavas are subaerially-erupted lava flows indicating that the volcano has subsided at least one kilometer since the eruption of the bottomhole lava at about 450 ka [*Brent Turrin*, written comm., March, 1995]. This subsidence is confirmed by the presence offshore of a drowned coral reef submerged to ~375 m.

A primary structural element in most of the volcanoes is the presence of two or more volcanic rift zones (delineated as short dashed lines on the geologic map, Plate 1, Figure 6). The rift zones radiate from the volcano summits across the shoreline to the deep ocean floor [*Fornari*, 1987] and serve as conduits for sub-horizontal transmission of magma from the sub-summit magma chamber to the flanks of the volcano. Generally, the volume of eruptions from the rift zones exceeds that from the summit, causing distinct ridges

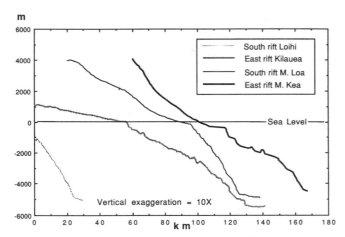

Fig. 10. Longitudinal profiles of the south rift zone of Loihi, east rift zone of Kilauea, south rift zone of Mauna Loa, and east rift zone of Mauna Kea. Profiles extend downward from the center of the summit or caldera to the bases of the volcanoes.

to mark the rift zones both above and, to an even greater degree, below sea level. The rift zones are apparently zones of extension which accommodate gravitational spreading of the edifice [*Fiske and Jackson*, 1972]. They are marked by eruptive vents and fissures, open cracks, pit craters and grabens. These features may be distributed over a zone up to 5 km wide; the dashed lines depicting the rift zones of the geologic map (Plate 1) mark only the axis or center of the rift zones. In addition to symmetrical rift zone spreading outward from their axes, the flanks of submarine rift zones may be carved by extensive landslides. Such landslides commonly cut back to the rift zone axes, where the reinforcement from dike complexes apparently inhibits further failure. Such a process is evident on the south side of the Hualalai northwest rift, the west side of the Mauna Loa south rift, both sides of the Loihi rifts, and possibly the north side of the Kilauea east rift (Plate 1).

The rift zone crests, like the flanks of the volcanoes, are steeper underwater, but the overall steepness of the longitudinal rift zone axes is less than that of volcano flanks both above and below sea level. The rift zones are, on average, longer than most volcano flanks since they generally cross the shoreline at island salients and they terminate at abyssal depths at salients of the volcano base. The steepening of the rift zone underwater cannot be due to all the processes that cause the volcano flanks to be steeper underwater because lava issues, over time, from innumerable vents spaced along the rift zone both above and below sea level, and is not primarily flowing from subaerial vents into the sea as is the case with other volcano flanks. Hence, the factors controlling the more gentle rift axis slope above sea level than below are related

to those that favor the subterranean movement of magma to subaerial vents and the extrusion of lava at subaerial vents. Included in these factors is the lower effective density difference between the magma and fluid (air or water) in which the volcano is growing, as well as the greater resistance to magma flow within the rift seaward of the coastline because of an increase in cooling rate [*Lonsdale*, 1989]. In addition, the expansion of volatiles at subaerial vents relative to submarine vents may also control the volume and rate of subaerial vs. submarine eruption.

The longitudinal profiles of rift zones from four volcanoes show a distal steepening of the submarine part of the rift zone below 2300-3200 m depth despite the variations in the overall subaqueous steepness between the profiles (Figure 10). One possible explanation for this increase in slope with depth is the effect of the critical pressure of water on eruptive processes. The critical constants of pure water are 374°C and 220 bars and of seawater ~405°C and ~300 bars [*Bishoff and Rosenbauer*, 1985]. The seawater critical pressure is equivalent to ~2900 m depth when the density of sea water is taken into account. Above this depth seawater heated to more than 405° C by lava will boil and the expanding steam will tend to clear the vent in the same (though much reduced) manner that phreatomagmatic eruptions near sea level are rendered explosive by the expansion of steam. Below 2900 m, seawater will not expand as much with the same amount of heating and will not promote eruption as vigorously.

The growth of Hawaiian rift zones may be examined by comparing the longitudinal rift zone profiles of four volcanoes, Loihi, Kilauea, Mauna Loa and Mauna Kea in order from youngest to oldest (Plate 1, Figure 10). The rift zones begin to form early in the history of a volcano and eventually grow above sea level as the volcano matures. The submarine south rift zone of the embryonic Loihi volcano is 25 km long, the east rift of Kilauea is 130 km long (55 % of which is submarine) the south rift of Mauna Loa is 108 km long (28 % of which is submarine), and the east rift zone of Mauna Kea (which is not exposed above sea level) is 107 km long (61% of which is submarine, Figure 10, Plate 1). The submarine part of all rift zones are convex upward, whereas the subaerial part only of Kilauea is convex upward. Mauna Loa's subaerial part is complex with two convex upward and two concave upward segments, and Mauna Kea's is entirely concave upward.

Loihi is the steepest, averaging 8-11°, all below sea level. Kilauea has developed the gentlest profile, perhaps because it is in a stage of most vigorous activity with the most frequent and largest eruptions occurring along the length of the rift zone establishing an equilibrium gradient of 1.2° from caldera to shoreline, and then 2.8° to 5.6°

below sea level. The older, larger, and higher Mauna Loa is now erupting a greater proportion of lava from its subaerial part, thus advancing its shoreline and increasing the proportion of its rift zone above sea level. The submarine part is substantially steepened, (relative to Kilauea) to 7.1°-10.3°. The Mauna Kea rift zone is obscured in its subaerial part, apparently by its cap of alkalic lava. These lava flows are more viscous causing them to be thicker which has substantially increased the upper subaerial slope making it distinctively concave upward [*Moore and Mark*, 1992].

Mauna Kea, and to a lesser extent Mauna Loa, show the subsidence of the slope change that marked the position of sea level when lava flows were last actively crossing the shoreline. This slope change is now at 150 m depth on Mauna Loa and about 400 m depth on Mauna Kea (Figure 11).

The major mapped fault systems on land, except for those bounding the summit calderas, were apparently generated in the upper, tensional regime of giant landslides that are mapped below sea level on the submarine flanks of the volcanoes and the surrounding abyssal plain [*Moore and Mark*, 1992]. The best example of faults related to landsliding is the Hilina fault system on the south side of Kilauea volcano, where a series of steep, concave-seaward normal fault scarps occurs near the upper part of the Hilina slump (Plate 1, Figure 8). Likewise, a major system of subaerial faults occurs on the upper reaches of the Punaluu landslide on the southeast flank of Mauna Loa. Most of the displacement on these faults apparently occurred at an earlier time when the Punaluu slump was unstable due to the absence of Kilauea and Loihi volcanoes which now buttress this flank of Mauna Loa. The Punaluu slide, however, may be still active since this region remains the most seismic part of Mauna Loa [*Klein and Koyanagi*, 1989], and Hawaii's biggest historic earthquake is attributed to seaward movement in this region [*Wyss*, 1988]. A third system of major subaerial faults (also with some historic activity) occurs at the upper reaches of the North Kona slump and the Alika debris avalanches on the west flank of Mauna Loa (Plate 1). The southern extension of the Kealakekua fault is covered by unbroken young lava cover.

The South Kona, Hilina, and Punaluu slumps are characterized by mid-slope benches. These benches are apparently related to major thrust faults in the lower, compressional part of the slumps. Thrusting elevates the seaward part of the bench, forms a sediment trap behind, and creates a steep scarp at the seaward edge of the bench. Borgia et al (1990) suggest that such thrusts are probably blind thrusts surmounted by folds.

DIRECTIONS FOR FUTURE WORK

Much work remains in unraveling the geologic processes and products that have shaped the submarine parts of the Hawaiian Ridge. The preparation of the geologic map of offshore Hawaii (Plate 1) has helped to identify many of the problems that should be explored, and some of these are mentioned in the text. More sampling and observations are needed to verify assignments of rock units and to test assumptions of important geologic processes.

Precisely-located multibeam bathymetry is essential in providing a key data source for evaluation of oceanfloor morphology and geology, and to serve as base maps for future studies. Marine studies can proceed in a systematic fashion only after the areas offshore from all the main Hawaiian Islands have been covered by such bathymetry.

The giant landslides are a critical element in understanding the volcanic processes that built these volcanoes. Efforts should continue to collect and analyze lava samples from the submerged landslide material in order to determine the host volcano and the environment of its eruption, whether above sea level or in shallow or deep water. Determination of the stage of magmatic evolution of the material may place better age constraints on landsliding. The landslide blocks can be regarded as samples plucked from the interior of the volcanoes and will provide information on the internal makeup and structure of the volcanoes.

The giant landslides constitute a potential hazard. Catastrophic movement of a giant debris avalanche would cause tsunamis to sweep the Pacific Basin. We need, therefore, to determine the time of failure of as many of the mapped landslides as possible so that the frequency of their generation can be estimated. Such studies could include radiometric dating of rock units involved in landsliding to provide maximum ages, and dating of units such as sediment or MnFe oxides covering slide material to determine minimum ages. Systematic sediment coring beyond the distal reaches of landslides may reveal evidence of landslide-induced disruption and deposition of turbidites within the sedimentary column that can be dated.

Geodetic methods need be expanded to monitor the non-catastrophic movement of landslides, particularly for the Hilina slump. The onland GPS measurements underway are an important element in such a program, but the area of coverage should be expanded to monitor west Hawaii. Acoustic geodetic measurements should be initiated on the submarine part of the Hilina slump, where movement rates are perhaps greater than on land. Ocean-bottom seismometers, tied into the on-land seismic net, should be placed so that the seismicity of the submarine part of the Hilina slump can be studied.

Acknowledgments. In addition to those cited, we wish to mention specifically some of the dedicated individuals from several institutions who spent weeks at sea in the mapped area collecting the data that is the essence of this report. These include scientists of the GLORIA program: R. H. Belderson, D.A. Clague, C. Gutmacher, R. T. Holcomb, M. Holmes, P. W. Lipman, W. R. Normark, R. Searle, A. Shor, M. E. Torresan, J. B. Wilson, and of other oceanographic cruises, W. B. Bryan, D.A. Clague, M. O. Garcia, J. P. Lockwood, W. R. Normark, and M. E. Torresan. Submersible diving programs were done under the auspices of NOAA's Hawaiian Undersea Research Laboratory, Woods Hole Oceanographic Institution, and the U. S. Navy: The scientific observers included W. B. Bryan, D.A. Clague, D. J. Fornari, M. O. Garcia, A. Malahoff, W. R. Normark, and J. R. Smith.

The NOAA multibeam data (areas 3 and 5 in Figure 1) was processed by D.R. Herlihy of NOAA's National Ocean Service. The digital data was compiled and our maps were produced at the NOAA Pacific Marine Environmental Laboratory, Newport, Oregon, with the assistance of C. G. Fox. WWC expresses thanks to R.W. Embley and S.R. Hammond for their support during this project. Thanks also to J.R. Smith, C.L. Wilkerson, and R.T. Holcomb for generously sharing their data for this compilation.

Plate 1 and Figure 5 were drafted by Ellen Lougee, and Glenn Schumacher provided computer support for printing of the slope and shaded relief maps. Critical reviews by Melvin Beeson, Michael Garcia, and Peter Lipman substantially improved the manuscript and are appreciated.

REFERENCES

Bischoff, J. L. and R. J. Rosengbauer, An empirical equation of state for hydrothermal seawater (3.2 percent NaCl): *Am. Jour. Science, 285,* 725-763, 1985.

Borgia, A., J. Burr, W. Moniero, L. D., Morales, and G. E. Alvarado, Fault propagation folds induced by gravitational failure and slumping of the Central Costa Rica volcanic range: implications for large terrestrial and Martian volcanic edifices, *J. Geophys. Res. 95,* 14,357-82, 1990.

Borgia, A., and Treves, B., Volcanic plates overriding the oceanic crust: Structure and dynamics of Hawaiian volcanoes: *Geol. Soc. London Special Publication 60,* 277-299, 1992.

Chadwick, W. W., Jr., Moore, J. G., Garcia, M. O., and Fox, C. G., Bathymetry of southern Mauna Loa volcano, Hawaii: *U. S. Geol. Survey Miscellaneous Field Studies Map MF-2233,* scale 1:150,000, 1993.

Chadwick, W. W., Jr., Smith, J. R. Jr., Moore, J. G., Clague, D. A., Garcia, M. O., and Fox, C. G., Bathymetry of south flank of Kilauea volcano, Hawaii: *U. S. Geol.Survey Miscellaneous Field Studies Map MF-2231,* scale 1:150,000, 1993.

Chadwick, W.W. Jr., Moore, J. G., and Fox, C. G., Bathymetry of the southwest flank of Mauna Loa Volcano, Hawaii: *U.S. Geol.Survey Miscellaneous Field Studies Map MF-2255,* Scale 1:150,000, 1994a.

Chadwick, W.W., Jr., Moore, J. G., and Fox, C. G., Bathymetry of the west-central slope of the Island of Hawaii: *U.S. Geol. Survey Miscellaneous Field Studies Map MF-2269,* scale 1:150,000, 1994b.

Clague, D. A. and R. P. Denlinger, The M7.9 1868 earthquake: Hawaii's active landslides (abs.), *Eos, 74,* 635, 1993.

Clague, D. A. and R. P.and Denlinger, Role of olivine cumulates in destabilizing the flanks of Hawaiian volcanoes: *Bull. Volcanol., 56,* p.425-434, 1994.

Clague, D. A., K. A. Hon, J. L. Anderson, W. W. Chadwick, Jr., and C. G. Fox, Bathymetry of Puna Ridge, Kilauea volcano, Hawaii: *U. S. Geol. Survey Miscellaneous Field Studies Map MF-2237,* 1:150,000, 1994.

Clague, D. A., J. G. Moore, J. E. Dixon, W. B. Friesen, Petrology of submarine lavas from Kilauea's Puna Ridge, Hawaii, *Jour. Petrology,* in press, 1995.

Delaney, P. T., R. S. Fiske, A. Miklius, A. T. Okamura, and M. K. Sako, Deep magma body beneath the summit and rift zones of Kilauea Volcano, Hawaii, *Science, 247,* 1265-1372, 1990.

Delaney, P. T., A. Miklius, T. Arnadottir, A. T. Okamura, and M. K. Sako, Motions of Kilauea volcano during sustained eruption from the Pu'u O'o and Kupai'anaha vents, 1983-1991, *J. Geophys. Res., 98,* 17,801-17,820, 1993.

Duffield, W. A., R. L. Christiansen, R. Y. Koyanagi, and D. W. Peterson, Storage, migration, and eruption of magma at Kilauea volcano, Hawaii, 1971-1972, *J. Volcan. Geoth. Res., 13,* 273-307, 1982.

Endo, E. T., Seismotectonic framework for the southeast flank of Mauna Loa volcano, Hawaii, *Ph.D. Thesis, University of Washington, Seattle, Washington,* 1985.

Fiske, R. S., and E. D. Jackson, Orientation and growth of Hawaiian volcanic rifts-The effects of regional structure and gravitational stresses, *Proc. Royal. Soc. London, A 329,* 299-326, 1972.

Fornari, D. J., Thle geomorphic and structural development of Hawaiian rift zones, *U. S. Geol. Survey Professional Paper 1350,* 125-132, 1987.

Fornari, D. J., J. P. Lockwood, P. W. Lipman, M. Rawson, A. Malahoff, Submarine volcanic features west of Kealakekua Bay, Hawaii, *J. Volc. Geotherm. Res., 7,* 323-327, 1980.

Fornari, D. J., J. G. Moore, L. Calk, A large submarine sand-rubble flow on Kilauea volcano, Hawaii: *Jour. of Volcan. and Geothermal Res., 5*, 239-256, 1979a.

Fornari, D. J., Peterson, D. W., Lockwood, J. P., Malahoff, A., and Heezen, B. C., Submarine extension of the southwest rift zone of Mauna Loa volcano, Hawaii: visual observations from U. S. Navy Deep submergence vehicle DSV *Sea Cliff* : *Geol. Soc. Am. Bull. 90*: 435-443, 1979b.

Garcia, M. O. and D. M. Hull, Turbidites from giant Hawaiian landslides: results from Ocean Drilling Program site 842, *Geology, 22,* 159-162, 1994.

Garcia, M. O., T. P. Hulsebosch, and J. M. Rhodes, Glass and mineral chemistry of olivine-rich submarine basalts, southwest rift zone, Mauna Loa volcano: implicatrions for magmatic processes, this volume.

Garcia, M. O., B. A. Jorgenson, J. J. Mahoney, E. Ito, and A. J. Irving, An evaluation of temporal geochemical evolution of Loihi summit lavas; results from Alvin submersible dives, *Jour. Geoph. Res., 98,* 537-550, 1993.

Holcomb, R. T., J. G. Moore, P. W. Lipman, and R. H. Belderson, Voluminous submarine lava flows from Hawaiian volcanoes, *Geology, 16,* 400-404, 1988.

Hon, K, T. Mattox, J. Kauahikaua, and J. Kjargaard, The construction of pahoehoe lava deltas on Kilauea volcano, Hawaii (abs.): *Eos, 74,* 616, 1993.

Klein, F. W. and R. Y. Koyanagi, The seismicity and tectonics of Hawaii, The Geology of North America, v. N, The eastern Pacific Ocean and Hawaii, *Geol.Soc. Am., Chap. 12,* 238-252, 1989.

Lipman, P. W., J. P. Lockwood, R. T. Okamura, D. A. Swanson, and K. M. Yamashita, Ground deformation associated with the 1975 magnitude-7.2 earthquake and resulting changes in activity of Kilauea volcano, Hawaii, U. S. Geol. Surv. Prof. Pap. 1276, 45pp, 1985.

Lipman, P. W., Declining growth of Mauna Loa during the last 1,000,000 years: rates of lava accumulation vs. gravitational subsidence, this volume,.

Lipman, P. W., Normark, W. P., Moore, J. G., Wilson, J. B., and Gutmacher, C. E., The giant submarine Alika debris slide, Mauna Loa, Hawaii: *Jour.Geophys. Res., 93,* 4279-4299, 1988.

Lipman, P. W., Rhodes, J. M. and Dalrymple, G. B., The Ninole basalt--Implications for the structural evolution of Mauna Loa volcano, Hawaii: *Bull. Volcan., 53,* 1-19, 1990.

Lonsdale, P., A geomorphological reconnaissance of the submarine part of the east rift zone of Kilauea volcano, Hawaii: *Bull. Volcan., 51,* 123-144, 1989.

Ludwig, K.R., Szabo, B. J., Moore,J. G.,and Simmons, K. R.,

Crustal subsidence rate off Hawaii, determined from $^{234}U/^{238}U$ ages of drowned coral reefs: *Geology, 19,* 171-174, 1991.

Malahoff, A., Geology of the summit of Loihi submarine volcano: *U. S. Geol. Survey Prof. Paper 1350,* 133-144, 1987.

Mark, R. K., and Moore, J. G., Slopes of the Hawaiian Ridge: *U. S. Geol. Survey Prof. Paper 1350,* 101-107, 1987.

Moore, G. W. and J. G. Moore, Large-scale bedforms in boulder gravel produced by giant waves in Hawaii, *Geol. Soc. Am. Spec. Pap. 229,* 101-110, 1988.

Moore, J. G., Petrology of deep-sea basalt, Hawaii, *Am. J. Sci, 263,* 40-52, 1965.

Moore, J. G., Bathymetry and geology-east cape of the island of Hawaii: *U. S. Geol. Survey Misc. Geol. Inves. Map I-677,* scale 1:62,500, 1971.

Moore, J. G., Subsidence of the Hawaiian Ridge, U. S. *Geol. Survey Prof.Paper 1350,* 85-100, 1987.

Moore, J. G., W. B. Bryan, M. H. Beeson, and W. R. Normark, Giant blocks in the South Kona Landslide, Hawaii: *Geology, 23,* 125-128, 1995.

Moore, J. G., W. B. Bryan, and K. R. Ludwig, Chaotic deposition by a giant wave, Molokai, Hawaii, *Geol. Soc. Am. Bull. 106,* 962-967, 1994a.

Moore, J. G.and J. F. Campbell, Age of tilted reefs, Hawaii: *Jour. Geoph.Res., 92,* 2641-2646, 1987.

Moore, J. G. and Clague, D. A., Coastal lava flows from Mauna Loa and Hualalai volcanoes, Kona, Hawaii, *Bull. Volcan., 49,* 752-764, 1987.

Moore, J. G., and Clague, D. A., Growth of the island of Hawaii: *Geol. Soc. Am. Bull., 104,* 1471-1484, 1992.

Moore, J. G., Clague, D. A., Holcomb, R. T., Lipman, P. W., Normark,W. R., and Torresan, M. E., Prodigious submarine landslides on the Hawaiian Ridge: *Jour. Geoph. Res., 94,* 17,465-17,484., 1989.

Moore, J. G. and Fiske, R. S., Volcanic substructure inferred from dredge samples and ocean-bottom photographs, Hawaii, *Geol. Soci. Am. Bull., 80,* 1191-1201, 1969.

Moore, J. G.; D. J. Fornari; and D. A. Clague, Basalts from the 1877 submarine eruption of Mauna Loa Volcano, Hawaii: new data on the variation of the palagonitization rate with temperature: *U.S. Geol. Survey Bull. 1663,* 1-11,1985.

Moore, J. G., and Mark, R. K., World Slope Map: *Eos, 67,* 1353-1362, 1986.

Moore, J. G. and R. K. Mark, Morphology of the island of Hawaii: *G S A Today, 2,* 257-259 & 262, 1992.

Moore, J. G., and G. W. Moore, Deposit from a giant wave on the island of Lanai, Hawaii, *J. Geoph. Res. 226,* 1312-1315, 1984.

Moore, J. G., W. R. Normark, and B. J. Szabo, Reef growth and volcanism on the submarine southwest rift zone of Mauna

Loa, Hawaii, *Bull. Volcan., 52,* 375-380, 1990.

Moore, J. G., W. R. Normark, and C. E. Gutmacher, Major landslides on the submarine flanks of Mauna Loa volcano, Hawaii, *Landslide News, 6,* 13-16, 1992.

Moore, J. G., W. R. Normark, and R. T. Holcomb, Giant Hawaiian landslides, *Annu. Rev. Earth Planet. Sci., 22,* 119-144, 1994b.

Moore, J. G., R. L. Phillips, R. W. Grigg, D. W. Peterson, and D. A. Swanson, Flow of lava into the sea 1969-1971, Kilauea Volcano, Hawaii, *Geol. Soc. Am. Bull., 84,* 537-546, 1973.

Owen, S., P. Segall, J. Freymueller, A. Miklius, R. Denlinger, T. Arnadottir, M. Sako, and R. Burgmann, Rapid deformation of the south flank of Kilauea volcano, Hawaii, *Science,* 267, 1328-32, 1995.

Peterson, D. W., R. B. Moore, Geologic history and evolution of geologic concepts, island of Hawaii, *U. S. Geol. Surv. Prof. Pap., 1350,* 149-189, 1987.

Sager, W. W. and M. S. Pringle, Paleomagnetic evidence for Cretaceous age of two volcanoes on the south flank of the island of Hawaii, *Geoph.Res. Letters, 17,* 2445-48. 1990,

Stearns, H. T., and G. A.Macdonald, Geology and ground-water resources of the island of Hawaii, *Hawaii Div. Hydrography, 9,* 1-363, 1946.

Stolper, E. M., D. M. Thomas, D. J. DePaolo, Hawaii scientific drilling project: baciground and overview of initial results (abs.), *Eos, 75,* 707, 1994.

Swanson, D. A., W. A. Duffield, and R. S. Fiske, Displacement of the south flank of Kilauea volcano: the result of forceful intrusion of magma into the rift zones, *U. S. Geol. Surv. Prof. Pap. 963,* 30pp., 1976.

Tilling, R. I., R. Y. Koyanagi, P. W. Lipman, J. P. Lockwood, J. G. Moore, and D. A. Swanson, Earthquake and related catastrophic events, Island of Hawaii, November 29, 1975, A preliminary report, *U. S. Geol. Survey Circular 740,* 1-33, 1976.

Waggoner, D. G., The age and alteration of central Pacific oceanic crust near Hawaii, Site 843, *Proc. Ocean Drill. Prog., Scientific Rusults, 136,* 119-132, 1993.

Wyss, M., A proposed source model for the great Kau, Hawaii, earthquake of 1868, *Bull. Seismological Soc. Am.,* 78, 1450-1462., 1988.

J. G. Moore, Branch of Volcanic and Geo-thermal Processes, U. S. Geological Survey, 345 Middlefield Road, MS 910, Menlo Park, CA 94025

W.W. Chadwick, Jr., Hatfield Marine Sciences Center, Oregon State University, Newport, OR 97365

Declining Growth Of Mauna Loa During The Last 100,000 Years: Rates of Lava Accumulation vs. Gravitational Subsidence

Peter W. Lipman

Branch of Volcanic and Geothermal Processes, U.S. Geological Survey, Menlo Park, CA 94025

Long-term growth rates of Hawaiian volcanoes are difficult to determine because of the short historical record, problems in dating tholeiitic basalt by K-Ar methods, and concealment of lower volcanic flanks by 5 km of seawater. Combined geologic mapping, petrologic and geochemical studies, geochronologic determinations, marine studies, and scientific drilling have shown that, despite frequent large historical eruptions (avg. 1 per 7 years since mid 19th century), the lower subaerial flanks of Mauna Loa have grown little during the last hundred thousand years. Coastal lava-accumulation rates have averaged less than 2 mm/year since 10 to 100 ka along the Mauna Loa shoreline, slightly less than recent isostatic subsidence rates of 2.4-2.6 mm/yr. Since 30 ka, lava accumulation has been greatest on upper flanks of the volcano at times of summit caldera overflows; rift eruptions have been largely confined to vents at elevations above +2,500 m, and activity has diminished lower along both rift zones. Additional indicators of limited volcanic construction at lower levels and declining eruptive activity include: (1) extensive near-surface preservation of Pahala Ash along the southeast coast, dated as older than about 30 ka; (2) preservation in the Ninole Hills of block-slumped ancestral Mauna Loa lavas erupted at 100-200 ka; (3) preservation low in the subaerial Kealakekua landslide fault scarp of lavas newly dated by K-Ar as 166±53 ka; (4) preservation of submerged coral reefs (150 m depth) dated at 14 ka and fossil shoreline features (as much as 350-400 m depth), with estimated ages of 130-150 ka, that have survived without burial by younger Mauna Loa lavas and related ocean-entry debris; (5) incomplete filling of old landslide breakaway scars; (6) limited deposition of post-landslide lava on lower submarine slopes (accumulation mostly <1,000 m depth); and (7) decreased deformation and gravitational instability of the volcanic edifice. In addition, the estimated recent magma-supply rate for Mauna Loa, about $28 \times 10^6 \, m^3/yr$ since 4 ka (including intrusions), is inadequate to have constructed the present-day edifice ($80 \times 10^3 \, km^3$) within a geologically feasible interval (0.6-1.0 m.y.); higher magma supply ($100 \times 10^6 \, m^3/yr?$, comparable to present-day Kilauea) must have prevailed during earlier times of more rapid volcano growth. Interpreted collectively, these features indicate that the emerged area of Mauna Loa and its eruptive vigor were greater in the past than at present. Volcanic growth due to lava accumulation has been offset by subsidence and by landsliding on the lower flanks of the volcano. Along with the apparent "drying up" of distal parts of the rift zones, these features suggest that Mauna Loa is nearing the end of the tholeiitic shield-building stage of Hawaiian volcanism.

Mauna Loa Revealed: Structure,
Composition, History, and Hazards
Geophysical Monograph 92
This paper is not subject to U. S. Copyright.
Published in 1995 by the American Geophysical Union

1. BACKGROUND

Despite rapid subsidence . . . Mauna Loa's robust activity. . . now in its vigorous shield building stage. . . has built the volcano probably higher now than ever before."

[*Moore and Clague*, 1992]

Long-term growth rates of Hawaiian volcanoes are poorly understood because of the short historical record, difficulties in dating relatively young tholeiitic basalt by isotopic methods, and concealment of the volcanoes' lower flanks by 4-5 km of seawater. Nevertheless, combined geologic mapping, ^{14}C and K-Ar determinations, marine studies, and scientific drilling have begun to provide partial constraints on the growth of Mauna Loa and other Hawaiian volcanoes over time periods of tens to hundreds of thousands of years.

Mauna Loa, the largest volcano on Earth rising 8.5 km from the Hawaiian Deep (Figures 1, 2), is a mature Hawaiian basaltic shield. Its morphology and structure are complex because of changing rates of lava accumulation and subsurface intrusion [*Lipman*, 1980a; *Lockwood and Lipman*, 1987], interference from adjacent volcanic edifices [*Fiske and Jackson*, 1972; *Lipman*, 1980b], isostatic subsidence of the edifice along with the entire Island of Hawaii [*Moore*, 1987], and lateral spreading of unstable flanks of the volcano accompanied by recurrent gravi-tational slumps and catastrophic landslides [*Lipman et al.*, 1988; *Moore et al.*, 1989; *Lipman et al.*, 1990; *Borgia and Treves*, 1992].

Evidence concerning the early growth of Mauna Loa is entirely concealed by young lavas, but general features of the Hawaiian Ridge and its component volcanoes suggest that the growth of Mauna Loa could not have begun before about one million years ago. Because the present-day axis of the Hawaiian Deep is about 100 km southeast of Mauna Loa's summit, average Pacific plate motion of 9-10 cm/yr during recent growth of the Hawaiian Ridge [*Clague and Dalrymple*, 1987] would place the locus of Mauna Loa approximately on the axis of the ancestral Hawaiian Deep at 1 Ma, suggesting that the volcano is unlikely to have begun growing any earlier. Since along each Hawaiian magmatic locus (Loa and Kea trends; e.g., *Moore and Clague*, 1992, Figure 1), tholeiitic shield-building activity appears to terminate at one volcano as it becomes established at the next (for example, Hualalai terminating at or shortly before inception of Loihi; Mauna Kea shield stage terminating as Kilauea developed), the 40-60-km spacing of volcanic centers along the Hawaiian Ridge also probably provides broad information on the inception and life span for individual volcanoes. The distance from Hualalai to Loihi, about 110 km, suggests a maximum span of approximately one million years for the shield-building stage of Mauna Loa. If the rate of propagation of the Hawaiian hot spot has accelerated in recent time, perhaps to as much as 13-15 cm/yr as proposed by some

[*Shaw*, 1973; *Moore and Clague*, 1992], then the inception age of Mauna Loa volcanism could be correspondingly more recent. Based in part on an inferred high recent rate of propagation, *Moore and Clague* [1992] estimate the inception of Mauna Loa at only about 600 ka; much of the present paper represents an attempt to evaluate and refine the insights of these authors concerning the evolution of Hawaii Island, utilizing diverse data that have recently become available for Mauna Loa.

The early growth and structural geometry of Mauna Loa was probably influenced by the buttressing effects of preexisting volcanoes (Mauna Kea, Hualalai) to the northeast and northwest [*Fiske and Jackson*, 1972; *Lipman*, 1980b] that have continued to grow concurrently with Mauna Loa, requiring that lavas from these volcanoes must interfinger complexly along their subsurface boundaries. Based on magma discharge rates [*Swanson*, 1972; *Dzurisin et al.*, 1984] and new volume estimates in this paper (Table 1), Kilauea is a much younger volcano, probably originating at about 150 ka and forming a veneer on the south flank of Mauna Loa. Development of some Kilauea structures is generally interpreted as having been controlled by Mauna Loa [*Fiske and Jackson*, 1972], but Kilauea in turn appears to have grown rapidly to a size large enough to influence rift-zone structures and flank stability on Mauna Loa [*Lipman*, 1980b; *Lipman et al.*, 1990]. A recent alternative suggestion by *Walker* [1990, Figure 13], that curvature along Hawaiian rift zones such as those on Mauna Loa is due to asymmetric injection of a wedge-like dike complex, seems unlikely to be broadly applicable; this hypothesis requires dikes to intrude preferentially on the buttressed side of rift zones and to cause extensional deformation of that side, rather than along seaward flanks that are tectonically mobile, as indicated by the extensional faults and geodetically determined displacements on the south flanks of Mauna Loa and Kilauea.

This paper reviews diverse evidence for estimating rates of lava accumulation on Mauna Loa, based on results from recent field and laboratory studies. Although many of the data are individually of limited resolution, together they provide coherent if imperfect constraints on magma-eruption and lava-coverage rates for the entire volcanic edifice. These data thus provide the framework for evaluating the growth history of Mauna Loa over the last few hundred thousand years, and lead to the conclusion that this large volcanic shield is no longer actively increasing in overall size above sea level,

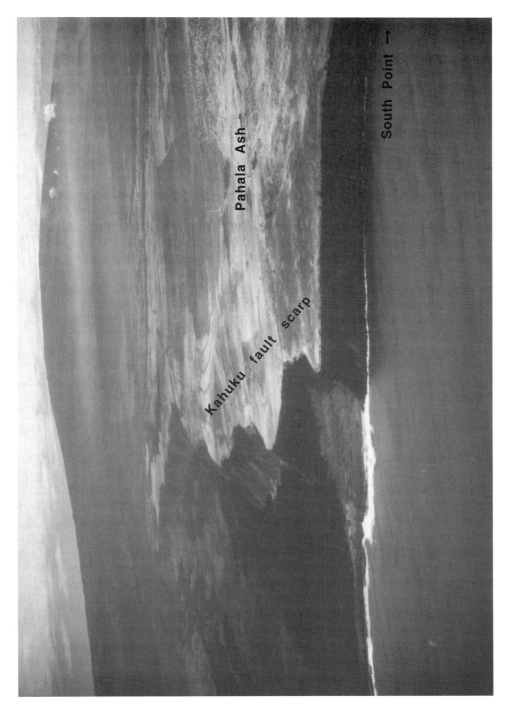

Fig. 1. Oblique north-looking aerial view of the south side of Mauna Loa. Erosionally modified Kahuku fault scarp trends southward, continuing under water at South Point (Ka Lae), to right of photograph. The southwest rift zone trends parallel and to the right (east) of the Kahuku fault on lower slopes, then at upper left of photo bends northeastward toward summit of Mauna Loa. Grass-covered lower slopes in South Point area are underlain by Pahala Ash and underlying Kahuku Basalt; little eruptive activity has occurred in this area during at least the last few tens of thousands of years. The subtle increase in slope angles high on Mauna Loa define an "upper shield," discussed in the text.

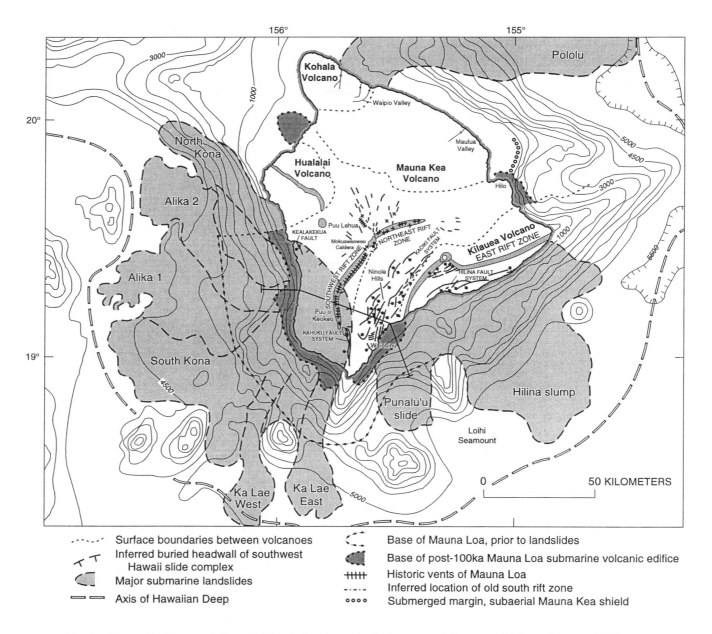

Fig. 2. Generalized map of Hawaii Island showing principal structural features of Mauna Loa and Kilauea volcanoes, locations of major submarine landslide terrains, and geographic features noted in text. Bathymetric contour interval is 500 m. Base of post-100-ka Mauna Loa edifice, demarking the major areas of post 100-ka submarine lava accumulation, is inferred from detailed bathymetry in *Chadwick et al.* [1994a, 1994b, 1994c]. *Dotted contact* at depth of about -360 m on Mauna Kea, east of Hilo, marks a change in bathymetric slope that is interpreted as the submerged coastal margin of Mauna Kea at the end of its shield-building stage. Narrow belts of *fine stipple* on land indicate general areas of exposed vents along rift zones; *bar and ball* indicate downthrown sides of faults or fault scarps. Modified from *Lipman et al.* [1988, 1990] and *Moore et al.* [1989].

because of the offsetting effects of isostatic subsidence and gravitationally driven landsliding and block slumping.

2. PRESENT SIZE OF MAUNA LOA

Interpretation of the growth history of Mauna Loa, Kilauea, and adjacent volcanic edifices requires valid information on the aerial extents and volumes of these volcanoes, as well as the times of initial growth (Table 1). High-resolution bathymetry, which has become available recently for much of the underwater slopes adjacent to Hawaii, provides new insights into volcanologic and structural processes but does not drastically modify previous estimates of the areas occupied by individual volcanoes. In contrast, the most

Table 1. Summary of areas, volumes, eruption rates, lava-coverage rates, and inferred inception ages, Mauna Loa and Kilauea volcanoes

	MAUNA LOA	KILAUEA
AREA (km^2):		
Subaerial	5,125	1,425
Submarine	4,000	4,700
Total	*9,125*	*6,125*
VOLUME: (x10^3 km^3)		
Bargar & Jackson	42.5	19.4
This study	80	15
		[E rift entirely Kilauea]
ERUPTION RATE (10^6 m^3/yr):		
Historical (incl. intracaldera)	28	40-50
Last 4000 yr	20	
	[based on coverage-rate ratio]	50?
LAVA-ACCUMULATION RATE since 750 yr bp, based on surface mapping (mm/yr):		
Coastline	0.9	6.4
3,000 m contour (upper shield)	3.7	-
LAVA-COVERAGE RATE (entire edifice, mm/yr):		
Eruption rate @ 10 x10^6 m^3/yr	-	17
@ 5 x10^6 m^3/yr	-	8.5
@ 2 x10^6 m^3/yr	2.2	-
MAGMA-SUPPLY RATE (10^6 m^3/yr):		
Eruption rate (last 4,000 yrs)	20	50
Rift-zone intrusions	4	30-50
Deep cumulates	4	10-15
Estimated total	*28*	*90-115*
SUBSIDENCE VOLUME (10^6 m^3/yr):		
Simple cone of depression	24	?
Seismic geometry	40	?
VOLUME-BASED INCEPTION AGE:		
@ 100 x10^6 m^3/yr	0.8 Ma	150 ka
@ 50 x10^6 m^3/yr	1.6 Ma	300 ka
@ 28 x10^6 m^3/yr	2.9 Ma	-

Sources of data and calculations methods discussed in text

A.

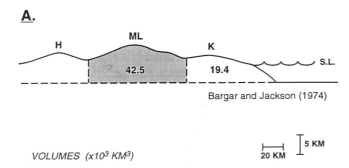

VOLUMES (x10³ KM³)

B.

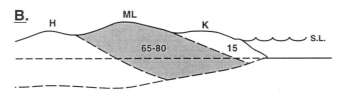

Fig. 3. Diagrammatic cross section, showing basis for alternative volume estimates for Mauna Loa (ML) and Kilauea (K) volcanoes, in relation to adjacent Hualalai (H): *A. Bargar and Jackson* [1974]; *B.* this paper, assuming isostatically subsided contact with pre-existing oceanic crust and sloping interfaces between volcanoes.

detailed previous estimates of the volumes of the Mauna Loa and Kilauea edifices, calculated by *Bargar and Jackson* [1974] as part of their evaluation of the entire Hawaiian Ridge (Table 1), are in need of revision because they assumed (1) vertical contacts between adjacent volcanoes and (2) horizontal contacts between the volcanic constructs and underlying sea floor (Figure 3).

Bargar and Jackson were fully aware that adjacent volcanoes along the Hawaiian Ridge grew sequentially against older edifices, implying complexly interfingering and sloping mutual contacts rather than vertical ones, but concluded that these effects would largely cancel out between opposing sides of each edifice. Such an approach, while broadly valid for Mauna Loa and older volcanoes along the Hawaiian ridge, led to a significant volume overestimate for Kilauea (19.4×10^3 km³, versus the present maximum estimate of 15×10^3 km³), which in geologically plausible cross section must be a relatively thin veneer on the south flank of Mauna Loa [e.g., perhaps 1 km thick at the summit, 2-3 km at the coast: *Stearns and Macdonald*, 1946, Plate 1; *Swanson et al.*, 1976, Figs. 15, 16; *Lipman et al.*, 1985, Figure 20]. If the east rift zone of Kilauea partly overlies a presently

inactive and buried eastward continuation of Mauna Loa's northeast rift zone, as suggested by some geophysical data [*Flannigan and Long*, 1987], then the volume of Kilauea eruptions could be substantially lower than estimated here. Similarly, Kilauea's volume would be smaller if some lower submarine parts of the Hilina block-slump system are surviving remnants of the formerly more mobile south flank of Mauna Loa, as speculatively discussed for many years [*Stearns and Clark*, 1930; *Lipman* 1980b, *Lipman et al.*, 1990].

Seismic-refraction and gravity studies demonstrate that much of the constructional Hawaiian Ridge is depressed below the mean depth of adjacent sea floor [*Eaton*, 1962; *Hill*, 1969; *Zucca and Hill*, 1980; *Zucca et al.*, 1982]. These studies, along with tide gauge and other geologic evidence for isostatic subsidence of the enormous Hawaiian volcanoes [*Moore*, 1970, 1987], demonstrate that the earlier volume estimates for individual volcanoes were too low by factors of as much as two. For Mauna Loa, which current-ly stands about 8.5 km above the adjacent sea floor, the seismic data document that the pre-existing oceanic crust has been depressed about 5 km at the west coastline and 8-9 km beneath the summit of this volcano (Figure 4). For a conservative model of geometrically simple conical depression of the ocean floor, the total volume of Mauna

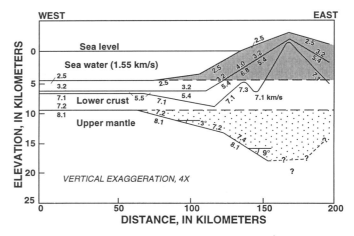

Fig. 4. Structural section, based on seismic profiling perpendicular to the Kona coast of Mauna Loa, documenting large-scale isostatic subsidence. Numbers indicate P-wave velocities, in kilometers per second. Upper shaded area is equivalent to volume of Mauna Loa edifice (42,000 km³) as modeled by Bargar and Jackson [1974]; lower stippled area denotes depressed oceanic crust (about 40,000 km³), which should equal the volume of the subsided Mauna Loa edifice and associated intrusions [modified from *Hill and Zucca*, 1987].

Loa would be about 65,000 km^3, rather than the 42,000 km^3 estimated by *Bargar and Jackson* [1974]. For the concave-downward geometry interpreted by Zucca, Hill, and associates from seismic profiling, the total volume is about 80,000 km^3 (Table 1). *Garcia et al.* [this volume] come up with a similar volume estimate, using a slightly different approach.

Alternatively, from calculations of crustal flexure beneath Mauna Loa, *Thurber and Gripp* [1988] infer only about 3 km of isostatic subsidence beneath Mauna Loa and that the 5-km difference with the seismic results is due to major local crustal underplating. By this structural interpretation, the bulk volume of Mauna Loa would be only about 60,000 km^3. Voluminous underplating of the oceanic crust by basaltic magma seems unlikely at Kilauea and Mauna Loa, however, based on buoyancy considerations and the subvolcanic distribution of seismic hypocenters [*Ryan*, 1987, 1988]. In addition, the flexural-subsidence calculation by *Thurber and Gripp* assumes a uniform volcano density of only 2.6 gm/cm^3, much lower than the 2.9 gm/cm^3 dominant density that has been modeled from combined gravity and seismic data by *Hill and Zucca* [1987] and is required by the voluminous picritic to dunitic cumulates inferred to be present low in Hawaiian volcanic edifices [*Jackson*, 1968; *Ryan*, 1988; *Clague and Denlinger*, 1994]. A purely flexural model also neglects effects of probable crustal fracturing associated with subsidence [*Ten Brink and Watts*, 1985], as well as the effects of lithospheric weakening due to heating and thinning by voluminous ascending hot-spot magma [*Detrick and Crough*, 1978; *McNutt*, 1984], which would decrease the effective elastic thickness of the oceanic lithosphere.

3. RATES OF SUBSIDENCE

Isostatic subsidence of Mauna Loa (and all of south Hawaii), produced by volcanic loading, is currently about 2.4 mm/yr at sea level, as indicated by long-term historical tide-gauge measurements at Hilo when normalized against similar data from other tide gauges in the Pacific to adjust for eustatic sea-level changes [*Moore*, 1970, 1987]. Similar integrated rates for late Quaternary time (as much as 2.6 mm/yr for North Kona since 475 ka) have been confirmed for several coastal areas of Hawaii Island by U-disequilibrium dates on submerged coral reefs [*Szabo and Moore*, 1986; *Ludwig et al.*, 1991]. More qualitatively, submerged Hawaiian recognized as evidence for ongoing subsidence along parts of the Mauna Loa coastline [*Apple and Macdonald*, 1966], and drowned lava deltas and shoreline topographic

inflections are especially clear on recent high-resolution bathymetric maps (*Chadwick et al.*, 1994a, 1994b, 1994c], especially along the Ka'u coast and from Kealakekua Bay to Honaunau. Part of the Mauna Loa coastal submergence reflects rapid postglacial eustatic rise of sea level, but no comparable submerged features occur along the Kilauea coast, where accumulation of lava has kept pace with, or exceeded rise of sea level. The submerged features along the Mauna Loa coast thus frame a primary concern of this paper: to what degree has recent lava accumulation been inadequate to compensate for sustained subsidence?

In addition, rates of isostatic subsidence must increase inland, reaching a maximum below the summit of Mauna Loa, which is also marked by concentration of high-density material at depth, presumably picritic or dunitic cumulate residues from crystallization in the summit magmatic system [*Jackson, 1968*; *Hill and Zucca*, 1987; *Ryan*, 1987; *Clague and Denlinger*, 1994]. Over the estimated maximum 1 m.y. lifespan of Mauna Loa, subsidence must have approximated 8 mm/yr at the summit, based on the seismic-refraction data indicating depression of the crustal basement by about 8 km [*Zucca et al.*, 1982], but the current rate of subsidence at the summit is unknown. Because the crust responds rapidly to changes in loading, equilibrating nearly completely within a few thousand years even for cold cratonic crust as indicated by glacial rebound rates (*Flint*, 1971), current volcanic-loading subsidence rates beneath Mauna Loa may have decreased appreciably from the integrated average rate, especially if geologically recent rates of shield building have slowed at Mauna Loa as proposed in this paper.

The relatively constant long-term subsidence rates obtained by dating submerged coral reefs (to 475 ka) may thus represent average rates that integrate somewhat more variable isostatic subsidence effects associated with shorter-term changes in loci of volcanic accumulation across Hawaii Island. In addition to relatively rapid subsidence due to volcanic loading, inactive northern volcanoes on Hawaii and older islands of the chain are also subject to sustained subsidence at slower rates as the Pacific lithosphere cools to equilibrate from thermal effects of passage over the Hawaiian hot spot [*Detrick and Crough*, 1978]. Such thermal equilibration, at an estimated rate of about 0.02 mm/yr for the Hawaiian chain since 40 Ma [*Clague and Dalrymple*, 1987], would be too slow to detect from existing tide-gauge records on the older islands. Current subsidence rates across upper parts of the Mauna Loa edifice should be determinable in future years by GPS

Table 2. K-Ar ages for Mauna Loa lava flows, Kealakekua and Kahuku fault scarps

Sample No.	Unit	K_2O wt%	K_2O/P_2O_5 #	Argon			Age ± S.D (ka)
				Weight (gms)	$^{40}Ar_{rad}$ (10^{-14} mol/g)	$^{40}Ar_{rad}$ (%)	
Kealakekua Pali:							
90L-10-A2	Flow	0.252	1.6	14.13	5.56	0.7	153 ± 126
				14.38	10.54	1.3	291 ± 137
				Weighted mean sample age =			*216 ± 93*
90L-10C	Flow	0.298	1.5	15.73	8.51	1.4	198 ± 91
				15.27	3.78	0.6	88 ± 90
				Weighted mean sample age =			*142 ± 64*
				Weighted mean site age =			*166 ± 53*
Kahuku Pali (subaerial):							
90L-12-B:	Flow	0.366	1.7	15.19	-5.66	-0.2	-108 ± 270
90L-12 E:	Flow	0.476	1.6	14.86	2.81	0.4	41 ± 65
				14.63	2.24	0.4	33 ± 62
				Weighted mean sample age =			*37 ± 45*
				Weighted mean site age =			*(no meaningful date)*
Kahuku Pali (submarine):							
183-15	Flow, -630 m (Glass S = 250 ppm*)	0.315	1.6	17.66	61.88	8.2	1370 ± 100
				16.82	41.73	6.5	920 ± 100
				Weighted mean sample age =			*1145 ± 71*
183-7	Flow, -960 m (Glass S = 230 ppm*)	0.294	1.6	20.39	66.04	5.8	1560 ± 130
				20.16	59.08	5.1	1390 ± 130
				Weighted mean sample age =			*1475 ± 92*
182-6B	Flow, -1310 m (Glass S + 470 ppm*)	0.336	1.7	9.72	74.78	13.7	1550 ± 150
				9.42	60.83	1.5	1260 ± 340
				Weighted mean sample age =			*1503 ± 186*
182-5	Flow, -1335 (Glass S = 200 ppm*)	0.405	1.8	19.83	7.10	0.2	122 ± 266
182-1	Flow, -1420 m (Glass S, ppm = n.d.)	0.428	1.6	14.12	40.88	1.2	664 ± 213
				13.80	31.54	1.0	512 ± 204
				Weighted mean sample age =			*585 ± 147*

K-Ar analyses by G. Brent Dalrymple, U.S. Geological Survey

*, Electron microprobe analysis provided by M.O. Garcia

#, K_2O/P_2O_5 ratios from whole-rock XRF analyses provided by J.M. Rhodes

Table 2: (continued)

Sample locations and descriptions [all chemical analyses by J.M. Rhodes, written commun., 1990, 1992]:

90L-10-A2: Kealakekua fault scarp (Pali Kapu o Keoua): 155°55.64'W, 19°28.98'N, collected along boulder beach, beneath prominent west-dipping ash layer: lowest exposed flow. Vesicular coarsely devitrified interior pahoehoe, containing about 5% small olivine phenocrysts and 2% plagioclase. Chemical composition: 50.75% SiO_2, 11.12% MgO, 1.55 K_2O/P_2O_5.

90L-10-C: Kealakekua fault scarp (Pali Kapu o Keoua): 155°55.56'W, 19°28.96'N. Pahoehoe flow above thin discontinuous ash lens that marks a local erosional or structural truncation; above 90L-10-A2, but below main ash layer. Vesicular completely crystallized dense interior of flow containing about 1% olivine phenocrysts. Chemical composition: 51.38% SiO_2, 7.12% MgO, 1.49 K_2O/P_2O_5.

90L-12-B: Kahuku Pali: 155°41.70'W, 18°57.69'N, collected along obscure old Hawaiian trail at elevation of about 125 m, about 50 m below scarp crest; dense interior of thick pahoehoe flow containing about 2% olivine phenocrysts and sparse plagioclase in a finely devitrified matrix with some surviving glass. Chemical composition: 51.55% SiO_2, 10.53% MgO, 1.69 K_2O/P_2O_5.

90L-12-E: Kahuku Pali: 155°41.67'W, 18°57.71'N, collected along obscure old Hawaiian trail at elevation of about 65 m; dense interior of thick flow containing about 10% large olivine phenocrysts in a coarsely devitrified microcrystalline matrix. Chemical composition: 49.60% SiO_2, 14.05% MgO, 1.65 K_2O/P_2O_5

183-15: Submarine Kahuku Pali: 155°41.4'W, 18°51.8'N. Sparsely vesicular finely devitrified interior of pillow fragment, containing 15.8% phenocrysts and 6.6% microphenocrysts of olivine (M. Garcia, written commun., 1992), Chemical composition: 48.01% SiO_2, 17.41% MgO, 1.63 K_2O/P_2O_5.

183-7: Submarine Kahuku Pali: 155°42.0'W, 18°51.8'N. Vesicular finely devitrified interior of pillow fragment, containing 10.2% phenocrysts and 4.0% microphenocrysts of olivine, 0.6% microphenocrysts of plagioclase (M. Garcia, written commun., 1992). Chemical composition: 50.12% SiO_2, 12.19% MgO, 1.61 K_2O/P_2O_5.

182-6B: Submarine Kahuku Pali: 155°42.6'W, 18°52.0'N. Vesicular finely devitrified interior of pillow fragment, containing 35.6% phenocrysts and 22.0% microphenocrysts of olivine (M. Garcia, written commun., 1992). Chemical composition: 45.01% SiO_2, 28.19% MgO, 1.68 K_2O/P_2O_5.

182-5: Submarine Kahuku Pali: 155°42.8W, 18°52.1'N. Sparsely vesicular coarsely devitrified interior of pillow fragment, containing 1.8% phenocrysts and 1.6% microphenocrysts of olivine, 0.6% microphenocrysts of augite (M. Garcia, written commun., 1992). Chemical composition: 49.88% SiO_2, 9.87% MgO, 1.76 K_2O/P_2O_5.

182-1: Submarine Kahuku Pali: 155°43.1'W, 18°52.2'N. Vesicular coarsely devitrified interior of pillow fragment, containing about 30% olivine phenocrysts and 1% plagioclase microphenocrysts. Chemical composition: 46.69% SiO_2, 21.84% MgO, 1.63 K_2O/P_2O_5.

measurements, however, as quantitative deformation data become sufficiently numerous and precise to see through effects of short-lived inflation-deflation cycles high on the volcano.

4. K-AR DATING OF OLD MAUNA LOA LAVAS

During the last 20 years radiocarbon methods have been spectacularly successful in dating prehistoric Hawaiian lava flows younger than about 35 ka [*Rubin et al.*, 1987], leading to the availability by 1995 of more than 350-age determinations for about 200 separate Mauna Loa lava flows [*Lockwood*, this volume]. In contrast, severe difficulties have been encountered in attempts to utilize K-Ar methods for dating young tholeiitic lavas of oceanic-island and submarine spreading-ridge tholeiites [*Dalrymple and Moore*, 1968; *McDougall and Swanson*, 1972; *McDougall*, 1979]. Such lavas are low in total potassium, which mostly resides in interstitial phases where it is mobilized by even incipient surficial weathering. In addition, more thoroughly crystallized dikes (and also submarine-erupted lavas) have been found to contain mantle-derived radiogenic argon as a magmatic volatile ("excess argon"), commonly in sufficient quantity to yield impossibly old apparent ages [*Lipman et al.,* 1990]. As a result, virtually all relatively precise and reliable K-Ar ages for late Pleistocene Hawaiian basalts obtained to date have been on late-stage alkalic lavas [*Clague and Dalrymple,* 1987], a stage which has not yet begun on Mauna Loa.

Because Mauna Loa lavas yielding ^{14}C ages greater than 30 ka are locally present on the surface or at shallow erosional depth and are underlain by thick erosional or fault-exposed sections, especially along the Kahuku and Kealakekua faults (Figure 2), additional K-Ar dating has been attempted while trying to develop strategies to overcome the problems of low K, alteration, and excess Ar. Determinations that yielded approximate ages of 100-200 ka for the Ninole Basalt on the south flank of Mauna Loa have already been reported [*Lipman et al.,* 1990].

For additional determinations on subaerial flows, 17 samples from the two fault-scarp sections were evaluated for minimum alteration, based on preserved K_2O/P_2O_5 ratios near 1.6-1.7 in bulk-rock samples (typical magmatic values for Mauna Loa [*Wright,* 1971; *Rhodes,* 1983; *Lipman et al.,* 1990]) and relatively high K_2O contents (>0.2). Olivine phenocrysts were separated from highly porphyritic samples (a common lithology in the Kahuku section) to increase the K content of the remaining groundmass fraction that was analyzed. Analytical methods were similar to those summarized in *Lipman et al.* [1990]. New determinations for three shoreline lava flows along the Kealakekua fault scarp, about 185 m below the top of the exposed section, yield a weighted mean age of 166 ± 53 ka (Table 2). These lava flows may thus be roughly correlative with the Ninole Basalt. Similar samples from the Kahuku fault scarp, 50-100 m below its top (capped by Pahala Ash), failed to yield distinct ages; the calculated maximum age for these samples, assuming that 2% Ar could be detected reliably, is about 200 ka (G.B. Dalrymple, written commun., 1993). The Kahuku section must have a minimum age of greater than 30 ka, based on the ^{14}C ages for lava flows overlying the Pahala Ash.

In addition, we attempted to date stratigraphically and petrologically well-constrained pillow lavas, collected by submersible from outcrop exposures of the underwater Kahuku fault at depths of 630-1420 m [*Garcia et al.,*

1993]. Thirty-eight samples of thoroughly devitrified pillow interiors were again screened, based on relatively high K_2O contents and magmatic K_2O/P_2O_5 ratios (much less variable in the submarine environment than subaerially), as well as low contents of S (M. Garcia, written commun., 1992) as determined by microprobe analysis of glassy rinds of the same pillow fragment. The approach was to attempt to identify pillow lavas that had degassed in a low-pressure environment, presumably as dikes propagated along the southwest rift zone from the summit magma reservoir. Such vigorous degassing uprift from active vents has been observed on Mauna Loa during several historical eruptions, most recently in 1984 when a larger gas plume was emanating from fissures at the 3300-m level than from the fountaining lava vents at 2,900 m [*Lipman et al.,* 1985; *Lockwood et al.,* 1987]. Despite S contents in some glassy pillow rinds as low as 200 ppm, similar to those of subaerially erupted tholeiitic lavas, the resulting K-Ar determinations (Table 2), ranging from 585 to 1,500 ka, are inconsistent with the sampled stratigraphic sequence. The youngest apparent ages are from the lowest two samples, and some stratigraphically higher samples seem implausibly old, even for the inception of Mauna Loa volcanism.

In light of these unsatisfactory results, an alternative interpretation is that the pillow lavas were erupted in a shallower marine environment where S was able to degas. Original depths would have been near sea level to perhaps 1,000 m, based on comparison with S contents of dredge samples from recent lavas along the east rift zone of Kilauea and young submarine Mauna Loa lavas (Figure 5). Significant mantle-derived radiogenic argon was apparently retained by some basalt magmas during the low-pressure subaqueous degassing of S. Such an approach suggests 500 to as much as 800 m net subsidence of this segment of the submarine southwest rift zone of Mauna Loa since the sampled pillow lavas were erupted (see also *Garcia et al.,* this volume). Calibrated to the modern coastal subsidence rate of 2.4-2.6 mm/yr as determined from the tide gauge at Hilo, submerged reefs, and eustatic sea-level changes [*Moore,* 1987; *Ludwig et al.,* 1991], ages of the sampled pillow lavas would be approximately 200-350 ka. Thus, even though our strategy to attempt to date these lavas by K-Ar methods has been unsuccessful, some useful age and paleodepth constraints have emerged.

5. RATES OF LAVA ACCUMULATION ON MAUNA LOA

Long-term rates of lava accumulation on Mauna Loa are gradually being constrained by mapping surface

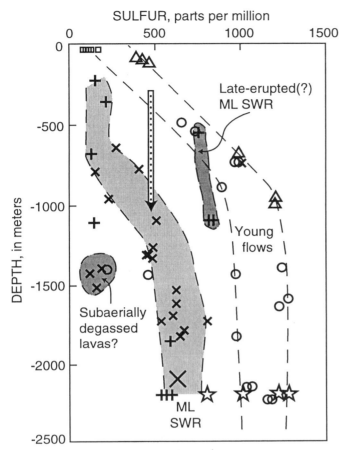

Fig. 5. Water depth versus sulfur contents for glassy margins of pillow lavas sampled along the submarine Kahuku fault scarp, in comparison with samples of young flows from Mauna Loa and from the east rift zone of Kilauea. Dashed lines enclose the bulk of the data for each volcano, indicating inferred typical ranges in S content with depth. Lower S contents for the Mauna Loa SWR submersible samples at any present-day depth range in comparison to the young flows are interpreted as evidence for 500-800 m of post-emplacement subsidence along Mauna Loa's submarine southwest rift zone, as indicated by horizontal arrow. Unpublished Mauna Loa SWR submersible-sample data (x) provided by M. Garcia [written commun., 1992]; Mauna Loa SWR dredge-sample data (X) from *Garcia et al.* [1989] and (+) from *Moore and Clague* [1992]. Dredge-sample data for young Kilauea flows (o) from *Clague et al.* [in press] and (stars) from Garcia *et al.* [1989]; for 1877 (squares) and late prehistoric (triangles) Mauna Loa flows from *Moore and Clague* [1987].

flows, ^{14}C and K-Ar dating of fault- and erosionally-exposed older lavas, marine studies, and results from scientific drilling near Hilo in 1993. Each type of data has interpretive problems: for example, the coastal-lava accumulation rates determined by surface mapping vary widely depending on local microtopography, relatively few sections have yielded reliable dates for lavas sufficiently old to represent a significant time interval, and only a single cored drill hole is available for Mauna Loa. Together, however, the combined data provide a fairly coherent pattern for coastal lava-accumulation rates. In this paper, dense-rock-equivalent (DRE) corrections for vesicularity of lavas have not been applied, because the purpose is to constrain the rate of growth of the volcanic edifice, which consists of a mix of porous surface flows and denser intrusions.

5.1. Surface Mapping

Mapping and dating of the subaerial lava flows of Mauna Loa [*Lipman and Swenson*, 1984; *Lockwood et al.*, 1988, this volume], while still incomplete for the huge Mauna Loa edifice, have progressed sufficiently that rough calculations of areas and rates of coverage are now available for most areas of the volcano during the last few thousand years [*Lipman*, 1980a; *Lockwood and Lipman*, 1987]; in-progress refinements to the published maps (J.P. Lockwood, unpubl. data, 1994) are unlikely to modify substantially the broad interpretations developed here. Combined with estimates of mean flow thickness, the map data can be used to place limits on eruption rates through time (Table 3). Direct measurement of areas covered and eruptive volume is relatively straightforward for historical flows, but becomes more difficult for older prehistoric flows that are increasingly buried by younger lavas. For the 150 years of historical activity [*Lockwood and Lipman*, 1987, table 18.1], the eruption rate on Mauna Loa has fluctuated substantially: 47×10^6 m^3/yr for the interval 1843-1877, but only 18×10^6 m^3/yr from 1877 to 1994, averaging 28×10^6 m^3/yr for the entire historical period. Geologic mapping of vent distributions and associated flows indicates that the historical period from 1868 to 1950 was atypical; eruption rates were lower for the late prehistoric record, averaging 20×10^6 m^3/yr back to at least 4 ka.

Evaluation of eruption rates for prehistoric time intervals subdivided in the mapping, where appreciable parts of the lava record are covered by younger flows, is facilitated by comparisons with cumulative areal-for geometrically simple uniform coverage per thousand years [*Lipman*, 1980a]. Such comparisons indicate that the latest prehistoric period mapped for Mauna Loa (0.15-0.75 ka) was a period of relatively low eruptive activity (about 18×10^6 m^3/yr), especially on the southwest side of Mauna Loa [*Lipman*, 1980a], in comparison to the historical period and the preceding 0.75-1.5 ka map unit (about 28×10^6 and 21×10^6 m^3/yr, respectively), but similar to the overall Mauna Loa eruptive rate since 4 ka [*Table 3; Lockwood and Lipman, 1987, Figure 18.11*].

Some fluctuations in coverage rate through time may reflect short-term variations in the magma supply to Mauna Loa, for example, the 50% decrease in eruption rates and changes in magma composition that occurred in the late 19th century [*Lockwood and Lipman, 1987; Tilling et al., 1987*]. Other periods of low prehistoric lava coverage may also relate to intermittent episodes of lava ponding within newly subsided summit calderas.

An overall eruption rate of 20×10^6 m^3/yr (no density correction) for the past 4,000 years could cover all of subaerial Mauna Loa (5,125 km^2) uniformly at a rate of 4 mm/yr thickness, nearly double the current coastal subsidence rate at Hilo (2.4 mm/yr; *Moore* [1987]). Alternatively, for the entire volcano, subaerial and submarine (about 9,000 km^2, excluding submarine landslide aprons), the uniform coverage rate would be about 2 mm/yr, slightly less than the present-day subsidence of the Island of Hawaii. Because erupted

lava accumulates preferentially high on the Mauna Loa edifice, near the summit caldera and flanking rift zones, more rigorous determinations of lava accumulation rates are needed at key elevations, especially along the coastline, in order to make valid comparisons with known subsidence rates.

5.2. Coastal Lava-Accumulation Rates

Valid coastal lava-accumulation rates are particularly useful for constraining growth of Mauna Loa, because reliable estimates for the offsetting effects of isostatic subsidence exist only along the coast. In addition, sea-cliff erosion and fault scarps provide exceptional coastal exposures of relatively old Mauna Loa lavas that can be dated under favorable conditions.

In order to quantify coastal lava-accumulation rates on a basis that would permit direct comparisons with other eruption-rate and subsidence data for Mauna Loa and Kilauea, the approximate areal percentages occupied by shoreline-crossing flows of historical (0-0.75 ka) and period 4 lava flows (0.15-0.75 ka) were determined for discrete segments of the Mauna Loa coastline, based on

Table 3. Lava-coverage and eruption rates for mapped age units of Mauna Loa

Map Unit & Time Interval	Estimated Coverage Rate (%/ka)[*]	Eruption Rate (x10^6/m^3/yr)	DRE[#] Eruption Rate (x10^6/m^3/yr)
Historical			
1843-1994	65	28	23
1877-1994	-	18	15
1843-1877	-	47	38
Group 4,			
0.15-0.75 ka	40	18	15
Group 3,			
0.75-1.5 ka	45	21	17
Group 2			
1.5-4.0 ka	40	18.5	15
Average			
0-4.0 ka	-	20	16.5

Data adjusted from *Lockwood and Lipman,* 1987

[*] Percentage of subaerial surface covered per thousand years; from plots in *Lipman* [1980a, Fig. 9] and *Lockwood and Lipman* [1987, *Fig. 18.11*].

[#] Based on assumed average lava density of 2.3 gm/cm^3, and dense-rock equivalent of 2.85 gm/cm^3.

the age-distribution maps of *Lockwood and Lipman* [1987] and *Lockwood et al.* [1988]. No significant segment of the coastline appears to have been crossed by more than a single flow since 0.75 ka. The areal percentages were multiplied by an estimated average flow thickness of 6 m, representative of maximum a'a accumulations along the coastline, to yield an average thickness for each coastal segment, then divided by 750 years to yield a covering rate in millimeters per year (Figure 6). It is not presently possible to determine such covering rates

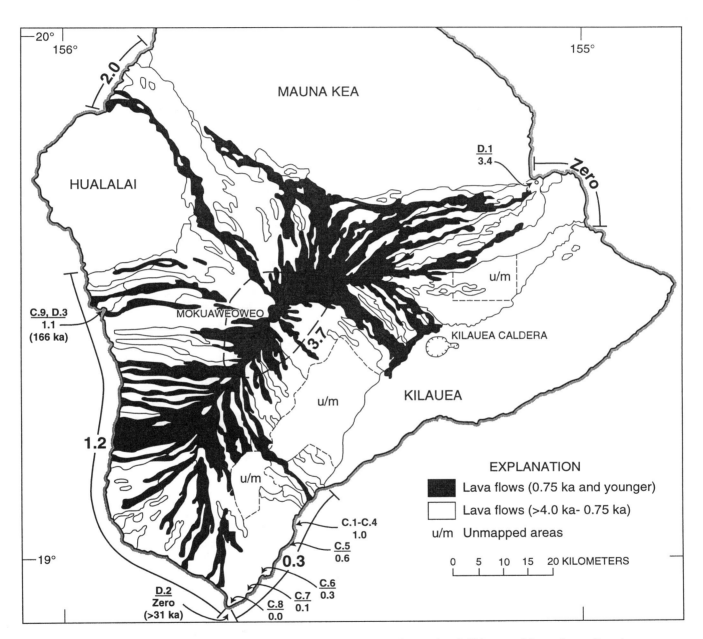

Fig. 6. Average rates of lava accumulation, in mm/yr, at coastline and at 3,000 m on Mauna Loa. Based on distributions of historical (0-150 yr bp) and age-group IV flows 150-750 yr bp); average flow thickness of 6 m is assumed except for dated sections (see text for discussion of calculation assumptions). Average rate is 0.9 mm/yr at the coast; 3.7 mm/yr at 3,000 m. Locations are also shown for sections listed in Table 3. Age-distribution map modified from *Lockwood and Lipman* [1987, Figure 18.2].

reliably for earlier time intervals, because of increasing complexity arising from repeated covering of coastline segments by multiple flows, and because of remaining uncertainties concerning the ages of some sizable flows, especially for the important South Kona segment.

The resulting calculated coastal accumulation rates since 0.75 ka (Table *4, A*) vary from zero in the Hilo area, to 2 mm/yr for North Kona where nearly a quarter of the coastline was covered by the 1859 flow (yet no other post-0.75 ka flows reached the coast). Intermediate values were obtained for the South Kona and Ka'u coastal segments, and the average calculated rate of covering for the entire Mauna Loa coastline is 0.9 mm/yr. These results are roughly consistent in their relatively limited

Table 4. Lava-coverage rates for Mauna Loa: summary of key locations

Location	Age (ka), Data source	Lava Thickness (m)	Accumulation Rate (mm/yr)
A. Coastal Lava Accumulation (percent cover)			
1. Hilo area (0)	0.75, map (1)	6	0.0
2. Ka'u coast (4.2 %)	0.75, map (1)	6	0.3
3. S. Kona-W. Ka'u (15 %)	0.75, map (1)	6	1.2
4. N. Kona (25 %)	0.75, map (1)	6	2.0
5. Entire coast (11.6 %)	0.75, map (1)	6	0.9
B. Up-slope (3000 m) Lava Accumulation (percent cover)			
1. Summit "shield" (47)	0.75, map (1)	6	3.7
C. Dated Section (Ka'u - S. Kona):			
1. Naalehu Pali	7.3, [14]C (2)	6	0.82
2. Naalehu Pali	13.2, [14]C (2)	6	1.02
3. Naalehu Pali	31.1, [14]C (2)	15	0.83
4. Naalehu Pali (cumul.)	31.1, [14]C (2)	27	0.87
5. Maniani Pali	9.1, [14]C (2)	5	0.55
6. Kahukupoko	11.8, [14]C (2)	3	0.25
7. Puu o Mahana	28.2, [14]C (2)	4	0.14
8. South Point	>30, [14]C (2)	0	0.0
9. Kealakekua Pali	166, K-Ar (3)	185	1.1 (0.6-1.6)
D. Accumulation Rate from Coastal-Subsidence Data:			
1. Hilo drill hole	0-290 (4)	239	2.4
2. Lower southwest rift	130-140 (5)	-	0.0
3. Kealakekua Bay (reef)	14 (6)	?	(low)

Locations shown on Figure 6; see text for discussion of assumptions and calculations

[1]References for data sources:

1. Computed from *Lockwood and Lipman* [1987, Fig. 18.2] and *Lockwood et al.* [1988]; percent cover is ratio of coastline-crossing flows to entire coastline.

2. *Rubin et al.* [1987, Table 10.1]: *Lipman and Swenson* [1984] and unpubl.

3. This report, Table 2; range in accumulation rate (brackets) reflects sizeable analytical uncertainty

4. Interpreted from core logs [*DePaolo et al.*, 1994] and ages in *Beeson et al.* [1994] and *Moore et al.* [1994]

5. Age interpreted from coastal subsidence rates [*Moore et al.*, 1990] and identification of submerged seacliff [P. Lipman, J.M. Rhodes, and M. Garcia, unpubl. data, 1992]

6. *Moore and Clague* [1987]

range and, as shown below, with estimates from dated lava sections.

The limitations inherent in the use of this relatively short time interval are well illustrated by the Hilo segment, which received no lava flows since 0.75 ka, yet was largely covered by thick flows of the Panaewa eruption at about 1.4 ka. Some effects tend to be counterbalancing. Both the Hilo and the North Kona segments are constricted segments where Mauna Loa flows are funneled between adjacent volcanoes. As a result, accumulation rates along these segments may be atypically high and irregular in timing compared to the volcano as a whole.

An important area where the 750-yr accumulation rates are likely to be misleadingly low is the South Kona-West Ka'u coast, where the shoreline has grown prominently westward, marking the focus of lava accumulation within the broad break-away zone of the composite South Kona slide complex (Figure 2). Although no 0-0.75 ka flows have been identified along this segment, much of it is surfaced by late lavas of the 0.75-1.5 ka age group, and the longer-term lava accumulation rate must have been the highest anywhere along the Mauna Loa coast in order to account for the shore-line bulge. This segment of the Mauna Loa coast reflects both infilling of the landslide break-away zone and also a favored site of lava accumulation downslope from the locus of maximum post-100-ka activity along the southwest rift zone, near Pu'u o Keokeo (Figure 2). If the large Hapaimanu flow in this segment [*Lipman and Swenson,* 1984] is younger than 0.75 ka, as suggested by in-progress studies (J.P. Lockwood, written commun., 1995), the overall rate of coastal lava accumulation for South Kona-West Ka'u would be 2.3 mm/yr, rather than the 1.2 mm/yr calculated from published mapping (Figure 6, Table 4.A.3). The rate for the active 30-km coastal segment of the Ka'u district northwest of South Point, which encompasses the Hapaimanu flow, would be 3.2 mm/yr, exceeding the average island sub-sidence rate.

5.3. Up-Slope Lava-Accumulation Rates

To provide insight into rates of lava accumulation high on the Mauna Loa edifice, accumulation rates since 0.75 ka (historical and period 4 flow units of *Lockwood and Lipman* [1987] were also calculated for the 3,000 m level (10,000 ft contour actually used) by the same approach as for the shoreline-crossing flows. The 3,000 m level marks the approximate base of a well-defined "upper shield" morphologic feature high on

Mauna Loa (Figure 1), marked by steeper slopes (greater than 10° over large areas: *Moore and Mark,* 1987, Figure 3.2), that defines a summit region of high lava accumulation (and intrusion?) during caldera overflow.

At this level, lava-accumulation rates have been higher than anywhere along the coast, calculated as averaging 3.7 mm/yr (Table 4, *B*); corrections were required for several segments where more than a single flow has crossed since 0.75 ka. Even this rate is probably somewhat low, because about a third of the perimeter of the "upper shield" has been blocked from summit lava eruptions by the walls of the summit caldera (Moku'aweoweo), which developed its present configuration sometime since 0.57 ka [*Lockwood and Lipman,* 1987]. This effect is in part offset by funneling of lavas ponded in the caldera along the upper northeast and southwest rift zones, where rates of lava accumulation are especially high. In the extreme (and demonstrably overstated) limiting assumption that no lava flow reached the base of the upper shield downslope from the sector below the walls of Moku'aweoweo since 0.75 ka, the lava accumulation rate for the remaining region would be 5.5 mm/yr.

5.4. Dated Lava-Flow Sections

Another approach to estimating lava-accumulation rates for Mauna Loa is from dated lava-flow sections along sea cliffs and fault scarps (Table 4, *C*), where the depth of exposure samples a longer time frame than can be obtained from surface mapping. The major shortcoming of this approach is that reliable ages have been determined for only a few sections. These are mostly in Ka'u, where coastal accumulation rates are known from surface-mapping relations to have been relatively slow in comparison to other sectors of Mauna Loa.

One informative site is along the coastal cliff (pali) southeast of Na'alehu, where ages have been obtained for three successive lava flows from a 27-m section near Ka'ukupoko Point (Figure 6). The resulting interpreted accumulation rates are only 0.8 to 1.0 mm/yr, with an average rate of 0.87 mm/yr for the entire section (Table 4, *C.* 1-4). Despite these low rates, this coastal segment is probably the locus for above-average accumulation rates along the Ka'u coast, because eruptions from the middle southwest rift zone would tend to be deflected to this area by the topographic high bounded by the Na'alehu fault [*Lipman and Swenson,* 1984].

Dates from three other sections further southwest along the Ka'u coast, each containing only a single dated flow, yield even lower accumulation rates (0.1-0.5 mm/yr: Table 4, *C.* 5-7). Consistent with these relatively low accumulation rates is the widespread surface exposure of thick Pahala Ash along this coastline, especially near South Point (Figure 1). The main body of Pahala Ash, although not dated directly, is overlain locally by lava flows dated by ^{14}C at greater than 30 ka [*Rubin et al.,* 1987; *Lipman et al.,* 1990], and the age of the main ash deposit may be beyond reliable resolution by the radiocarbon method.

An important constraint on rates of lava accumulation is the newly determined K-Ar age for lavas at the base of Kealakekua Pali in South Kona (Table 4, *C.* 9). The relatively large analytical uncertainties for K-Ar age at this site (166±53 ka) allow accumulation rates in the range 0.6-1.7 mm/yr (average of 1.1 mm/yr) for this 185-m fault-scarp section, the thickest exposed subaerial section available for study. This is the only currently available determination for the west flank of Mauna Loa; it represents a site where only flows from radial vents and overflows of the summit caldera could accumulate, but where lava has been deflected by proximity to the Hualalai edifice. Relatively rapid accumulation rates during the sustained growth of Mauna Loa are indicated for this coastal segment by the complete filling of any broad embayment in the coastline adjacent to Hualalai, in contrast to the surviving embayments where Mauna Loa lavas are funneled between Hualalai and Mauna Kea in North Kona, and between Kilauea and Mauna Kea near Hilo (Figure 6). In the North Kona sector, sidescan-sonar data also indicate that recent Mauna Loa flows have flowed only short distances over the sea floor from the shoreline [*Moore and Clague,* 1992].

The highest rates of coastal lava accumulation and shoreline growth on Mauna Loa are almost certainly along the westward-flaring coast in that reflects infilling of the composite South Kona slide complex (Figure 2). Unfortunately, no dateable sections have been located along this coastal segment, and the age control on flow distri-butions remains limited. Newly available high-resolution bathymetry for the underwater slopes of Mauna Loa offshore of South Kona [*Chadwick et al.,* 1994b, 1994c] strongly suggests, however, that even along this segment lava accumulated only on upper parts of the volcanic edifice, within the large break-away scar generated by the South Kona slide complex at 100 ka or earlier. All the submarine slopes below about 1,000 m depth remain morphologically dominated by block-slump and debris-avalanche features, indicating that post-slide lava accumulation has been mainly subaerial and in shallow water. Such an interpretation is supported by the still-incomplete burial of the bounding faults (Kealakekua, Kahuku) of the slide-breakaway scar and by anomalously steep lava-draped slopes within the partly infilled scar. Indeed, the anomalous topography on the South Kona flank of Mauna Loa was the basis for initial inference of submarine sliding on this flank of Mauna Loa, providing the impetus for the marine studies that identified the underwater Alika slide complex [*Normark et al.,* 1979].

5.5. Accumulation Rates from Drill-Hole Data

Important additional data on rates of lava accumulation for Mauna Loa are emerging from scientific drilling near Hilo [*Stolper et al.,* 1994; *Lipman and Moore,* in press]. The upper part of this hole, spudded at an elevation of only 4.2 m above sea level, penetrated 280 m of Mauna Loa lavas, close to their lap-out against the south flank of Mauna Kea as the flows were deflected from the northeast rift zone toward Hilo Bay (Table 5). The Mauna Loa-Mauna Kea boundary is marked by a clay-rich soil [*DePaolo et al.,* 1994] and by a compositional change to alkalic basalt [*Rhodes and Sweeney,* 1994]. Virtually all the Mauna Loa lavas are subaerial, but carbonate-rich sediment, beach sand, hyaloclastite, and other littoral and shallow-water deposits were encountered at six interflow horizons (Table 5). Accumulation of Mauna Loa lavas kept pace with isostatic subsidence and eustatic sea level rise at this site, in part reflecting the relative ease of flow across gently dipping coastal slopes of the Mauna Kea subaerial shield, which remains incompletely buried by Mauna Loa flows even though now submerged to a water depth of about 350 m (Figure 2). The gentle submerged slopes of Mauna Kea in Hilo Bay have also permitted large shoreline displacements by Mauna Loa flows.

Depths to initial beach deposits of successive Mauna Loa submergence sequences can be converted to age estimates and in turn to lava-accumulation rates for the Hilo area (Table 5), using eustatic global sea level curves and the isostatic subsidence rate for Hawaii Island (Figure 7) constrained by published isotopic ages [*Buchanan-Banks,* 1993; *Beeson et al.,* 1994; *Moore et al.,* 1994],. Excellent agreement between the average ^{14}C age (39 ka) from a boggy littoral deposit at -178 m [*Beeson et al.,* 1994] and the depth predicted from the submergence curve (Figure 7) confirms the validity of the calibration, at least for upper parts of the drill-hole

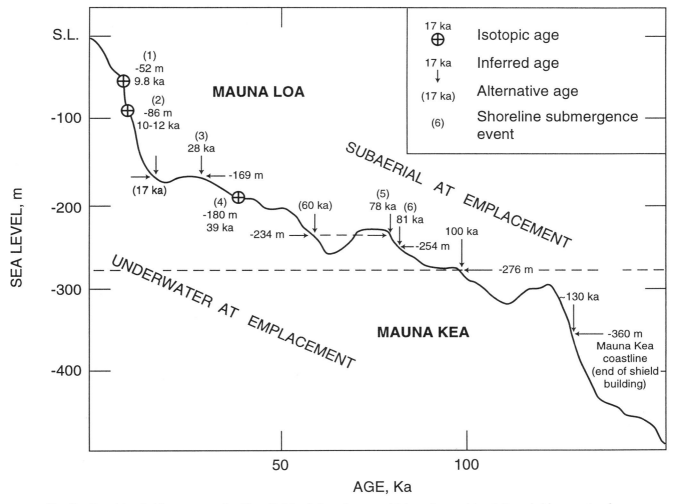

Fig. 7. Coastal subsidence curve for Hawaii Island, based on combining observed isostatic subsidence rates for the island (2.4 mm/yr) with eustatic sea-level changes [from *Moore et al.,* 1990], showing interpreted age-subsidence relations for intervals of coastal submergence documented in the Hilo drill hole (Table 5). Contact between Mauna Loa and underlying Mauna Loa lavas is at -276 m depth in the drill hole; all Mauna Kea and virtually all Mauna Loa lavas are subaerial.

section. The age of the lowest Mauna Loa flows must be at least 95 ka based on the submergence curve; because these flows are all subaerial, they could be older, but no older than the estimated 130-ka age of the last shield-stage lavas of Mauna Kea. For the accumulation-rate estimates (Table 5), an approximate age of 100 ka has been assumed for initial deposition of Mauna Loa lavas at the drill site.

Accumulation rates for Mauna Loa flows, as sampled in the drill hole, range widely for short time intervals (Table 5), but are broadly similar overall to those documented elsewhere on this volcano since 100 ka.

The overall accumulation of 239 m of lava at the drill site since about 100 ka (2.4 mm/yr) is only slightly higher than rates of 1-2 mm/yr determined elsewhere along the Mauna Loa coast (Table 4). Integrated accumulation rates for the four deepest lava intervals in the Mauna Loa section (0.4-4.0 mm/yr, average 1.3 mm/yr), interpreted to represent 28 to about 100 ka (Figure 8), are similar to coastal lava-accumulation rates elsewhere on the volcano (Table 4) and only about half the measured coastal subsidence rate for Hawaii Island.

The variably higher accumulation rates for lava packages in the range 0-28 ka (4.3-68 mm/yr; average 5.1

Table 5. Rates of Mauna Loa lava accumulation at the Hilo Scientific Drill Site

D.H., depth (m)	Depth, -S.L. (m)	Lithology & submergence event (no.)	Age (ka)	Lava thickness (m)	Lava thickness (cummul.)	Time span (ka)	Interval accum. rate (mm/yr)	Cumulative accum. rate (mm/yr)
0	0	—	0		(239)	1.3	0	2.4
0-31	+4.2-26.5	Lava (all Panaewa?)	1.3*	31	(239)	1.3	22.	2.4
31-56.5	27-52	Carbonate sediments	1.3-9.8	0	(208)	8.5	0	2.1
56.5	52	*Shoreline submerged (1)*	9.8[&,#]					
56-90.7	52-86.5	Subaerial lava	9.8-10.3?	34	(208)	0.4	68.	2.3
90.7	86.5	*Thin silt w/shells (2)*	10.3?[#]					
91-167	87-163	Subaerial lava	10-28	77	(174)	18	4.3	1.9
167-173	163-169	*Black-sand beach(3)*	28[&,x]					
173-182	169-178	Subaerial lava	28-39	9	(97)	11	0.8	1.3
182-184	*178-180*	*Ash w/littoral sandstone (4)*	39[@,&]					
184-230	180-226	Subaerial lava	39-60	46	(88)	21	2.2	1.4
230-238	226-234	Hyaloclastite	60-78	8	(42)	18	0.4	1.1
238	234	*Shoreline submerged (5)*	78[&,+]					
238-250	234-246	Subaerial lava	78-81?	12	(34)	3	4.0	1.5
250-258	246-254	Beach sand (Kilauea?)						
258	254	*Shoreline submerged (6)*	81[&]					
258-280	254-276	Subaerial lava	81-100+	22	(22)	19	1.2	1.2
280	276	Near shore(?)	100+[&,o]					
280-	276-	Mauna Kea flows	>130					

Sources of age data and interpretive problems

&, Inferred from sea-level subsidence curve [*Moore et al.*, 1990]

*, Radiocarbon age [calibrated from data in *Buchanan-Banks*, 1993]

#, U-disequilibrium age determinations [*Moore et al.*, this volume]; the 10.3 ka age may be unreliable (see text discussion)

@, Radiocarbon age [*Beeson et al.*, this volume]

x, The 28-ka age is for initial submergence; the beach-sand deposit may record relatively stable sea level until 17 ka, which would indicate a higher accumulation rate for the overlying flow sequence (11 mm/yr)

+, Because of the gentle slope of the sea-level curve and calibration uncertainties, submergence could have been as young as 60 ka; shallow marine or litoral deposition in the interval 78-70 ka, followed by emergence before 60 ka, is supported by inferred subaerial weathering of the hyalolastite at this depth [*DePaolo et al.*, 1994].

o, Minimum age to remain subaerial at this depth; because of gently sloping sea-level curve, might be as young as 95 ka. Preliminary identification of the 120-ka Blake polarity excursion near the base of the Mauna Loa section (at 260 m depth: J. Holt and J. Kirschvink, written commun., 1995) suggests a somwhat earlier age for inception of Mauna Loa lava deposition at the drill site mm/yr) are thought to reflect construction of only a few thickly inflated coastal lava deltas [*Hon et al.*, 1994], emplaced while postglacial sea level was rising rapidly after 18 ka (Figure 7). The upper 31 m of lava in the hole is interpreted, based on the lack of weathering between flow units, as consisting entirely of lavas from the Panaewa eruption, which has been dated at the surface at 1.4 ka [*Buchanan-Banks*, 1993; *DePaolo et al.*, 1994]. The next underlying 34-m lava package similarly appears to have accumulated within a brief time interval, probably during another single eruptive episode at about 10-12 ka [*Moore et al.*, 1994]. The recent relatively high rates of Mauna Loa lava accumulation in the Hilo area are thus considered atypical, as indicated by the longer-term record from the drill hole and by incomplete coverage of the gently inclined Mauna Kea surface by

Mauna Loa lava in Hilo Bay (Figure 2). Mauna Loa tholeiitic flows apparently do not interleave with the Mauna Kea lavas deeper in the drill hole section [*Rhodes and Sweeney*, 1994], indicating that earlier flows from Mauna Loa were able to flow into the ocean at lower elevations and that this part of the Mauna Kea shield was probably well above sea level during its subaerial growth. Perhaps initial growth of Kilauea, beginning at about 150 ka (Table 1), played a critical role in channeling Mauna Loa lavas higher against the flank of Mauna Kea. Despite the subsequent accumulation of Mauna Loa flows in the Hilo area, the record of recurrent submergence documented by the drill core suggests that long-term future problems for coastal Hilo may be as great from continuing isostatic subsidence of the island as from coverage by lava.

5.6. Accumulation Rates from Marine Data

Additional evidence on lava-accumulation rates in comparison with amounts of coastal subsidence, and resulting coastline growth or retreat, can be derived locally from marine oceanographic surveys and direct observation of underwater features.

Previous inference, from on-land geologic mapping [*Lipman*, 1980a; *Lipman and Swenson*, 1984], that the lower southwest rift zone of Mauna Loa has been relatively inactive in comparison with upper parts of this rift zone and with coastal parts of Kilauea rift zone, is confirmed by new bathymetric data and submersible observations. Although the coverage remains incomplete, multiple submersible dives have failed to disclose any young-appearing lava or eruptive vents along the crest of the underwater southwest rift zone of Mauna Loa down to a depth of 2000 m [*Fornari et al.*, 1979; *Moore et al.*, 1990; *Garcia et al.*, 1993, this volume]. A drowned coral reef at 150-160-m depth, sampled by submersible and dated by radiocarbon and U-disequilibrium methods at 14 ka [*Moore et al.*, 1990], is not crossed by any subsequent lava flows along the rift crest. The break in bathymetric slope at this depth is traceable for about 10 km to the northeast [*Chadwick et al.*, 1994b], indicating little or no modification by eruptive activity or structural events until proximity to the active Wai'ohinu fault, which funnels lava flows from upper parts of the southwest rift zone. In comparison, no comparable drowned shorelines are present along the Kilauea coast [Chadwick et al., 1994a], indicating that this volcano has been growing sufficiently vigorously to offset effects of isostatic and eustatic submergence.

Deeper along the submarine southwest rift zone, drowned boulder beach deposits have been identified by

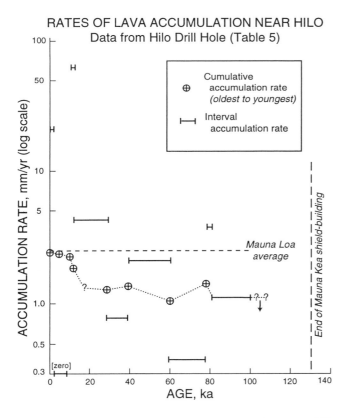

Fig. 8. Rates of lava accumulation on Mauna Loa near Hilo, based on data from the Hilo scientific drill hole (Table 5). Note log scale for lava accumulation rate. Mauna Loa lava accumulation rate for entire hole averages 2.4 mm/yr, about the same as the current rate of island subsidence. Lava accumulation rates in the drill hole vary widely for short time intervals; as discussed in text, higher coastal rates since about 20 ka may be related to rapid eustatic rise of sea level.

submersible at 350-400 m depth [*Moore et al.*, 1990; *Garcia et al.*, 1993]. Based on comparison with the submergence curve for Hawaii Island (Figure 7), the age of this drowned paleo-coastline is estimated at 130-150 ka, providing a limiting minimum age for activity along this sector of the rift zone, and documenting 8 km of subsequent shore-line retreat at this site. The ending of peak magmatic activity along the submarine southwest rift zone by about 300-350 ka is implied by the evidence for 500-800 m of cumulative subsidence since deposition of S-poor pillow lavas that are the dominant component exposed in the submarine Kahuku fault (Figure 5), as discussed earlier. Thus, a variety of evidence indicates that the lower southwest rift zone has been relatively inactive since at least 100 ka, and perhaps several times as long.

Farther north at Kealakekua Bay, a coral reef at a depth of 150 m was dated by U-disequilibrium at about 13 ka, showing that it is correlative with the one at South Point [*Moore and Clague*, 1987]. In Kealakekua Bay, this reef segment is crossed by only a few younger Mauna Loa flows, and none of the three largest flows occupying the present shoreline reach the reef front, indicating low rates of lava accumulation (Table 4, D.3) and net shoreline retreat of several kilometers since formation of the reef. This reef was traced across the Kealakekua fault without apparent offset, demonstrating limited recent movement along this still seismically active structure [*Moore and Clague*, 1987], and about 2 km of shoreline retreat in Kealakekua Bay since 14 ka. Similarly, the inferred depth to the paleo-shoreline during deposition of the 166-ka lavas in the Kealakekua scarp (Table 2), based calibration to the submergence curve (Figure 7), suggests about 500 m of subsequent submergence and 3 km of shoreline retreat along this steep slope.

Deeper on the underwater flanks of Mauna Loa, no sizable young lavas erupted from submarine vents have been reliably identified by GLORIA sidescan sonar surveys, even though the sonar images have been effective for mapping young flows elsewhere along the Hawaiian Ridge [*Holcomb et al.*, 1988; *Lipman et al.*, 1989]. One small area of possible young lava, low on the east side of the southwest rift zone [*Holcomb et al.*, 1988], is indistinct on GLORIA images and remains unsubstantiated by bottom photography or dredging. The only confirmed young underwater Mauna Loa flows, other than clastic debris aprons from subaerially erupted a'a flows that crossed the shoreline, are the diminutive 1877 lavas, which cover less than 1 km² and were erupted from three isolated vents at depths of 100-1,000

m in Kealakekua Bay [*Normark et al.*, 1978; *Fornari et al.*, 1980; *Moore et al.*, 1985]. On a regional scale, the Kealakekua Bay sector of Mauna Loa is currently shielded from summit-erupted lavas by the high west wall of Moku'aweoweo caldera, and only relatively infrequent flank eruptions from radial vents can reach the shoreline. As a result in the Kealakekua Bay area, the distribution of Hualalai lava flows has migrated southward since formation of the 13 ka reef, covering older Mauna Loa lavas [*Moore and Clague*, 1987]. These observations further confirm that recent Mauna Loa eruptive activity is largely confined to subaerial vents on upper flanks of the volcano.

5.7. Caldera Formation and Intrusion

Because the geometry of summit calderas such as Moku'aweoweo influences the distribution of downslope lavas, mechanisms of caldera evolution are important for understanding the rates of lava accumulation on Hawaiian volcanoes. The absence of sizable young lava flows on submarine flanks of Mauna Loa, along with evidence of inactivity along lower parts of both subaerial rift zones, provides a key constraint for the poorly understood processes of caldera formation. Geologists have long inferred that caldera collapses on basaltic volcanoes are caused by voluminous submarine eruptions that drain magma chambers beneath their summits [*Macdonald*, 1965; *Holcomb*, 1987], and identification of voluminous young submarine lava fields surrounding the deep-water termination of the east rift zone of Kilauea seemingly provides support for such an interpretation [*Holcomb et al.*, 1988]. In contrast, at Mauna Loa the absence of large lower-rift lavas, either underwater or on land, that could be associated with formation of the present caldera suggests that other processes must cause at least some caldera subsidence on Hawaiian-type volcanoes.

At Mauna Loa, the present summit caldera (Moku'aweoweo) formed at least in part since about 0.6 ka, as indicated by truncation of a dated flow on its rim [*Lockwood and Lipman,* 1987]. The volume of the caldera at that time must have been substantially greater than 2 km³: the present volume is about 1 km³, and an additional 1.2 km³ of lava is estimated to have accumulated within Moku'aweoweo just during the 150-yr historical period, leading to overflow in 1914 [*Lockwood and Lipman*, 1987]. Perhaps long-lived eruptions higher on a volcano of the size of Mauna Loa can also lead to caldera collapse, but no single eruption since 0.75 ka (beginning of age period 4) with a volume greater than about 1 km³ has been identified by geologic mapping [*Lockwood and Lipman*, 1987; J.P. Lockwood, oral commun., 1994].

Alternatively, rather than representing a single large collapse event, Moku'aweoweo may have enlarged incrementally, by repeated collapse of smaller craters during modest-volume eruptions, and perhaps in association with seismically induced extension and seaward motion of its southeast flank with or without concurrent flank eruptive activity. Sustained seaward spreading, which is especially clear for Kilauea [*Swanson et al.*, 1976; *Borgia and Treves*, 1992; *Delaney et al.*, 1993], has the overall effect of splitting a volcano into two subequal blocks, separated by the extending rift zones. Such interpretation of the structure of Kilauea and Mauna Loa as dominated by a single "break-away rift zone" [*Fiske and Swanson*, 1992] downplays the overall tectonic role of the summit caldera, which could largely represent passive responses to extension along the crest of the rift, with or without associated eruptive activity.

Caldera collapse and rifting of the summit region must, at times, have been sufficiently inactive to permit complete filling of the caldera, burial of upper-rift fissures, and overflow enlargement of the "upper shield" above 3,000 m elevation (Figures 1, 6). The upper shield of Mauna Loa, with slopes locally steeper than 10° in striking contrast with the summit morphology of Kilauea, must have grown during repeated periods of small-volume summit lava-lake overflows between caldera-collapse events, most recently during the 0.75-1.5 ka time interval [*Lipman*, 1980b; *Lockwood and Lipman*, 1987; *Lockwood*, this volume]. Loading by dense intrusions concentrated in the summit area may further encourage intermittent caldera subsidence not directly related to large-volume eruptions [*Walker*, 1988]. Quantifying intrusive volumes within Mauna Loa is also a critical problem for estimating magma-supply rates, as discussed in a later section.

5.8. Evidence from Ninole Hills

Lava flows of the Ninole Hills, the oldest subaerial rocks exposed on the south side of Hawaii Island, provide key evidence on the evolution of Mauna Loa and the southeastward progression of Hawaii volcanism. Recent interpretation of these rocks as faulted remnants of the south flank of Mauna Loa at approximately 100-200 ka [*Lipman et al.*, 1990] provides a framework for interpreting subsequent depositional and structural events.

Deep canyons, interpreted as having eroded in Ninole rocks in response to massive landslide failure of the south flank of Mauna Loa at 100 ka or earlier to produce the Punalu'u slide (Figures 2, 9), have provided efficient channels for younger Mauna Loa lavas to reach the Ka'u coastline. Nevertheless, this coastline remains embayed because the volume of the younger flows has been insufficient to infill completely the break-away scar of the Punalu'u slide, providing further evidence of relatively low lava accumulation rates along the Ka'u coastline of Mauna Loa since 100 ka.

Subaerial presence of the Ninole Basalt also documents that the southern flank of Mauna Loa had grown to much of its present size above sea level by 100-200 ka. Flat-topped primary depositional surfaces are preserved at the crests of hills in the Ninole area (Kaiholena, Pu'u Enuhe, Pu'u Makanau, etc.: see USGS Punalu'u 7.5' quadrangle), although modestly jostled by subsequent faulting along the Wai'ohinu-Kao'iki fault zone (Figure 1). These flat summit areas stand as much as several hundred meters above more recent Mauna Loa lavas that have infilled around their bases but are well aligned with surfaces on Mauna Loa upslope from the Ninole Hills, especially above about 1,500 m elevation (Figure 9) This relation suggests that upper slopes of the middle southwest rift zone of Mauna Loa had grown to much of their present size by 100-200 ka, the time of accumulation of the lavas now exposed in the Ninole Hills, and that only modest growth of the rift zone has taken place subsequently.

6. LANDSLIDING AND GRAVITATIONAL FLANK FAILURE

In addition to isostatic subsidence, growth of Hawaiian volcanoes is modulated by lateral gravity-driven spreading of their oceanward flanks, as well documented by seismic and geodetic studies, especially for the south flank of Kilauea [*Swanson et al.*, 1976; *Lipman et al.*, 1985; *Borgia and Trewes*, 1992; *Denlinger and Okubo*, in press]. The spreading is accommodated by recurrent low-angle thrusting at the base of the volcanic edifice, shallower high-angle normal faults, and catastrophic debris-avalanche failures. These faults, driven by gravity on steep submarine slopes, by dike injection high along rift zones, and probably by ultramafic cumulates deeper in the magma syste n [*Clague and Denlinger*, 1994], have produced block slumps on land and underwater along the active Hilina system on Kilauea (about one M 7-8 earthquake/100 yr) and along the Wai'ohinu-Kao'iki and Kealakekua-Kahuku fault zones on Mauna Loa.

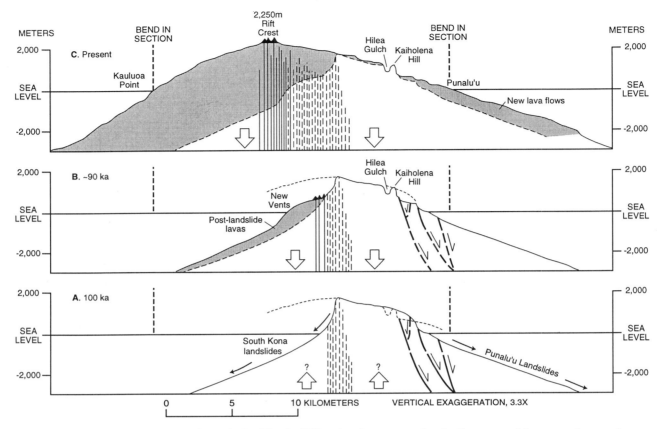

Fig. 9. Schematic sections through the Ninole Hills, showing pre-erosional alignment with upper slopes of southwest rift zone of Mauna Loa and evolution of the southwest rift zone in response to landsliding and rift migration (from *Lipman et al.* [1990, Figure 6]). Line of section shown on Fig. 2. The evolution of this segment of Mauna Loa is discussed in more detail in *Lipman et al.* [1990].

Rapidly emplaced giant submarine landslides have catastrophically removed entire flanks of Hawaiian volcanoes; the resulting deposits have moved as much as 250 km on the sea floor and morphologically resemble subaerial sector-collapse debris-avalanche deposits [*Moore et al.*, 1989; *Normark et al.*, 1993]. The Alika and Ka Lae slides on the west flank of Mauna Loa, the youngest large submarine debris avalanches in Hawaii (about 100 ka?), have large scarps at their heads [*Lipman et al.*, 1988]. The seemingly older Punalu'u slide on the southeast flank probably triggered rapid canyon cutting in its headwall, generating the present morphology of the Ninole Hills [*Lipman et al.*, 1990]. Conditions that trigger such prodigious slides, the ultimate geologic hazard in Hawaii, remain poorly understood, but the net effect is to further reduce the rate of upward growth of mature Hawaiian volcanoes. For Mauna Loa, the cumulative volume of the South Kona slide complex has been estimated at 1,500-2,000 km^3 [*Lipman et al.*, 1988], equivalent to the volcano-wide lava supply for 75-100 ka at the recent eruption rate of 20×10^6 m^3/yr (Tables 1, 3), although much of the slide material still resides high on the submarine flank of the volcanic edifice. The volumes of large slides on the southeast flank of Mauna Loa, including Punalu'u, are poorly known. A further complication is the possibility that much early slide material from the south flank of Mauna Loa, associated with gravitational instabilities along the Wai'ohinu-Kao'iki fault zone, may lie buried beneath present-day Kilauea. Such a relation would further add to the volume of Mauna Loa and decrease the volume of Kilauea as estimated in Table 1.

In contrast to its earlier history of flank instability and despite continuing frequent eruptions (historically about every 8 years), Mauna Loa's seaward flanks appear to have increasingly stabilized since at least 30 ka and perhaps 100 ka, in comparison with its dynamic neighbor Kilauea. Sustained displacement of the south flank of Kilauea, punctuated by failure events such as the 1975 Kalapana earthquake, is well documented by historical deformation data, fault offsets along the Hilina system, and underwater block slumps [*Swanson et al.*, 1976; *Lipman et al.*, 1985; *Delaney et al.*, 1993; *Denlinger and Okubo*, in press]. On Mauna Loa, analogous flank failures are recorded in the Kao'iki-Wai'ohinu and Kealakekua-Kahuku fault systems (and underwater by long-traveled debris avalanches), but many of the faults are draped by >30-ka ash deposits without major recent ffset despite continuing seismicity [*Lipman*, 1980b; *Lipman et al.*, 1990]. The South Kona block slumps are apparently presently inactive, because their benches lack the large closed depressions that characterize the Hilina lump [*Chadwick et al.*, 1993, 1994b; 1995]. Historical geodetic data, while less complete than for Kilauea, also document recent relative stability on middle slopes of Mauna Loa during inflation-deflation cycles at the summit, even during the major 1984 rift eruption [*Lockwood et al.*, 1987; *Dvorak et al.*, 1993]. Some slow spreading of the massive lower flanks of Mauna Loa could be taking place by sustained creep, without the large stress accumulation along faults and catastrophic release in large earthquakes that characterizes the oceanward south flank of Kilauea; such a possibility for Mauna Loa should be testable in future years by regional GP measurements that are currently being initiated. The apparent increased recent stability of Mauna Loa may be due to combined effects of (1) the great areal extent of this mature shield and flanking landslide debris that buttress further spreading, (2) the cumulative isostatic subsidence beneath its summit (about 8 km) that requires increasingly "uphill" spreading at the base of the volcanic edifice along the old sea floor, (3) growing volcanic buttresses against its south flank--first Kilauea, and more recently Loihi, and (4) the declining growth of Mauna Loa itself since 100-200 ka, leading to cooling and partial stagnation of its cumulus dunite core.

7. MAGMA-SUPPLY RATES

Estimation of magma-supply rates for Mauna Loa and other Hawaiian volcanoes through time provides a valuable framework to interpret the present state of the volcanoes, their growth in relation to isostatic subsidence and times of flank failure and sliding, their inception ages, and the general evolution of hot-spot volcanoes. Eruption rates for the 150 year historical record of Mauna Loa (1843-present) are known quite well from contemporary observations and recent field studies [28×10^6 m^3/yr: *Lockwood and Lipman*, 1987]; eruption rates since about 4 ka have also been determined with fair accuracy from geologic mapping and ^{14}C geochronology, averaging 20×10^6 m^3/yr (Table 1). These values are derived from prehistoric surface-covering rates, compared to the historical covering and eruption rates. Because the historical rates include a substantial volume (about 25%) of lava ponded within the summit caldera (Moku'aweoweo), a similar factor is incorporated in the prehistoric rates. In discussing magma-supply rates, the 4-ka average is assumed to represent recent Mauna Loa eruptive activity. Determining magma-supply rates is more difficult because of the need to include (1) effects of summit inflation/deflation associated with filling and draining of the summit magma reservoir, (2) high-level dike intrusion along rift zones, and (3) magma accumulation deeper within the volcanic edifice.

The amount of any sustained inflation high on Mauna Loa related to growth of the summit magma reservoir is poorly defined by available geodetic data, even for the last few decades. At Kilauea, the summit region has been shown by leveling and tilt measurements to have fluctuated during successive inflation-deflation cycles associated with eruptive activity during the last 40 years, but little or no net vertical uplift has accumulated [e.g., *Tilling and Dvorak*, 1993, Figure 5]. Seismic and deformation data show that Mauna Loa has been less active than Kilauea between eruptions [*Klein and Koyanagi*, 1989], and deformation observations (especially microgal gravity) have documented that inflation and uplift premonitory to the 1984 eruption at Mauna Loa were largely offset by eruption-related subsidence [*Lockwood et al.*, 1987]. Accordingly, cumulative summit inflation is assumed to have contributed negligibly to the growth of both Mauna Loa and Kilauea, at least in recent time.

In contrast, dike emplacement and lateral spreading due to intrusion have been important for endogenous growth of both volcanoes. At Kilauea, recurrent dike propagation along rift zones, both during rift eruptions and as frequent intrusions that could be tracked seismically and geodetically, has been associated with meter-scale lateral seaward displacement of the south flank [*Swanson et al.*, 1976; *Klein et al.*, 1987; *Tilling and Dvorak*, 1993]. For Mauna Loa, 15 sizable eruptions have occurred along its rift zones during the 150-yr historical period, or about 1 per 10 years, but only a single dike intrusion without eruption has been recorded since seismic monitoring became feasible for the upper slopes of Mauna Loa in mid century. The locus of this intrusion, into the northeast rift zone after the 1975 summit eruption [*Lockwood et al., 1987*], later became the site for the vent of the 1984 rift eruption, which may have followed and reused the 1975 dike system. Non-eruptive dike intrusions appear to occur much less frequently on Mauna Loa than at Kilauea, as indicated both by the imperfect historical seismic and deformation record, and also by geologic evidence of limited recent seaward mobility for the south flank of Mauna Loa in comparison with Kilauea [*Lipman*, 1980b; *Lipman et al.,* 1990]. Certainly, summit seismicity and deformation at Mauna Loa were remarkably quiet for the 10-year period of good-quality data that are available prior to the precursors of the 1975 eruption [*Lockwood et al.*, 1987]. Accordingly, most intrusions high in the Mauna Loa edifice are interpreted to occur as dikes along the rift zones in association with rift eruptions; non-eruptive intrusions may have been more common early during the growth of Mauna Loa but have become less frequent as the mobility of Mauna Loa's south flank has been impeded by the growth of Kilauea.

The following estimate of cumulative volume for shallow dike emplacement on Mauna Loa, associated with eruptions for the historical period, is based on an average dike distance of about 20 km for historical vents along the rift zones, an average dike height of 3 km (from the top of the shallow summit magma reservoir [*Decker et al.*, 1983], and an average dike thickness of 0.5 m (maximum measured dilations associated with dike emplacement during the 1975 and 1984 eruptions were 0.7 and 0.4 m across the summit caldera, where maximum dilations would be expected). The inferred 3-km dike height is in accord with maximum estimates for Kilauea dikes based on seismic and modeling data [*Klein et al.*, 1987; *Rubin and Pollard*, 1987], and the inferred average width for Mauna Loa dikes is similar to the 0.53 m median width measured by *Walker* [1987] for several thousand dikes erosionally exposed at Ko'olau Volcano on Oahu. The resulting average dike volume would be 30×10^6 m^3, with an emplacement recurrence for the historical period of eight years, indicating a contribution to the overall magma supply of Mauna Loa of about 4×10^6 m^3/yr. The potential for modest additional contributions by infrequently emplaced non-eruptive dikes along Mauna Loa rift zones is perhaps balanced by the probability that the historical period has been more active than the longer-term eruptive record since 4 ka. This estimate brings the high-level magma supply at Mauna Loa up to 24×10^6 m^3/yr.

A final component is accumulation of magma at depth in Mauna Loa without erupting. From density and buoyancy relations, sizable volumes of sparsely porphyritic melt seem unlikely to accumulate at or near the base of a Hawaiian volcano, but rather would rise to the shallow summit magma chamber that is well defined from seismic and deformation monitoring [*Decker et al.*, 1983]. Similarly to Kilauea, Mauna Loa magma is thought to have fractionated olivine as it rises and accumulates in the summit reservoir prior to eruption [*Wright*, 1971; *Rhodes*, 1987, this volume]. Olivine-rich (picritic) lavas are notably more commonly from vents low along Mauna Loa rift zones than from the summit region [*Garcia et al.*, 1993, this volume]. For Kilauea, picritic lavas and Mg-rich liquid compositions are especially prevalent along submarine segments of the lower east rift zone, and mass-balance calculations indicate that about 14 volume percent of Kilauea magma must accumulate beneath the summit reservoir as picritic or dunitic cumulate intrusions [*Clague and Denlinger*, 1994]. If a similar proportion of magma is inferred to accumulate low within the Mauna Loa edifice, it would add about 3.4×10^6 m^3/yr to the overall magma supply. Based on these admittedly rough estimates, the combined processes of deep cumulates and shallower dike intrusion may currently be contributing 25-30 percent to the total magma accumulation at Mauna Loa, and the total magma supply for the past few thousand years would be about 28×10^6 m^3/yr (Table 1). The uncertainties embedded in this value are difficult to quantify, as indicated by the above discussion, but the magma-supply estimate is considered likely valid within ±10 percent, and certainly within ±20 percent.

At a sustained magma-supply rate of 28×10^6 m^3/yr, the present Mauna Loa edifice (80,000 km^3) would have required almost 3 m.y. to accumulate (Table 1). In light of an approximate limiting age of 1 Ma for inception of Mauna Loa, the magma-supply rate must have been substantially higher during early growth of the volcano. A plausible growth scenario, as developed more fully later, could be magma accumulation at a rate of 100×10^6 m^3/yr for about half a million years, followed by gradually declining magma supply to the present rate (28×10^6 m^3/yr) since 0.2-0.5 Ma. An analogous decline in magma output has been inferred for the late evolution of Mauna Kea, based on results from the Hilo scientific drill hole [*DePaolo and Stolper*, 1994].

In addition to the inferred volumetric decline in magma supply at Mauna Loa, helium isotopic systematics for accessible Mauna Loa lavas (to about

200 ka?) show significantly lower ^{3}He/^{4}He ratios (and parallel changes in strontium and lead isotopic compositions) for the younger samples, interpreted as indicating decreasing contributions with time from an undegassed source, probably isotopically enriched deep-plume material [*Kurz and Kammer*, 1991; *Kurz et al.*, this volume]. Helium compositions of the oldest Mauna Loa lavas are isotopically similar to historical Kilauea flows and dredged samples from the active Loihi seamount, while younger and historical Mauna Loa lavas are isotopically similar to mid-ocean ridge basalts (MORB) derived from depleted asthenospheric sources (Figure 10). Similar decreases in helium isotopic ratios have

been documented late during the tholeiitic shield-building cycle and the transition to alkalic volcanism at Haleakala (East Maui) and Mauna Kea volcanoes [*Kurz et al.*, 1987; *Kurz et al.*, 1994], suggesting that helium composition is a sensitive indicator for changing conditions of magma generation during the late stages of volcanic activity. Notably, the shift in helium isotopic compositions at Mauna Loa has taken place late during the tholeiitic shield-building stage, while the comparable compositional changes at Haleakala and Mauna Kea appear to have occurred approximately concurrently with the change to alkalic volcanism. While these results may document subtle contrasts among the evolutionary trends of different Hawaiian volcanoes, the decreasing involvement of isotopically enriched magma sources in generation of the younger Mauna Loa magmas is another indicator that shield-building activity has begun to wane at this large volcano.

Rare-earth-element modeling of lavas from several Hawaiian volcanoes, documenting depletion in incompatible elements for some historical and late prehistoric Mauna Loa lavas relative to those predicted from primitive mantle sources, has also been interpreted recently as indicating that this volcano is near the end of its shield-building stage [*Watts*, 1993]. The elemental concentrations fail to define systematic temporal trends for Mauna Loa lavas, however, and end-member compositions were erupted in only 44 years, between 1843 and 1887 [*Rhodes and Hart*, this volume], suggesting that these data may not be reliable for forecasting the demise of tholeiitic volcanism at Mauna Loa.

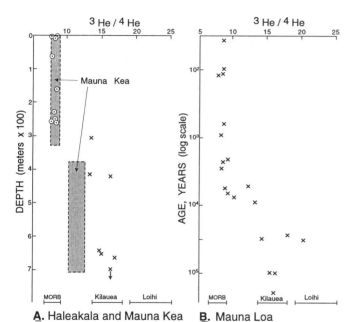

A. Haleakala and Mauna Kea   **B.** Mauna Loa

Fig. 10. Variations in helium isotopic ratios with age and stratigraphic position for Hawaiian volcanoes studied to date: *A.* Haleakala and Mauna Kea [*Kurz et al.*, 1987, 1994] , *B.* Mauna Loa [*Kurz and Kammer* , 1991]. Stratigraphically controlled samples of Mauna Loa tholeiitic lavas show a trend of decreasing radiogenic ratios with time, similar to those accompanying termination of the shield-building stage at Haleakala and Mauna Kea. O, alkalic basalt flows of Haleakala (Hana and Kula Volcanics); X, tholeiitic flows of Haleakala (Honomanu Basalt) and Mauna Loa (Ka'u, Kahuku, and Ninole Basalts). The Haleakala sample with attached arrow is a dredged basalt, which is inferred to be from deeper in the section. The Mauna Kea samples (stippled rectangular fields) are from the Hilo drill hole, consisting of interlayered alkalic and tholeiitic flows in the upper 200 m of section, underlain by 500 m of tholeiitic basalts to the bottom of the hole.

8. COMPARISONS WITH KILAUEA AND LOIHI

Comparisons with Kilauea and Loihi, as analogs for earlier growth of Mauna Loa, provide additional insights into the long-term eruption, magma-supply, and growth rates of Hawaiian volcanoes. Kilauea appears to be characterized by higher and more variable lava-eruption and intrusion rates than Mauna Loa, although direct comparisons of eruption and lava-coverage rates between the two volcanoes are difficult because of their disparate surface areas. Subaerial Kilauea covers about 1,425 km^2 (Table 1), only about one-fourth the size of Mauna Loa, and overall surface coverage rates and coastline crossing rates are thus expectedly higher for Kilauea. Since 0.75 ka, more than 80% of subaerial Kilauea but only 29% of Mauna Loa has been covered by lava flows [*Holcomb*, 1987, Table 12.2; *Lockwood and Lipman*, 1987, Figure 18.10], and the average coastal lava-accumulation rate is 7 times greater for Kilauea (Table 1).

Higher eruption rates for Kilauea are also clearly documented by its 750-year lava-accumulation rate in comparison with the "upper shield" of Mauna Loa (3,000 m level, 3.7 mm/yr: Figure 6), which has an area of about 700 km^3--only about half that of subaerial Kilauea. Based on the same calculation approach and the age-distribution map of *Holcomb* [1987, Figure 12.12], 78% of the Kilauea coastline has been crossed by separately mapped age units since 0.75 ka, many segments more than once for the earlier age units. The resulting minimum average coastal accumulation rate for Kilauea is 6.4 mm/yr, nearly double that at the base of the Mauna Loa "upper shield" (3,000 m level), despite the much larger size of subaerial Kilauea.

The historical magma-supply rate for Kilauea has been closely determined from total lava output during several long-lived eruptions, during which shallow intrusive activity has been limited, at about 100x10^6 m^3/yr [*Swanson*, 1972; *Wolfe et al.*, 1987]; this is more than three times the estimated rate for Mauna Loa (28x10^6 m^3/yr: Table 1). Between prolonged eruptions, lower average eruption rates have been partly offset by increased shallow intrusions (summit inflation, dike intrusion into rift zones), estimated to constitute as much as 45 percent of the total Kilauea magma supply between 1956 and 1983 [*Dzurisin et al.*, 1984; *Dvorak and Dzurisin*, 1993]. During earlier historical activity (1820-1950), when eruptions were largely confined to the summit region and there were long periods of inactivity, the Kilauea magma supply may have been lower than in the late 20th century [*Peterson and Moore*, 1987, Table 7.3; *Dvorak and Dzurisin*, 1993; *Wright and Fiske*, in press], although determining the 19th century magma supply for Kilauea is complicated by uncertainties about rates of intracaldera lava accumulation, summit inflation, dike intrusion, and mobility of its south flank. At times, large volumes of magma may have accumulated within the south flank without associated eruptions, in conjunction with sustained seaward spreading, punctuated by increased movements during large earthquakes [*Delaney et al.*, 1993]. Rough estimates, based on the observed average 6 cm/yr seaward displacement of Kilauea's south flank between 1983-1991 [*Delaney et al.*, 1993], and the greater intermittent displacements associated with south-flank earthquakes [*Lipman et al.*, 1985], suggest that as much as two-thirds of Kilauea's magma supply may accumulate within the extending region. Such effects on overall magma supply should be much less significant for Mauna Loa because of the lower

present mobility of its south flank, now buttressed by Kilauea and Loihi [*Lipman*, 1980b].

At an average magma supply rate of 100x10^6 m^3/yr, the entire estimated volume of Kilauea (15,000 km^3) could have accumulated since 150 ka (Table 1). Initial growth of Kilauea could have begun as early as about 300 ka if the sustained magma supply rate were as low as 50x10^6 m^3/yr [*Dzurisin et al.*, 1984]. Alternatively, a model of gradually increasing magma supply, similar to that developed for Mauna Loa later in this paper, would indicate an inception age of about 200 ka for Kilauea. If sizable parts of Kilauea are constructed over old block slumps along Mauna Loa's south flank, and/or if the east rift zone of Kilauea partly overlies an inactive eastward continuation of the northeast rift zone of Mauna Loa as discussed earlier, the resulting smaller total volume of Kilauea could have accumulated in less time.

With similar assumption of Kilauea rates, Loihi (400 km^3) could have grown to its present size since as recently as 4 ka. If early magma supply at Loihi were closer to that for late-stage alkalic volcanism (perhaps 10x10^6 m^3/yr) than to the current Kilauea rate, as suggested by the eruption of alkalic basalt as well as tholeiite, activity at this volcano could have begun as early as 40 ka. More plausibly, the current magma supply at Loihi is inferred to be between late-stage alkalic activity and present Kilauea, perhaps 30-50x10^6 m^3/yr, with an inception age of 10-15 ka.

9. DISCUSSION AND SUMMARY

The combined effects of isostatic subsidence, lateral spreading along low-angle faults, and landsliding appear to have narrowly dominated over volcanic growth during the last few hundred thousand years at Mauna Loa. Despite frequent historical rift and flank eruptions (avg. 1 per 8 yrs since mid-19th century), the lower subaerial flanks of Mauna Loa have grown little during this time interval. Since 10-100 ka, lava-accumulation rates along most of the coast have averaged only 1±1 mm/yr (Figure 11), although rates have been much higher for short time intervals and locally where topography has deflected and ponded lava. Coastal lava-accumulation rates thus have not kept pace with isostatic subsidence of Hawaii Island (2.4 mm/yr), and in conjunction with postglacial eustatic sea-level rise, subaerially emplaced lavas are now submerged below sea level along much of the Mauna Loa coastline. The only coastal segment characterized by significant growth since 100 ka is the large composite lava delta that infilled the landslide scar in South Kona

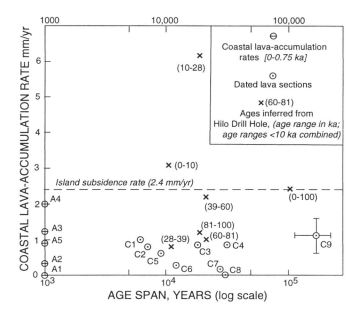

Fig. 11. Summary of rates of coastal lava accumulation on Mauna Loa, based on dated lava sections, coastal-lava accumulation rates from mapping, and coastal-subsidence ages inferred from Hilo drill-hole data. Lava accumulation rates are lower than island-subsidence rates except for post 20-ka lava near Hilo.

and generated a prominent bulge in the coastline. Even here, rates of lava accumulation have been inadequate to fill in the landslide headwalls, and lava accumulation has only modestly exceeded subsidence rates, based on data for the last 750 years (Figure 6). Lava accumulation rates have been greater high on the volcano, averaging nearly 4 mm/yr at 3,000-m elevation for the past 750 years, but are several times less than those for Kilauea at comparable distances from its summit. Based on published geologic mapping, overall lava accumulation rates are interpreted as averaging 2 mm/yr for subaerial Mauna Loa, 4-8 times lower than for subaerial Kilauea.

Eruptions of Mauna Loa are also less frequent than at Kilauea, and none historically have continued as long as several east rift eruptions of Kilauea. In contrast, eruption rates and volumes during historical Mauna Loa eruptions have been higher than for Kilauea. During the last major eruption of Mauna Loa in 1984, the volume of erupted lava exceeded 1×10^6 m^3/hr during the first days of the three-week eruption; in contrast, eruption rates during brief episodic activity at Kilauea typically have been in the range $0.1-0.5 \times 10^6$ m^3/hr, as during the Pu'u

O'o eruptions [*Wolfe et al.*, 1987], and have been only about $0.2-0.4 \times 10^6$ m^3/day during recent sustained activity [*Heliker et al.*, 1993]. As a result of the large volumes and high eruption rates, the relatively infrequent Mauna Loa rift eruptions present continuing major hazards to the Island of Hawaii, covering 35-40% of the volcano surface area every 1,000 years [*Lockwood and Lipman*, 1987; *Trusdell*, this volume], even though this volcano appears to be less active than in the past. Lava has flowed into the area of present-day Hilo at least 10 times since 10 ka [*Buchanan-Banks*, 1993], most recently in 1880-81, and much of the city is built on the 1.4-ka Panaewa flow, which is among the largest-volume eruptive deposits exposed on Mauna Loa. Volcanic hazards are especially severe on the South Kona side of the southwest rift zone, where increasing development along the crest of the rift zone and steep adjacent slopes could permit inundation by rapidly flowing lava with little lead time for evacuation (flow advance rates of up to 10 km/hr in 1950: see *Finch and Macdonald* [1953]). Even though the lowermost part of the southwest rift zone now erupts only infrequently, much of this area is directly downslope from the active middle sector of the rift, and only the elevated areas along the east side of the Kahuku fault (Figure 2) are shielded from lavas erupted further upslope.

Inferred rates of magma supply and lava accumulation since 100 ka seem too low by a factor of 3-4 to account for the growth of Mauna Loa since its estimated inception at about 1 Ma or even more recently; early rates of eruption and/or intrusion must have been greater. Diminished recent eruptive rates are also indicated by eruptive inactivity along both rift zones below elevations of about 2,500 m, and by diverse other subaerial and marine geologic features discussed in this paper and summarized in Table 6. The apparent "drying up" of distal parts of both rift zones, restriction of underwater lava accumulation to upper slopes of the submarine volcanic edifice, and decreases in radiogenic helium-isotope ratios suggesting a declining deep mantle component in recent Mauna Loa lavas, all provide strong additional indications that Mauna Loa is near the end of the tholeiitic shield-building stage of Hawaiian volcanism. The declining shield activity at Mauna Loa also suggests that rates of isostatic subsidence should have slowed along central and northern parts of Hawaii Island from peak rates before several hundred thousand years ago.

Overall growth of Mauna Loa seems most appropriately modeled by magma supply at rates similar

Table 6. Summary: Evidence for Decline in Mauna Loa Eruptive Activity

---

Low coastal lava-accumulation rates regionally, in dated sections, and in Hilo drill hole (less than subsidence)

Eruptive inactivity on lower southest & northeast rifts in Holocene and late Pleistocene time

Ninole Hills: surviving remnants of Mauna Loa at 150-200 ka

Widespread near-surface exposures of Pahala Ash (>31 ka)

Small volumetric proportion of historical eruptions emplaced beyond shoreline (8%), in relation to submarine
     area (45%)

Limited extent of post-landslide (post-100 ka?) submarine lava accumulation: mostly at depths <1000 m

Limited coverage of Mauna Kea lavas on the Kohala shelf and near Hilo

Incomplete infilling of landslide break-away scars, especially Punalu'u and Alika slides

Preserved -150 m coral reefs (14 ka): Kealakekua Bay, Ka Lae, Hilo Bay(?)

Preserved -350-400 m boulder beach at Ka Lae (100-130 ka)

Shoreline retreat: 2 km since 14 ka at Kealakekua Bay, 3-4 km since 166 ka at Kealakekua Bay,   8 km since 140
     ka at South Point

Decreased deformation and gravitational flank instability on lower parts of the volcanic edifice

Recent magma supply rates (since 4 ka) inadequate to generate edifice

Helium isotopic evidence of decreased involvement of enriched plume-like mantle source for young Mauna Loa
     lavas

---

to (or greater than) present-day Kilauea (100×10^6 m^3/yr) during peak shield growth when Mauna Loa most directly overlay the Hawaiian hot spot, followed by a prolonged period of gradual decline in magma supply to the present estimated rate (28×10^6 m^3/yr) as plate motion has carried the volcano away from the hot-spot locus (Figure 12). This quantitave model is constrained by the inferred lifespan of about one million years for Mauna Loa, somewhat longer than the 600,000 years estimated for Hawaiian volcanoes in general by Moore and Clague [1992], and the total volume of about 80×10^3 km^3 estimated for present-day Mauna Loa. Inception of Mauna Loa may have included an early alkalic stage, similar to that well documented at Loihi, and must have

been marked by rapidly increasing magma-supply rates. The stage of peak shield-building, at magma-supply rates at least as great as those for present-day Kilauea, is inferred to have begun to decline no later than about 200 ka, as evidenced by subsequent inactivity along the lower southwest rift zone (and probably a previously more extensive northeast rift zone as well) and also by preservation of the Ninole Basalt in high-standing hills on the south flank of Mauna Loa. The apparent absence of major landsliding from Mauna Loa's flanks since about 100 ka and increased stability along the block-faulted Wai'ohinu-Kao'iki zone are also inferred to reflect declining vigor of shield-building volcanism at Mauna Loa, as well as buttressing effects from the growth of

Kilauea and Loihi.   Broadly similar time-volume relations for volcano growth, including rapid inception and more gradual decline of tholeiitic shield building, have previously been inferred for Mauna Kea [*Wise,* 1982] and for a generalized 5-m.y. Hawaiian evolutionary cycle [*Clague,* 1987, Figure 3], although these models were based on fewer constraints for eruption rates, edifice volumes, and inception of shield building than the Mauna Loa interpretation in Figures 12 and 13.

Based on the estimated inception age and variations in magma supply (Figure 12), and a highly simplified geometric conical model for the Mauna Loa edifice, it is possible to approximate other aspects of its growth history with time, such as height of the summit above the sea floor, subsidence component of the total height, time of initial rise to the water surface, and subaerial growth

for the summit area (Figure 13). The geometric model is a double-sided cone, with one apex projecting upward from the sea floor (Cretaceous oceanic crust) to the summit of the vol-cano as defined by bathymetric and topographic data, and a second cone of isostatic subsidence facing downward as con-strained by the seismic data (Figure 4).   Distinction is made between total height of the present edifice (H_T, 17 km), height above the regional sea floor on which the volcano grew (H_A, 9 km), the subsidence component of total height (H_S, 8 km), and height above water (H_W, 4 km).   A simple relation of height above the sea floor (H_A) to average edifice radius (R), $R = 6\ H_A$, is used to constrain volcano size; this ratio is reasonable for the present average shape of Mauna Loa (R=55 km, $H_A = 9$ km), and even fairly good for the steeper and more

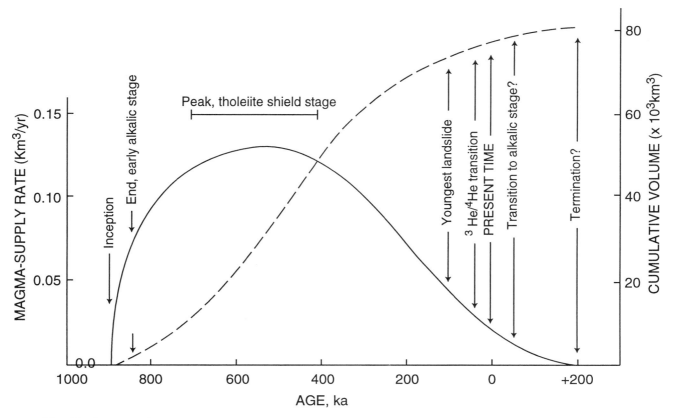

Fig. 12. Interpreted growth history of Mauna Loa, showing inferred magma-supply rates (*solid line*) and cumulative volumes (*dashed line*) with time. Curves are constrained by: (1) present total volume and magma-supply rates from Table 1, (2) requirement for higher magma-supply rates during peak of tholeiitic shield-building stage as documented in this paper, (3) lower rates during probable early alkalic stage at Mauna Loa and analogous to Loihi, (4) indication from the declining recent eruption rates and helium-isotope evidence that tholeiitic shield building is nearing completion and alkalic volcanism should be expected at Mauna Loa in the not-too-distant geologic future, and (5) overall estimated lifespan of about 1 m.y. for Hawaiian volcanoes.

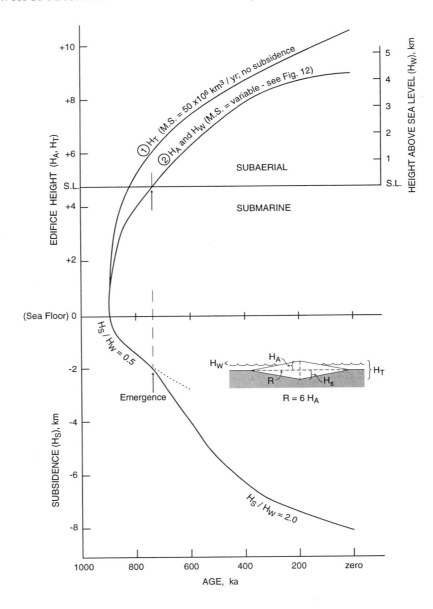

Fig. 13. Model growth rates for the summit of a geometrically simple conical volcano (see inset diagram): (1) Simple growth at constant magma-supply rate (M.S. = 50x10^6 m³/yr), without accompanying subsidence. (2) Growth of an edifice more like Mauna Loa: (a) volume of subsidence that nearly equals the volume above the sea floor; (b) magma-supply rates that vary as estimated from Figure 12, initially increasing rapidly to a maximum of about 130x10^6 m³/yr, then declining to present-day rate of 28x10^6 m³/yr; and (c) ratio of subsidence to growth above sea floor that increases when the edifice grows above sea level. Growth heights plotted (and present-day Mauna Loa values) are: *total height* of the present edifice (H_T, 17 km), *height above the sea floor* on which the volcano grew (H_A, 9 km), *subsidence component* of total height (H_S, 8 km), and *height above water* (H_W, 4.2 km).

elliptical Loihi (R=10-15 km, H_A = 2.5 km). Calculated model volumes are low because both upward and downward facets of the actual Mauna Loa shield are convex outward, but this limitation is unimportant for inferring the growth history of the summit region. Also, the model illustrates growth of the volcano only qualitatively because, rather than starting volcanic growth on flat sea floor, Mauna Loa probably began as a submarine vent on the lower south flank of Hawaii Island near the junction between already-large subaerial shields of Mauna Kea and Hualalai, much like the subsequent growth of Kilauea and Loihi from sites on the south flank of Mauna Loa.

Two time-height curves constructed from the model illustrate general aspects of shield-stage growth: (1) simple growth at a constant magma-supply rate, and (2) growth of an edifice more like Mauna Loa: the volume of subsidence nearly equals the volume above the sea floor, magma-supply rates vary as estimated in Figure 12, and the ratio of subsidence to growth above sea floor increases as the edifice grows above sea level. Curve (1) shows that as the shield grows, the rate of increase in height diminishes, based solely on the simple cone geometry. With the modest complexities added in curve (2), an initially rapid increase in magma supply tends to counteract the declining rate of increase in height, generating a flattened growth curve; the later decline in magma supply causes the growth curve to flatten rapidly. The rate of subsidence starts slowly, because the strength of the crust is significant when the volcano is small; then increases rapidly (along with cone height) because the base of early shield is small. As the basal area of the shield increases and rate of upward growth declines, so does the rate of subsidence. Subsidence is assumed to be minor until the edifice reached an initial height of 1 km above the sea floor, then is inferred to be geologically rapid (mostly within 10-20 ka of loading, based on analog with glacial rebound). The relation between net growth in edifice height (H_A) versus edifice subsidence (H_S) must change further as the volcano becomes subaerial, because of the greater loading by rock above sea level. The submarine volcano is modeled to grow upward relative to subsidence at a ratio of 2:1; in contrast, the subaerial upper part of the edifice is modeled to grow at the reduced ratio of 2:3. In addition to the modeled parameters, such a change in upward growth rate above sea level would be in accord with the increased tendency of the volcanic edifice to spread laterally as it becomes larger [*Fiske and Jackson*, 1972; *Borgia and Treves*, 1992].

The resulting model values are only approximations, and alternative scenarios could be developed. The internal geometry of Mauna Loa is far more complex than the model (e.g., onlap and interfingering with Mauna Kea and Hualalai, elliptical geometry imposed by rift zones, and variable angles of submarine and subaerial slopes), and the model is dependent on the uncertain values for changing eruption rates and total volume of the volcano (Figure 12). Because the subsidence model is based solely on the eruptive volume of Mauna Loa, subsidence rates are probably underestimated: subsidence of Mauna Loa would have been augmented early in its evolution by vigorous continuing growth of adjacent Hualalai and Mauna Kea, while subsidence of Mauna Loa in more recent time should be augmented by rapid growth of Kilauea. More broadly, long-term subsidence rates for Hawaii Island (consistently 2.3-2.6 mm/yr), as determined from dated submerged coral reefs [*Moore et Moore et al.*, 1990, 1994; *Ludwig et al.*, 1991], likely record integrated effects of the growth of multiple adjacent volcanoes, while subsidence related to a single isolated volcano should vary widely during growth of its edifice.

A further complication is the increasing concentration of Mauna Loa lavas on upper parts of the edifice, since at least 100 ka (Figure 2). The effect of such reduced areal extent of lava accumulation is to increase loading on upper proximal slopes relative to lower (mainly submarine) slopes, and to further deviate from the simple conical model geometry. The model predicts continued slow increase in the summit height of Mauna Loa, as long as lava erupts there, even if net subsidence dominates on lower slopes, as documented in this paper for many coastal regions. The current vertical velocity of Mauna Loa's upper slopes, which may be determinable in the near future by GPS methods, should record interplay between volcano growth, by lava accumulation and intrusive inflation, versus isostatic subsidence associated with both Mauna Loa and adjacent volcanoes.

Because Mauna Loa likely began to grow on the flank of Mauna Kea or Hualalai, rather than directly on oceanic crust, the model overestimates the time for the edifice to reach sea level. The modeled time of emergence (740 ka, 160 k.y. after inception) is much longer than that predicted for Loihi, which is estimated to have grown to its present size (within 950 m of sea level) within 8-12 k.y. after inception, and would reach the surface in about an additional 10-30 k.y. This discrepancy results largely from the modeled inception of volcano growth on the oceanic crust at a water depth of about 5 km, in contrast to the actual location of Loihi on the flank of Mauna Loa at shallower ocean depths, and also reflects the relatively gentle slope on the modeled cone (9.5°) in comparison to the steep submarine slopes of Loihi (as much as 25° over several kilometers vertically). Thus, the early upward growth of Mauna Loa may be better approximated by a curve similar to (1) in Figure 13, with an estimated elapsed time to reach sea level of a few tens of thousands of years.

Analogies with other volcanoes also provide some insight into the potential future alkalic stage of Mauna Loa. The transition from a dominant isotopically undegassed deep magma source to increasing proportions of degassed magma (asthenospheric source?), as tracked by helium-isotope ratios, suggests by analogy with Mauna Kea and Haleakala that alkalic magmatism should be expected at Mauna Loa in the not-too-distant geologic future (Figure 12). The decline in magma-supply rate at Mauna Loa prior to this predicted transition, as documented in this paper, is closely comparable to the decreased rates of lava accumulation for late stages in shield building at Mauna Kea, as interpreted from initial results of the Hilo scientific drilling project [*DePaolo and Stolper*, 1994]. Comparisons of lava accumulation rates in relation to compositional evolution at several such volcanoes promises to provide much new information about overall growth of Hawaiian volcanoes and magma production in their plume source.

Volume comparisons with Mauna Kea suggest that future alkalic volcanism will probably occur over a several hundred thousand year period at Mauna Loa, but will generate only a few thousand cubic kilometers of additional magma. At Mauna Kea, the transition to an alkalic stage was marked by compositional transitions in late-erupted tholeiitic basalts as well: submarine lavas dredged from low along Mauna Kea's east rift zone are typical tholeiites that are difficult to distinguish in elemental ratios from Mauna Loa or Hualalai tholeiites (although isotopically distinct), while late Mauna Kea tholeiites sampled at the surface and in the Hawaii drill hole are transitional toward alkalic compositions [*Frey et al.*, 1990; *Moore and Clague*, 1992; *Yang et al.*, 1994a, 1994b]. The spatial and probable temporal variations among Mauna Kea tholeiites suggest progressive restriction of magmatism to more proximal sites and abandonment of lower reaches of its east rift zone, analogous to that documented for Mauna Loa's southwest rift. Accordingly, future eruptions of alkalic lava on

Mauna Loa probably will take place high on the volcanic edifice, perhaps permitting the summit height of Mauna Loa to continue to increase slowly, even as its above-water surface area decreases due to coastal subsidence. As the island continues to be loaded by further growth of Kilauea and Loihi, eventually growth rates even for the summit of Mauna Loa will fail to keep pace with subsidence.

Acknowledgments. Thought-provoking discussions over many years with Jack Lockwood, Jim Moore, Mike Garcia, Mark Kurz, and Mike Rhodes have contributed much to my evolving ideas concerning the volcanic and petrologic evolution of Mauna Loa, but none of these associates should be held responsible for the ideas and interpretations expressed here. For the present study, Brent Dalrymple attempted valiantly to obtain meaningful K-Ar ages from recalcitrant Mauna Loa tholeiites; Mike Rhodes provided XRF analyses to aid in sample selection for age-determination efforts, and Mike Garcia shared microprobe determinations for glassy pillow lavas that we had collected together. Jim Moore has been my long-term sounding board and critic for interpretation of shore-line and marine volcanic processes, and Jack Lockwood and associates have generated much of the primary data on map distribution and ages of subaerial Mauna Loa lava flows. Thanks also to Dave Clague, Dick Fiske, Mike Garcia, Jack Lockwood, Jim Moore, Mike Rhodes, and Don Swanson for thoughtful comments on earlier versions of this paper.

REFERENCES

Apple, R.A., and G.A. Macdonald, The rise of sea level in contemporary time at Honaunau, Kona, Hawaii, *Pacific Sci.*, *20*, 125-136, 1966.

Bargar, K.E., and E.D. Jackson, Calculated volumes of individual shield volcanoes along the Hawaiian-Emperor chain, *U.S. Geol. Surv. J. Res.*, *2*, 545-550, 1974.

Beeson, M.H., D.A. Clague, and J.P. Lockwood, Origin of clastic deposits in the Hilo drill hole, Hawaii [abs.], *EOS*, *75*, 712, 1994.

Borgia, A. and B. Treves, Volcanic plates override the ocean crust: Structure and dynamics of Hawaiian volcanoes, *Geol. Soc. Spec. Publ. London*, *60*, 277-299, 1992.

Buchanan-Banks, J.M., Geologic map of the Hilo 7 1/2' quadrangle, Hawaii, *U.S. Geol. Surv. Map I-2274*, 1993.

Chadwick, W. W. Jr., J.R. Smith Jr., J.G. Moore, D.A. Clague, M.O. Garcia, and C.G. Fox, Bathymetry of south flank of Kilauea volcano, Hawaii, *U.S. Geol. Survey Misc. Field Studies Map MF-2231*, 1993.

Chadwick, W. W. Jr., J.G. Moore, M.O. Garcia, and C.G. Fox, Bathymetry of southern Mauna Loa, Hawaii, *U.S. Geol. Survey Misc. Field Studies Map MF-2233*, 1994a.

Chadwick, W. W. Jr., J.G. Moore, and C.G. Fox, Bathymetry of the southwest flank of Mauna Loa volcano, Hawaii, *U.S. Geol. Survey Misc. Field Studies Map MF-2255*, 1994b.

Chadwick, W. W. Jr., J.G. Moore, and C.G. Fox, Bathymetry of the west-central slope of the Island of Hawaii, *U.S. Geol. Survey Misc. Field Studies Map MF-2269*, 1994c.

Clague, D.A., Hawaiian xenolith populations, magma supply rates, and development of magma chambers, *Bull. Volc., 49,* 577-587, 1987.

Clague, D.A., and G.B. Dalrymple, The Hawaiian-Emperor volcanic chain: Part I. Geologic evolution, *U.S. Geol. Surv. Prof. Pap. 1350,* 5-54, 1987..

Clague, D.A., and R.P. Denlinger, The role of cumulus dunite in destabilizing the flanks of Hawaiian volcanoes, *Bull. Volc., 56,* 425-434, 1994.

Clague, D.A., J.G. Moore, J.E. Dixon, and W.B. Friesen, Petrology of submarine lavas from Kilauea's Puna Ridge, Hawaii, *J. Petrol.,* in press.

Dalrymple, G.B., and J.G. Moore, Argon-40: Excess in submarine pillow basalts from Kilauea Volcano, Hawaii, *Science, 161*, 1132-1135, 1968.

Decker, R.W., R.Y. Koyanagi, J.J. Dvorak, J.P. Lockwood, A.T. Okamura, K.M. Yamashita, and W.R. Tanigawa, Seismicity and surface deformation of Mauna Loa volcano, Hawaii, *EOS, 64,* 545-547, 1983.

Delaney , P.T., A. Miklius, T. Arnadottir, A.T. Okamura, and M.K. Sako, Motions of Kilauea volcano during sustained eruption from the Pu'u O'o and Kupai'anaha vents, 1983-1991, *J. Geophys. Res., 98,* 17,801-17,820, 1993

Denlinger R.P., and P. Okubo, A huge landslide structure on Kilauea Volcano, Hawaii, *J. Geophys. Res.,* in press.

DePaolo, D.J., E.M. Stolper, and D.M. Thomas, Core logs, Hawaii Scientific Drilling Project, 79 pp & unpaged, 1994.

DePaolo, D.J., and E.M. Stolper, Accumulation rate of Mauna Kea lavas at Hilo: Implications for volcano construction times, plume structure, and plate velocity [abs.], *EOS, 75,* 708, 1994.

Detrich, R.S., and S.T. Crough, Island subsidence, hot spots, and lithospheric thinning, *J. Geophys. Res., 83,* 1236-1244, 1978.

Dvorak, J.J., and D. Dzurisin, Variations in magma-supply rate at Kilauea Volcano, Hawaii, *J. Geophys. Res., 98,* 22,255-22,268,1993.

Dvorak, J.J., A.T. Okamura, and A. Miklius, Mauna Loa geodesy: Variable filling of the summit reservoir and spreading of the rift zones, *EOS, 74,* 641, 1993.

Dzurisin, D., R.Y. Koyanagi, and T.T. English, Magma supply and storage at Kilauea volcano, Hawaii, 1956-1983, *J. Volc. Geotherm. Res., 21,* 177-206, 1984.

Eaton, J.P., Crustal structure and volcanism in Hawaii, *Am Geophys. Union Mon. 6*, 13-29, 1962.

Finch, R.H., and G.A. Macdonald, 1953, Hawaiian volcanoes during 1950, *U.S. Geol. Surv. Bull. 996-B*, 89 pp, 1953.

Fiske, R.S., and E.D. Jackson, Orientation and growth of Hawaiian volcanic rifts - The effects of regional structure and gravitational stresses, *Prof. R Soc. London, A 329,* 299 - 326, 1972.

Fiske, R.S., and D.A. Swanson, One-rift, two-rift paradox at Kilauea Volcano, Hawaii, *EOS, 73,* 506, 1992.

Flannigan, V.J., and C.L. Long, Aeromagnetic and near-surface electrical expression of the Kilauea and Mauna Loa rift systems, *U.S. Geol. Surv. Prof. Pap 1350,* 935-946, 1987.

Flint, R.F., Glacial and Quaternary geology, *John Wiley and Sons,* 892 p., 1971.

Fornari, D.J., D.W. Peterson, J.P. Lockwood, A. Malahoff, and B. Heezen, Submarine extension of the southwest rift zone of Mauna Loa volcano, Hawaii: Visual observations from the U.S. Navy deep submersible vehicle DSV *Sea Cliff, Geol. Soc. Am. Bull., 90,* 435-443, 1979.

Fornari, D.J., J.P. Lockwood, P.W. Lipman, M. Rawson, A. Malahoff, Submarine volcanic features west of Kealakekua Bay, Hawaii, *J. Volc. Geotherm. Res., 7,* 323-337, 1980.

Frey, F.A., W.S. Wise, M.O. Garcia, A. Kennedy, P. Gurriet, and F. Albarede, The evolution of Mauna Kea volcano, Hawaii,: Petrogenesis of tholeiitic and alkalic basalts, *J. Geophys. Res., 96,* 14,347-14,375, 1990.

Garcia, M.O., D.W. Muenow, K.E. Aggray, and J.R. O'Neil, Major element, volatile, and stable isotope geochemistry of Hawaiian submarine tholeiitic glasses, *J. Geophys. R, 94,* 10,525-10,538, 1989.

Garcia, M.O, J.M. Rhodes, P. Lipman, and M. Kurz, Submarine Mauna Loa: Opportunities for investigating the volcano's petrologic history and structure [abs.], *EOS, 74,* 628-629, 1993.

Garcia, M.O., T.P. Hulsebosch, and J.M. Rhodes, Glass and mineral chemistry of olivine-rich submarine basalts, southwest rift zone, Mauna Loa volcano: Implications for magmatic processes, *Am. Geophys. U. Mon.,* this volume.

Heliker, C.C., T.N. Mattox, M.T. Mangan, and J. Kauahikaua, Kilauea volcano update: Eleven-year-long eruption continues [abs.], *EOS, 74,* 644, 1993.

Hill, D.P., Crustal structure of the Island of Hawaii from seismic-refraction measurements, *Bull. Seis. Soc. Am., 59,* 101-130, 1969.

Hill, D.P., and J.J. Zucca, Geophysical constraints on the structure of Kilauea and Mauna Loa volcanoes, and some implications for seismomagmatic processes, *U.S. Geol. Surv. Prof. Pap. 1350,* 903-917, 1987.

Holcomb, R.T., Eruptive history and long-term behavior of Kilauea volcano, *U.S. Geol. Surv. Prof. Pap. 1350,* 261-350, 1987.

Holcomb, R.T., J.G. Moore, P.W. Lipman, and R.H. Belderson,

Voluminous submarine lava flows from Hawaiian volcanoes, *Geology, 16,* 400-404, 1988.

Hon, K., J. Kauahikaua, R. Denlinger, and K. MacKay, Emplacement and inflation of pahoehoe sheet flows: Observations and measurements of active flows on Kilauea volcano, Hawaii, *Geol. Soc. Am. Bull., 106*, 351-370, 1994.

Jackson, E.D., The character of the lower crust and upper mantle beneath the Hawaiian Islands, *23rd Internat. Geol. Congress (Prague), 1*, 135-140, 1968.

Klein, F.W., R.Y. Koyanagi, J.S. Nakata, and W.R. Tanigawa, The seismicity of Kilauea's magma system, *U.S. Geol. Surv. Prof. Pap. 1350*, 1019-1185, 1987.

Klein, F.W., and R.Y. Koyanagi, The seismicity and tectonics of Hawaii, *in* The Geology of North America, Vol., N, The Eastern Pacific Ocean and Hawaii, *Geol. Soc. Am.,* 238-252, 1989.

Kurz, M.D., M.O. Garcia, F.A. Frey, and P.A. O'Brien, Temporal helium isotopic variations within Hawaiian volcanoes: basalts from Mauna Loa and Haleakala, *Geochim. Cosmochim. Acta, 51,* 2905-2914, 1987.

Kurz, M.D., and D.P. Kammer, Isotopic evolution of Mauna Loa volcano, Hawaii, *Earth and Plan. Sci. Lett., 103,* 257-269, 1990.

Kurz, M.D., T.C. Kenna, D.P. Kammer, J.M. Rhodes, and M.O., Garcia, Isotopic evolution of Mauna Loa volcano: A view from the submarine southwest rift, *Am. Geophys. U. Mon.,* this volume

Kurz, M.D., J.K. Lassiter, B.M. Kennedy, D.J. DePaolo, J.M. Rhodes, and F.A. Frey, Helium isotopic evolution of Mauna Kea volcano: first results from the 1 km drill core [abs.], *EOS, 75,* 711, 1994.

Lipman, P.W., Rates of volcanic activity along the southwest rift zone of Mauna Loa, Hawaii, *Bull. Volc, 43,* 703-725, 1980a.

Lipman, P.W., The southwest rift zone of Mauna Loa-- Implications for structural evolution of Hawaiian volcanoes, *Am. J. Sci, 280-A,* 752-776, 1980b.

Lipman, P.W., N.G. Banks, and J.M. Rhodes, Gas-release induced crystallization of 1984 Mauna Loa magma, Hawaii, and effects on lava rheology, *Nature, 317,* 604-607, 1985.

Lipman, P.W., D.A. Clague, J.G. Moore, and R.T. Holcomb, South Arch volcanic field - Newly identified young lava flows on the sea floor south of the Hawaiian Ridge, *Geology, 17,* 611-614, 1989.

Lipman, P.W., J.P. Lockwood, R.T. Okamura, D.A. Swanson, and K.M. Yamashita, Ground deformation associated with the 1975 magnitude-7.2 earthquake and resulting changes in activity of Kilauea volcano, Hawaii, *U.S. Geol. Surv. Prof. Pap. 1276*, 45 pp, 1985.

Lipman, P.W., and J.G. Moore, Mauna Loa lava-accumulation rates at the Hilo drill site: Formation of lava deltas during a period of declining overall volcanic growth, *J. Geophys. Res.,* in press.

Lipman, P.W., W.R. Normark, J.G. Moore, J.B. Wilson, and C.E. Gutmacher, The giant submarine Alika debris slide, Mauna Loa, Hawaii, *J. Geophys.. Res., 93,* 4279-4299, 1988.

Lipman, P.W., J.M. Rhodes, and G.B. Dalrymple, The Ninole Basalt - Implications for the structural evolution of Mauna Loa volcano, Hawaii, *Bull. Volc., 53,* 1-19, 1990.

Lipman, P.W., and A. Swenson, Generalized geologic map of the southwest rift zone of Mauna Loa Volcano, Hawaii, *U.S. Geol. Surv. Misc. Invest. Map I-1323*, 1984.

Lockwood, J.P., Mauna Loa eruptive history -The radiocarbon record, *Am. Geophys. U. Mon.*, this volume.

Lockwood, J.P, and P.W. Lipman, Holocene eruptive history of Mauna Loa Volcano, *U.S. Geol. Surv. Prof. Pap. 1350*, 509-536, 1987.

Lockwood, J.P., J.J. Dvorak, T.T. English, R.Y. Koyanagi, A.T. Okamura, M.L. Summer, and W.R. Tanigawa, Mauna Loa 1974-1984: A decade of intrusive and extrusive activity, *U.S. Geol. Surv. Prof. Pap. 1350*, 537-570, 1987.

Lockwood, J.P., P.W. Lipman, L.D. Petersen, and F.R. Warshauer, Generalized ages of surface lava flows of Mauna Loa volcano, Hawaii, *U.S. Geol. Surv. Misc. Invest. Map I- 1908*, 1988.

Ludwig, K.R., B.J. Szabo, J.G. Moore, and K.R. Simmons, Crustal subsidence rate off Hawaii determined from $^{234}U/^{238}U$ ages of drowned coral reefs, *Geology, 19,* 171-174, 1991.

Macdonald, G.A., Hawaiian calderas, *Pac. Sci., 19,* 320-333, 1965.

McDougall, I., Age of shield-building volcanism of Kauai and linear migration of volcanism in the Hawaiian Island chain, *Earth and Plan. Sci. Let., 46,* 31-42, 1979.

McDougall, I., and D.A. Swanson, Potassium-argon ages of lavas from the Hawi and Polulu volcanic series, Kohala Volcano, Hawaii, *Geol. Soc. Am. Bull., 83,* 3731-3738, 1972.

McNutt, M.K., Lithospheric flexure and thermal anomalies, *J. Geophys. Res., 89,* 11,180-11,194, 1984.

Moore, J.G., Relationship between subsidence and volcanic load, Hawaii, *Bull. Volc., 34,* 562-576, 1970.

Moore, J.G., Subsidence of the Hawaiian Ridge, *U.S. Geol. Surv. Prof. Pap. 1350*, 85-100, 1987.

Moore, J.G., W.B. Bryan, M.H. Beeson, and W.R. Normark, Giant blocks in the South Kona landslide, Hawaii, *Geology, 23,* 125-128, 1995.

Moore, J.G., and D.A. Clague, Coastal lava flows from Mauna Loa and Hualalai volcanoes, Kona, Hawaii, *Bull. Volc., 49,* 752-764, 1987.

Moore, J.G., and D.A. Clague, Volcano growth and evolution of the island of Hawaii, *Geol. Soc. Am. Bull., 104,* 1471-1484, 1992.

Moore, J.G., D.A. Clague, R.T. Holcomb, P.W. Lipman, W.R. Normark, and M.E. Torresan, Prodigious submarine landslides on the Hawaiian Ridge, *J. Geophys. Res., 94,* 17,465-17,484, 1989.

Moore, J.G., D.J. Fornari, and D.A. Clague, Basalts from the 1877 submarine eruption of Mauna Loa, Hawaii: new data on the variation of palagonitization rate with temperature, *U.S. Geol. Survey Bull. 1663,* 11 pp., 1985.

Moore, J.G., B.L. Ingram, and K.R. Ludwig, Coral ages and island subsidence, Hilo Drill Hole [abs.], *EOS, 75,* 711, 1994

Moore, J.G., and R.K. Mark, Slopes of the Hawaiian Ridge, *U.S. Geol. Surv. Prof. Pap. 1350,* 101-107, 1987.

Moore, J.G., W.R. Normark, and B.J. Szabo, Reef growth and volcanism on the submarine southwest rift zone of Mauna Loa, Hawaii, *Bull. Volc., 52,* 375-380, 1990.

Normark, W.R., P.W. Lipman, J.P. Lockwood, and J.G. Moore, Bathymetric and geologic maps of Kealakekua Bay, Hawaii, *U.S. Geol. Survey Misc. Field Invest. Map MF-986,* 1978

Normark, W.R., P.W. Lipman, and J.G. Moore, Regional slump structure on the west flank of Mauna Loa volcano, Hawaii [abs.], *Hawaii Symp. on Intraplate Volcanism and Submarine Volcanism, Abstract Vol.* 72, 1979.

Normark, W.R., J.G. Moore, and M.E. Torresan, Giant volcano-related landslides and the development of the Hawaiian Islands, *U.S. Geol. Surv. Bull. 2002,* 184-196, 1993.

Peterson, D.W., and R.B. Moore, Geologic history and evolution of geologic concepts, Island of Hawaii, *U.S. Geol. Surv. Prof. Pap. 1350,* 149-189, 1987.

Rhodes, J.M., Homogeneity of lava flows; Chemical data for historic Mauna Loa eruptions, *J. Geophys. Res., 93,* 4453-4466, 1983.

Rhodes, J.M., The 1852 and 1868 picrite eruptions: Clues to parental magma compositions and the magmatic plumbing system, *Am. Geophys. U, Mon.,* this volume.

Rhodes, J.M., How Mauna Loa works: A geochemical perspective [abs.], *Hawaii symposium on How Volcanoes Work, Abs. Vol.,* 208, 1987.

Rhodes, J.M., and S.R. Hart, Episodic trace element and isotopic variations in historic Mauna Loa lavas: Implications for magma and plume dynamics, *Am. Geophys. U. Mon.,* this volume.

Rhodes, J.M., and J. Sweeney, Geochemical stratigraphy of the Hawaii scientific drill hole [abs.], *EOS, 75,* 708, 1994.

Rubin, A.M, and D.D. Pollard, Origins of blade-like dikes in volcanic rift zones, *U.S. Geol. Surv. Prof. Pap. 1350,* 1449-1470, 1987.

Ryan, M.P., Elasticity and contractancy of Hawaiian olivine tholeiite and its role in the stability and structural evolution of subcaldera magma reservoirs and rift systems, *U.S. Geol. Surv. Prof. Pap. 1350,* 1395-1448, 1987

Ryan, M.P, The mechanics and three-dimensional internal structure of active magmatic systems: Kilauea volcano, Hawaii, *J. Geophys. Res., 93,* 4213-4248, 1988.

Shaw, H.R., Mantle convection and volcanic periodicity in the Pacific: evidence from Hawaii, *Geol. Soc. Am. Bull., 84,* 1505-1526, 1973.

Stearns, H.T., and G.A. Macdonald, Geology and ground water resources of the Island of Hawaii: *Hawaii Div. Hydrol. Bull. 9,* 363 pp., 1946.

Stearns, H.T., and W.O. Clark, Geology and water resources of the Kau District, Hawaii: *U.S. Geol. Surv. Water-Supply Pap. 616,* 194 pp., 1930.

Stolper, E.M., D.M. Thomas, and D.J. DePaolo, Hawaii Scientific Drilling Project: background and overview of initial results [abs.], *EOS, 75,* 707, 1994.

Swanson, D.A., Magma supply rate at Kilauea volcano, 1952-1971, *Science, 175,* 169-170, 1972.

Swanson, D.A., W.A. Duffield, and R.S. Fiske, Displacement of the south flank of Kilauea volcano: The result of forceful intrusion of magma into the rift zones, *U.S. Geol. Surv. Prof. Pap 963,* 30 pp., 1976.

Szabo, B.J., and J.G. Moore, Age of -360 m reef terrace, Hawaii, and the rate of late Pleistocene subsidence of the island, *Geology, 14,* 967-968, 1986.

Ten Brink, U.S., and A.B. Watts, Seismic stratigraphy of the flexural moat flanking the Hawaiian Islands, *Nature, 317,* 421-424, 1985.

Thurber, C.H., and A.E. Gripp, Flexure and seismicity beneath the south flank of Kilauea volcano and tectonic implications, *J. Geophys. Res.,* 93, 4271-4278, 1988.

Tilling R.I., J.M. Rhodes, J.W. Sparks, J.P. Lockwood, and P.W. Lipman, Disruption of the Mauna Loa magma system by the 1868 Hawaiian earthquake: geochemical evidence, *Science,* 235, 196-199, 1987.

Tilling, R.I., and J.J. Dvorak, Anatomy of a basaltic volcano, *Nature, 363,* 125-133, 1993

Trusdell, F.A., Lava flow hazards and risk assessment on Mauna Loa volcano, Hawaii, *Am. Geophys. U.,* this volume.

Walker, G.P.L., The dike complex of Koolau Volcano, Oahu: Internal structure of a Hawaiian rift zone, *U.S. Geol. Surv. Prof. Pap. 1350,* 961-993, 1987.

Walker, G.P.L, Three Hawaiian calderas: an origin through loading by shallow intrusions, *J. Geophys. Res., 93,* 773-784, 1988.

Walker, G.P.L., Geology and volcanology of the Hawaiian Islands, *Pacific Sci., 44,* 315-347, 1990.

Watson, S., Rare earth element inversions and percolation models for Hawaii, *J. Petrology, 34,* 763-783, 1993.

Wise, W.S., A volume-time framework for the evolution of Mauna Kea volcano, Hawaii [abs.]: *EOS, 63,* 1137, 1982.

Wolfe, E.W., M.O Garcia, D.B. Jackson, R.Y. Koyanagi, C.A.

Neal, and A.T. Okamura, The Puu Oo eruption of Kilauea volcano, episodes 1-20, January 3, 1983, to June 8, 1984, *U.S. Geol. Surv. Prof. Pap. 1350*, 471-508, 1987.

Wright, T.L., Chemistry of Kilauea and Mauna Loa in space and time, *U.S. Geol. Surv. Prof. Pap. 735*, 40pp, 1971.

Wright, T.L., and R.S. Fiske, Magma tectonics and the recent spreading history of Kilauea volcano, Hawaii, *Geol. Soc. Am. Bull.,* in press.

Yang, H-J., F.A. Frey, M.O. Garcia, and D.A. Clague, Submarine lavas from Mauna Kea volcano, Hawaii: Implications for Hawaiian shield stage processes, *J. Geophys. Res., 99,* 15,577-15,594, 1994a.

Yang, H-J., F.A. Frey, J.M. Rhodes, and M.O. Garcia, The temporal evolution of Mauna Kea volcano: Geochemical comparisons of basalts from the Hawaii scientific drilling project with dredged submarine lavas and subaerial surface lavas [abs.], *EOS, 75,* 708, 1994b.

Zucca, J.J., D.P. Hill, and R.L. Kovach, Crustal structure of Mauna Loa volcano, Hawaii, from seismic refraction and gravity data, *Bull. Seis. Soc. Am., 72,* 1535-1550, 1982.

Zucca, J.J., and D.P. Hill, Crustal structure of the southeast flank of Kilauea volcano, Hawaii, from seismic fraction measurements, *Seis. Soc. Am. Bull., 70,* 1149-1159, 1980.

Peter W. Lipman, U.S. Geological Survey, 345 Middlefield Road, MS 910, Menlo Park, CA 94025

Mauna Loa Eruptive History - The Preliminary Radiocarbon Record

John P. Lockwood

U.S. Geological Survey, Hawaiian Volcano Observatory, Hawaii

Radiocarbon dating of charcoal from beneath lava flows of Mauna Loa has provided the most detailed prehistoric eruptive chronology of any volcano on Earth. Three hundred and fifty-five ^{14}C dates have been reviewed, stratigraphically contradictory dates have been rejected, and multiple dates on single flows averaged to give "reliable" ages on 170 separate lava flows (about 35% of the total number of prehistoric Mauna Loa flows mapped to date). The distribution of these ages has revealed fundamental variations in the time and place of Mauna Loa eruptive activity, particularly for Holocene time. As lava flow activity from Mauna Loa's summit waxes, activity on the rift zones wanes. A cyclic model is proposed which involves a period of concentrated summit shield-building activity associated with long-lived lava lakes and frequent overflows of pahoehoe lavas on the north and southeast flanks. At this time, compressive stresses across Mauna Loa's rift zones are relatively high, inhibiting eruptions in these areas. This period is then followed by a relaxation of stresses across Mauna Loa's rift zones and a long period of frequent rift zone eruptions as magma migrates downrift. This change of eruptive style is marked by summit caldera collapse (possibly associated with massive eruptions of picritic lavas low on the rift zones). Concurrent with this increased rift zone activity, the summit caldera is gradually filled by repeated summit eruptions, stress across the rift zones increases, magma rises more easily to the summit, rift activity wanes, and the cycle repeats itself. Two such cycles are suggested within the late Holocene, each lasting 1,500-2,000 years. Earlier evidence for such cycles is obscure. Mauna Loa appears to have been quiescent between 6-7 ka, for unknown reasons. A period of increased eruptive activity marked the period 8-11 ka, coincident with the Pleistocene-Holocene boundary. Other volcanoes on the Island of Hawaii for which (limited) radiocarbon dating are available show no evidence of similar cyclicity or repose. Mauna Loa may be presently nearing the end of a thousand-year-long period of increased rift zone activity, and sustained summit eruptions may characterize the volcano's most typical behavior in the millennium to come. Such a shift could eventually alter the nature of volcanic risk for future populations on Hawaii.

1. INTRODUCTION

Mauna Loa, the largest volcano on Earth, has grown rapidly over its short 0.6-1.0 m.y. history [*Moore and Clague*, 1992; *Lipman*, this volume] and later eruptive products have buried all traces of the volcano's earlier eruptions. In contrast to the deeply buried lavas of Mauna Loa's youth, Holocene lavas are well preserved, as about 98% of Mauna Loa's subaerial surface is covered by lava flows and ash deposits younger than 10 ka [The abbreviation "ka" is used throughout this paper to describe time in thousands of "radiocarbon years" (before the

Mauna Loa Revealed: Structure,
Composition, History, and Hazards
Geophysical Monograph 92

calendar year datum of A.D. 1950)]. Dating of these surface lavas is relatively easy, as several factors have combined to make Mauna Loa an ideal volcano for the formation and preservation of organic material suitable for radiocarbon dating: (1) plants grow quickly in the tropical climates of Mauna Loa's lower slopes, flows vegetate within a few decades, and provide material for carbonization by subsequent flows. (2) Mauna Loa tholeiites are highly fluid, and plants are enveloped quickly by advancing flows, conditions which are conducive to charcoal preservation. (3) Due to the large size of Mauna Loa (>5,000 km^2) relative to individual lava flows (typically 10-50 km^2 area), most flows reaching lower areas of the volcano will be in lateral contact with older, forested substrates where charcoal preservation is likely.

In order to systematically recover datable material from beneath lava flows on the Island of Hawaii, [*Lockwood and*

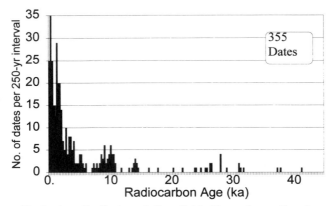

Fig 1. Age distribution of all available Mauna Loa radiocarbon dates, from "modern" to 49 ka, plotted as the number of dates per 250-year interval. These 355 dates include a large number of geologically unreasonable dates, especially younger values, which result from post-eruptive wildfire charcoal contamination.

Lipman, 1980] investigated the formation and preservation of charcoal beneath modern flows. The resulting techniques for charcoal recovery have proven to be extremely useful for deciphering Mauna Loa's Holocene eruptive history. This paper is a report on the results of the Mauna Loa radiocarbon dating efforts and on the patterns of varying eruptive activity now emerging from the large set of chronologic data.

2. THE MAUNA LOA RADIOCARBON RECORD

355 radiocarbon dates have now been obtained on organic material collected beneath (or, rarely, within) Mauna Loa lavas (Fig. 1). Of these 355 dates, 273 were considered "reasonable" in terms of stratigraphic position or degree of weathering; 82 were rejected. Most of the rejected ages were "too young" (Fig. 2) and resulted from sample contamination by later wildfire charcoal, some were obtained from dating sooty soils and were "too old," owing to relict soil carbon, and a few were simply "anomalous," although obtained from apparently ideal charcoal. Of the 273 "reasonable" dates (Fig. 3), 102 were multiple dates on single flows. In these cases, weighted averages of all dates for each flow were calculated according to reported lab precisions, using the statistical formulas provided by *Bevington* [1969, p. 70-71]. These formulas are shown in Appendix "A" with examples, as there appears to be a lack of uniformity on how "average" ages of multiply-dated flows should be calculated. The radiocarbon ages of 170 individual Mauna Loa flows were thus established (Fig. 4; Table 1), through "reasonable" single dates (110 flows) or averaged multiple dates (60 flows). The distribution of these lava flow ages in time and space form the basis of this paper.

Ages (radiocarbon or historical) are now available on about 35% of the approximately 500 separate lava flows mapped so far on Mauna Loa. The distribution of these flow ages in time is providing an increasingly representative picture of Mauna Loa's later eruptive history. This picture is most comprehensive for the recent past, however, and becomes less clear with increasing age. As younger flows are better exposed than older ones, and since recovery of datable material from beneath them is easier (and more critical for hazards evaluations), more than 85% of flows younger than 1,500 years are now dated. In contrast, ages for fewer than 25% of surface flows believed to be more than 5,000 years old are presently available.

Charcoal has now been recovered and dated from all parts of Mauna Loa, from sea level to 2,750 m elevation (Fig. 5). Many of the dated flows have been mapped upslope from the vegetated areas where charcoal was collected; hence, the geographic distribution of dated flows is representative of

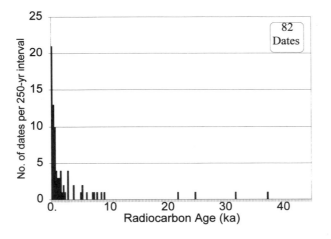

Fig. 2. Age distribution of 82 "unreasonable" dates, plotted as the number of dates per 250-year interval. These dates were rejected from the Mauna Loa data set because of discrepancies with stratigraphic position or weathering characteristics.

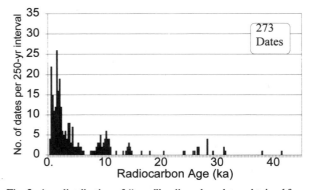

Fig. 3. Age distribution of "good" radiocarbon dates obtained from beneath Mauna Loa lava flows, plotted as the number of dates per 250 year interval. Of these 273 dates, 104 were multiple dates on individual flows.

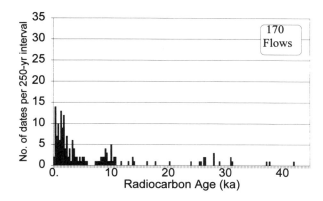

Fig. 4. Age-frequency distribution of all dated flows on Mauna Loa. Distribution of 170 flow ages, plotted as the number of flows per 250-year interval. Weighted averages of all multiple dates on individual flows were calculated according to laboratory precision to determine average ages for the flows plotted in this figure.

most of the volcano's subaerial surface. Mauna Loa has been divided into five subdivisions to evaluate geographic variations in eruptive activity for discussion in this paper (Fig. 6). Most Mauna Loa eruptions occur in the summit region (defined following *Lockwood and Lipman* [1987] as the area above 3,660 m (12,000') elevation). "*Summit lava flows*" are thus either restricted to the summit caldera (Moku`aweoweo) or to the northwest (MKN) or southeast (MKS) flanks. These flows are fed from vents which initially erupt within Moku`aweoweo but which commonly migrate beyond the caldera into the upper few kilometers of the rift

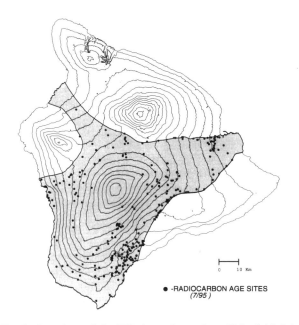

Fig. 5. Locations of the 273 charcoal samples which yielded the "good" dates discussed in this paper.

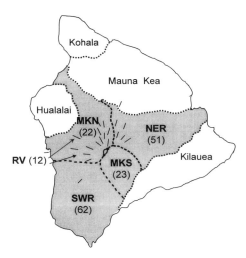

Fig. 6. Island of Hawaii, showing location of volcanoes (Mauna Loa shaded), structural subdivisions of Mauna Loa (boldface), and numbers of dated lava flows within each subdivision (parentheses) [NER = Northeast Rift Zone, SWR = Southwest Rift Zone, MKS = Moku`aweoweo overflows - southeast flank, MKN = Moku`aweoweo overflows - northwest flank, RV = Radial vent flows on north and northwest flanks (vents are shown schematically)].

zones. *Rift zone lava flows* (NER or SWR) are defined as those derived from vents below 3,660 m elevation, located along the well-defined, narrow northeast or southwest rift zones. Another class of lava flows, *radial vent flows* (RV), were erupted from the north through west flanks of Mauna Loa from vents oriented radially to the summit. These radial vents are mostly located in subdivision MKN where buttressing by older volcanoes results in the radial orientation of maximum principal stress axes and eruptive fissures.

3. VALIDITY OF FLOW AGE VARIATIONS

The early record of Mauna Loa activity is fragmentary and only 24 lava flows older than 12,000 years have been dated. The distribution of these older flows suggests a possible increase in flow activity between 25-32 ka, but the limited number of dated flows may reflect sampling bias rather than eruptive temporal variations. The remaining discussion will therefore be focused on flows younger than 12 ka (here loosely termed "Holocene"), as the abundant data in this age range are more statistically significant and the shorter time interval allows better plotting of temporal variations.

Radiocarbon dates are traditionally referenced to the year A.D. 1950 ("zero" age). In order to discuss the age distribution of all <u>dated</u> Mauna Loa lavas erupted prior to this zero age, historical lavas with ages prior to 1950 must also be included in the prehistoric data set for plotting purposes. In order to be comparable to the radiocarbon-dated flows, however, they must be flows which would be mapped and radiocarbon-dated as individual flows if no historical

TABLE 1. Mauna Loa radiocarbon-dated flows, showing age, precision, and
number of ^{14}C dates per flow.

AGE (Yrs. b.p.)	PRECISION		DATES (No. / flow)	SUBDIVISION	FLOWID[1] No	UNIT NAME
140	+/-	15	1	NER	497	MANUKA
200	+/-	100	1	NER	330	
265	+/-	46	2	NER	800	KEAMOKU-KKK
265	+/-	43	2	RV	0	KEAMUKU
278	+/-	46	3	NER	0	LZ FLOW
297	+/-	40	3	NER	801	MAUNAIU
300	+/-	50	1	NER	796	KUPANAHA
392	+/-	27	2	SWR	502	KIPAHOEHOE
410	+/-	80	1	SWR	510	
410	+/-	60	1	NER	0	
430	+/-	50	2	NER	838	KEAMOKU-KLW
437	+/-	46	2	NER	793	KULANI
468	+/-	36	3	NER	799	KEAMOKU-KLE
476	+/-	42	3	MKS	274	KE A POOMOKU
480	+/-	60	1	NER	794	ML BOYS' SCHOOL
490	+/-	60	1	SWR	495	
521	+/-	78	3	RV	412	PUU O UO
570	+/-	60	1	NER	791	KULALOA
580	+/-	80	1	NER	802	KEAMOKU-KKL
590	+/-	50	1	NER	797	KEAWEWAI CAMP
640	+/-	45	1	SWR	496	POHINA
640	+/-	80	1	RV	0	
750	+/-	60	1	SWR	627	KIPUKA KANOHINA
760	+/-	70	1	NER	833	KEAUHOU RANCH
780	+/-	70	1	NER	840	
799	+/-	34	4	SWR	547	KAWA BAY
860	+/-	20	1	SWR	626	KIPUKA NOA
880	+/-	60	1	NER	795	OLAA UKA
910	+/-	70	1	NER	837	PUU WAHI
940	+/-	100	1	RV	0	
942	+/-	59	2	SWR	551	HALEPOHAHA
980	+/-	80	1	SWR	549	
1000	+/-	150	1	RV	442	
1020	+/-	100	1	RV	0	
1111	+/-	51	2	MKN	55	PUUHONUA O HONAUNAU
1139	+/-	120	2	MKN	56	
1150	+/-	200	1	MKN	57	
1151	+/-	43	2	MKS	0	AINAPO TRAIL
1190	+/-	150	1	SWR	542	
1292	+/-	32	4	MKS	298	PAHUAMIMI
1333	+/-	34	5	NER	829	KUKUAU

TABLE 1. (Con't.)

AGE (Yrs. b.p.)	PRECISION		DATES (No. / flow)	SUBDIVISION	FLOWID[2] No	UNIT NAME
1382	+/-	35	4	MKS	288	KAALAALA GULCH
1438	+/-	37	2	MKS	297	KEAKAPULU FLAT
1456	+/-	58	2	RV	439	PUU KINIKINI
1470	+/-	33	2	MKS	344	
1470	+/-	50	1	NER	828	PANAEWA PICRITE
1480	+/-	60	1	RV	0	
1500	+/-	100	1	MKN	0	
1500	+/-	30	1	MKS	273	
1524	+/-	63	2	MKS	301	KAPAPALA DRIVEWAY
1534	+/-	71	3	NER	0	REDLEG TRAIL
1632	+/-	120	2	MKN	144	POOPAAELUA
1634	+/-	65	2	MKS	336	EAST WOOD VALLEY
1640	+/-	150	1	MKS	343	HALFWAY HOUSE
1720	+/-	65	1	RV	0	
1730	+/-	60	1	SWR	545	PUU O KEOKEO
1737	+/-	70	3	NER	864	UPPER WAIAKEA PC AA
1740	+/-	250	1	MKN	0	
1810	+/-	105	1	RV	0	PUU 6995
1819	+/-	43	2	MKN	679	HOOKENA
1827	+/-	17	6	SWR	540	MOAULA
1838	+/-	94	2	RV	866	PUU KAHILIKU
1860	+/-	200	1	SWR	354	
1863	+/-	46	3	MKS	289	MAKAKUPU
1880	+/-	200	1	NER	0	AINAKAHIKO
1883	+/-	43	3	MKS	290	KAPAPALA RANCH
1891	+/-	99	1	NER	0	
1910	+/-	120	1	MKS	349	
1960	+/-	90	1	SWR	674	
1993	+/-	43	3	NER	883	UPPER STRIP ROAD
2010	+/-	200	1	SWR	590	
2050	+/-	160	1	MKN	145	
2068	+/-	85	1	MKN	0	
2075	+/-	36	3	SWR	295	RED CONE
2312	+/-	58	2	SWR	638	NINOLE GULCH
2330	+/-	60	1	SWR	498	
2330	+/-	60	1	SWR	498	
2365	+/-	30	1	SWR	637	NINOLE A'A
2457	+/-	38	2	SWR	644	HOKUHANO
2480	+/-	35	2	SWR	544	PUU 2847
2497	+/-	63	2	SWR	680	KALAHIKI
2902	+/-	40	3	MKS	346	PUU ELEELE

TABLE 1. (Con't.)

AGE (Yrs. b.p.)	PRECISION	DATES (No. / flow)	SUBDIVISION	FLOWID[3] No	UNIT NAME
2920	+/- 60	1	SWR	634	
2940	+/- 200	1	MKN	138	PALEMANO POINT
3070	+/- 60	1	MKN	0	
3270	+/- 180	1	MKN	686	KAUNENE
3280	+/- 142	2	MKN	0	KANIKU
3330	+/- 80	1	MKN	141	
3347	+/- 28	6	NER	863	PUNAHOA
3393	+/- 57	3	SWR	624	KAUAHAAO CHURCH
3480	+/- 90	1	MKS	355	UPPER WAIHAKA GULCH
3548	+/- 46	2	SWR	0	
3680	+/- 25	1	SWR	646	HIONAA
3710	+/- 74	2	NER	892	OHAIKEA
3750	+/- 70	1	NER	0	
3790	+/- 200	1	NER	868	
3850	+/- 38	2	SWR	641	WAILAU
4080	+/- 60	1	SWR	623	PUU AKIHI
4122	+/- 32	5	NER	954	WILDER ROAD
4324	+/- 79	2	MKS	0	
4575	+/- 35	1	SWR	555	KAUMAIKEOHU
4695	+/- 58	2	NER	876	
4770	+/- 90	1	SWR	751	
5140	+/- 80	1	SWR	396	
5160	+/- 200	1	SWR	0	
5319	+/- 86	2	NER	947	PUU MAKALA PICRITE
5460	+/- 60	1	SWR	745	
5534	+/- 70	2	NER	976	MOUNTAIN VIEW
5903	+/- 42	2	SWR	548	ALAPAI
7360	+/- 70	1	SWR	739	WAIKAPUNA PALI
7750	+/- 70	1	SWR	747	PUU POO PUEO
7960	+/- 110	1	NER	962	
8030	+/- 70	1	NER	952	ALAWAENA ROAD
8365	+/- 140	1	NER	948	KINOOLE STREET
8370	+/- 200	1	MKS	0	
8550	+/- 90	1	MKS	0	
8645	+/- 71	2	NER	978	PUU ULAULA
8910	+/- 210	1	MKN	218	NAPOOPOO AA
9000	+/- 100	1	MKN	219	
9020	+/- 88	2	NER	953	AINAOLA DRIVE
9080	+/- 80	1	SWR	0	
9170	+/- 100	1	SWR	743	
9250	+/- 80	1	MKN	225	

TABLE 1. (Con't.)

AGE (Yrs. b.p.)	PRECISION	DATES (No. / flow)	SUBDIVISION	FLOWID[4] No	UNIT NAME
9680	+/- 70	1	MKS	338	
9760	+/- 90	1	SWR	0	
9960	+/- 70	1	NER	0	
10010	+/- 70	1	MKS	299	
10030	+/- 85	1	MKS	388	
10048	+/- 54	3	NER	957	WAIAKEA HOMESTEADS
10140	+/- 300	1	SWR	0	
10199	+/- 41	5	NER	950	ANUENUE
10400	+/- 150	1	NER	974	
10600	+/- 150	1	MKS	386	
10660	+/- 400	1	NER	0	
10820	+/- 400	1	SWR	0	
10887	+/- 33	3	SWR	758	HONUAPO
11780	+/- 100	1	SWR	749	
13210	+/- 300	1	SWR	0	
13800	+/- 300	1	MKN	471	
13940	+/- 110	1	SWR	748	
14219	+/- 64	7	NER	949	WAIANUENUE
16390	+/- 80	1	SWR	757	
17910	+/- 810	1	SWR	0	MOHOKEA
20370	+/- 250	1	SWR	0	
24076	+/- 384	2	NER	0	WAIPAHOEHOE STREAM
25570	+/- 600	1	SWR	0	
25970	+/- 220	1	SWR	0	
26410	+/- 390	1	SWR	0	
26450	+/- 120	1	SWR	0	
26630	+/- 130	1	SWR	0	
26650	+/- 570	1	SWR	0	
28140	+/- 590	1	SWR	735	
28145	+/- 333	2	NER	960	KALUAIKI STREAM
28150	+/- 800	1	SWR	752	
29010	+/- 160	1	SWR	0	
31020	+/- 3100	1	SWR	0	
31100	+/- 900	1	SWR	0	
31400	+/- 2000	1	MKN	0	
37430	+/- 1050	1	NER	0	HILO DRILL HOLE
38000	?	1	NER	0	
42030	+/- 1800	1	NER	0	HILO DRILL HOLE

1 "FLOWID" numbers are the identifiers by which the hundreds of mapped flows on Mauna Loa are known. These are the "common denominator" numbers used to link various databases and are the reference numbers to be cited if more information is required about any particular flow. Many flows do not yet have FLOWID numbers assigned.

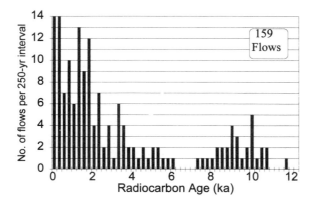

Fig. 7. Age distribution of all dated Mauna Loa lava flows (to 12.5 ka), plotted as the number of flow ages per 250-year age interval. Ages of 159 flows, plotted as number of flow dates per 250-year age interval. The plot includes 12 historically dated lava flows judged to be comparable to the data set of 147 radiocarbon-dated flows.

records were available. For this reason, only historical flows which reached vegetated areas were counted (summit-erupted historical lava flows were thus excluded). Of the 17 historical flank eruptions [*Lockwood and Lipman,* 1987, Table 1], only 12 were included with the radiocarbon-dated samples in the following age plots (1984 lavas were excluded as the eruption occurred after 1950; 1919 and 1916 were excluded as they would have been mapped as part of the 1926 eruption without historical records; 1899 was excluded as it would have been mapped with the 1880 flow, and 1877 was excluded as it is submarine). The historical flows included in the following plots are, from the NER: 1843, 1852, 1855-56, 1880-81, 1935 and 1942; and from the SWR: 1868, 1887, 1907, 1926, and 1950; and from the RV: 1859.

4. THE HOLOCENE RECORD

An age-frequency distribution plot of the 159 Mauna Loa flows with ages < 12 ka shows a general decrease in numbers of flows with time (Fig. 7). This decrease in dated flows over time is a simple function of the progressive burial of older flows by younger ones, a process which amounts to about 40% coverage of the Mauna Loa edifice per millennium [*Lipman,* 1980; *Lockwood and Lipman,* 1987; Fig. 7]. The assumption of a uniform burial rate is, of course inadequate, as the coverage rate actually varies inversely with the distance from source vents over the Mauna Loa edifice (older flows are thus preferentially preserved on lower flanks of the volcano). The absence of dated flows in the interval 6-7.4 ka and an abundance of flows between 8-11 ka represent a significant excursion from a uniform burial rate curve (Fig. 8). Major departures from this generalized burial curve, as seen in Figures 4 and 7, might be explained by random sampling variation, but if this were true, different random variation patterns should

be expected from different parts of the volcano. Yet a plot of lava flow ages from each rift zone (Fig. 9) shows very similiar distribution variations, although these flows are from widely separated areas. Clearly, these similar distribution patterns suggest underlying variations in eruptive frequency that affected both rift zones of Mauna Loa. I conclude that the variations from a uniform burial curve to be described in this paper are not results of sampling bias but represent actual variations in the distribution of Mauna Loa lava flows in space and time.

Much of the flow age variation indicated on Figure 7 results from superimposed age distribution patterns of flows from different parts of the volcano. By separating lava flows erupted from the rift zones from those erupted at the summit, an antithetic eruption pattern emerges. The period 0.9 ka to present has been a time of frequent rift zone eruptions but was marked by only one overflow from Moku`aweoweo (Fig. 10). Relatively few lava flows were erupted from the rift zone during the preceding period 0.9 -1.7 ka, a time of frequent overflows of lava from the summit (Fig. 10). In the plot of all Mauna Loa flows (Fig. 7), this period of lowered rift zone activity is partially obscured by an abundance of summit overflows of 1.1-2.0 ka age. Evidently this long-lived period of heightened summit activity, probably marked by near-continuous lava lakes and voluminous overflows, was a time of rapid shield-building and coincided with a time of greatly restricted activity on both Mauna Loa rift zones. The extensive overflows from this period are clearly delineated on the highly generalized geologic map of *Lockwood et al.* [1988] and on the more detailed map of the Island of Hawaii [*Wolfe and Morris,* in press].

The period of prolonged summit lava lake activity was also a time of frequent "radial vent" eruptions on the north and northwest flanks of Mauna Loa (Figs. 6, 11). Radial vent lavas of this period probably represent the lateral draining of magma

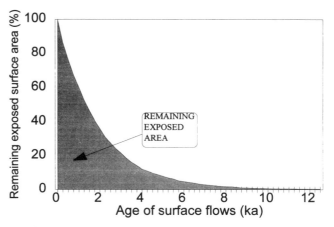

Fig. 8. Holocene burial of Mauna Loa's surface. This plot shows the amount of an original 10,000 year-old surface which remains with the passage of time, assuming burial by lava flows at a uniform rate of 40%/1,000 years.

stored high within the Mauna Loa edifice (possibly in lava lakes). These lavas were mostly erupted as very fluid, degassed pahoehoe lavas from fissures along which normal spatter deposits were not formed [*Lockwood and Lipman*, 1987, Fig. 18.19]. In contrast, radial vent lavas erupted after 1 ka are "fountain-fed" lavas [*Swanson*, 1973] and consist mostly of a`a [*Lockwood and Lipman*, 1987, Fig. 18.18]. More than 50 radial vent sources have been mapped to date on Mauna Loa, but only 14 (including two historical vents) have been dated. Older radial vent flows are difficult to recognize or date, as they are usually overlain by younger summit-derived flows, and only the spatter cone source may be preserved. Some older lava flows exposed along lower flanks could be derived from "spatterless" radial vents, but they are indistinguishable from summit overflows if their source vents are later buried. The areas covered by individual radial vent flows are also very small relative to the sizes of individual summit overflows. For these reasons, the numbers of older radial vent flows are seriously underrepresented by radiocarbon age distribution plots (Fig. 11).

The increase in rift zone activity as summit overflows ceased at about 1 ka (Fig. 10) was possibly accompanied by the formation of an early Moku`aweoweo caldera. The resumption of rift zone activity may also have been accompanied by a lower NER eruption of the voluminous basalt flow informally known as the "Panaewa picrite" [Lockwood, 1984]. The "Panaewa picrite" forms a prominant lava delta east of Hilo and is 30 m thick where penetrated by a research drill hole on the edge of Hilo Bay [*Stolper et al.*, 1994]. This vast outpouring of basalt (minimum volume, one cubic kilometer) may be genetically related to the formation of Moku`aweoweo Caldera. The most "reasonable" date on this flow is 1.47 ka, however (Table 1), which may be too old.

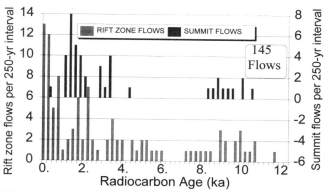

Fig. 10. Comparison of age distribution of lava flows erupted from Mauna Loa's summit (43 flows) *vs.* rift zone flows (102) in the past 12 ka. Eleven of the 13 ages plotted in the 0-250 bracket are from historical flows. Note the inverse relationship between ages of flows from the two areas.

The minimum of rift zone activity seen between 2.6 -3.5 ka (Fig. 10) was also accompanied by an increase in summit activity. This may represent another period of extended lava lake activity prior to formation of a now obscured summit caldera, possibly the buried pre-Moku`aweoweo caldera proposed by *Holcomb* [1980, Fig. 6]. Voluminous picritic basalts currently being mapped by USGS geologists on Mauna Loa's lower southwest rift zone, with ages around 2.5 ka ("Pu`u 2847 picrite", Table 1), may have accompanied the formation of this caldera and heralded the increase in rift zone eruptive activity that occurred at this time. The increase in summit activity at about 3.25 ka coincides with a decrease in rift activity, but evidence for antithetic relationships between summit and rift zone activity is faint and unconvincing before this time. A very large eruption of picritic basalt from the lower northeast rift zone ("Pu`u Maka`ala picrite" - 5.3 ka, Table 1) may have preceded the apparent increase in rift zone activity at 4.5-5.25 ka. This eruption could have also followed caldera formation as proposed for the previous two minima, but evidence for any preceding period of increased summit activity lies buried beneath voluminous summit overflows of younger age.

The cause of the abrupt transition from summit shield-building to prolonged periods of dominant rift zone activity are not known but must be related to changed stress regimes within Mauna Loa. The periods of summit-dominated eruptive activity imply that the rift zones are relatively "closed" to the injection and eruption of lava from Mauna Loa's shallow central magma storage reservoir. This reservoir, whose location and dimensions are becoming well-known from seismic, gravity, and deformation studies (See papers by *Okubo, Johnson,* and *Miklius et al.*, this volume), must preferentially send magma upward to summit vents during periods of regional compression, and laterally to rift vents during periods of relaxed stress across the rift zones. *Miklius et al.,* this volume, have shown that Mauna Loa deformation is asymmetrical across

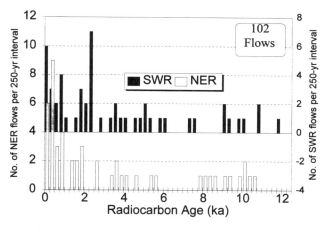

Fig. 9. Age distribution of all dated flows (to 12.5 ka) on Mauna Loa's Northeast Rift Zone (NER - 50 flows) and Southwest Rift Zone (SWR - 52 flows), plotted as the number of flow dates per 250-year interval for each rift zone. Five historically dated lava flows are included in the SWR data set, six in the NER data set. Note the similar distribution of flow ages from each rift zone.

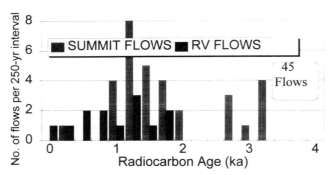

Fig. 11. Comparison of ages of dated flows erupted from "radial vents" (13), with ages of dated flows derived from Mauna Loa summit overflows (32) during the past 4 ka. One historical radial vent flow is included in the data set. Radial vent lava flows older than 1 ka are difficult to date, as they are mostly buried by younger summit flows.

the summit and rift zones, so that the stress changes which control the alternation between summit- and rift-dominated activity are ultimately related to the mobility of Mauna Loa's south flank. This, in turn, may be related to the mobility of Kilauea and the entire southern flank of Hawaii Island.

The absence of dated lava flows at 6-7 ka may represent a period of eruptive repose for all Mauna Loa. Of the 170 prehistoric lava flows dated so far, only a single one (5.9 ka) falls in the range 5.6-7.3 ka, although 12 flows have been dated in the previous 1.7 ka. In the 1.7 ka following the 5.6-7.3 ka period, 10 flows have been dated. The period of apparent eruptive repose suggests a cessation of magmatic production, but the possibility exists that magma was being produced and was either injected into the volcanic edifice without surface eruption or was erupted in areas not presently exposed for sampling. Evidence of caldera in-filling and shield-building during this period *could* be completely buried beneath younger flows, but if so the 6-7 ka flows did not travel far from the summit area.

The pronounced increase of eruptive activity between 8-11 ka indicated in Figure 7 is clearly "real," as lava flows in this age range are found on all regions of Mauna Loa (Figs. 9, 10). The cause of this heightened activity is conjectural, but it is interesting to note that this period coincides with the Pleistocene-Holocene boundary. Could there be any relationship to climatic changes at the end of Pleistocene time? Mauna Loa must have been covered by a Late Pleistocene ice cap, as has been documented on Mauna Kea [*Porter*, 1979]. All traces of this presumed ice cap, which disappeared from Mauna Kea by 9 ka [*Porter et al.*, 1977] are now buried by younger lavas. Although the mass of ice which covered Mauna Loa at the end of Pleistocene time would have been relatively small, could the sudden melting of this ice cap at the beginning of Holocene time have been a de-pressurization "trigger" which initiated a period of increased production from a delicately balanced magma system? Could the period of eruptive quiescence after 7 ka be a delayed response to this period of

increased magma production - a period when the perturbed magmatic system restored itself to equilibrium? Or could rapidly rising sea levels in early Holocene time have influenced magma equilibrium? *Kurz et al.* [this volume] and *Kurz and Kammer* [1991] have also noted major shifts in isotopic compositions of Mauna Loa at this time and have called for changes in magmatic production rates around 10 ka.

5. REGIONAL RELATIONSHIPS

Are the variations in eruptive frequency documented for Mauna Loa restricted to this single volcano or did such cyclic patterns affect adjacent volcanoes on the Island of Hawaii? Could regional variations in magma production affect more than one volcano on Hawaii? Holocene lava flows have been dated for three other volcanoes on the Island of Hawaii (Kilauea, Hualalai, and Mauna Kea), but the data are unfortunately sparse. Age distribution plots of radiocarbon-dated flows on these volcanoes are instructive in showing pronounced contrasts between the eruptive histories of these volcanoes (Figs. 12a-c), but no indication of inter-volcano temporal relationships is apparent from the small data sets.

Kilauea Volcano is more frequently active than Mauna Loa and, because Kilauea's surface area is less than one-third that of Mauna Loa, its surface is buried much more rapidly (Fig. 12a). Insufficient evidence is preserved for meaningful comparison with Mauna Loa's record, although *Holcomb* [1987] does state that a major hiatus in overflows from Kilauea's summit caldera occurred between 1.1 - 1.5 Ka, which happens to coincide with the period of maximum overflow activity from Mauna Loa's summit (Fig. 10).

Hualalai Volcano, in contrast to Kilauea, has a much *lower* lava production rate than Mauna Loa [*Moore et al.*, 1987], so that the Holocene record remains well preserved. A plot of all dated flows on Hualalai (Fig. 12b) shows a fairly uniform age

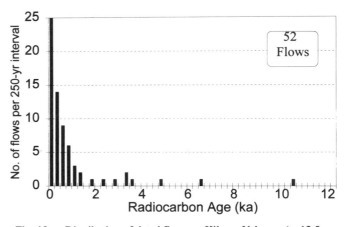

Fig. 12a. Distribution of dated flows on Kilauea Volcano (to 12.5 ka), plotted as number of flow dates per 250-year age interval (52 flows). Data from *Holcomb* [1987]; *Rubin et al.* [1987].

distribution over the Holocene, with no obvious variation patterns.

Mauna Kea Volcano had only limited activity in Holocene time, with dated flows restricted to the interval 4.5-7.2 ka (Fig. 12c). The only other Hawaiian volcano with dated Holocene flows is Haleakala on the island of Maui [*Crandell*, 1983], but the sparse chronology and limited mapping make a comparison with Mauna Loa's record impossible.

6. CONCLUSIONS

Radiocarbon dating of Mauna Loa lava flows has provided the most refined eruptive chronology of any volcano on Earth. An analysis of 170 well-dated prehistoric Mauna Loa lava flows reveals variations in eruptive frequency not explicable by random sampling, but which reflect instead systematic changes in time and place of eruptive activity. Pronounced decreases in eruptive activity on Mauna Loa's rift zones correlate with periods of increased lava lake activity and shield-building at the volcano's summit. Voluminous eruptions of olivine-rich lavas low on the rift zones may have heralded the cessation of summit overflows and the resumption of increased rift zone activity. These picrites were possibly erupted at a time when picritic melts had risen higher than normal into shallow supply conduits [*Rhodes*, this volume], and their eruption low on Mauna Loa's flanks may have initiated summit caldera collapse.

Radiocarbon dating, combined with detailed geologic mapping in progress, suggests the following cyclic model for this summit-flank alternation of eruptive activity, cycles that may have been repeated numerous times in Mauna Loa's past. A cycle begins during a period of high-standing magma at Mauna Loa's summit. This is reflected by near-continuous lava lake activity and shield-building, from which vast sheets of pahoehoe blanket the northwest and southeast flanks. The

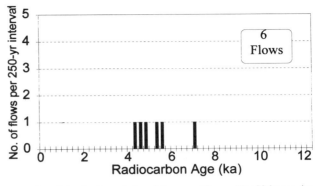

Fig. 12c. Distribution of dated flows on Mauna Kea Volcano (to 12.5 ka), plotted as number of flow dates per 250-year age interval (6 flows). Data from *Wolfe and Morris* [in press].

last such period of summit activity lasted for about 800 years between 1.2-2.0 ka, and was accompanied by frequent small eruptions from radial fissures on the northwest flank. This period of summit overflows may end when a large flank eruption occurs low on one of the rift zones, initiating caldera collapse at Mauna Loa's summit. Such flank eruptions are probably triggered by, or at least may accompany, substantial dilation of the rift zones. Dilation facilitates eruptive activity lower on the flanks and robs Mauna Loa of the high-standing magma required for shield building. The sudden lowering of the magma column associated with caldera collapse results in a major shift of surface activity as both rift zones become characterized by frequent eruptive activity (an estimated average of one rift zone eruption each 20-25 years over the past millennium). Summit activity probably continues during times of increased rift activity (as it has during the 19-20th centuries - *Barnard*, this volume), but all erupted lavas are trapped within the summit caldera, and no trace of this activity remains at the surface. As stresses increase across the rift zones over time, magma again rises more easily within the edifice, and the caldera is filled. A lava lake appears at Mauna Loa's summit, pahoehoe sheets are able to overflow the caldera walls, and the cycle begins again.

The record of repetitive cycles of rift zone activity (Fig. 10) suggests that these cycles each last about 2,000 years. The earliest written descriptions of Moku`aweoweo Caldera [*Barnard*, this volume] showed that the caldera was more than 300 m deep in 1794 but has been steadily infilling ever since. The caldera is as full as possible at this time, as caldera floor lavas now spill out to the southwest. Since the last period of summit overflows began about 2,000 years ago, I suggest that Mauna Loa may be on the verge of shifting to a period of long-lived lava lake activity, shield-building, increased summit overflows, and diminished rift zone eruptions. The sharp increase in historical rift zone activity, noted by *Lipman* [1980] and conspicuous for flows younger than 0.5 ka on Figure 7, may be a "finale" related to an impending shift in

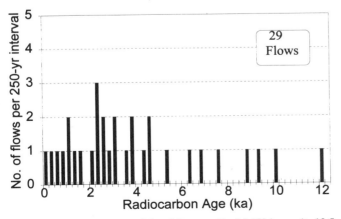

Fig. 12b. Distribution of dated flows on Hualalai Volcano (to 12.5 Ka), plotted as number of flow dates per 250-year age interval (29 flows). Data from *Moore et al.* [1987] and *Moore and Clague* [1991].

eruptive style. Such a transition will obviously have impact on the nature of volcanic risk on the Island of Hawaii, as the incidence of lava flow activity increases to the northwest and southeast of the summit (subdivisions MKN and MKS of Fig. 6) and lessens in the areas downslope from the rift zones (subdivisions NER and SWR). *Lipman* [this volume] has shown convincing evidence to suggest that in a long-term context Mauna Loa may be nearing the end of its tholeiitic shield-building phase. This does not contradict my suggestion that another period of summit activity is nearing, but might suggest that future shield-building episodes will be less vigorous than those of the past.

Radiocarbon dating of Mauna Loa lavas has yielded preliminary information on the distribution of eruptive activity in time and space, but much needs to be done to further refine this record. Only about 35% of surface lava flows have been dated so far, and most of these have single dates. Many flows have been dated in areas where geologic mapping is not complete, areas where stratigraphic frameworks to evaluate flow ages are inadequate. Mauna Loa's radiocarbon record continues to improve, however, as more charcoal samples are collected for dating and as new flows are mapped and relationships clarified between already dated lavas. The number of radiocarbon-dated Mauna Loa lava flows reported in this paper (170) is not a firm number but depends on partly subjective decisions about how individual dates will be assigned to particular lava flows. Where ages are obtained from beneath similiar lava flows in distantly separated outcrop areas, a decision must be made as to whether a correlation can be made between the two lava flow exposures or not. These decisions to "lump" or "split" flows greatly affect the number of resultant dated flows and, thus, the age-frequency distribution patterns.

Consider the problem of similar lava flow exposures on opposite sides of Mauna Loa's rift zones. We know from historical examples (e.g., 1880, 1926, 1950) that large eruptions on Mauna Loa's rift zones can send flows in opposite directions down different flanks of the rift zones. Prehistoric radiocarbon ages obtained from "similar" lava flow remnants on opposite flanks of rift zones *may* date the same event, but how does one decide to average the two dates as *one* flow or to treat them as *two* eruptive events? Paleomagnetic analysis of dated Mauna Loa lava flows is critical to correlation efforts of this sort and, as paleomagnetic secular variation is better understood [*Hagstrum and Champion*, in press], these techniques are becoming more appropriate for evaluation of questionable radiocarbon dates and for "permissible" (not absolute) age dating. The temporal variation of minor element abundances [*Rhodes and Hart*, this volume; *Rhodes*, 1983] provides another valuable tool for correlation of isolated flow remnants through geochemical analysis. The tools of remote sensing have also proven valuable for correlation of younger flows [*Kahle et al.*, this volume; *Kahle et al.*, 1988], and quantitative analyses of forest ages developed on separate flows [*Vitousek et al.*, this volume] can provide further evidence. None of the above techniques can individually provide proof of correlation between isolated flow outcrops, but together they can provide guidance on flow age assignments.

Ages reported in this paper are derived from laboratory radiocarbon dates, and no effort has been made at "calibration" (the conversion of ^{14}C dates to calendar equivalents). Calibration is a complex matter, as no single equivalent calibrated age can be given for a particular ^{14}C date. Temporal variations in atmospheric ^{14}C content mean that a single radiocarbon date has a *range* of calendar equivalents, regardless of laboratory precision [*Stuiver and Reimer*, 1993]. Sophisticated statistical calibrated age analyses of the large Mauna Loa radiocarbon age data set could eventually yield great improvements in the quantitative understanding of eruptive temporal patterns on Mauna Loa.

The record of prehistoric activity, derived from geologic mapping and radiocarbon dating, is the basis of volcanic hazards and risk assessments on the flanks of Mauna Loa [*Trusdell*, this volume]. Improvements in our ability to predict future activity are a most important element of the Mauna Loa Decade Volcano Program, and will depend on an improvement of our understanding of Mauna Loa's past.

APPENDIX A.

Formulas for calculating weighted averages (X) and standard errors (σ) for multiply-dated lava flows [*Bevington*, 1969].

$$\bar{X} = \frac{\sum_{i=1}^{n} \left\{ \frac{x_i}{\sigma_i^2} \right\}}{\sum_{i=1}^{n} \left\{ \frac{1}{\sigma_i^2} \right\}}$$

$$\bar{\sigma} = \sqrt{\frac{1}{\sum_{i=1}^{n} \frac{1}{\sigma_i^2}}}$$

Where x_i = reported radiocarbon age; σ_i = laboratory standard error

Example: Pu'u O Uo flow [FLOWID **412**], Reported ^{14}C dates (yrs. b.p.): 1) 450 +/- 140, 2) 510 +/- 150, 3) 580 +/- 120[1].

$$\bar{X} = \frac{\left\{ \frac{410}{140^2} + \frac{510}{150^2} + \frac{580}{120^2} \right\}}{\left\{ \frac{1}{140^2} + \frac{1}{150^2} + \frac{1}{120^2} \right\}} = 521$$

$$\bar{\sigma} = \sqrt{\dfrac{1}{\left\{\dfrac{1}{140^2} + \dfrac{1}{150^2} + \dfrac{1}{120^2}\right\}}} = 78$$

∴ Weighted average age of Pu'u O Uo flow: **521 +/- 78** yrs. b.p.

---

[1]Dates from [*Rubin et al.*, 1987]

Acknowledgments. Recovery of the several hundred charcoal samples reported on in this paper represents a huge investment of time and labor, supported by the U.S. Geological Survey. These samples have been collected by many individuals over the past three decades, and to each of them, especially to Peter Lipman and Frank Trusdell, much is owed for their labors under the sun and in the mud over the years. Norm Banks, Jane Buchanan-Banks, Marie Jackson-Delaney, and Ed Wolfe also collected more than five charcoal samples each. Meyer Rubin and his colleagues have dated most of these samples, although AMS dating at the Universities of Arizona, California, and Toronto is becoming ever more important. Deborah Trimble's precise conventional radiocarbon dates are also greatly appreciated. Discussions with Dave Clague, Bob Decker, Peter Lipman, Jim Moore, Mike Ryan, and Frank Trusdell are always stimulating, and their thoughts have undoubtedly crept into this paper. Ed Wolfe provided valuable information on radiocarbon dating for Mauna Kea, as did Dick Moore for Hualalai. Paula Reimer provided important insights on radiocarbon calibration systematics. Ed Wolfe, Peter Lipman, and Tom Simkin each provided extremely conscientious reviews; this paper has been greatly improved by their many thoughtful suggestions. Above all, I thank my wife Martha, who has assisted in recovery of several score of the charcoal samples reported on in this paper.

REFERENCES

Barnard, W.M., Mauna Loa Volcano: historical eruptions, exploration, and observations (1779-1910), *this volume.*

Bevington, P.R., Data reduction and error analysis for the physical sciences, McGraw-Hill, New York, 298 pp., 1969.

Crandell, D.R., Potential hazards from future volcanic eruptions on the island of Maui, Hawaii, *U.S. Geol. Surv. Misc. Investig. Ser. Map I-1442*, scale 1:100,000, 1983.

Hagstrum, J. T., and D. E. Champion, Late Quaternary geomagnetic secular variation from historical and ^{14}C-dated lava flows on Hawaii, *J. Geophys. Res.* (submitted).

Holcomb, R.T., Kilauea Volcano, Hawaii: Chronology and morphology of surficial lava flows, *Stanford Univ. Ph. D. Dissert.*, 1980.

Holcomb, R.T., Eruptive history and long-term behavior of Kilauea Volcano, in Decker, R.W., T.L. Wright, and P.H. Stauffer, eds., Volcanism in Hawaii, *U.S. Geol. Surv. Prof. Pap. 1350*, 1, 12, 261-350, 1987.

Johnson, D.J., Gravity changes on Mauna Loa Volcano, *this volume.*

Kahle, A. B., M.J. Abrams, E.A. Abbott, P.J. Mouginis-Mark, and V.J. Realmuto, Remote sensing of Mauna Loa, *this volume.*

Kahle, A.B., A.R. Gillespie, E.A. Abbott, M.J. Abrahms, R.W. Walker, and G. Hoover, Relative dating of Hawaiian lava flows using multispectral thermal infrared images: A new tool for geologic mapping of young volcanic terranes, *J. Geophys. Res.*, 93, 15239-15251, 1988.

Kurz, M.D. and D. P. Kammer, Isotopic evolution of Mauna Loa Volcano, *Earth Planet. Sci. Lett.*, 103, 257-269, 1991.

Kurz, M.D., T.C. Kenna, D.P. Kammer, J.M. Rhodes, and M.O. Garcia, Isotopic evolution of Mauna Loa Volcano: a view from the submarine southwest rift zone, *this volume.*

Lipman, P.W., Declining growth of Mauna Loa during the last 100,000 years: rates of lava accumulation vs. gravitational subsidence, *this volume.*

Lipman, P.W., Rates of volcanic activity along the southwest rift zone of Mauna Loa Volcano, Hawaii, in Garcia, M.O., and R.W. Decker, eds., G.A. Macdonald Special Memorial Issue, *Bull. Volcanologique*, 43, 4, 703-725, 1980.

Lockwood, J.P., Geologic map of Mauna Loa Volcano-Kulani Quadrangle, Hawaii (preliminary), with preliminary map legend, *U.S. Geol. Surv. Open-File Rep. 84-12*, 8 pp., scale 1:24,000 1984.

Lockwood, J.P., and P.W. Lipman, Recovery of datable charcoal from beneath young lava flows—lessons from Hawaii, *Bull. Volcanologique*, 43, 609-615, 1980.

Lockwood, J.P., and P.W. Lipman, Holocene eruptive history of Mauna Loa Volcano, in Decker, R.W., T.L. Wright, and P.H. Stauffer, eds., Volcanism in Hawaii, *U.S. Geol. Surv. Prof. Pap. 1350*, 1, 18, 509-535, 1987.

Lockwood, J.P., P.W. Lipman, L. Petersen, and F.R. Warshauer, Generalized ages of surface lava flows of Mauna Loa Volcano, Hawaii, *U.S. Geol. Surv. Misc. Investig. Ser. Map I-1908*, scale 1:250,000, 1988.

Miklius, A., M. Lisowski, P.T. Delaney, R. Denlinger, J. Dvorak, A.T. Okamura, and M.K. Sako, Recent inflation and flank movement of Mauna Loa Volcano, *this volume.*

Moore, R.B., D.A. Clague, M. Rubin, and W.A. Bohrson, Hualalai Volcano: A preliminary summary of geologic, petrologic, and geophysical data, in Decker, R.W., T.L. Wright, and P.H. Stauffer, eds., Volcanism in Hawaii, *U.S. Geol. Surv. Prof. Pap. 1350*, 1, 20, 571-585, 1987.

Moore, J.G., and D.A, Clague, Volcano growth and evolution of the island of Hawaii: *Geol. Soc. Amer. Bull.*, 104, 1471-1484, 1992.

Okubo, P.G., A seismological framework for Mauna Loa Volcano, Hawaii, *this volume.*

Porter, S.C., Hawaiian glacial ages, *Quaternary Res.*, 12, 161-187, 179.

Porter, S.C., M. Stuiver, and I.C. Yang, Chronology of Hawaiian glaciations, *Science*, 195, 61-63, 1977.

Rhodes, J.M., Homogeneity of lava flows: Chemical data for historic Mauna Loa eruptions, *J. Geophys. Res.*, 93, 4453-4466, 1988.

Rhodes, J.M., The 1852 and 1868 Mauna Loa picrite eruptions: Clues to parental magma compositions and the parental magma system, *this volume.*

Rhodes, J.M., and S.R. Hart, Episodic trace element and isotopic variations in historical Mauna Loa lavas: Implications for magma and plume dynamics, *this volume.*

Rubin, M., L.K. Gargulinski, and J.P. McGeehin, Hawaiian radiocarbon dates, in Decker, R.W., T.L. Wright, and P.H. Stauffer, eds., Volcanism in Hawaii, *U.S. Geol. Surv. Prof. Pap. 1350*, 1, 10, 213-242, 1987.

Stolper, E.M., D.M. Thomas, and D.J. DePaolo, Hawaii Scientific Drilling Project: Background and overview of initial results (abs.),

Eos Trans. AGU, 75, 707, 1994.

Stuiver, M., and P.J. Reimer, Extended ^{14}C data base and revised CALIB 3.0 ^{14}C age calibration program. *Radiocarbon*, 35, 215-230, 1993.

Swanson, D.A., Pahoehoe flows from the 1969-1971 Mauna Ulu eruption, Kilauea Volcano, Hawaii, *Geol. Soc. Amer. Bull.*, 84, 615-626, 1973.

Trusdell, F.A., Lava flow hazards and risk assessment on Mauna Loa Volcano, Hawaii, *this volume*.

Vitousek, P.M., G.H. Aplet, J.W. Raich, and J.P. Lockwood, Biological perspectives on Mauna Loa Volcano: A model system for ecological research, *this volume*.

Wolfe, E.W., and J. Morris, Geologic map of the island of Hawaii: *U.S. Geol. Surv. Misc. Investig. Ser. Map I-xxx*, scale 1:100,000, 3 sheets (in press).

Wolfe. E.W., W.S. Wise, and G.B. Dalrymple, The geology and petrology of Mauna Kea Volcano, Hawaii: A study of post-shield volcanism, *U.S. Geol. Surv. Prof. Pap. 1557* (in press).

Quiescent Outgassing of Mauna Loa Volcano 1958-1994

Steven Ryan

Mauna Loa Observatory, Climatic Monitoring and Diagnostics Laboratory, National Oceanic and Atmospheric Administration, Hilo, Hawaii

A continuous 37 year record of the quiescent CO_2 outgassing of Mauna Loa volcano was derived from atmospheric measurements made 6 km downslope of the summit caldera at Mauna Loa Observatory. The volcanic plume is sometimes trapped in the temperature inversion near the ground at night and transported downslope to the observatory. The amount of volcanic CO_2 was greatest shortly after the 1975 and 1984 eruptions and then decreased exponentially with decay constants of 6.5 and 1.6 years respectively. Between 1959 and 1973 the decay constant was 6.1 years. The total reservoir mass of CO_2 during each of the three quiescent periods was similar and estimated to be between 2×10^8 kg and 5×10^8 kg (0.2 Mt to 0.5 Mt). The 1975 eruption may have been preceded by a small increase in CO_2 emissions. A similar increase has occurred since early 1993. Condensation nuclei (CN), presumably consisting of sulfate aerosol, were measured in the volcanic plume throughout the 1974 to 1994 record. The post-1975 period had consistently high levels of CN. Between 1977 and 1980, light-scattering aerosols were detected, coincident with a period of visible fuming at the summit. CN levels after the 1984 eruption were greatly reduced. Two brief periods of low CN emissions during this time correlate with temporary halts or reductions in the rate of summit expansion. These temporary reversals in the inflation of the mountain did not affect the steady exponential decline of the CO_2 emissions rate. Upper limits were set on the amounts of H_2O, O_3, CH_4, SO_2, aerosol carbon, radon, CO, and H_2 present in the plume at various periods between 1974 and 1993. The ratio of SO_2 to CO_2 was less than 1.8×10^{-3} between 1988 and 1992.

1. INTRODUCTION

Mauna Loa Observatory (MLO) has been an important site for the continuous climatological monitoring of atmospheric CO_2 levels since 1958 [*Pales and Keeling*, 1965], and of a growing number of aerosol and trace gas species since 1974 [*Ferguson and Rosson*, 1991; *Peterson and Rosson*, 1993]. The observatory is at an elevation of 3400 m on the northern flank of 4169 m Mauna Loa Volcano, 6 km from the summit caldera.

Vented gas from the nearby Mauna Loa summit is sometimes transported downslope at night and detected by the CO_2

analyzers [*Pales and Keeling*, 1965; *Miller and Chin*, 1978] and aerosol monitors [*Bodhaine et al.* 1980] at MLO. At a remote location such as MLO, the background air is normally well mixed and exhibits a steady hour-to-hour CO_2 concentration. Plumes from the summit caldera, a nearby source of CO_2, are poorly mixed with the background air upon reaching MLO and can easily be identified by their highly variable CO_2 concentration. Previous studies have been concerned with identifying and eliminating this volcanic contamination from the climatological record [e.g. *Keeling et al.*, 1976; *Thoning et al.*, 1989]. The present study is the first to use the suite of MLO trace-gas data sets to monitor the long-term outgassing behavior of Mauna Loa volcano.

The history of published volcanic gas measurements on Mauna Loa is brief and intermittent. These have included fumarole SO_2 emission estimates for 1978-1979 [*Casadevall and Hazlett*, 1983], five months of continuous in-situ fumarole temperature and reducing gas activity measurements taken just

Mauna Loa Revealed: Structure, Composition, History, and Hazards
Geophysical Monograph 92

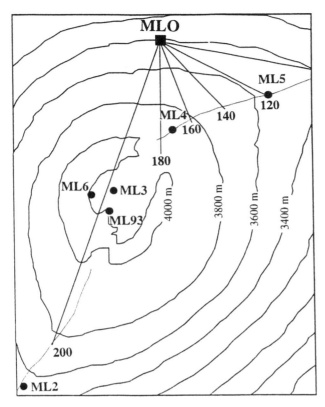

Fig. 1. Map of Mauna Loa volcano summit vicinity. Radial lines from the location of MLO give bearings in degrees from true north. The locations of identified thermal features are shown and described in the text.

west side to near zero where it intersects the rifts to the NNE ("North Pit") and SSW ("South Pit"). This topography is a likely factor in the transport of volcanic fume from sources inside the caldera. Air trapped near the surface might "drain" out of the caldera at the locations of North Pit and South Pit. The observatory is almost directly downslope from North Pit, a favorable location for detecting air exiting the caldera at this point.

Outgassing sites in the summit region were visually identified by *Casadevall and Hazlett* [1983]. Five of these are located on Figure 1 and identified as ML2 - ML6. The primary source of outgassing before 1984 was at location ML3. This feature consisted of seven 120°- 362° C active fumaroles located on the 1975 eruptive fissure, which produced 0.5 to 5 tons of SO_2 per day [*Casadevall and Hazlett*, 1983]. This fissure was a source of reducing gas activity [*Lockwood et al.*, 1985] and visible fume which sometimes produced a dense blue haze on the mountain's upper flank [*Lockwood et al., 1987*]. The remaining features in Figure 1 are comparatively minor gas sources and will not be discussed further.

The 1984 eruption covered a large portion of the caldera floor and formed new fissures. Visible fuming from ML3 ceased following this eruption [*Lockwood et al.*, 1987]. Visible fuming in the post-1984 era is greatly reduced and comes from a location marked ML93 in Figure 1 [J. Sutton, pers. communication, 1993]. No thermal inventories of Mauna Loa have been published since the 1984 eruption.

2.2 Site Meteorology

The volcanic plume only reaches MLO under certain meteorological conditions [*Price and Pales*, 1963; *Garrett*, 1980; *Hahn et al.*, 1993]. On clear nights radiative cooling produces a temperature inversion near the ground. Gravity pulls this cool, dense air down the mountain slope at speeds of several m/s in a thin layer tens of meters thick. If the free tropospheric winds are light, the volcanic plume remains trapped beneath the inversion layer to potentially be transported in the downslope wind to MLO. This condition is often disrupted by strong easterly or westerly tropospheric winds, which break up the surface temperature inversion and steer the plume away from the direct downslope path to MLO.

Meteorological variables have been measured at MLO since 1958 and include wind, temperature, humidity, and pressure. Before 1977, data were recorded on chart records and hand-scaled to obtain hourly averages with a 45-degree directional resolution. After this date, a computer data acquisition system was installed along with improved sensors, some of which have been subsequently upgraded. The wind direction data after 1977 has 1 degree resolution. Details of this program are given in *Herbert et al.* [1981].

before the 1984 eruption [*Lockwood et al.*, 1985; *Sutton and McGee*, 1989], a series of samples taken during and immediately after the 1984 eruption for eight gases including CO_2 [*Greenland*, 1987], and satellite and aircraft measurements of the SO_2 and CO_2 emissions during the 1984 eruption [*Casadevall et al.*, 1984].

2. ENVIRONMENT

2.1 Site Description

MLO is located relative to the summit features in Figure 1. The summit caldera is 6 km from MLO at a bearing of 190°, with a floor elevation of 4000 meters. The caldera is 3 km by 5 km in diameter, elongated along two major rift zones extending to the northeast and southwest. The northeast rift lies at bearings of 100° to 180° from MLO, at a minimum distance of 4 km. The southwest rift lies at bearings between 200° and 210° at distances of 10 km and greater.

The height of the caldera rim varies from 180 meters on the

3. CO$_2$ MEASUREMENTS

3.1 Measurement Methodology

The original CO$_2$ monitoring program was started in 1958 by the Scripps Institute of Oceanography (SIO) and has produced the longest continuous record of atmospheric CO$_2$ available in the world. It has provided a textbook example of the effect of fossil-fuel burning on the global atmosphere. Complete details of this program are given in *Pales and Keeling* [1965] and *Keeling et al.* [1976, 1982, 1987]. Measurements were made using an Applied Physics Corporation (APC) dual detector non-dispersive infrared analyzer [*Smith*, 1953]. An air stream was sampled alternately from two separate intake lines for 10 minutes each followed by a 10 minute flow of reference gas. Water vapor was removed by passing the air through a freezer trap at -60 to -80 °C. The output of the analyzer was recorded on a chart record from which data was hand-scaled. If the variability of the trace during each 10-minute interval was visually judged to be significantly greater than that of the reference gas trace, the interval was flagged as "variable." These subjective flags were the basis for identifying the presence of volcanic plume in the SIO data set. Data used in the present study were obtained through the Carbon Dioxide Information Center [*Keeling*, 1986]. The precision of the SIO system in measuring reference gases was between 0.1 and 0.2 ppm. The precision of monthly baseline averages was approximately 0.5 ppm, increasing to as much as 1.0 ppm from mid-1964 to late 1968 [*Keeling*, 1986].

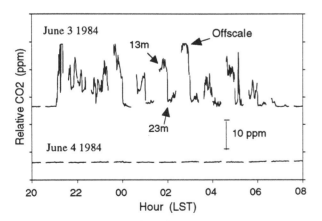

Fig. 2. Minute average atmospheric CO$_2$ concentrations for two consecutive nights in June 1984. The top trace illustrates the effects of the volcanic plume and the bottom trace shows the uncontaminated free tropospheric background. The two traces are offset by 20 ppm. Air was being sampled from heights of 13 meters and 23 meters respectively for 25 minutes each hour. A periodic 10 minute break occurs when the system is sampling reference gas. The data labeled "Offscale" are caused when atmospheric CO$_2$ concentrations exceed the range of the analyzer.

In May 1974, a second continuous CO$_2$ monitoring program was established by what is now the National Oceanic and Atmospheric Administration (NOAA). Details are given in *Komhyr et al.* [1983, 1989]. This program began using a Hartmann and Braun URAS-2 non-dispersive infrared analyzer until August 1987, when it was replaced by a Siemens Ultramat-3 analyzer. Two separate intake lines and two reference gases were sampled every hour. The air stream was dehumidified by passing it through a cold trap at -60 to -80 °C. The two reference gases were calibrated weekly by comparison with a suite of five standard gases. Beginning in 1976, 1-minute averages of the analyzer output voltage were recorded by a computerized data acquisition system [*Herbert et al.*, 1986]. Variability in these data is flagged by computer.

These data were recently re-processed for this study to obtain 1-minute average concentrations following the calibration methodology of *Komhyr et al.* [1989]. The 1-minute data were then visually edited to remove periods when the system malfunctioned (brief power outages, freezer trap blockages, air pump failures, etc.)

3.2 Variability in CO$_2$ Concentration

The presence of a volcanic plume caused an increase in the level of minute-scale variability in the CO$_2$ record. Hawaii is centrally located in the Pacific Ocean, far from continental CO$_2$ sources, so that background air at MLO is well mixed and has a steady hourly concentration. Since the Mauna Loa volcanic source is only a few km away, the plume gas is poorly mixed with background air upon reaching MLO, resulting in a large increase in the minute-scale variability of the CO$_2$ concentration. An illustrative example of background data and volcanically disturbed data taken shortly after the 1984 eruption is shown in Figure 2. Background CO$_2$ levels have risen from 315 ppm in 1958 [*Keeling et al.*, 1976] to 358 ppm in 1993 with an annual cycle averaging 6-7 ppm and an average diurnal cycle of about one ppm [Thoning et al., 1989].

Minute-scale variability in the CO$_2$ record had other potential causes. These fall into two categories: sources of "noise" that caused both positive and negative changes in the background concentration and had a long-term sum of zero, and true CO$_2$ sources that caused only positive changes in the background concentration.

The identified or potential nearby, nighttime sources of CO$_2$, in approximate order of their influence were:

1. Volcanic emissions from the Mauna Loa summit. These were the primary CO$_2$ sources, typically producing increases of several ppm.

2. Volcanic emissions from Kilauea volcano. CO$_2$, SO$_2$, and other volcanic emissions came from the nearby Kilauea region [*Greenland et al.*, 1985; *Connor et al.*, 1988] southeast of Mauna Loa at altitudes between sea level and 1200 m. This

source was active intermittently in the 1960s and 1970s, and was virtually continuous after 1982. The emissions usually remained trapped in the marine boundary layer, below 2000 m. In the afternoon, upslope winds commonly brought air from the marine boundary layer up to MLO. This air sometimes contained fume from Kilauea volcano. *Luria et al.* [1992] showed that this was the principal daytime source of SO_2 at MLO in 1989 (at concentrations of up to 50 ppb), and estimated a corresponding upper limit daytime CO_2 increase of 0.9 ppm. The possibility exists that some Kilauea-contaminated marine boundary layer air was occasionally caught up in a large-scale circulation pattern and became entrained in the downslope winds at night. Dilution would probably have reduced this nighttime excess CO_2 concentration to less than 0.1 ppm with variability less than that of the Mauna Loa plume because of greater travel times (about ten hours) and distances (at least 100 km).

3. Respired CO_2 from island ecosystems and island anthropogenic CO_2. Most vegetation on the island of Hawaii is at elevations below 2000 m and most of the human population lives below 500 m. The daytime photosynthetic uptake of CO_2 by island vegetation at low altitudes caused decreases of up to several ppm when the air was blowing upslope in the afternoon [e.g. *Pales and Keeling*, 1965]. Because these sources were widely dispersed at distances of many tens of km, the CO_2 was more thoroughly mixed and had a correspondingly small minute-scale variability [*Thoning et al.*, 1989]. Nighttime contamination from these sources would have been minimal for the same reason as for the Kilauea plume. There is no vegetation on the barren lava at elevations above MLO.

4. Contamination from the vicinity of MLO. Events of this type were noted by *Keeling et al.* [1976] for certain daytime periods before mid-1971, and were attributed to local automobile traffic, which was absent at night. A diesel generator on the site provided station power and a local source of CO_2 from 1958 to early 1967. During this period, air was selected according to wind direction from two out of four orthogonal lines located on the corners of the site to maintain an upwind sampling of air. The steadiness of the nighttime downslope wind would have made inadvertent contamination from the generator a rare occurrence. Another potential problem was the leakage of room air through air lines or leaking diaphragm pumps. These episodes were presumed to be identified by the observers and flagged as instrument malfunctions.

The identified or potential sources of within-hour "noise" variability in the CO_2 record, in approximate order of their importance were;

1. Changes in the background concentration. Smooth within-hour variations (typically a few tenths of a ppm) occasionally occurred in the background CO_2 concentration. These could be caused by the synoptic movement of airmasses

having differing CO_2 concentrations, or result from a change in the vertical circulation of free tropospheric air near the mountain at times when a large vertical gradient of tropospheric CO_2 was present. For this study it was necessary to optimize the ability to distinguish changes in background concentration from volcanic plume events. The use of an hourly standard deviation as a measure of variability [e.g. *Thoning et al.*, 1989] was not satisfactory for this purpose because it often failed to discriminate between the high-frequency variability characteristic of plume events and the low-frequency changes in background concentration. An alternative measure was developed that more effectively separated out higher-frequency variability; the variability index (VI), defined as the average absolute difference between successive 1-minute average concentrations during each measurement interval.

2. Instrument noise. The SIO analyzer output was subject to occasional periods of excessive noise and drift primarily due to ageing and deterioration of vacuum tubes in the power supply, amplifier, and thermal regulation circuits. Locations of the analyzer and room temperature control apparatus were changed several times during the program to reduce the thermal drift of the analyzer. The decision to flag suspect periods as either an instrument malfunction or due to natural variability was made by the observer based on his experience and daily monitoring of the analyzer. The more modern NOAA analyzers had solid-state electronics and were much more stable. Two measures of the stability of the NOAA analyzers were obtained from analysis of the 1-minute data. First, the hourly standard deviation was calculated for the last two minutes of the two five-minute reference gas runs (n=4 each hour). The URAS-2 analyzer had an average reference gas standard deviation of 0.015 ppm from 1976 to 1980 and 0.03 ppm from 1981-1987. The Ultramat-3 analyzer had an average reference gas standard deviation of 0.009 ppm. Next, the drift in instrument output voltage between successive hourly calibration runs was calculated. The voltage drift in 30 minutes multiplied by the instrument scale factor ranged between 0.04 ppm and 0.1 ppm for the URAS-2 analyzer and between 0.01 ppm and 0.02 ppm for the Ultramat-3 analyzer.

3. Line voltage and frequency fluctuations. These caused a corresponding shift in the analyzer output that could appear as an abrupt or gradual drift, or a high frequency noise in the SIO data. Most events were presumed to be recognized by the observer and flagged as an instrument malfunction.

4. Radio frequency noise. In the 1960's, radio transmitters at the observatory site occasionally produced a high frequency noise on the CO_2 trace. These events were presumed to be flagged as an instrument malfunction by the observer and were so infrequent as to be of minor consequence.

5. Physical vibration of the analyzer. Vibration of either analyzer produced a brief "spike" in the instrument output.

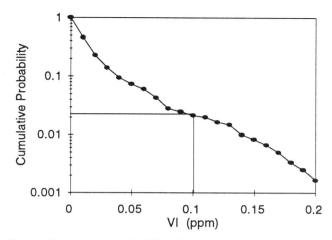

Fig. 3a. The cumulative probability of occurrence of hourly variability index values above a given threshold for presumably uncontaminated background air. Air was assumed to be uncontaminated between 1990 and 1992 when winds blew from 235 to 245 degrees. The sample size was 639 hours.

This was quite uncommon and is assumed to be of little consequence to this study.

3.3 Calculation of ΔCO_2

ΔCO_2 was defined as the difference between the measured concentration and the estimated background concentration during periods of high CO_2 variability. ΔCO_2 was calculated as follows:

1. Only nighttime data between 2000 LST and 0759 LST (Local Standard Time) were analyzed. This was when the plume was most likely to be transported down the slope beneath the temperature inversion.

2. Every hour the average CO_2 concentration measured from each of two intake lines, L1 and L2, was calculated. Each line was sampled for either 20 minutes (SIO data) or 25 minutes (NOAA data). The first three minutes of every sample was rejected to allow for a complete flushing of the lines. This produced two independent "measurements" per hour.

3. Each measurement was flagged as either variable or background.

   a. For the SIO hand-scaled data, the variability flag in the original data set was used.

   b. For the NOAA computerized data, an objective algorithm was used. Measurements in which the variability index (VI) exceeded a threshold value were flagged.

4. For each measurement flagged as variable, a corresponding background CO_2 concentration was estimated by linear interpolation between the nearest non-flagged measurement before and after that hour.

5. ΔCO_2 was calculated as the difference between the average CO_2 concentration and the estimated background.

As stated earlier, the NOAA data were flagged using a "variability index" (VI), which was the average absolute difference between successive 1-minute CO_2 concentrations during a measurement interval. The choice of an optimum threshold value of VI involved a subjective compromise between including relatively small-amplitude, presumably volcanic events, and excluding relatively large amplitude noise as described earlier. In the SIO data, this decision had already been made by the observer in assigning the variability flags.

For the NOAA data, the distribution of hourly VI for presumed background air is shown in Figure 3a. Background air was presumed to be present when winds came from a 10° sector centered on 240° between 1990 and 1992. It will be shown later that volcanic CO_2 was at a minimum in this sector, and that the level of CO_2 outgassing was at a minimum during this period. Figure 3a shows that a VI threshold of 0.1 ppm excluded about 98% of the presumed background data.

The effect of varying the VI threshold on the distribution of ΔCO_2 for the entire NOAA record is shown in Figure 3b. The distributions were made up of the two components identified earlier; one arising from sources having only positive ΔCO_2, and the other arising from "noise" which had ΔCO_2 of both signs and a net sum of zero. Reducing the VI threshold from 0.5 ppm to 0.05 ppm greatly increased the number of small-magnitude ΔCO_2 events detected from both components. At a VI threshold of 0.5 ppm, there was essentially no detectable noise component, but also a greatly reduced population of less than four ppm ΔCO_2 events.

The areas under the distributions shown in Figure 3b are plotted in Figure 3c as a function of VI. Since the noise component of the distribution had a net sum of zero, the area under the distribution gave the sum of the source component. Figure 3c suggests that a VI of 0.1 ppm detected 95% of the total source component, whereas a VI of 0.5 ppm detected only

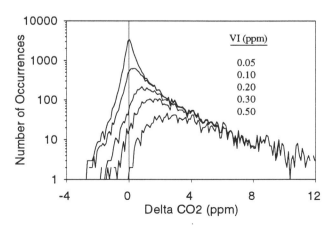

Fig. 3b. The effect of varying the variability index threshold on the distribution of hourly delta CO_2. The five labeled VI thresholds correspond vertically to the five distributions.

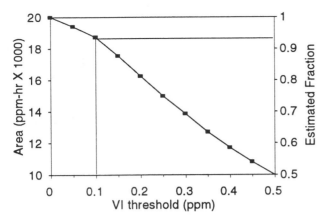

Fig. 3c. The areas under the distributions shown in Figure 3b. The "estimated fraction" gives the fraction of the total source component of CO_2 that would be detected at the given VI threshold, assuming an extrapolated total component of 2×10^4 ppm-hr.

50% of the total source component. Based on these considerations, the VI threshold was chosen to be 0.1 ppm.

The next step in calculating ΔCO_2 was to apply a correction to those periods in the NOAA 1-minute data when the analyzer signal went off-scale, as illustrated in the top trace of Fig. 2. The NOAA analyzers had a range of about 50 ppm, and a manual offset adjustment was periodically made to keep the output voltage approximately centered in this range during background CO_2 conditions. A strong volcanic plume having excess CO_2 greater than 25 ppm above background caused the analyzer output to saturate at a constant maximum voltage. This occurred during 72 measurements in 1984, 23 in 1985, 17 in 1986, 5 in 1987, and one each in 1976 through 1981. The curve in figure 4 was used to extrapolate for ΔCO_2 at these times. This curve was obtained from over 3000 within-range measurements in 1985 when ΔCO_2 was greater than zero. It gives the average fraction of the total ΔCO_2 represented by a cumulative number of ascending-ordered 1-minute values. The extrapolation was applied according to the following example. Suppose that only 15 minutes of a measurement interval had within-range ΔCO_2 and that the average ΔCO_2 of these was 10 ppm. From Figure 4, this represents 0.5 of the extrapolated 22-minute average ΔCO_2, which would therefore be 20 ppm. Of the 121 measurements in which one or more minutes were over-range, nine had an extrapolated average ΔCO_2 of 100 ppm or greater. The maximum extrapolated 22-minute average ΔCO_2 was 690 ppm.

3.4 Vertical ΔCO_2 Gradient

The height above the ground from which CO_2 was sampled varied between 7 meters and 40 meters [*Keeling et al.*, 1982; *Komhyr et al.*, 1989]. Both the SIO and NOAA programs

sampled air from two separate lines each hour. The average hourly ΔCO_2 ratio between these lines was calculated (Table 1). When the lines were at the same height, this ratio was within a few percent of 1.00. Ratios from lines at different heights were used to derive the average vertical ΔCO_2 profile (Figure 5). This shows that the volcanic plume near MLO was trapped beneath the temperature inversion near the ground [e.g. *Hahn et al.*, 1992; *Lee et al.*, 1993]. The evolution of this phenomena throughout the night is shown in Figure 6. From 2000 LST to about 0100 LST the concentration of the volcanic plume measured at MLO gradually increased. This was likely caused by the strengthening of the surface temperature inversion and downslope wind as the sun-warmed lava slope underwent radiative cooling throughout the evening. Meteorological conditions stabilized after 0100 LST, and the average concentration of the plume was steady until the breakup of the temperature inversion after sunrise.

3.5 Long-term ΔCO_2 Record

The long-term ΔCO_2 record is shown in Figure 7. It was derived from SIO data between 1958 and 1975 and NOAA data from 1976 to 1994. Each ΔCO_2 measurement (two per hour from separate intake lines) was normalized to a standard sampling height of 23 meters using the ratios given in Table 1 and was used to calculate monthly averages. Hours in which the plume was absent (ΔCO_2 = zero) were included in the averages.

The period between August 1968 and April 1971 had

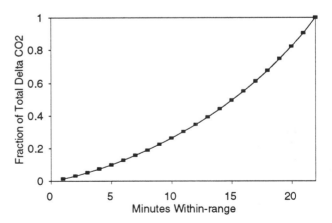

Fig. 4. For each measurement interval, the 1-minute-averaged delta CO_2 values were arranged in 22 ascending order bins by concentration. The average concentration of each bin was calculated for over 3000 within-range measurement intervals in 1985. This figure shows the cumulative fraction of the total delta CO_2 represented by a cumulative number of ascending order bins. This relationship was used to estimate delta CO_2 for those measurement intervals in which one or more minutes had CO_2 concentrations above the range of the analyzer.

anomalously high nighttime ΔCO_2 concentrations. *Keeling et al.* [1982] report that the sampling lines were found broken near the ground several times during this period, possibly contributing to a dramatic increase in daytime CO_2 "peaks." In the present study, it was assumed that air was sampled near ground level throughout this period, and a 23 meter normalization factor of 3.0 (extrapolated from Figure 5) was applied.

Figure 7 shows a strong association between ΔCO_2 and the volcanic activity of Mauna Loa volcano. Eruptions of Mauna Loa occurred in 1950, 1975, and 1984. ΔCO_2 increased abruptly shortly after the 1975 and 1984 eruptions and decreased systematically after that.

The month-to-month variability in ΔCO_2 was primarily caused by variations in the frequency and efficiency of plume transport to MLO, as suggested by Figure 8. An annual cycle was seen in the monthly frequency of volcanic plume episodes at MLO, with the minimum occurring in winter/spring. Strong free tropospheric winds occurred more frequently during these seasons [e.g. *Harris and Kahl,* 1990] which tended to prevent the plume from reaching MLO, as discussed in section 2.2.

The distribution of ΔCO_2 with wind direction is shown in Figure 9. The distribution peaked in the 180° to 190° direction.

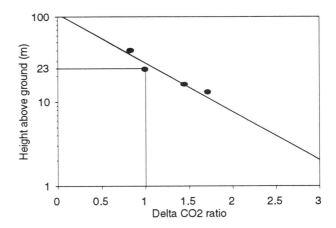

Fig. 5. The vertical profile of delta CO_2 based on the results in Table 1, and normalized to a standard sampling height of 23 meters. The fit is a logarithmic regression.

This is the bearing to North Pit, identified in section 2.1 as the most likely location for volcanic plume trapped in the surface inversion layer to emerge from the caldera at night. The distribution peak had a full width at half-maximum of about 40°. Nighttime wind directions east of 110° or west of 260° occurred less than 20 hours per year.

4. AEROSOL MEASUREMENTS

4.1 Measurement Methodology

Continuous measurements of atmospheric aerosol particles were begun in January, 1974 when a Meteorology Research Inc. four-wavelength nephelometer was installed to monitor the aerosol light scattering extinction coefficient, σ_{sp}, at wavelengths of 450, 550, 700, and 850 nm [*Bodhaine*, 1978]. The

TABLE 1. ΔCO_2 Ratios at Various Sample Heights

	L1 (m)	L2 (m)	Period	Months	ΔCO_2 L1/L2
SIO	7	7	05/58 - 02/71	143	0.97
	23	23	04/71 - 09/73	16	1.09
	23	16	10/73 - 12/86	140	0.69
NOAA	13	13	06/77 - 01/80	31	0.98
	23	13	02/80 - 11/84	55	0.58
	23	23	12/84 - 04/88	38	0.97
	40	23	05/88 - 10/93	65	0.87

Ratios Normalized to 23m

Height	ΔCO_2 Ratio
40	0.87
23	1.00
16	1.45
13	1.72
7	2.0 (est)
1	3.0 (est)

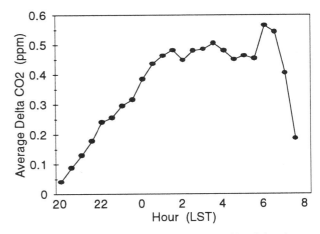

Fig. 6. The average delta CO_2 as a function of local time between 1977 and 1980. All sampling was done from a height of 13 meters during this period.

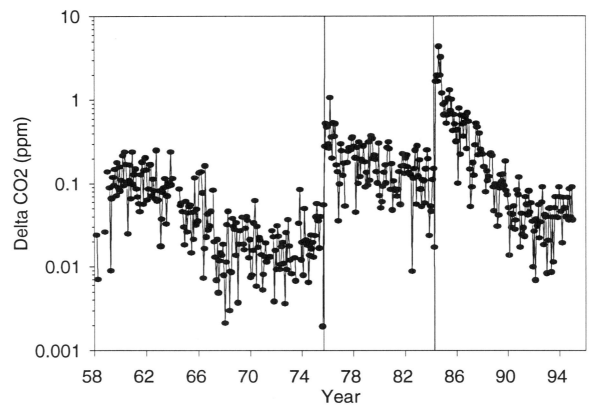

Fig. 7. Monthly average delta CO_2 between 2000 LST and 0759 LST, normalized to a standard sampling height of 23 meters. The 1975 and 1984 eruptions are denoted by vertical lines. Data before 1976 were derived from hand-scaled data obtained by the Scripps Institute of Oceanography and the rest were derived from computer digitized NOAA data.

molecular component of light scattering is typically one to one hundred times greater than the background aerosol component of light scattering. The aerosol component is isolated by real-time subtraction of the signal measured from a filtered air sample, using a 45 minute averaging time constant to obtain a suitable signal to noise ratio. Details of the nephelometer program are given by *Bodhaine* [1983] and *Massey et al.* [1987]. The nephelometer measures particles in the size range of 0.1 to 1.0 µm. In sufficiently large numbers these appear as visible haze.

In May 1975, continuous measurements of condensation nuclei (CN) were begun using a General Electric CN counter. In 1991, this was replaced by an alcohol-based instrument manufactured by Thermo Systems, Inc. (TSI). Details of the CN program are given by *Bodhaine* [1983] and *Massey et al.* [1987]. In the General Electric CN counter, humidified air undergoes a rapid adiabatic expansion, which creates a supersaturation of water vapor. The water vapor condenses around aerosol nuclei, forming a cloud that diminishes the amount of light reaching a photodetector. The attenuated signal minus a dark background signal is used to calculate the number

density of CN. In the modern TSI instrument, alcohol droplets are formed around each CN and are counted individually as they interrupt a laser beam. The condensation nuclei counter responds to particles in the 0.002 to 0.1 µm range. Oxidation of SO_2 forms sulfate aerosols in this size range. There have been no previous studies of Mauna Loa volcanic aerosols. The Mauna Loa volcanic aerosol measured at MLO is young (typically 0.5 hours) and is transported in dry air (relative humidity typically less than 20%).

All aerosol sampling was done from a height of 13 meters. Data were recorded by a computer and on a chart record. The chart record data were visually checked to remove periods of instrument malfunction and local contamination from the final digital data record.

4.2 Nearby Aerosol Sources and Variability

Variations in background aerosols were large compared to variations in background CO_2. The annual cycle in CN varied by 50% from monthly averages of 218 cm^{-3} to 326 cm^{-3}. The 550 nm σ_{sp} varied annually by a factor of five, from an average

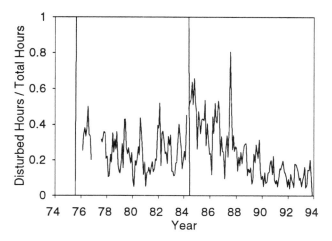

Fig. 8. Monthly fraction of hours between 2000 LST and 0759 LST in which the variability index was greater than 0.1.

November low of 3.1×10^{-7} m^{-1} to an April high of 1.7×10^{-6} m^{-1} [*Massey et al.*, 1987].

Sources of aerosols on Hawaii were identified by *Pueschel and Mendonca* [1972]. These sources can be divided into two categories: those at lower altitudes (in the marine boundary layer), and those at or above MLO.

Aerosols from sources in the marine boundary layer were frequently transported to MLO by the daytime upslope winds, but were mostly absent in the nighttime downslope wind. These included Kilauea volcano (the largest source of condensation nuclei, frequently accompanied by visible haze), combustion from forest fires and sugar cane fires, combustion from anthropogenic activity in coastal towns, and sea salt aerosols. *Pueschel and Mendonca* [1973] combined visual observations, aerosol measurements at MLO, and thermal energy calculations to show that Kilauea aerosols could penetrate the trade wind inversion (at an altitude of 1700 m) during an episode of active fountaining, but not during a subsequent period of flowing surface lava. This implied that direct injection of Kilauea aerosols above the inversion into the free troposphere was only possible on those rare occasions of active fountaining.

Aerosols from the vicinity of MLO came primarily from infrequent automobile traffic, which occurred almost entirely during the day. The only identified nearby source of aerosols from altitudes above MLO was Mauna Loa volcano itself [*Bodhaine et al.*, 1980].

4.3 ΔCN and $\Delta \sigma_{sp}$ Records

Most of the aerosol data were recorded by computer as 10 minute averages and processed as hourly averages. It was therefore impossible to identify aerosol plume events on the basis of minute-scale variability in aerosol data. Because of this, variability in CO_2 (as described earlier) was used to

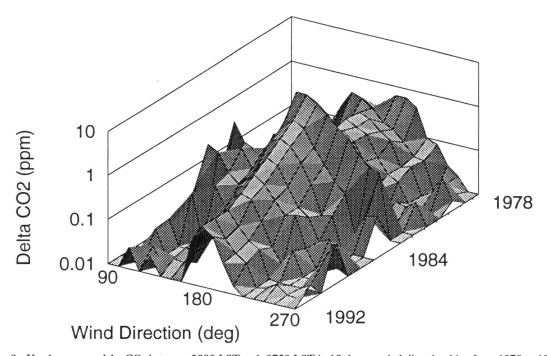

Fig. 9. Yearly average delta CO_2 between 2000 LST and 0759 LST in 10 degree wind direction bins from 1978 to 1992. Delta CO_2 was normalized to a standard sampling height of 23 meters. Winds outside of 110 to 260 degrees blew too infrequently to yield statistically significant delta CO_2 averages.

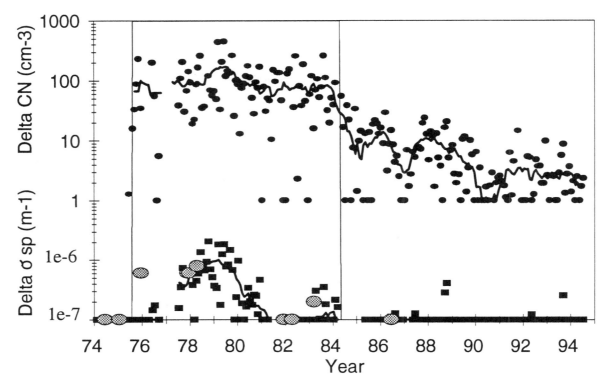

Fig. 10. Monthly average delta aerosols between 0000 LST and 0759 LST measured from a sampling height of 13 meters. Vertical lines denote the 1975 and 1984 eruptions. The fits are 12 month running means. Delta CN less than 1 cm^{-3} is plotted as 1 cm^{-3}. Delta 550nm light scattering less than 1×10^{-7} m^{-1} is plotted as 1×10^{-7} m^{-1}. The solid circles with the light scattering data indicate the visual thickness of fume at the caldera vents estimated from aerial photographs on a scale of zero to five and then linearly scaled between 1×10^{-7} m^{-1} and 1×10^{-6} m^{-1}.

identify the presence of volcanic plume or the presence of background conditions. Calculations of ΔCN and $\Delta \sigma_{sp}$ were then made in the same way as for ΔCO_2.

Bodhaine [1978] noted that the best time for sampling background aerosols was between 0100-0700 LST. On average, the downslope wind pattern did not become fully developed until after midnight (as suggested by Figure 6). Unlike CO_2, the background aerosol concentration at MLO typically had a factor of 10 diurnal variation, which was due to the upslope transport of aerosol-rich marine boundary layer air in the afternoon. Evening hours between 2000 LST and 0000 LST commonly had steady background levels of CO_2 while the corresponding aerosol concentrations were still decreasing from high afternoon levels [*Clarke and Bodhaine*, 1993]. This was more evident in the nephelometer data (which had a 45 minute averaging time constant) than the CN data (which had a 0.2 second response time). When contaminated evening hours were misidentified as having background aerosol levels, the interpolated background later in the night was over-estimated, frequently resulting in negative ΔCN and $\Delta \sigma_{sp}$ values. Inadvertent background contamination was minimized

by restricting delta aerosol calculations to the stable period between 0000 LST and 0759 LST.

The complete ΔCN and $\Delta \sigma_{sp}$(550 nm) record is shown in Figure 10. Volcanic aerosols behaved differently from volcanic CO_2. The post-1975 quiescent period had higher levels of ΔCN and $\Delta \sigma_{sp}$ than the post-1984 period. Since all aerosol data were measured from the same height above the ground, the vertical distribution of the aerosol plume could not be derived.

To find out if $\Delta \sigma_{sp}$ was a measure of the intensity of visible fume from the volcano, a comparison was made with a series of aerial photographs taken of the summit caldera area (J. Lockwood, pers. com., 1993). In each photograph the relative size and opacity of the visible plumes, which emanated from location ML3 (Figure 1), were estimated on a scale of zero to five, with five being the most intense. The estimates were not corrected for the effects of wind speed and relative humidity on the opacity of the visible fume. These estimates were compared with the $\Delta \sigma_{sp}$ data in Figure 10 (where a value of zero was scaled to 1×10^{-7} m^{-1} and a value of five was scaled to 1×10^{-6} m^{-1}). The subjective photographic data gave supporting evidence that $\Delta \sigma_{sp}$ was a measure of the visible fume from

Mauna Loa. It showed the high levels of 1978, the cessation of visible fuming in 1981-82, and a return to low levels of visible fuming in 1983. The two records only disagreed once out of ten times, in late 1976, when the photographic evidence suggested a greater degree of fuming than the light scattering data.

At those times when measurable fume was present, $\Delta \sigma_{sp}$ measured by the four channels was systematically greater at shorter wavelengths. This showed that the peak of the aerosol size distribution occurred at a particle size smaller than about 0.3 µm.

5 TRACE SPECIES MEASUREMENTS

5.1 Measurement Methodology

The following species were analyzed using the method outlined in the preceding sections: H_2O, CO, H_2, SO_2, O_3, CH_4, $Radon_{222}$, and aerosol Black Carbon.

Water vapor was measured from a height of 2 m using a dew cell from 1974 to 1981, and a dew point hygrometer after 1981. Temperature was measured by a thermograph before 1975, and after that by an aspirated, shielded thermocouple at a height of 2 m [*Herbert et al.*, 1987]. Hourly water vapor mixing ratios were calculated from measured dew point, temperature, and pressure observations.

Carbon monoxide and hydrogen were sampled from a height of 40 m using a Trace Analytical Reduction Gas Analyzer gas chromatograph [*Novelli et al.*, 1991]. The instrument precision for CO was approximately 0.5 ppb [*Ferguson and Rosson*, 1991; *Novelli et al.*, 1991]. The data used here were unedited preliminary results from 1992 and 1993 [P. Novelli, pers. com., 1994]. The chromatograms have a hydrogen peak that was not analyzed as part of the climatological monitoring program (the concentration of hydrogen in the reference tank was not measured). Hourly hydrogen concentrations provided for the present study were based on an arbitrary assignment of 100 ppb to the reference tank H_2 concentrations, and may have been systematically low by a factor of five.

Sulfur dioxide was measured by a Thermo Environmental Instruments (TEI) model 43S pulsed florescence analyzer. From December 1988 to November 1989 an instrument was operated by the NOAA Air Resources Laboratory from a sampling height of 13m, with a 1-hour detection limit of 41 ppt [*Luria et al.*, 1992]. From September 1991 to August 1992, an identical instrument was operated intermittently as part of the MLOPEX-II experiment from a sampling height of 7 m [*Hubler*, 1993, Hubler pers. com., 1993]. A program designed specifically to detect SO_2 in the Mauna Loa plume was started in June 1994 and continues to the present. It also uses a TEI model 43S analyzer, sampling alternatively from heights of 4 m and 34 m. Zero-SO_2 measurements are made twice per hour

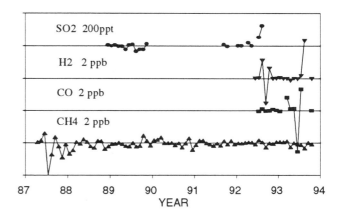

Fig. 11a. Monthly average delta values of four trace gas species between 0000 LST and 0759 LST from 1987 and 1993. The scale interval for the upper trace (SO_2) is 200 ppt. The scale interval for the lower three traces is 2 ppb.

for 10 minutes each. A 10 ppm reference gas is injected into the high-volume sampling line to obtain a 1.2 ppb calibration twice daily with a 120 ppt to 5 ppb six-point calibration made every 10 days. The 95% confidence detection limit for a 20 minute measurement is 30 ppt.

Radon (Rn_{222}) was measured from a height of 40 m by an instrument built by the DOE Environmental Research Labs [*Thomas and LeClare*, 1970; *Negro*, 1979]. The half-hour detection limit was 70 mBq m^{-3} in 1991 and 1992, and was 30 mBq m^{-3} in 1993. Radon measurements at MLO are discussed by *Whittlestone et al.* [1992].

Ozone was measured from a height of 13 m by an electrochemical concentration cell [*Komhyr*, 1969] from 1974 to 1976, and by a Dasibi ultraviolet photometer from 1976 to 1993 [*Oltmans*, 1981; *Oltmans and Komhyr*, 1986].

Methane was measured from a height of 23 m between 1987 and 1991 and from 40 m after that using a Carle Series 400 gas chromatograph with flame ionization detection [*Ferguson and Rosson*, 1992; *Masarie et al.*, 1991].

Aerosol black carbon was measured from a height of 13 m starting in 1990 using an aethaelometer [*Hansen et al.*, 1984; *Gundel et al.*, 1984].

5.2 Delta Analysis for Trace Species

Hourly delta values were calculated for the eight trace species listed using the method described in section 4.3 for aerosols. Calculations were restricted to the period between 0000 LST and 0759 LST to reduce the possibility of inadvertent contamination from residual marine boundary layer air which occasionally persisted into the late evening hours. Results are shown in Figure 11 and Table 2. The long-term average monthly delta values for seven of the species

TABLE 2.   Monthly Average Delta Trace Species

Species	Period	Months	Avg. Δ Species		2 σ	
H_2O	1974-1993	196	-7	ppm	32	ppm
CO	1992-1993	14	4	ppt	1.6	ppb
H_2	1992-1993	16	0.2	ppb	1.5	ppb
SO_2	1988-1992	20	6	ppt	64	ppt
SO_2	1994-1995	8	2.8	ppt	4.4	ppt
O_3	1974-1993	204	-0.3	ppt	215	ppt
CH_4	1987-1993	79	70	ppt	640	ppt
Carbon	1990-1993	33	-0.3	ng m^{-3}	3.5	ng m^{-3}
Radon	1991-1993	27	1.8	mBq m^{-3}	3.5	mBq m^{-3}

(excluding radon) were not significantly different from zero. Three species with long data records, H_2O, O_3, and CH_4, showed no trends or systematic changes associated with the eruptive cycle of Mauna Loa.

The detection limit at the 95% confidence level for each species was taken as two standard deviations about the mean, and is given in Table 2. This represented an upper limit to the volcanic contamination potentially present in an unedited, monthly averaged climatological baseline data set.

Delta radon averaged 1.8 mBq m^{-3} (with a monthly standard deviation of 1.7 mBq m^{-3}) between 1991 and 1993. Since radon is known to emanate from Mauna Loa lavas [*Wilkening*, 1974], a rough calculation was made to determine if the upper slopes of Mauna Loa could have been the source of this excess radon rather than the volcanic plume. *Wilkening* [1974] reported an average radon flux of 0.012 atoms cm^{-2} sec^{-1} for Mauna Loa and Cape Kumukahi lavas. If all the radon emanating from the slope above MLO was trapped in the nighttime inversion layer and uniformly mixed to a height of 50 m at an average downslope wind speed of 3.4 m s^{-1}, the resultant radon activity at MLO (6 km from the summit) would be 30 mBq m^{-3}. This is significantly greater than 1.8 mBq m^{-3}, the average delta radon activity. Conditions that favored the transport of volcanic plume (i.e. light winds and a strong surface temperature inversion) would have also created the greatest atmospheric concentrations of ground-emanated radon at MLO [*Whittlestone et al.*, 1993]. This effect could easily account for the small positive delta radon observed. It is therefore concluded that no volcanic radon was present in the plume at a detection limit of 3.5 mBq m^{-3}.

6.   INTERPRETATION OF RESULTS

6.1   Short-term Variations in ΔCO_2

Variations in ΔCO_2 on short time-scales (hours-to-weeks) before and after the 1975 and 1984 eruptions were examined. The amount of CO_2 reaching MLO depended on two factors; the volcanic emissions rate and the airflow pattern between the point(s) of emission and the observatory. To reduce the effects of airflow variations, hourly data were selected in which the observatory wind direction was within a 45° sector centered on 180° and the wind speed was between 2 and 5 m s^{-1}. These conditions were most favorable for plume transport to MLO (Figure 9). These data are shown for 2-year periods centered on the 1975 and 1984 eruptions (Figure 12). Data was missing for 35 days following the start of the 1984 eruption because the lava flow cut the power to MLO.

No significant increase in hourly $\Delta CO2$ averages occurred in the twelve months preceding either the 1975 or 1984 eruption. The probability that a random, short-duration outgassing event would have been detected at MLO depends upon the frequency of wind conditions favorable for plume transport. Detection probabilities were calculated for a 1-week period prior to each eruption under the assumption that a random event would be detected only if it occurred between 2000 LST and 0659 LST when the hourly average wind direction was within a 45-degree sector centered on 180°. No events would have been detected for 24 hours before the 1975 eruption, which started at 2342 LST on July 5 [*Lockwood et al.*, 1987]. In the seven days preceding this eruption, the probabilities of detecting events with durations of 1 hour, 10 hours, and 24 hours were 21%, 63%, and 89% respectively. No events would have been detected for 90 hours prior to the 1984 eruption, which began at 0125 LST on March 25 [*Lockwood et al.*,

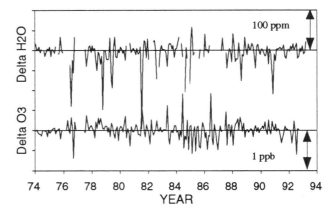

Fig. 11b. Monthly average delta values for water vapor (top, scale interval of 100 ppm) and ozone (bottom, scale interval of 1 ppb) between 0000 LST and 0759 LST from 1974 to 1993.

1987]. The probabilities of detecting events with durations of 1 hour, 10 hours, and 24 hours in the week before this eruption were 15%, 32%, and 46% respectively.

After the 1975 eruption, there was a period of 65 days in which every wind-selected hour had ΔCO_2 = zero. On day 65, there was an hour in which ΔCO_2 was 1.7 times greater than any hourly value that occurred in the year before the eruption. This shows that enhanced outgassing was delayed by about 65 days following the end of the 1975 eruption. The eruption ended with magma venting at an elevation of 3700 m [*Lockwood et al.*, 1987]. If the primary source of CO_2 was a recently recharged magma reservoir at 3 km depth [*Decker et al.*, 1983] (equivalent to an elevation of 1000 meters above sea level), it would follow that the newly exsolved bubbles would have to rise through a 2700 m column to reach the surface. A bubble rising 2700 m in 65 days would have an average ascent rate of 1.7 m hr^{-1}.

Following the end of the 1984 eruption, MLO was without power for 14 days. For 6 days after this, every wind selected hour had ΔCO_2 = zero. Then on May 6, there were several hours with an elevated ΔCO_2, the highest being 1.7 times greater than any hourly value that occurred in the year preceding the 1984 eruption. This shows that enhanced outgassing was first observed 21 days after the end of the 1984 eruption, although the power outage caused a data black-out for the first 14 days. During this time, atmospheric air samples were collected almost daily in glass flasks for later analysis. Most were exposed during periods when the winds brought clean, baseline air to MLO. Fortunately, four flask samples were collected while MLO was in "heavy fumes" (according to the observers logbook) on the morning of April 24. These flasks had CO_2 concentrations averaging 4.7 ppm above a baseline concentration estimated from clean air samples taken on April 20 and April 25. This is 1.2 times greater than the maximum hourly ΔCO_2 measured in the year preceding the eruption. It suggests that enhanced outgassing of CO_2 was present 9 days after the end of the 1984 eruption. Direct measurements of vented gas by *Greenland* [1987] show that CO_2 was becoming enriched relative to SO_2 on April 18, three days after the end of the eruption.

6.2 Pre-Eruption Trends in ΔCO_2

Variations in ΔCO_2 on time-scales of months-to-years were examined before the 1975 and 1984 eruptions. *Gerlach* [1986] suggested that, for Kilauea volcano, monitoring summit emissions might show variations in the rate of supply of parental magma to the summit magma chamber and provide a tool for eruption forecasting. This idea was tested for Mauna Loa using the CO_2 outgassing record.

Examination of Figure 7 shows that the exponential decrease of ΔCO_2 that occurred throughout the 1960's leveled off

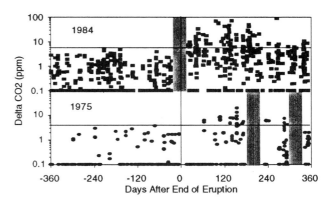

Fig. 12. Hourly delta CO_2 between 0000 LST and 0659 LST selected for winds between 158 and 202 degrees with wind speeds between 2 and 5 m/s. Hours with delta CO_2 less than 0.1 ppm (primarily including delta CO_2 = 0 ppm) are plotted as 0.1 ppm. Periods with missing CO_2 or wind data are left blank. The horizontal lines give the maximum hourly delta CO_2 which occurred during the 360 days before the start of each eruption.

sometime after 1970. There may have been a slight increase of about 0.015 ppm in the trend of ΔCO_2 beginning two to three years before the 1975 eruption. As mentioned in section 3, the data taken before 1976 were recorded on a first-generation analyzer, were hand scaled, and were subjectively selected for variability. These and other factors may have contributed to drifts in ΔCO_2 that were not related to changes in volcanic emissions, so caution must be exercised in drawing conclusions from the pre-1976 data. There was no apparent increase before the 1984 eruption, but ΔCO_2 at this time was about a factor of 10 greater than in 1972-73. An increase of 0.015 ppm would represent a 10% change in pre-1984 levels and may not have been detectable.

An increasing trend in ΔCO_2 began in early 1993 and continued up through the most recent data available for this paper, January 1995 (Figure 7). Based on a 1-year running mean, ΔCO_2 increased by almost 0.02 ppm, from 0.034 ppm to 0.053 ppm. The distribution of ΔCO_2 with wind direction (Figure 13) changed dramatically between 1992 and 1993-1994. The height of the peak in the distribution near 180° decreased by a factor of two while there was a large increase in ΔCO_2 from both the southeast and southwest directions, being greatest at 230°. The broadening of the ΔCO_2 distribution observed in 1994 was unprecedented in the 37-year record. Annual average ratios of ΔCO_2 were calculated between the 135°±22.5° (southeast) and 180°±22.5° (south) sectors, and between the 225°±22.5° (southwest) and 180°±22.5° (south) sectors. A flat distribution would have a ratio near 1.0 and a distribution sharply peaked near 180° would have a ratio approaching 0.0. In every year from 1958 to 1992, the calculated ratios were all less than 0.35, with an average of

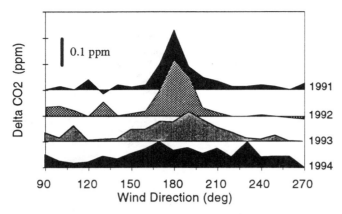

Fig. 13. Average delta CO_2 as a function of wind direction in 10 degree bins for 1991-1994. The distributions are each displaced by 0.1 ppm for clarity.

0.15. In 1994, the southeast to south ratio was 0.7 and the southwest to south ratio was 0.9.

The changes in ΔCO_2 observed in 1993-94 were not due to changes in wind patterns or the performance of the CO_2 analyzer. There were no significant changes in the annual MLO wind direction frequency distributions, the average MLO wind speeds partitioned by wind direction, or the noise level of the analyzer reference gas signal between 1991 and 1994. Increases in ΔCO_2 from the southeast and southwest could have potentially come from a source other than Mauna Loa volcano, but this is considered unlikely as described in section 3.2. Kilauea volcano emissions reaching MLO, which typically have SO_2 to CO_2 ratios greater than 0.1, would have been accompanied by a large increase in SO_2, which was not observed.

There have been no detailed studies of the airflow patterns around the summit of Mauna Loa that would allow the location of a new source(s) to be identified based on the ΔCO_2 distribution. The most likely location(s) for a volcanic source outside the caldera are one or both of the rifts [*Casadevall and Hazlett*, 1983]. The simplest interpretation is that the northeast rift was the source of increased ΔCO_2 from the southeast and the southwest rift was the source of ΔCO_2 from the southwest. From a historical perspective, this is perhaps unlikely, since one or the other, but not both, rifts tend to be active at one time [*Lockwood and Lipman*, 1987]. A second possibility is that there is a single source on the southwest rift. When the free-tropospheric winds blow from the southwest, MLO is on the leeward side of the mountain. Under these conditions, a plume originating on the southwest rift would travel equal distances around the mountain to arrive at MLO from either the southwest or the southeast (Figure 1). The northeast rift is unlikely to be the only source since the plume would have to travel clockwise about 330 degrees to arrive at MLO from the southwest.

In summary, it appears that the outgassing behavior of Mauna Loa has undergone an unprecedented transition during the last two years. The CO_2 emissions coming from the summit have continued to decline, while CO_2 emissions coming from a source or sources located high on the southeast rift (or possibly both rifts) have apparently increased. This has resulted in a net increase in ΔCO_2 measured at MLO. Although the size of this increase has thus far been small (0.02 ppm), it is similar in size to an increase that may have preceded the 1975 eruption. This activity could be an early precursor to the next eruption. The magma responsible for the increase in CO_2 must be at a depth great enough not to cause increases in either SO_2 or sulfate (CN), which have not been observed.

6.3 Mass Estimate of CO_2 Emissions

The annual CO_2 mass emission rate of Mauna Loa volcano was estimated based on the observatory measurements of ΔCO_2 as a function of wind direction (Figure 9) and height above the ground (Figure 5) shown earlier. The following assumptions were made:

1. The plume measured at MLO was fully trapped in the surface temperature inversion between the hours of 0000 LST and 0759 LST (Figure 6).

2. Variations in ΔCO_2 caused by changing meteorological transport conditions could be eliminated by taking yearly averages.

3. At night, all of the CO_2 in the plume was trapped in the surface temperature inversion and transported down the slope. This assumption had no supporting evidence. The degree to which the plume was trapped in the inversion layer has never been measured. To the extent that part of the plume may have escaped directly into the free troposphere, the CO_2 emissions estimate based on this assumption would represent a lower limit.

4. The average normalized vertical profile of the plume CO_2 6 km downslope from the summit was given by Figure 5. The integrated area under this curve is equivalent to a 79 meter column having a uniform ratio of 1.0.

5. The distribution of ΔCO_2 with wind direction was observed to have a full width at half-maximum of 40 degrees (section 3.5). This was assumed to be the azimuthal extent ("width") of the plume 6 km downslope from the summit.

6. The average speed of the plume 6 km downslope from the summit was 3.4 m/s. This was the climatological average of the nighttime downslope component of wind velocity measured at a height of 8.5 meters.

CO_2 mass emissions were calculated as follows. The plume was contained in a three-dimensional pie-shaped segment originating at the summit with a radius of 6 km, an angle of 40 degrees (from assumption 5), and a scale height of 79 meters (from assumption 4). The downslope vertical face of the

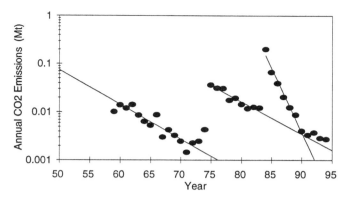

Fig. 14. Estimated annual output of CO_2 from the summit area of Mauna Loa volcano using the transport model described in the text. Yearly averages are taken for calendar years except for the years occurring on either side of an eruption. These are the 365 day intervals before and after the date of the start of the eruption. The fits are logarithmic regressions to the 1960-1973, 1975-1983, and 1984-1989 points respectively. One Mt equals 10^9 kg.

segment had an area of 3.3 X 10^5 m^2. The volume of air moving at 3.4 m s^{-1} (from assumption 6) through this face was 1.1 X 10^6 m^3 s^{-1}. At the 680 mb average atmospheric pressure of MLO, one ppm of CO_2 is equivalent to 1.06 X 10^{-6} kg m^{-3} CO_2. The total mass of CO_2 emerging from the segment face, equivalent to the emission rate of the source, is therefore 1.2 kg s^{-1}ppm^{-1}. Over one year, an average MLO plume concentration of one ppm is equivalent to an emission of 3.7 X 10^7 kg CO_2. The plume concentration was taken as the annual average ΔCO_2 between 0000 LST and 0759 LST (assumption 1) when the wind was in a 45° sector centered on 180°.

Annual CO_2 emission estimates from the Mauna Loa summit between 1959 and 1994 are shown in Figure 14. Logarithmic regressions were calculated for the three post-eruptive periods. The fit to the 1960-1973 period was extended back in time to obtain an estimated emissions of 7.4 X 10^7 kg for 1950, the year of the previous eruption. The area under each fit, integrated from the year of the eruption to T = infinity, was

taken as an estimate of the total mass of CO_2 in each quiescent reservoir. These results are listed in Table 3 along with the eruptive volume of the preceding eruption, taken from *Lockwood and Lipman* [1987].

The emissions estimated in Table 3 can be compared with measurements of the CO_2 emission rate during the 1984 eruption [*Casadevall et al.*, 1984], which ranged between 2.4 X 10^5 kg day^{-1} and 1.4 X 10^6 kg day^{-1}. If the average emission rate was the mean of these extreme values, the total mass of CO_2 produced during the 21 days of the 1984 eruption would have been 1.7 X 10^7 kg. This represents about 5% of the average quiescent reservoir mass of CO_2 from Table 3, suggesting that much more CO_2 is degassed during quiescent periods than during eruptions.

The volume of magma required to supply a given reservoir quantity of CO_2 can be estimated using the CO_2 barometer of *Harris* [1981] and the model of *Gerlach* [1986]. The dissolved component of CO_2 as a function of pressure is taken to be 5.9 X 10^{-4} wt % MPa^{-1} and the magma density is assumed to be 2.6 X 10^3 kg m^{-3}. The 1984 eruption produced 2.2 X 10^8 m^3 of lava [*Lockwood et al.*, 1987], equivalent to 5.7 X 10^{11} kg of magma. The average quiescent reservoir mass of CO_2 (from Table 3) was 3.5 X 10^8 kg. A mass of magma equivalent to that erupted in 1984 would lose 0.061 wt % CO_2 in outgassing the mass of CO_2 lost during quiescence. This represents a magma decompression of 103 MPa, equivalent to an ascent of 4.1 km, which is comparable to the vertical scale size of the magma system beneath Mauna Loa based on seismic evidence [e.g. *Lockwood et al.*, 1987]. The mass estimates of quiescent CO_2 emissions reported here are therefore consistent with the view that bodies of magma degas in a shallow summit chamber before being erupted.

The quiescent reservoir mass of CO_2 was similar for all three periods, yet the volume of the 1975 eruption was much less than the volumes of the 1950 and 1984 eruptions (Table 3). This suggests that a large fraction of the 1975 magma did not erupt, consistent with seismic evidence of magma intrusion into the northeast rift during and following that eruption [*Lockwood et al.*, 1987].

TABLE 3.   Estimated CO_2 Emissions From Mauna Loa Summit

Period Fit	r^2	1/e (years)	$\sum$ Mass (10^8 kg CO_2)	Previous Eruption	Initial Rate (10^7 kg CO_2 yr^{-1})	Lava Volume (10^6 m^{-3})
1960-73	0.86	6.1	4.8 (est) ??	1950	7.4 (est) ??	376
1975-83	0.89	6.5	2.4	1975	3.5	30
1984-89	0.97	1.6	3.3	1984	15.0	220

TABLE 4. Ratio of Gas Species to CO_2 in the Plume

Period	H_2O	CO	H_2	SO_2
01/74 - 12/74	< 390			
01/78 - 12/78	< 350			
06/84 - 05/85	< 47			
12/88 - 11/89	< 270			< 1.0 X 10^{-4}
09/91 - 08/92	< 540			< 1.8 X 10^{-3}
01/92 - 12/93	< 470	< 3.1 X 10^{-2}	< 3.0 X 10^{-2}	
06/94 - 01/95				7 X 10^{-5}
4/18/84 vent sample [*Greenland*, 1987]	21.4	1 X 10^{-3}	1.6 X 10^{-2}	1.65

An exponential regression provided an excellent fit ($r^2 > 0.85$) to the CO_2 emissions data for all three quiescent periods (Figure 14). This characteristic is predicted by *Johnson* [this volume], who suggests that Mauna Loa's summit reservoir is rapidly resupplied with a large influx of fresh magma from a deeper source while an eruption is in progress. The new magma enters the reservoir from below but does not mix quickly enough to be part of the eruption. If the rate of magma resupply to the summit reservoir during the subsequent repose is low, this fresh body of magma would be the primary source of quiescent CO_2 emissions. It would degas as a single batch with a characteristic exponentially decaying rate.

Following both the 1975 and 1984 eruptions, there were periods of several months when ΔCO_2 was greater than the subsequent exponential decay rate would predict (Figure 7). This could mean that the period of rapid refilling of the summit reservoir proposed by *Johnson* [this volume] continued for several months beyond the end of the eruption.

Between 1984 and 1989, CO_2 emissions decreased at a rate that should have resulted in an estimated output of 3 X 10^5 kg CO_2 by 1994. The observed emissions in 1994 were almost 10 times greater than this, so an additional source must have been present. It is interesting that the fit to the 1975-1983 data comes close to fitting the 1990-1994 data points as well (Figure 14). This is unlikely to be a coincidence. It suggests that the excess emissions observed after 1990 came from the same source that was outgassing between 1975 and 1984, and that the 1984 eruption did not affect the exponentially decaying CO_2 emissions rate of this source. This implies that the post-1975 magma body remained physically separate from and was not disrupted by the emergence of the post-1984 magma body. Therefore, the post-1975 magma was not the source of the 1984 eruption, in agreement with the conclusions of *Lockwood et al.* [1987] and *Rhodes* [1988] based on lava chemistry evidence. After 1984, there were two independent magma bodies degassing CO_2 from the vicinity of the summit caldera.

6.4 Gas Ratios in the Plume

For gas species that do not react or fractionate during atmospheric transport in the plume, the ratio of the delta values as measured at MLO is equivalent to the emission ratio at the source. Changes in this ratio over time may be directly related to the volcanic processes that produce the gases.

Four of the trace gases examined in section 5 were measured in vent samples taken by *Greenland* [1987] four days after the end of the 1984 eruption. In Table 4, the mole percent ratios for four gases measured in a vent sample taken on April 18, 1984 are compared to the delta ratios obtained at MLO for various periods between 1974 and 1993. The detection limit ratios of ΔH_2O, ΔCO, and ΔH_2 to ΔCO_2 were all significantly greater than ratios of these gases measured in the post-eruption vent sample. Although the relative abundances of these gases could not be measured in the MLO data, the upper limits show that they did not increase greatly between 1984 and 1992.

The detection limited ΔSO_2 to ΔCO_2 ratio between 1991 and 1993 was over three orders of magnitude less than the post-eruptive vent sample ratio, suggesting that volcanic SO_2 should have been easily detected in the MLO measurements. *Gerlach* [1986] predicted a total S to CO_2 mole fraction exsolution ratio near 0.2 for reservoir-equilibrated magma ascending through a depth of one to three thousand meters (the presumed depth of the top of the Mauna Loa summit magma chamber from *Decker et al.*, [1983]). Airborne measurements of non-eruptive degassing at the Kilauea summit caldera gave SO_2 to CO_2 ratios of about 0.1 [*Greenland et al.*, 1985]. From this evidence, the quiescent Mauna Loa SO_2 to CO_2 ratio might be expected to be on the order of 0.1. This is over 100 times greater than the detection limit of the MLO measurements, yet essentially no SO_2 was present in the plume.

The loss of a large fraction of SO_2 during transport between the vent(s) and the observatory could potentially account for the discrepancy, so this was investigated. The two principal processes in atmospheric SO_2 removal expected for the nighttime summit environment are liquid phase oxidation, and dry deposition to the ground. *Moller* [1980] gave a typical SO_2 liquid-phase mean residence time in dry air (characteristic of the Mauna Loa summit environment) of 30 hours. Dry deposition velocities reported in the literature for SO_2 range from 0.04 to 7.5 cm s^{-1} [*Sehmel*, 1980]. The dry deposition velocity depends on the surface moisture content and surface roughness. *Lee et al.* [1993] reported dry deposition velocities for HNO_3 at MLO of 0.27 to 4 cm s^{-1}. A reasonable, conservative estimate of the SO_2 dry deposition velocity at MLO is therefore 1 cm s^{-1}. Using this value and assuming a uniform downslope mixing depth of 74 m (obtained in section 6.3), the dry deposition mean residence time would be two hours, making this the predominant loss term. The average plume transit time between the vent(s) and the observatory (at a distance of 6 km and speed of 3.4 m s^{-1}) is only 0.5 hours. This suggests that less than 20% of the vented SO_2 was lost during transit to MLO. It is thus likely that there was no appreciable atmospheric loss of SO_2 and that the ΔSO_2 to ΔCO_2 ratios in Table 3 represent the emissions at the vent(s).

In 1978 Mauna Loa was visibly fuming. ΔCN was 10 to 50 times greater than during the 1988 to 1993 period (Figure 10). SO_2 measurements were made at MLO in 1978 using a chemical method [*Bodhaine et al.*, 1980; Komhyr, pers. com., 1993]. The data were recorded on chart records that were never reduced. It was found that nighttime episodes of 0.5 to >10 ppb SO_2 (above a < 0.1 ppb background) often occurred, usually coinciding with periods of high CN associated with the presence of volcanic plume. Hourly ΔCO_2 in 1978 was typically in the range of 0.5 to 10 ppm, suggesting that the ΔSO_2 to ΔCO_2 ratio during this period was about 1 X 10^{-3}.

The fact that there is over 100 times less SO_2 relative to CO_2 in the quiescent Mauna Loa plume compared to the quiescent Kilauea plume shows that there must be a major difference between these two systems in the chain of events from exsolution at depth to plume dispersal. A continuous monitoring program for SO_2 was established in early 1994 at MLO to look more closely at this problem. Initial results for the first eight months of this program show that the average SO_2 to CO_2 ratio was approximately 7 X 10^{-5}.

6.5 Ratios of Aerosols to CO_2

Besides CO_2, the only observatory-monitored species detected in measurable quantities in the Mauna Loa plume were CN (0.002 to 0.1 μm particles) and light scattering (0.1 to 1 μm) particles. The CN in the volcanic plume presumably consisted primarily of sulfate aerosol produced by gas-to-particle conversion of SO_2. Sub-micron sulfate particles have

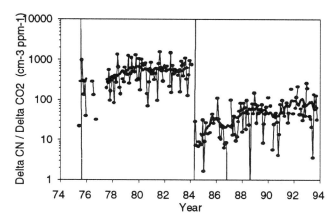

Fig. 15 The ratio of monthly average delta CN and delta CO_2 between 0000 LST and 0759 LST. Vertical lines denote the 1975 and 1984 eruptions. The fit is a 12 month running mean.

a much smaller deposition velocity than does SO_2 [*Fisher*, 1978], suggesting that losses of volcanic CN during atmospheric transit between the vent and observatory were negligible (i.e. much less than 20%). Therefore, ratios of ΔCN to ΔCO_2 measured at MLO are most likely the same as those present at the vent(s).

The monthly average ΔCN to ΔCO_2 ratio is shown in Figure 15 along with a 12-month running mean. In units of cm^{-3} ppm^{-1} the ratio increased from about 300 to 600 between 1974 and 1979 and remained near 600 until the 1984 eruption. Following the 1984 eruption, the ratio gradually climbed from 15 to 100, with two notable dips in 1986-87 and 1991. *Gerlach* [1986] has shown that the S to CO_2 ratio of exsolved gas increases as the depth of the source magma decreases. The high ΔCN to ΔCO_2 outgassing ratio during the post-1975 period compared to the post-1984 period would be expected if magma were emplaced at relatively shallow depths following the 1975 eruption, as suggested by *Lockwood et al.*, [1987].

6.6 Relationship of Emissions to Summit Inflation

Geodetic monitoring of Mauna Loa has shown that the summit has been gradually inflating since the 1984 eruption. There was one reported period of non-inflation in 1990 [*Okamura et al.*, 1991; *Miklius et al.*, 1993] which correlated with a large decrease in ΔCN and the ΔCN to ΔCO_2 ratio (Figures 10 and 15). Semi-annual EDM measurements of the distance across the summit caldera made by the Hawaii Volcanoes Observatory [Miklius, pers. com., 1994] showed a second period in 1986-1987 in which the rate of summit expansion was slowed or briefly reversed. This also correlated with a temporary decrease in both ΔCN and the ΔCN to ΔCO_2 ratio. These observations suggest that the short-term changes observed in the rate of CN (sulfate) production are related to changes in summit inflation.

The two brief halts in the rate of summit expansion in 1986-1987 and 1990 did not measurably affect the rate of CO_2 outgassing (Figures 7 and 14), which underwent steady exponential decay. If inflation of the summit is caused by refilling of the summit reservoir [Decker et al., 1983] with CO_2-rich parental magma [Gerlach, 1986], it follows that changes in the rate of inflation should be accompanied by changes in CO_2 emissions. Two presumed temporary halts in the magma supply rate did not produce measurable changes in the CO_2 emissions measured at MLO. Either (1) the new magma responsible for inflation was already depleted in CO_2 or (2) the quantity of new magma was so small that the CO_2 emissions from it were insignificant compared to those from the existing reservoir. This second possibility was not supported by the results of a comparison between the calculated volume of magma supplied to the reservoir and concurrent estimates of CO_2 emissions. Okamura et al. [1991] used geodetic measurements to calculate that 1.1×10^7 m^3 of magma was added to the summit reservoir in 1991. This is equivalent to 2.9×10^{10} kg of magma having a density of 2.6×10^3 kg m^3. Exsolution of CO_2 at the 0.06 wt % ratio of the existing quiescent reservoir (from section 6.3) would have produced 1.7×10^7 kg CO_2. Subsequent outgassing with a 1.6 year exponential decay rate (characteristic of the post-1984 reservoir) would have resulted in CO_2 emissions of 8×10^6 kg during the first year (1991). The observed CO_2 emissions in 1991 were only 3.3×10^6 kg (Figure 14). Most of the 1991 emissions presumably came from the gradual degassing of the pre-existing reservoir; therefore, the incremental CO_2 added by recently injected magma should have been much less than 3.3×10^6 kg. This raises the possibility that, by 1991, the new magma responsible for summit inflation may have already been depleted in CO_2 by the time it reached the reservoir.

6.7 Eruptive Degassing

The 1975 eruption lasted less than 19 hours during a period of unfavorable winds, and no trace of the eruptive plume was present in the data.

The 1984 eruption began at 0125 LST on March 25. Observatory winds throughout the night were from 140° to 160° at over 10 m s^{-1} and the plume was not detected. By 0600 LST, the wind speed fell below 6 m s^{-1} and the CO_2, CN, and σ_{sp} concentrations began to rise above background levels. Between 0700 and 0800 LST, the plume was most intense, with $\Delta CO_2 = 1.80$ ppm, $\Delta CN = 136,000$ cm^{-3}, and $\Delta \sigma_{sp} = 6.5 \times 10^{-6}$ m^{-1}. The ΔCN to ΔCO_2 ratio was 75,600 cm^{-3} ppm^{-1}. The CN levels recorded during this hour were the highest ever measured at MLO.

Compared to average quiescent conditions just before the eruption, the period in which the eruptive plume was most intense had similar amounts of CO_2, fifty times more light

scattering particles, and over one thousand times more CN. At this time, the eruption was emanating from the north-east rift zone near the caldera at 3700 meters elevation [Lockwood et al., 1987].

By 1300 LST on March 25, ΔCO_2 had returned to zero and ΔCN was down to 1300 cm^{-3}, presumably due to a shift to more northerly winds. The nephelometer had been turned off at 0800 LST, and MLO was completely shut down by a power failure the next morning. The analyzers were without power for the rest of the eruption. Manual CN readings taken at MLO between April 4 and April 16 (when lava was flowing from vents on the northeast rift between 2770 m and 2930 m) ranged between 1400 cm^{-3} and 31,000 cm^{-3}.

These observations show that the early eruptive magma was mostly depleted of CO_2, in agreement with Gerlach [1986] (for Kilauea), Greenland [1987], and Johnson [this volume]. The early eruptive plume was very rich in small particles (sulfate), and moderately enriched in large (0.1 to 1.0 µm) particles.

7. CONCLUSIONS

Atmospheric trace gas and aerosol measurements made at Mauna Loa Observatory were used to characterize the quiescent volcanic plume coming from the 6 km distant summit of Mauna Loa volcano. Minute-scale variability in the atmospheric CO_2 concentration was used to identify the presence of the plume at night in the downslope wind. The excess concentration of CO_2 above background levels was calculated for each hour in which the plume was present.

Excess CO_2 was greatest when winds blew from the direction of the summit caldera (180° to 190°). The distribution of excess CO_2 with wind direction had a full-width at half-maximum of about 40°. The plume was trapped in the nighttime surface temperature inversion layer with an average scale height of tens of meters. The strength of the plume at MLO followed the evolution of the temperature inversion, forming after sunset, gradually intensifying, and reaching a stable maximum between 0100 LST and 0600 LST.

Excess CO_2 was measured in the plume throughout the 1958 to 1994 period of record. The amount of volcanic CO_2 was greatest shortly after the 1975 and 1984 eruptions and decreased exponentially in the following years. Enhanced outgassing was delayed by 65 days following the 1975 eruption, and by less than 9 days following the 1984 eruption. The 1975 delay time implies a bubble ascent rate through a presumed 2700 m magma column of approximately 2 meters per hour.

From 1959 to 1994 the total annual mass of vented CO_2 was estimated, based on a simple model of plume dispersal. The total mass of the post-1950, post-1975, and post-1984 CO_2 reservoirs was estimated at 4.8×10^8 kg, 2.4×10^8 kg, and 3.3×10^8 kg respectively. This mass of CO_2 would require eruptive-scale volumes of magma (on the order of 10^8 m^3) to ascend several km. The three reservoirs had exponential decay constants of 6.1 years, 6.5 years, and 1.3 years respectively. The 1984 eruption apparently did not affect the outgassing rate of the post-1975 reservoir. After 1984, CO_2 was presumably being produced by both the post-1975 and post-1984 reservoirs.

The 1975 eruption was preceded by a three-year period in which the average excess CO_2 in the plume at MLO increased by 0.015 ppm. There was no measurable increase preceding the 1984 eruption, although an increase of this size would not have been observable due to higher average plume concentrations at this time. An increase of 0.02 ppm occurred from early 1993 to late 1994. During this time, excess CO_2 continued to decrease when the winds blew from the direction of the summit, while there was an unprecedented increase in excess CO_2 when winds blew from the southeast and southwest. This was most likely caused by a new source, possibly located on the southwest rift. This activity may be an early precursor to the next eruption.

Excess aerosol particles were measured in the plume through-out the record between 1974 and 1994. Condensation nuclei (particle size of 0.002 µm to 0.1 µm, presumably sulfate aerosol) were present in large numbers throughout the post-1975 period, decreased by a factor of five soon after the 1984 eruption, and gradually decreased by a further factor of five between 1984 and 1994. The post-1984 decrease was punctuated by two brief dips in 1986-87 and 1990-91 which correlate with temporary halts or reductions in the rate of summit expansion measured by Hawaii Volcanoes Observatory. These changes in the rate of summit expansion did not measurably affect the steady exponential decrease of CO_2 emissions. Particles that scatter light (0.1 µm to 1µm) were present in detectable quantities only between 1977 and 1980, and to a lesser degree in 1983. These data were consistent with estimates of the visual thickness of fume at the vents obtained from photographs.

Eight additional observatory data sets were examined for a volcanic plume component. These were H_2O (1974-93), O_3 (1974-93), CH_4 (1987-93), SO_2 (1988-92), aerosol carbon (1990-93), radon (1991-93), CO (1992-93), and H_2 (1992-93). None of these species were present in the plume to the detection limits of the analysis technique. The upper limit of the ratios of H_2O, CO, and H_2 to CO_2 was much greater than the ratios of these gases measured at the vent shortly after the 1984 eruption. The upper limit to the SO_2 to CO_2 ratio was 10^{-3}, approximately two orders of magnitude less than that reported for the quiescent Kilauea plume. Recent measurements show a ratio of 7 X 10^{-5}.

The 1984 eruptive plume was sampled at MLO early in the first day of the eruption. Compared to levels measured during the quiescent period before the eruption, the eruptive plume had similar concentrations of CO_2 and a thousand times greater number density of condensation nuclei.

Acknowledgments. The efforts of the all the MLO staff over the years is gratefully acknowledged, in particular J.F.S. Chin, who for thirty years operated the MLO CO_2 program begun in 1958 by C.D. Keeling. NOAA/CMDL data were obtained through the efforts of P. Tans and K. Thoning (CO_2), B. Bodhaine (aerosols), G. Herbert and M. Bieniulis (meteorology), S. Oltmans (ozone), E. Dlugokencky and P. Lang (CH_4), and P. Novelli and W. Coy (CO and H_2). SO_2 data for 1989 were provided by M. Luria of NOAA/ARL. SO_2 data for 1991-92 were provided by G. Hubler of NOAA/AL. Thanks to J.P. Lockwood of Hawaiian Volcanoes Observatory for encouragement and the use of his collection of Mauna Loa summit photographs. A. Miklius of HVO provided EDM data for the Mauna Loa summit caldera. The detailed review by A. J. Sutton of HVO is greatly appreciated.

REFERENCES

Bodhaine, B. A., The Mauna Loa four wavelength nephelometer: instrument details and three years of observations, *NOAA Tech. Report ERL 396-ARL5*, Boulder, Colorado, 1978.

Bodhaine, B. A., J. M. Harris, G. A. Herbert, W. D. Komhyr, Identification of volcanic episodes in aerosol data at Mauna Loa Observatory, *J. Geophys. Res.*, *85(C3)*, 1600-1604, 1980.

Bodhaine, B. A., Aerosol Measurements at four background sites, *J. Geophys. Res.*, *88(C15)*, 10753-10768, 1983.

Casadevall, T. J., and R. W. Hazlett, Thermal areas on Kilauea and Mauna Loa volcanoes, Hawaii, *J. Volcan. Geotherm. Res.*, *16*, 173-188, 1983.

Casadevall, T. J., A. Krueger, B. Stokes, The volcanic plume from the 1984 eruption of Mauna Loa, Hawaii (abstract), *EOS*, *45 No. 5*, 1984.

Clarke, A. and Bodhaine, B., A comparison of aerosol size distributions and nephelometer measurements at Mauna Loa Observatory, in *Climate Monitoring and Diagnostics Laboratory No. 21 Summary Report, 1992*, 93-96, Boulder, CO, 1993.

Connor, C. B., R. E. Stoiber, L. L. Malinconico, Jr., Variations in Sulfur Dioxide emissions related to earth tides, Halemaumau Crater, Kilauea Volcano, Hawaii, *J. Geophys. Res.*, *93(B12)*, 14867-14871, 1988.

Decker, R. W., R. Y. Koyanagi, J. J. Dvorak, J. P. Lockwood, A. T. Okamura, K. M. Yamashita, and W. R. Tanigawa, Seismicity and surface deformation of Mauna Loa volcano, Hawaii, *EOS*, *64 No. 37*, 545-547, 1983.

Ferguson, E. E., R. M. Rosson, (eds.), *Climate Monitoring and Diagnostics Laboratory No. 20: Summary Report 1991*, 131 pp., NOAA Environmental Laboratories, Boulder, Colorado, 1991.

Fisher, B. E. A., Long-range transport and deposition of sulfur oxides, in *Sulfur in the Environment Part 1*, J. O. Nriagu, ed., 245-295, John Wiley & Sons publisher, 1978.

Garrett, A. J., Orographic cloud over the eastern slopes of Mauna Loa Volcano, Hawaii related to insolation and wind, *Monthly Weath. Rev. 108 No. 7*, 1980.

Gerlach, T. M. Exsolution of H_2O, CO_2, and S during eruptive episodes at Kilauea Volcano, Hawaii, *J. Geophys. Res., 91(B12)*, 12177-12185, 1986.

Greenland, L. P., W. P. Rose, J. B. Stokes, An estimate of gas emissions and magmatic gas content from Kilauea volcano *Geochim. Cosmochim. Acta, 49*, 125-129, 1985.

Greenland, L. P., Composition of gases from the 1984 eruption of Mauna Loa Volcano, in Decker, R. W. et al. (eds), *Volcanism in Hawaii*, Chapter 30, U. S. Geological Survey Professional Paper 1350, 781-791, 1987.

Gundel, L. A., R. L. Dod, H. Rosen, T. Novakov, The relationship between optical attenuation and black carbon concentration for ambient and source particles, *Sci. Total Environ., 36*, 197-202, 1984.

Hahn, C. J., J. T. Merrill, B. G. Mendonca, Meteorological influences during MLOPEX, *J. Geophys. Res., 97(D10)*, 10291-10309, 1993.

Hansen, A. D. A., H. Rosen, T. Novakov, The aethaelometer - an instrument for the real-time measurement of optical absorption by aerosol particles, *Sci. Total Environ., 36*, 191-196, 1984.

Harris, D. M., The concentration of CO_2 in submarine tholeiitic basalts, *J. Geol., 89*, 689-701, 1981.

Harris, J. M. and J. D. Kahl, A descriptive atmospheric transport climatology for the Mauna Loa Observatory, using clustered trajectories. *J. Geophys Res., 95(D9)*, 13651-13667, 1990.

Herbert, G. A., J. M. Harris, M. S. Johnson, J. R. Jordan, The acquisition and processing of continuous data from GMCC observatories, *NOAA Tech. Memo. ERL ARL-93*, Air Resources Laboratories, Silver Spring, Maryland, 1981.

Herbert, G. A., E. R. Green, J. M. Harris, G. L. Koenig, S. J. Roughton, K. W. Thaut, Control and monitoring instrumentation for the continuous measurement of atmospheric CO_2 and meteorological variables, *J. Atm. and Ocean. Tech., 3 No.3*, 414-421, 1986.

Herbert, G. A., E. R. Green, G. L. Koenig, K. W. Thaut, Monitoring instrumentation for the continuous measurement and quality assurance of surface weather observations, *Sixth Symposium on Met. Obs. and Instrumentation*, 467-470, A.M.S., Boston, Mass, 1987.

Hubler, G. NO_y and SO_2 measurements at the Mauna Loa Observatory during 1991-92 (abstract), *EOS, 74 No. 43*, 119, 1993.

Johnson, D. J., Gravity changes on Mauna Loa volcano, this volume.

Keeling, C. D., R. B. Bacastow, A. E. Bainbridge, C. A. Ekdahl, Jr., P. R. Guenther, L. S. Waterman, and J. F. S. Chin, Atmospheric carbon dioxide variations at Mauna Loa Observatory, Hawaii, *Tellus, 28(6)*, 538-551, 1976.

Keeling, C. D., R. B. Bacastow, and T. P. Whorf, Measurements of the concentration of carbon dioxide at Mauna Loa Observatory, Hawaii., in W. C. Clark, Ed. *Carbon Dioxide Review: 1982.*, Oxford University Press, New York, 377-385, 1982.

Keeling, C. D., Atmospheric CO_2 concentrations - Mauna Loa Observatory, Hawaii 1958-1986. NDP-001/R1, Carbon Dioxide Information Center, Oak Ridge National Laboratory, Oak Ridge, Tennessee., 1986.

Keeling, C. D., D. J. Moss, T, P. Whorf, Measurements of the concentrations of atmospheric carbon dioxide at Mauna Loa Observatory, Hawaii 1958-1986, Final report for the Carbon Dioxide Information Center, Oak Ridge National Laboratory, Oak Ridge, Tennessee, 1987.

Komhyr, W. D., Electrochemical concentration cells for gas analysis, *Ann. Geophys., 25(1)*, 203-210, 1969.

Komhyr, W. D., L. S. Waterman, and W. R. Taylor, Semiautomatic nondispersive infrared analyzer apparatus for CO_2 air sample analyses., *J. Geophys. Res., 88*, 3913-3918, 1983.

Komhyr, W. D., T. B. Harris, L. S. Waterman, J. F. S. Chin, and K. W. Thoning, Atmospheric Carbon Dioxide at Mauna Loa Observatory 1. NOAA Global Monitoring for Climatic Change Measurements with a nondispersive infrared analyzer, 1974-1985, *J. Geophys. Res., 94(D6)*, 8533-8547, 1989.

Lee, G., L. Zhuang, B. J. Huebert, T. P. Meyers, Concentration gradients and dry deposition of nitric acid vapor at the Mauna Loa Observatory, Hawaii, *J. Geophys. Res., 98(D7)*, 12661-12671, 1993.

Lockwood, J. P., N. G. Banks, T. T. English, L. P. Greenland, D. B. Jackson, D. J. Johnson, R. Y. Koyanagi, K. A. McGee, A. T. Okamura, J. M. Rhodes, The 1984 Eruption of Mauna Loa Volcano, Hawaii, *EOS, 66 No. 16*, 169-171, 1985.

Lockwood, J. P., and P. W. Lipman, Holocene eruptive history of Mauna Loa volcano, in Decker, R. W. et at. (eds.) *Volcanism in Hawaii*, Chapter 18, U. S. Geological Survey Professional Paper 1350, 509-535, 1987.

Lockwood, J. P., J. J. Dvorak, T. T. English, R. Y. Koyanagi, A. T. Okamura, M. L. Summers, W. R. Tanigawa, Mauna Loa 1974-1984: A decade of intrusive and extrusive activity, in Decker, R. W. et al. (eds.), *Volcanism in Hawaii*, Chapter 19, U. S. Geological Survey Professional Paper 1350, 537-570, 1987.

Luria, M., J. F. Boatman, J. Harris, J. Ray, T. Straube, J. Chin, R. L. Gunter, G. Herbert, T. M. Gerlach, C. C. Van Valin, Atmospheric sulfur dioxide at Mauna Loa Hawaii, *J. Geophys. Res., 97(D5)*, 6011-6022, 1992.

Masarie, K. A. , L. P. Steele, P. M. Lang, A rule-based expert system for evaluating the quality of long-term, in situ, gas chromatographic measurements of atmospheric methane, *NOAA TM ERL CMDL-3*, 37 pp, Boulder, Colorado, 1991.

Massey, D. M., T. K. Quakenbush, B. A. Bodhaine, Condensation nuclei and aerosol scattering extinction measurements at Mauna Loa Observatory: 1974-1985, *NOAA Data Report ERL ARL-14*, Silver Spring, Maryland, July 1987.

Miklius, A., A. T. Okamura, M. K. Sako, J. Nakata, Current state of geodetic monitoring of Mauna Loa Volcano (abstract) 1993 Fall AGU Meeting, 1993.

Miller, J. M. and J. F. S. Chin, Short-term disturbances in the carbon dioxide record at Mauna Loa Observatory, *Geophys. Res. Lett.*, *5 No. 8*, 669-671, 1978.

Moller, D., Kinetic model of atmospheric SO_2 oxidation based on published data, *Atmos. Environ.*, *14 (No. 9)*, 1067-1076, 1980.

Negro, V. Environmental radon monitor, *USDOE Report EML-367*, 244-247, 1979.

Novelli, P. C., J. W. Elkins, L. P. Steele, The development and evaluation of a gravimetric reference scale for measurements of atmospheric carbon monoxide, *J. Geophys. Res.*, *96(D7)*, 13109-13121, 1991.

Okamura, A. T., A. Miklius, M. K. Sako, J. Tokuuke, Evidence for renewed inflation of Mauna Loa Volcano, Hawaii, (abstract), 1991 AGU Fall Meeting, 1991.

Oltmans, S. J. Surface ozone measurements in clean air *J. Geophys. Res.*, *86*, 1174-1180, 1981.

Oltmans, S. J. and W. D. Komhyr Surface ozone distributions and variations from 1973-1984 measurements at the NOAA Geophysical Monitoring for Climatic Change baseline observatories, *J. Geophys. Res.*, *91(D4)*, 5229-5236, 1986.

Pales, J. C. and C. D. Keeling, The concentration of atmospheric carbon dioxide in Hawaii, *J. Geophys. Res.*, *70*, 6053-6076, 1965.

Peterson, J. T., and R. M. Rosson (eds.) *Climate Monitoring and Diagnostics Laboratory No. 21 Summary Report 1992*, 131 pp., NOAA Environmental Research Laboratories, Boulder, CO, 1993.

Price, S., and J. C. Pales, Mauna Loa Observatory: the first five years, *Monthly Weath. Rev.*, 665-680, December, 1963.

Pueschel, R. F., and B. G. Mendonca, Sources of atmospheric particulate matter on Hawaii, *Tellus*, *24*, 139-148, 1972.

Pueschel, R. F., and B. G. Mendonca, Dispersion into the higher atmosphere of effluent during an eruption of Kilauea volcano, *J. de Recherches Atmospheriques*, 439-446, 1973.

Rhodes, J. M., Geochemistry of the 1984 Mauna Loa eruption: implications for magma storage and supply, *J. Geophys. Res.*, *93(B5)*, 4453-4466, 1988.

Sehmel, G. A., Particle and gas dry deposition: a review, *Atmos. Environ.*, *14*, 983-1011, 1980.

Smith, V. N., A recording infrared analyzer., *Instruments*, *26*, 421-427, 1953.

Sutton, A. J. and McGee, K. A., A multiple-species volcanic gas sensor- Testing and applications (abstract) *IAVCEI Continental Magmatism General Assembly*, Santa Fe, N.M. abstract volume, 262, 1989.

Thomas, J. W., and P. C. LeClare, A study of the two-filter method for radon-222, *Health Phys.*, *18*, 113-122, 1970.

Thoning, K. W., P. P. Tans, W. D. Komhyr, Atmospheric carbon dioxide at Mauna Loa Observatory 2. Analysis of the NOAA GMCC Data, 1974-1985, *J. Geophys. Res.*, *94(D6)*, 8549-8565, 1989.

Whittlestone, S., E. Robinson, S. Ryan, Radon at the Mauna Loa Observatory: transport from distant continents, *Atmos. Environ.*, *26A No. 2*, 251-260, 1992.

Whittlestone, S., S. D. Schery, Y Li, Separation of local from distant pollution at MLO using Pb-212, *Climate Monitoring and Diagnostics Laboratory No. 21 Summary Report 1992*, 116-118, Boulder, CO, 1993.

Wilkening, M. H., Radon-222 from the island of Hawaii: deep soils are more important than lava fields or volcanoes, *Science*, *183*, 413-415, 1974.

S. Ryan, Mauna Loa Observatory, P.O. Box 275, Hilo, Hi 96720. (808) 933-6965 ryan@mloha.mlo.hawaii.gov

Biological Perspectives on Mauna Loa Volcano: A Model System for Ecological Research

Peter M. Vitousek, Gregory H. Aplet[1], James W. Raich[2], and John P. Lockwood[3]

Department of Biological Sciences, Stanford University, Stanford, California

As a result of evaluations of volcanic hazards, most of the surface lava flows of Mauna Loa have been mapped and dated. Each of these flows represents a valuable resource for ecological studies – a single-age, single-substrate transect reaching from near the summit towards the sea, often spanning a range of nearly 20°C in mean annual temperature. The set of flows on a particular flank of the mountain represents an age sequence of parallel transects, and the influence of precipitation can be assessed separately by examining flows of comparable age on different aspects of Mauna Loa. We evaluated the development of plant communities and the functioning of ecosystems across portions of the age-elevation-precipitation matrix on Mauna Loa. On the wet east flank, plant communities develop more slowly at high elevation, although the composition of the vegetation of young flows is similar at all elevations. However, rainforest ultimately develops on older flows below 1800 m elevation, while open woodlands dominate old flows at higher elevations. Rates of both plant production and decomposition increase with decreasing elevation (increasing temperature) on a given flow – but production increases linearly while decomposition increases exponentially. Consequently, soil carbon turnover and rates of nutrient cycling increase progressively from high to low elevation. These examples of ecological research illustrate how the relative simplicity of the biological systems on Mauna Loa, in combination with their relatively well-understood geology, allow us to evaluate processes that are difficult to study in more complex continental ecosystems.

1. INTRODUCTION

Most earth scientists view Mauna Loa from perspectives of geology, geochemistry, or geophysics, and the name of this great volcano conjures up thoughts of magma genesis, eruptive processes, distribution of surface lava flows, and perhaps the associated hazards to humanity. To biologists as well, Mauna Loa is a place of global significance – not for its rocks, but for the organisms and ecosystems that occupy its surface.

Much of the progress in science is made by interdisciplinary research, but collaboration between the earth and life sciences is uncommon. In this paper we report on one such collaboration on Mauna Loa, where an understanding of substrate ages (derived from volcanic hazard studies) has opened fundamentally new avenues for biological research. This paper does not attempt to summarize the very wide spectrum of biological research being carried out on Mauna Loa, but instead describes how the broad yet well-defined range of soils and environments present on Mauna Loa Volcano make it an extraordinarily valuable resource for ecological studies, one that can contribute substantially to our understanding of the basic processes regulating ecological systems.

[1]The Wilderness Society, 900 17th Street, N.W., Washington, D.C. 20006.

[2]Department of Botany, Iowa State University, Ames, Iowa 50011.

[3]Hawaii Volcano Observatory, U.S. Geological Survey, Hawaii National Park, Hawaii 96718.

Mauna Loa Revealed: Structure,
Composition, History, and Hazards
Geophysical Monograph 92

On a simple level, we think of the characteristics and dynamics of terrestrial ecosystems as being controlled by a relatively small number of factors, the most important of which are climate, the parent material in which soils and ecosystems develop, relief or topography, organisms (defined as the regional flora and fauna), and time or substrate age [*Jenny*, 1941, 1980]. Ideally, the ways that each of these factors (termed 'state factors') control soils and ecosystems can be determined by locating a sequence of sites along which one factor varies, but all of the others can be held constant. This provides a powerful conceptual framework for ecosystem analysis – but with the practical difficulty that is rare that all the factors but one can be held truly constant, especially across a wide range of variation in one particular factor. Mauna Loa, however, provides an extraordinarily well-defined natural system in which several of the state factors are, or can be held, remarkably constant, while others vary widely. Patterns of variation in the major state factors in the Hawaiian Islands were reviewed by Vitousek (1995); these apply to Mauna Loa as follows:

1.1. *Climate*

Mauna Loa supports extraordinary variation in climate in a very small area. Temperature decreases with increasing elevation at an environmental lapse rate of 6.4°C for every 1000 m [*Juvik and Nullet*, 1994], from mean annual temperatures of 24°C at sea level to near 0°C at the summit. Variation in precipitation is a function of elevation and of exposure to the prevailing northeast trade winds; the windward east flank receives very high rainfall, up to 6000 mm/y, while nearby rain shadow areas on the northwest flank average as little as 200 mm/y. The east slope of Mauna Loa is almost always wet and the northwest almost always dry; elsewhere, there are distinct wet and dry seasons. Kona, on the southwest slope of Mauna Loa, is sheltered from the trade winds and would appear to be in a rain shadow. However, the mass of Mauna Loa and the overhead summer sun combine to establish a diurnal land-sea breeze cycle that brings substantial summer precipitation to Kona. In contrast, the southeast-facing flank receives most of its annual precipitation from the southeast winds of winter storms (Figure 1). Finally, high-elevation climates are shaped by the trade-wind inversion, which ranges from 1900 m upwards; areas above this inversion are exposed to very dry air, high radiation input, and elevated evaporative demand [*Juvik et al.*, 1978; *Kitayama and Mueller-Dombois*, 1992; *Juvik and Nullet*, 1994]. The net result of these processes is a fine-scale matrix of temperature and precipitation that encompasses nearly 95% of the climatic variation

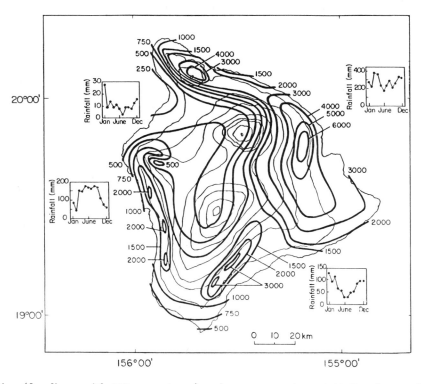

Fig. 1. Elevation (fine lines, with 500 m contours) and mean annual precipitation (coarse lines, in mm/y) on the Island of Hawaii. From Giambelluca et al. (1986).

in Earth's tropics [*George et al.*, 1987].

1.2 *Time*

Mauna Loa also supports an extraordinary range of substrate ages in a small area. Its frequent eruptions give rise to lava flows and tephra deposits that have been mapped in detail, and dated using historical records (since 1840) and ^{14}C dating of buried charcoal [*Lockwood and Lipman*, 1980]. J. P. Lockwood and collaborators have identified and mapped >600 individual flows with very fine resolution; these are available on file, and are now being digitized and entered into an ARC-INFO geographic information system (GIS) data base (Kauahikaua, this volume). For relatively young substrates (to several thousand years), this information provides an exceptionally detailed record of substrate ages – one that is unmatched for any volcano on Earth.

1.3. *Organisms*

Hawaii is the most isolated archipelago on Earth, and relatively few species have been able to colonize it naturally. Some of these successful colonists have radiated through evolution into a variety of quite different species that occupy very different environments, making the Hawaiian Islands a natural laboratory for evolutionary studies [*Carson and Kaneshiro*, 1976; *Carlquist*, 1980]; *Carr et al.*, 1989]. Other colonists have not diversified into many species, but still cover a much broader range of environments than do most continental species. One striking example of the latter is the native ohia tree, *Metrosideros polymorpha* (Myrtaceae), which is the dominant tree in natural ecosystems from treeline to sea, and from extremely wet (>10,000 mm/yr) to (on young soils) quite dry (<500 mm/y) sites. It is often the first woody plant to occupy young volcanic sites, and in wet areas it remains the most abundant tree on the oldest substrates in Hawaii (>4,000,000 y) [*Dawson and Stemmermann*, 1990]. *Metrosideros* dominates most natural systems below 2500 m elevation on Mauna Loa, making the organism state factor nearly constant across a very wide range of environments. A consequence of its broad range is that *Metrosideros* and other species offer unusual opportunities for determining how tradeoffs inherent to carbon fixation, nutrient use, and water loss vary within a species across a range of environments [*Britten*, 1962; *Pearcy and Caulkin*, 1983; *Robichaux and Canfield*, 1985; *Vitousek et al.*, 1990; *Meinzer et al.*, 1992]. Moreover, the widespread biological invasions of Hawaii by human-transported exotic species represent a fundamental alteration of the organism factor, one that has been shown to alter ecosystem-level properties such as nitrogen budgets [*Vitousek and Walker*, 1989] and fire frequencies [*Hughes et al.*, 1991] in areas of the Hawaiian Islands.

1.4. *Relief*

Relief can be held remarkably constant across a very broad range of environments. Rapidly aggrading shield volcanoes are built up of frequent and relatively fluid lava flows with gentle angles of repose; their surfaces are porous, with few surface streams, little surface runoff, and very little erosion [*MacDonald et al.*, 1983]. Mauna Loa in particular reaches from sea level to 4168 m with little coarse-scale topography across most of its surface.

1.5. *Parent Material*

Essentially all of the parent material of Hawaii is volcanic, and the chemistry of material produced during the shield-building stage is relatively constant over both short and long time scales [*Wright and Helz*, 1987]. There is some minor variation within and between eruptions, particularly in MgO content, but Mauna Loa is all tholeiitic basalt [*Wright*, 1971; *Rhodes*, 1983]. In contrast, the texture of the parent material varies widely, from smooth, massive pahoehoe flows through coarse aa to finer cinder and ash, and this variation can affect the development of biological systems substantially [*Smathers and Mueller-Dombois*, 1974].

1.6 *Interactions*

Overall, three of the major ecosystem state factors can be held remarkably constant on Mauna Loa, while climate (temperature and precipitation separately) and time vary spectacularly – but in continuous, well-defined, and largely independent gradients. For example, any one of the >600 mapped and dated lava flows on Mauna Loa is an extraordinary ecological resource – a single age, single substrate transect reaching from near the summit towards, and often to, the sea. The set of flows on a particular flank of the mountain is still more useful – an age sequence of parallel transects, each well-defined, consistent, and comparable to the others across a wide range of temperature and precipitation. Moreover, the effects of temperature and precipitation can be separated by comparing sites of similar age and temperature across aspects of Mauna Loa that differ in precipitation. For example, ecosystem properties on ~140 y old aa and pahoehoe flows can be determined under very wet (3000-6000 mm/y) versus dry (400-550 mm/y) conditions on the east versus northwest flanks of Mauna Loa [*Vitousek et al.*, 1992] (Figure 2). In all of these cases, sites that differ widely can be compared – but more importantly, the existence of complete gradients between the extremes allows the relationships as well as the differences among sites to be investigated.

Moreover, the older volcanoes on the Island of Hawaii, and on the older islands to the northwest, provide opportunities to evaluate ecosystem dynamics on older soils; these can greatly increase the scope and generality of

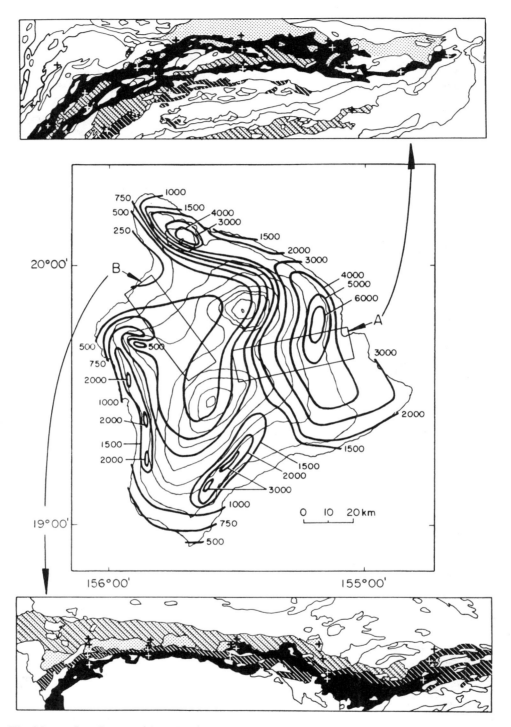

Fig. 2. The Mauna Loa flow age/elevation/precipitation matrix. A young (<140 y) versus an old (2800–4000 y) flow of aa versus pahoehoe lava were sampled across a wide range of elevations on the wet east (rectangle A) versus the dry northwest (rectangle B) flanks of Mauna Loa. The young flows sampled are mapped in black, while the old flows are stippled; aa flows are indicated with diagonal lines. Locations where soils were sampled are marked with a +; *Metrosideros* foliage was sampled at more frequent intervals. From Vitousek et al. (1992).

biological research on Hawaiian volcanoes [*Riley and Vitousek*, 1995; *Crews et al.*, in press; *Kitayama and Mueller-Dombois*, in press]. In this paper, we will describe results of two lines of research on Mauna Loa – one concerned with understanding the composition and dynamics of plant communities as they develop on new volcanic substrates, and the other concerned with understanding how temperature regulates carbon balance and biogeochemical cycling in montane rainforests.

2. PLANT COMMUNITIES

The development of complex biological communities on entirely new substrates such as recent lava flows is termed "primary succession", to distinguish it from the "secondary succession" that occurs on previously vegetated sites that have been cleared (for example, by fire, windthrow, or human land use) but retain the influences of earlier biological occupation in their soils. Primary succession has long received attention out of proportion to its current areal extent [cf *Cowles*, 1899; *Clements*, 1916; *Cooper*, 1919], in large part because an understanding of the processes operating during the development of communities on wholly new substrates provides insight into the general workings of biological communities. However, primary succession is relatively difficult to study. Its time frame is relatively long (generally several decades to several millenia), and the entire pathway can be evaluated only by piecing together sequences of sites of different ages, and assuming or demonstrating that they differ only or primarily in age.

The Hawaiian Islands in general, and Mauna Loa in particular, would appear to be a logical place to examine primary succession, and considerable research has been carried out on Hawaiian ecosystems [cf *Forbes*, 1912; *MacCaughey*, 1917; *Skottsberg*, 1941; *Atkinson*, 1970; *Eggler*, 1971; *Uhe*, 1988]. Until recently, however, the lack of information on ages and extents of prehistoric lava flows has left only the small historical proportion of the overall developmental sequence accessible to quantitative studies. Detailed dating of prehistoric flows was made possible by the recovery of charcoal beneath flows [*Lockwood and Lipman*, 1980], and extensive mapping and ^{14}C dating was then undertaken in order to determine volcanic hazards [*Lockwood et al.*, 1988]. A byproduct of this effort has been an extension of the time over which primary succession can be studied.

The east flank of Mauna Loa provides a useful array of lava flows for this purpose. Drake and Mueller-Dombois [1993] determined forest structure on eight aa lava flows from 47- to 3400 y old, all at 1200 m elevation (with ~4000 mm/y annual precipitation). They demonstrated that 1) the biomass and stature of the dominant *Metrosideros polymorpha* increased with age, although at a decreasing rate; and 2) colonizing varieties of *Metrosideros* with pubescent leaves were replaced by later-successional glabrous varieties on the oldest flow (see also *Stemmermann*, 1983). Kitayama et al. (in press) further demonstrated that tree ferns (*Cibotium* species) and a number of other native plants increased in relative abundance through time, and that the later deposition of volcanic ash onto some of these flows increased the biomass but did not alter the composition of developing forests.

Aplet and Vitousek (1994) extended these analyses by evaluating primary succession on an age-elevation matrix on the east flank of Mauna Loa; they sampled five aa flows from five to 3400 y old at each of six elevations from 915–2400 m. In this matrix, variation in the vegetation of each flow as a function of elevation can be viewed as representing the influence of climate on a particular substrate, and the change in vegetation as a function of flow age at each elevation represents a primary successional sequence. This matrix can be used to address how climate affects the rate and pathway of primary succession, under well-defined conditions that would be difficult to duplicate elsewhere.

Not surprisingly, plant biomass accumulates more rapidly in warm, wet, low elevation sites, and reaches higher equilibrium levels (Figure 3). More interestingly, while the composition of plant communities on young flows is relatively similar at all elevations, community composition on older flows diverges sharply above versus below the trade wind inversion near 1900 m elevation (Figure 4). Rainforest develops at lower elevation, while open woodlands dominate higher sites.

Patterns of succession appear to be rather different in drier sites on the leeward side of Mauna Loa. *Metrosideros* remains the initial colonizer, but it is replaced on older flows by a variety of tree species that are better able to cope with drought [*Stemmermann and Ihsle*, 1993].

The Mauna Loa system could be used further to investigate interactions of temperature, precipitation, and substrate age with the seasonal timing of precipitation, the texture of substrates, and the extent of biological invasions by exotic species. The results of these and similar studies would of course be specific to Mauna Loa, but an understanding of the underlying mechanisms involved should be more generally applicable.

3. ECOSYSTEM FUNCTION

The Mauna Loa system also is valuable for determining how ecosystems function – what controls production, decomposition, biogeochemical cycling of nutrients, fluxes of water – as well as for determining their composition. Generally, climate exerts an overriding influence on the structure and funtioning of terrestrial ecosystems. The effects of temperature have been addressed through broad comparisons across continental scales [*Holdridge*, 1971; *Jordan and Murphy*, 1978; *Post*

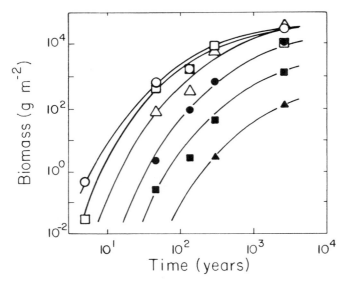

Fig. 3. Above-ground plant biomass as a function of age and elevation on the wet east flank of Mauna Loa, showing more rapid vegetation development at lower elevation. Five aa flows (then 5, 47, 137, ~300, and ~3400 y old) were sampled at six elevations; biomass at 914 m elevation is indicated by □, at 1219 m by ○, at 1524 m by △, at 1829 m by ●, at 2134 m by ■, and at 2438 m by ▲. From Aplet and Vitousek (1994).

et al., 1982; *Raich and Schlesinger*, 1992], through experimental studies using microcosms [*Billings et al*, 1982], and through ecosystem models that build on physiological information derived from plants and microorganisms [*Rastetter et al.*, 1991; *Parton et al.*, 1994]. The environmental matrix on Mauna Loa permits studies on an intermediate level of resolution between microcosms and continents; the effects of temperature on ecosystems can be determined directly by measuring ecosystem function across elevation on a single lava flow, often with similar dominant vegetation at all elevations. On the wet east flank of Mauna Loa, variation in elevation is primarily a gradient in temperature, up to near the elevation of the trade wind inversion.

Raich et al. [submitted] determined aboveground net primary production (ANPP) at four sites on young pahoehoe flows on the east flank of Mauna Loa. Net primary production is defined as total photosynthesis minus plant respiration; it represents the energy available for plant growth and for the respiration of animals and microbes. It is measured by determining net plant growth, consumption of live plants (where this is important), and the turnover of plants and plant parts. Measurements were carried out at 290 m on the 1881 flow, and 700, 1130, and 1660 m on the 1855 flow. All of these sites were dominated by the same major species, including most importantly *Metrosideros polymorpha* (ohia)

and *Dicranopteris linearis* (uluhe or false staghorn fern); their mean annual temperature ranged from 13°–22°C. Raich et al. found that while production by individual species varied differently along the elevational gradient, overall ANPP decreased linearly by 36 g·m^{-2}·y^{-1} for every 100 m gain in elevation ($r^2 = .99$) (Figure 5).

Vitousek et al. [1994] determined rates of decomposition at four elevations on the same flows; they measured mass loss over a two-year period of *Metrosideros* leaf litter collected at each site, and of *Metrosideros* litter collected at one of the sites that was distributed to and decomposed at all four elevations. Rates of decomposition increased exponentially with decreasing elevation (increasing temperature) for all of these substrates (Figure 5). The net result is that increasing temperature increases both ANPP and decomposition – but across this range of sites, temperature affects decomposition more than it does production. Similar patterns

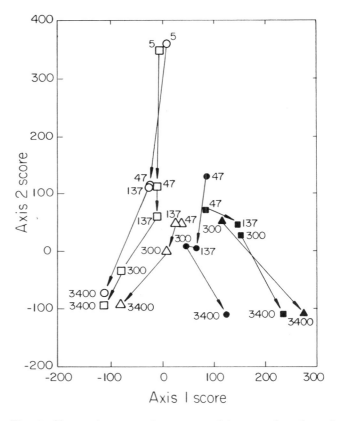

Fig. 4. Change in vegetation communities as a function of lava flow age and elevation on Mauna Loa. Canonical correspondence analysis was used to extract major axes of variation in species composition across the matrix described in Fig. 3; changes in vegetation with primary succession are plotted in the space defined by the first two axes. The numbers refer to lava flow ages at the time of sampling; symbols as in Fig. 3. From Aplet and Vitousek (1994).

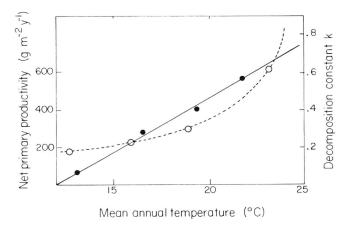

Fig. 5. Above-ground net primary production (solid line and symbols, from Raich et al. [submitted]) and the decomposition of *Metrosideros* leaf litter (dashed line and hollow symbols, from Vitousek et al. [1994]) as a function of mean annual temperature on the 1855 and 1881 pahoehoe lava flows, east flank of Mauna Loa. The decomposition constant k is defined as the coefficient of the first-order decomposition equation $dX/dt = -kX$, where X is the mass of decomposing substrate.

were observed on an older (3400 y) Mauna Loa flow, although absolute rates of both ANPP and decomposition were greater on the old flow – and only two sites were sampled for ANPP [*Raich et al.*, submitted; *Vitousek et al.*, 1994]. These results support a possibility suggested by a number of global models, that anthropogenic climatic warming could cause accelerated decomposition in excess of production, and thereby cause a net release of carbon dioxide to the atmosphere and hence a positive feedback to warming [*Townsend et al.*, 1992; *Melillo et al.*, 1993].

A further implication of the pattern in Figure 5. is that the availability of nutrient elements to plants could decline systematically with increasing elevation. Plant production requires substantial quantities of N, P, and other nutrients, and the decomposition of organic material is the major source of supply of these elements on a year-to-year basis. Since decomposition is slowed disproportionately (relative to production) in cold, high elevation sites, nutrient supply could be reduced to a greater extent than is nutrient demand. Consistent with this suggestion, concentrations of N and P in *Metrosideros* foliage decrease significantly with increasing elevation on east-flank Mauna Loa lava flows (Figure 6) [*Vitousek et al.*, 1992]. A broader implication of this temperature/nutrient supply interaction is that increased global temperatures could enhance nutrient release from soil organic matter, with consequences for plant growth, element leaching, and atmospheric chemistry [*Schimel et al.*, 1990; *Melillo et al.*, 1993].

Additional studies have evaluated the effects of substrate age and of precipitation on soil and plant nutrients. Concentrations of biologically available N and P increase with increasing flow age across the spectrum of ages present on Mauna Loa, as nutrients from the atmosphere and from mineral weathering accumulate in developing ecosystems [*Vitousek et al.*, 1988, 1992]. Ultimately, however, there is a substantial decrease in biologically available P on much older substrates within the Hawaiian Islands [*Crews et al.*, in press; *Vitousek et al.*, in press].

In contrast to temperature, evidence from Mauna Loa suggests that decreased precipitation reduces nutrient demand by plants to a greater extent than it reduces supply from decomposers – so that nutrient availability is relatively greater on the dry 1859 pahoehoe flow on northwest Mauna Loa than it is at similar elevations on the very wet 1855 pahoehoe flow on the east flank. Concentrations of both N and P in foliage are higher in dry sites on Mauna Loa, and the decomposability of tissue produced in such sites is substantially greater (Figure 6) [*Vitousek et al.*, 1992, 1994].

4. CONCLUSIONS

The range of environments on Mauna Loa, coupled with the quantity and quality of information on the geology of Mauna Loa, make it an unmatched resource for studies of plant communities and ecosystem function. Where comparisons are possible, patterns of ecosystem structure and function along the environmental gradients studied are consistent with those in continental areas. However, the relative simplicity of biological communities on Mauna Loa, and the tight control over a

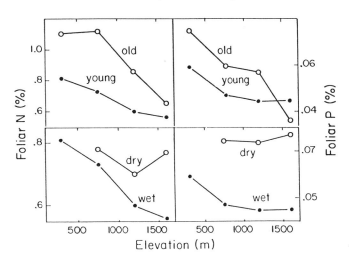

Fig. 6. Concentrations of N and P in full-sun leaves of *Metrosideros polymorpha* collected on young versus old pahoehoe flows on the wet east flank of Mauna Loa. Data from Vitousek et al. (1992).

broad range of environmental factors, makes Mauna Loa much more suitable for detailed studies of biological mechanisms than are most continental ecosystems. A detailed knowledge of lava flow ages has enabled us to understand controls over the rates at which plant communities develop and ecosystems function with a precision difficult to achieve elsewhere. Moreover, the ecological interplay between lava flows and lifeforms may not be limited only to the *passive* effect of substrate ages. Carson et al. [1990] have suggested that the frequent eruptive activity of Mauna Loa may play an *active* role in the acceleration of evolution itself, through the isolation of populations and habitat stress.

The benefits of interdisciplinary collaborative research of the sort we describe here do not accrue only to ecology. Refinement of our understanding of forest succession and ecosystem function may allow geologists to constrain the ages of young vegetated lava flows undatable by other means, and also yield quantitative information of biosphere-atmosphere exchange processes of vital concern to atmospheric scientists and climatologists. The effort that has gone into mapping and dating Mauna Loa lava flows for geological and hazard-evaluation purposes has paid and will continue to pay diverse scientific dividends.

Acknowledgements. This research was supported by NSF grants BSR-8718003 and BSR-891832 to Stanford University. C. P. Stone and the research staff of Hawaii Volcanoes National park provided advice and laboratory space, and C. Nakashima prepared the manuscript for publication.

REFERENCES

Aplet, G. H., and P. M. Vitousek, An age-altitude matrix analysis of Hawaiian rain-forest succession. *Journal of Ecology, 82*, 137-147, 1994.

Atkinson, I. A. E., Successional trends in the coastal and lowland forest of Mauna Loa and Kilauea volcanoes, Hawaii. *Pacific Science, 24*, 387-400, 1970.

Billings, W. D., J. O. Luken, D. A. Mortensen, and K. M. Peterson, Arctic tundra: a source or a sink for atmospheric carbon dioxide in a changing environment? *Oecologia, 53*, 7-11, 1982.

Britten, E. J., Hawaii as a natural laboratory for research on climate and plant response. *Pacific Science, 16*, 160-169, 1962.

Carlquist, S. C., *Hawaii, A Natural History. Pacific Tropical Botanical Garden*, 1980.

Carr, G. D., R. H. Robichaux, M. S. Witter, and D. W. Kyos, Adaptive radiation of the Hawaiian silversword alliance (Compositae-Madiinae), a comparison with Hawaiian picture-wing *Drosophila*, in *Genetics, Speciation, and the Founder Principle*, edited by L. V. Giddings, K. Y. Kaneshiro, and W. W. Anderson, pp. 79-97, Oxford University Press, New York, 1989.

Carson, H. L., and K. Y. Kaneshiro, *Drosophila* of Hawaii,

systematics and ecological genetics. *Annual Review of Ecology and Systematics, 7*, 311-345, 1976.

Carson, H. L., J. P. Lockwood, and E. M. Craddock, Extinction and recolonization of local populations on a growing shield volcano. *Proceedings of the National Academy of Sciences, 87*, 7055-7057, 1990.

Clements, F. E., Plant succession, an analysis of the development of vegetation. Carnegie Institution of Washington, Washington, D.C., 1916.

Cooper, W. S., Ecology of the strand vegetation of the Pacific coast of North America. *Carnegie Institution of Washington Yearbook, 18*, 96-99, 1919.

Cowles, H. C., The ecological relations of the vegetation on the sand dunes of Lake Michigan. *Botanical Gazette, 27*, 95-117, 167-202, 281-308, 361-391, 1899.

Crews, T., J. Fownes, R. Riley, D. Herbert, K. Kitayama, D. Mueller-Dombois, and P. Vitousek. Changes in soil phosphorus fractions and ecosystem dynamics across a long soil chronosequence in Hawaii. *Ecology*, in press, 1995.

Dawson, J. W., and L. Stemmermann, *Metrosideros* (Myrtaceae), in *Manual of the Flowering Plants of Hawaii*, edited by W. L. Wagner, D. R. Herbst, and S. H. Sohmer, Bernice P. Bishop Museum, Honolulu, Hawaii, 1990.

Drake, D. R., and D. Mueller-Dombois, Population development of rain forest trees on a chronosequence of Hawaiian lava flows. *Ecology, 74*, 1012-1019, 1993.

Eggler, W. A., Quantitative studies of vegetation on sixteen young lava flows on the island of Hawaii. *Tropical Ecology, 12*, 66-100, 1971.

Forbes, C. N., Preliminary observations concerning the plant invasion in some of the lava flows of Mauna Loa, Hawaii. *Occasional Papers of the Bishop Museum, 5*, 15-23, 1912.

George, T., B. B. Bohlool, and P. W. Singleton, *Bradyrhizobium japonicum*-environment interactions, nodulation and interstrain competition in soils along an environmental gradient. *Applied and Environmental Microbiology, 53*, 1113-1117, 1987.

Giambelluca, T. W., M. A. Nullet, and T. A. Schroeder, *Rainfall Atlas of Hawaii*. Department of Land and Natural Resources, State of Hawaii, Honolulu, 1986.

Holdridge, L. R., W. C. Grenke, W. H. Hatheway, T. Liang, and J. A. Tosi, *Forest Environments in Tropical Life Zones, A Pilot Study*. Pergamon Press, New York, 1971.

Hughes, R. F., P. M. Vitousek, and T. Tunison, Alien grass invasion and fire in the seasonal submontane zone of Hawai'i. *Ecology, 72*, 743-746, 1991.

Jenny, H., *Factors of Soil Formation*, McGraw-Hill, New York, 1941.

Jenny, H., *Soil Genesis with Ecological Perspectives*, Springer-Verlag, New York, 1980.

Jordan, C. F., and P. G. Murphy, A latitudinal gradient of wood and litter production, and its implications regarding competition and diversity in trees. *American Midland Naturalist, 99*, 415-434, 1978.

Juvik, J. O., and D. Nullet, A climate transect through tropical montane rain forest in Hawai'i. *Journal of Applied Meteorology, 33*, 1304-1312, 1994.

Juvik, J. O., D. C. Singleton, and G. G. Clarke, Climate and water balance on the Island of Hawaii. Mauna Loa Observatory 20th Anniversary Report, pp. 129-139, 1978.

Kitayama, K., and D. Mueller-Dombois, Vegetation of the wet windward slope of Mt. Haleakala, Maui, Hawaii. *Pacific Science, 46*, 197-220, 1992.

Kitayama, K., and D. Mueller-Dombois. Vegetation changes during long-term soil development in the Hawaiian montane rainforest zone. *Vegetatio*, in press, 1995.

Kitayama, K., D. Mueller-Dombois, and P. M. Vitousek, Primary succession of the Hawaiian montane rainforest on a chronosequence of eight lava flows. *Journal of Vegetation Science*, in press, 1995.

Lockwood, J. P., and P. W. Lipman, Recovery of datable charcoal beneath young lavas, lessons from Hawaii. *Bulletin Volcanologique, 43*, 609-615, 1980.

Lockwood, J. J., P. W. Lipman, L. D. Peterson, and F. R. Warshauer, Generalized ages of surface lava flows of Mauna Loa Volcano, Hawaii. Washington, D.C., U.S. Government Printing Office, 1988.

MacCaughey, V., Vegetation of Hawaiian lava flows. *Botanical Gazette, 64*, 386-420, 1917.

MacDonald, G. A., A. T. Abbot, and F. L. Peterson, Volcanoes in the Sea, The Geology of Hawaii. University of Hawaii Press, Honolulu, 1983.

Meinzer, F. C., P. W. Rundel, G. Goldstein, and M. R. Sharifi, Carbon isotope composition in relation to leaf gas exchange and environmental conditions in Hawaiian *Metrosideros polymorpha* populations. *Oecologia, 91*, 303-311, 1992.

Melillo, J. M., A. D. McGuire, D. W. Kicklighter, B. Moore, C. J. Vorosmarty, and A. L. Schloss, Global climate change and terrestrial net primary production. *Nature, 363*, 234-240, 1993.

Parton, W. J., D. S. Schimel, and D. S. Ojima, Modeling environmental change in grasslands. *Climatic Change, 28*, 142-160, 1994.

Pearcy, R. W., and H. Calkin, Carbon dioxide exchange and growth of C3 and C4 tree species in the understory of a Hawaiian forest. *Oecologia, 58*, 19-25, 1983.

Post, W. M., W. R. Emanuel, P. J. Zinke, and A. G. Stangenberger, Soil carbon pools and world life zones. *Nature, 298*, 156-159, 1982.

Raich, J. W., A. E. Russell, and P. M. Vitousek. Primary production and ecosystem development along elevational and age gradients in Hawaii. *Ecological Monographs*, submitted for publication.

Raich, J. W., and W. H. Schlesinger, The global carbon dioxide flux in soil respiration and its relationship to climate. *Tellus, 44B*, 81-99, 1992.

Rastetter, E. B., M. G. Ryan, G. R. Shaver, J. M. Melillo, K. J. Nadelhoffer, J. E. Hobbie, and J. D. Aber, A general biogeochemical model describing the responses of the C and N cycles in terrestrial ecosystems to changes in CO2. climate, and N deposition. *Tree Physiology, 9*, 101-126, 1991.

Riley, R. H., and P. M. Vitousek, Nutrient dynamics and trace gas flux during ecosystem development in Hawaiian montane rainforest. *Ecology, 76*, 292-304, 1995.

Rhodes, J. M., Homogeneity of lava flows, Chemical data for historic Mauna Loa eruptions. *Journal of Geophysical Research Supplement, 88*, A869-A879, 1983.

Robichaux, R. H., and J. E. Canfield, Tissue elastic properties of eight Hawaiian *Dubautia* species that differ in habitat and diploid chromosome number. *Oecologia, 66*, 77-80, 1985.

Schimel, D. S., W. J. Parton, T. G. F. Kittel, D. S. Ojima, and C. V. Cole, Grassland biogeochemistry, links to atmospheric processes. *Climatic Change, 17*, 13-25, 1990.

Skottsberg, C., Plant succession on recent lava flows in the island of Hawaii. Gotesborgs Kungl. Vetenskapsoch Vitterhets-samhalles Handlingar. *Sjatte Folden, Series B*, 1941.

Smathers, G. A., and D. Mueller-Dombois, Invasion and Recovery of Vegetation after a Volcanic Eruption in Hawaii. Natural Park Service Monograph, Washington, D.C., 1974.

Stemmermann, L., Ecological studies of Hawaiian *Metrosideros* in a successional context. *Pacific Science, 37*, 361-373, 1983.

Stemmermann, L., and T. Ihsle, Replacement of *Metrosideros polymorpha*, Ohia, in Hawaiian dry forest succession. *Biotropica, 25*, 36-45, 1993.

Townsend, A. R., P. M. Vitousek, and E. A. Holland, Tropical soils dominate the short-term carbon cycle feedbacks to atmospheric carbon dioxide. *Climatic Change, 22*, 293-303, 1992.

Uhe, G., The composition of plant communites inhabiting the recent volcanic deposits of Maui and Hawaii, Hawaiian Islands. *Tropical Ecology, 29*, 26-47, 1988.

Vitousek, P. M., The Hawaiian Islands as a model system for ecosystem studies. *Pacific Science, 49*, 2-16, 1995.

Vitousek, P. M., and L. R. Walker, Biological invasion by *Myrica faya* in Hawai'i, Plant demography, nitrogen fixation, ecosystem effects. *Ecological Monographs, 59*, 247-265, 1989.

Vitousek, P. M., P. A. Matson, and D. R. Turner, Elevational and age gradients in Hawaiian montane rainforest, foliar and soil nutrients. *Oecologia, 77*, 565-570, 1988.

Vitousek, P. M., C. B. Field, and P. A. Matson, Variation in $\delta^{13}C$ in Hawaiian *Metrosideros polymorpha*, A case of internal resistance? *Oecologia, 84*, 362-370, 1990.

Vitousek, P. M., G. Aplet, D. Turner, and J. J. Lockwood, The Mauna Loa environmental matrix, foliar and soil nutrients. *Oecologia, 89*, 372-382, 1992.

Vitousek, P. M., D. R. Turner, W. J. Parton, and R. L. Sanford, Litter decomposition on the Mauna Loa environmental matrix, Hawaii, Patterns, mechanisms, and models. *Ecology, 75,* 418-429, 1994.

Vitousek, P. M., D. R. Turner, and K. Kitayama. Foliar nutrients during long-term soil development in Hawaiian montane rain forest. *Ecology,* in press, 1995.

Wright, T. L., Chemistry of Kilauea and Mauna Loa lavas in space and time. *U. S. Geological Survey Professional Paper 735,* 1-49, 1971.

Wright, T. L., and R. T. Helz, Recent advances in Hawaiian petrology and geochemistry, in *Volcanism in Hawaii,* edited by R. W. Decker, T. L. Wright, and P. H. Stauffer, pp. 625-640, U.S. Geological Survey Professional Paper, Washington, D.C., 1987.

P. M. Vitousek, Department of Biological Sciences, Stanford University, Stanford, CA 94305

Gravity Changes on Mauna Loa Volcano

Daniel J. Johnson

Seattle, Washington

Abstract. Gravity observations made on Mauna Loa Volcano Hawaii, before and after the March 25-April 14, 1984 eruption indicate that a magma reservoir, centered 3630±200 m below the summit area, lost 136(±50)x10^9 kg of magma mass during the event. Comparison of the reservoir mass loss figure (ΔM) with the volume change by surface subsidence of the edifice (ΔV_e) gives $\Delta M/\Delta V_e$=2033 kg/m^3, consistent with the ratio predicted for magma withdrawal from a reservoir containing degassed, CO_2-poor magma. The net reservoir mass loss is insufficient to entirely account for the mass of erupted lava and dike intrusion. A proposed explanation is that a pulse of magma flow from depth, concurrent with the eruption, may have replaced reservoir magma lost to eruption. With this model, magma resupply to the shallow Mauna Loa reservoir is episodic and is associated with eruption; during repose, extensive CO_2 degassing and a low rate of magma resupply minimizes the CO_2 content of stored reservoir magma.

INTRODUCTION

Accelerating rates of seismicity and ground surface displacement observed at Mauna Loa Volcano led to a published statement in September 1983 of an increased probability of eruption within the following 2 years [*Decker et al.*, 1983]. Baseline gravity observations on Mauna Loa were made February 13-14, 1984 in anticipation of the eruption. Fortuitously an eruption began on March 25 at 01:25 HST [*Lockwood et al.*, 1985] within just 6 weeks of the initial gravity observations. Additional measurements were begun 8 hours after the first sighting of lava, and continued at intervals of 1-6 days for the duration of the 3 week eruption. Analysis of these data give a unique perspective on how Mauna Loa works.

A classic view of the eruptive behavior of Hawaiian volcanoes is that they contain a shallow magma reservoir, located below the summit, that gradually fills with mantle-derived magma during repose. Then, on eruption, previously stored magma is rapidly expelled from the reservoir to the surface or into rift zone intrusions [*Dzurisin, et al.*, 1984; *Decker et al.*, 1983]. This view is reinforced by observation of surface uplift during periods of repose and subsidence of the edifice during eruption, indicating filling and draining of a subsurface magma storage zone.

An unresolved issue of the 1984 eruption is a disparity between the volume of edifice contraction and the volume of erupted lava [*Dvorak et al*, 1985]. Previous estimates of the volume of edifice collapse are 110x10^6 m^3 [*Okamura et al.*, 1984], 100(±30)x10^6 m^3 [*Lockwood et al.*, 1985], and 55(±15)x10^6 m^3 [*Dvorak et al.*, 1985]. Approximately 220x10^6 m^3 of lava, which has an estimated density of 2000 kg/m^3, reached the surface during the eruption [*Lipman and Banks*, 1987]. This is equivalent to 170x10^6 m^3 of magma with density 2600 kg/m^3 - still more than the volume of collapse. Consider also that a significant volume of magma was delivered to the inferred 22 km-long intrusive dike that bisected the summit and rift zones. Perhaps 75x10^6 m^3 of magma (dike roughly estimated 0.75 m wide, 5000 m high, and 20 km long) ended up stored within the rift zone dike. The total of dike and lava flow volumes gives 245x10^6 m^3, far in excess of the subsidence volumes given above.

Okamura et al. [1984], states that the volume discrepancy between subsidence and erupted lava might be due to: (1) eruption of magma stored within the rift zone since the previous eruption in 1975, (2) subsidence restricted by crustal rigidity, and (3) vesiculation of stored reservoir magma. The first process may be minor, as geochemical analyses of 1984 lava samples presented by *Rhodes* [1988] do not indicate a significant proportion of rift zone-derived

Mauna Loa Revealed: Structure,
Composition, History, and Hazards
Geophysical Monograph 92

lavas. The remaining two processes are shown by *Johnson* [1992] to be important at neighboring Kilauea Volcano during a recent phase of frequent, low-volume eruptions. The idea is that the shear strength of the edifice limits the amount of downward sagging of the crust overlying the draining magma reservoir, while the space left by the expelled magma is claimed by decompressional expansion of magma and CO_2, as well as CO_2 exsolution.

Dvorak et al. [1985] proposed that additional magma reservoirs may also have contributed to the eruption, explaining limited subsidence with respect to the volume of lava observed at the surface. Such reservoirs may have been beyond the perimeter of the geodetic network, or possibly located deep enough that surface displacement was not detectable.

Johnson [1992] presented theoretical arguments and a suite of gravity and geodetic observations from Kilauea Volcano that show that it is not strictly necessary for the collapse volume to equal the volume of magma removed. This is because the volume change observed at the surface is the sum of the volume change due to removal of mass (i.e. magma), bulk compression of the magma resident in the reservoir, volatile (mainly CO_2) compression, exsolution, and migration, and lastly volume change due to density redistribution of the crust. For example, while magma is being removed from the reservoir during eruption, decreasing internal pressure causes exsolution of CO_2 plus volumetric decompression of exsolved gas and magma. All of these factors mitigate reservoir contraction [*Johnson*, 1992, equation 8]. Concurrently, the volume change of the edifice is 1.5 times the change in size of the reservoir cavity as a consequence of the crustal density change associated with deformation [*Johnson*, 1992, equation 9, with a Poisson's ratio of 0.25 typical of crustal material]. While the processes internal and external to the reservoir have opposite influence on the ratio of edifice volume change to internal reservoir mass change, *Johnson* [1992] shows that at Kilauea the internal processes may at times dominate. Comparison with observations from Kilauea are useful in the analysis of Mauna Loa.

The purpose of this paper is to use the gravity data collected before and after the 1984 eruption to examine the observed volume disparity between edifice contraction and lava flow at Mauna Loa. The utility of the gravity method with respect to monitoring a subsurface magma reservoir is that it is sensitive to *mass* change, whereas geodetic methods (such as leveling, tilt, trilateration, GPS) detect surface displacement only. An apparent volume change detected by geodetic methods may reflect expansion/contraction of existing crust and reservoir material as well as addition/subtraction of magma from the system. Analysis of gravity data may thus help sort out these kinds of volume ambiguities. The first goal is to determine the actual mass of magma removed from the known summit magma reservoir and determine if this amount is sufficient to explain the mass of the eruptive products. Secondly, the relationship between mass removed and the resulting summit collapse will be analyzed to learn more about the shear strength of the edifice and the compressibility of the magma reservoir itself.

THE 1984 ERUPTION

Mauna Loa, like neighboring Kilauea Volcano, contains a central subcaldera magma reservoir which is recharged by magma during repose [*Decker et al.*, 1983; *Rhodes*, 1988]. With time, this filling produces a measurable distention of the edifice. Analysis of surface displacement patterns prior to the 1984 eruption by *Decker et al.* [1983] placed the region of filling roughly 3 km below the southeast rim of the summit caldera Mokuaweoweo.

The events of the March 1984 eruption have been described by *Lockwood et al.* [1985]. The first phase of eruption saw the propagation of an eruptive fissure to the floor of Mokuaweoweo at 01:25 HST on March 25. Over the next several hours the eruptive fissures migrated out of the caldera, into both the southwest and the northeast rift zones. By 07:00 HST fountaining was restricted to a portion of the upper northeast rift zone at an elevation of 3700 m. The eruption migrated down the northeast rift zone through the first day in a series of jumps; as new fountains appeared downrift, activity farther uprift waned. At 16:41 HST venting began near 2900 m elevation and continued in that vicinity for the remainder of the 3-week eruption.

Observations

As the eruption progressed, subsidence of the ground surface above the reservoir was monitored by frequent geodetic surveys [*Lockwood et al.*, 1985; *Dvorak and Okamura*, 1987]. Subsidence is attributed to magma removal from the reservoir; some of this magma was intruded as dikes into both of Mauna Loa's rift zones while a large volume was erupted to the surface [*Lockwood et al.*, 1985]. Locations of geodetic and gravity observation sites on Mauna Loa are shown in Figure 1, and measured tilt and leveling changes are illustrated in Figure 2. The area of maximum subsidence, as indicated by the orientation of ground tilting and the vertical movement of leveling benchmarks, was located southeast of Mokuaweoweo Caldera (Figure 2), at a location similar to the area of previous uplift [*Decker et al.*, 1983].

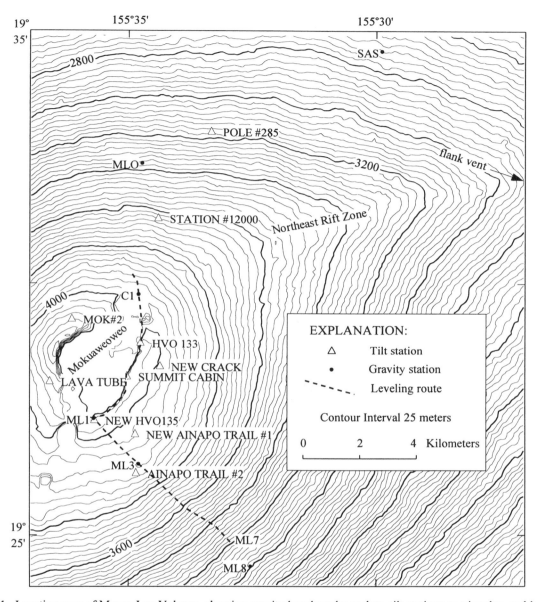

Fig. 1. Location map of Mauna Loa Volcano, showing gravity benchmarks as dots, tilt stations as triangles, and leveling route as dashed line. Site SAS is gravity base station and leveling data are referenced to ML7. Eruptive fissures migrated from the summit caldera down the Northeast Rift Zone during the first day of the 1984 eruption and became confined to region marked "vent site" for the remaining 3 weeks of the eruption.

Leveling surveys to third-order standards were done on June 27, 1983 and May 7-8, 1984 [*Okamura et al.*, 1984; *Dvorak, et al.*, 1985] on a route that traverses the summit area of Mauna Loa. A maximum subsidence of 574 mm relative to site ML7 was measured along the southeast rim of Mokuaweoweo caldera. Most likely neither end of the leveling traverse was distant enough from the apex of subsidence to escape subsidence. An estimate of the amount of subsidence of the reference benchmark, or "float" of the level line, will thus be made in the following section.

Occupation of the entire inventory of spirit-level tilt sites located around the rim of Mokuaweoweo caldera and the upper slopes of Mauna Loa was done between July 12-August 24, 1983 and April 23-27, 1984. These data are considered to have a precision of ±12 μrad [*Dvorak and Okamura*, 1987]. The changes (Figure 2) define an inward tilt, towards a common focus at the southeast rim of

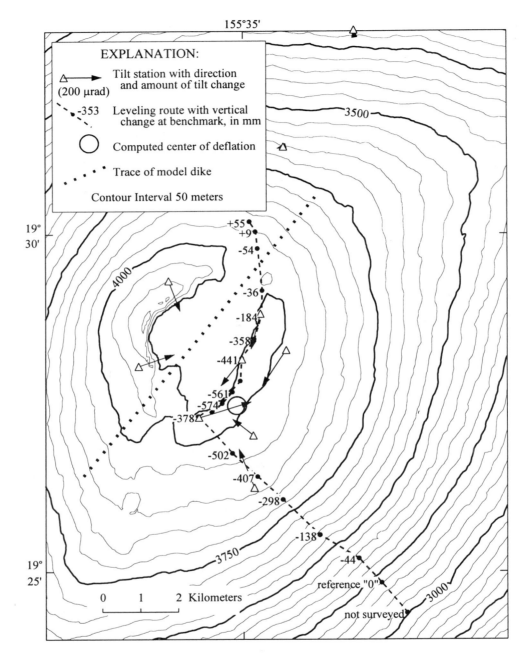

Fig. 2. Mauna Loa summit area tilt and level observations spanning March-April 1984 eruption. Tilt changes [*Lockwood et al.*, 1985] measured between July 12-August 24, 1983 survey and post-eruption survey April 23-27, 1984 drawn as vectors. Precision of each tilt measurement, determined by measuring changes in relative elevation along 40-m-long triangles, is ±12 μrad [*Dvorak and Okamura*, 1987]; measurement sites are shown by triangles at the tail of each arrow; arrows point in direction of relative downward deflection. Numbers indicate relative elevation changes of benchmarks, shown as solid dots, measured between third-order leveling surveys [*Okamura, et al.*, 1984; *Dvorak, et al.*, 1985]. Vertical movement of leveling benchmarks relative to ML7 (labeled as "0") between June 27, 1983 and May 7-8, 1984 labeled on map adjacent to benchmark location. Point of maximum subsidence near the southeast rim of Mokuaweoweo indicated as open circle.

Mokuaweoweo.

A complete survey of gravity sites C1, ML1, ML3, and ML8 was done on February 13-14 and May 2, 1984. Gravity readings are corrected for tidal effects [*Longman,* 1959]. Calibration functions with linear and periodic terms were determined from calibration ranges and applied to the data. Gravity data were reduced using JOSH v. 3 [unpublished, 1994] which inputs data from an unlimited number of individual runs and calculates a least squares solution of second-order polynomials to approximate time-dependent changes in the reading level of the gravimeters (gravimeter drift), offsets of the reading level (tares) as needed, and relative gravity *g* at each surveyed station. A run comprises a sequence of gravity readings made using a particular gravimeter. Separate runs, which may be made during the same day or on multiple days, are combined to make a survey. In this study, gravity surveys were done using two gravimeters run over closed loops between the base station SAS and monitoring sites using helicopter transport. The February survey comprised two loops over both days and reduced values have standard errors of from ±8 to ±12 µGal. The May survey comprised three loops on the same day and values have std. errors of about ±7 µGal. Observed gravity changes between the complete surveys, bracketing the 1984 eruption, are given in Table 1 along with corresponding elevation changes.

During the course of the 1984 eruption gravity measurements were occasionally made at station ML1 to record the chronology of gravity change. These surveys were accomplished by closing two loops between SAS and ML1 with two gravimeters. The exception was the March 26 survey, when only one loop could be completed because helicopter support was unavailable. A time plot of ML1 gravity is given in Figure 3. First impressions of the ML1 data are that the pattern of change closely follows the exponentially diminishing rates of tilt change and horizontal strain [*Lockwood, et al.,* 1985]. Also, the positive sign of the change is consistent with a strong contribution of an increase due to the decreased height of the observation point (the free-air change), which is both greater and of opposite sign to the component due to the subsurface magma mass loss.

As a part of the gravimeter calibration procedure, a gravity tie was made between base station SAS and GC7, located 16 km north of SAS on the lower slope of neighboring Mauna Kea Volcano. Measured gravity change at SAS between pre- and post-eruption surveys on February 21 and May 15, 1984 is +1.6±11.2 µGal relative to GC7. The absence of a significant gravity change at SAS diminishes the probability that this station moved up or down.

TABLE 1. Observed Gravity and Height Changes: February 13-14 to May 2, 1984

Station	Observed Δg (µGals)	Gravity std. error (µGals)	Observed Δh (mm)
C1	56.6	±10.4	-54
ML1	135.7	±10.9	-378
ML3	150.7	±13.4	-407
ML7	-	-	reference 0
ML8	13.5	±13.3	-
SAS	reference 0	±8.9	-

REVIEW OF MODEL EQUATIONS

Model for Deformation from Reservoir Volume Change

The principles of deformation and gravity analysis presented here are a foundation for the analysis that follows. These generalized equations enable inferences to be made about the nature of mass and volume changes at depth associated with the gravity and surface displacement anomalies. However, because of the dramatic topography of Mauna Loa which is not anticipated in the derivation of the principles, they should be used with some caution.

Deformation resulting from the inflation and deflation of the summit reservoir of Mauna Loa is simulated using a model first applied to volcanology by *Mogi* [1958]. This model gives deformation of an elastic body having one free surface as a function of pressure change within a spherical cavity inside the body. Surface uplift is given as

$$\Delta h = \frac{3\Delta P V_r (1-\nu)}{4\pi\mu} \cdot \frac{Z}{(Z^2 + X^2)^{3/2}} \qquad (1)$$

where Z is the source depth, X is the radial distance of the point from the source epicenter, ΔP is the pressure change, V_r is the volume of the source and ν and μ are the Poisson's ratio and shear modulus of the body [modified from *Hagiwara*, 1977]. The change in radius, Δa, of the source of radius a is given by *Hagiwara* [1977] as:

$$\Delta a = \frac{a\Delta P}{4\mu} \qquad (2)$$

As long as a is large relative to Δa, the volume change ΔV_r of the spherical source may be estimated as the surface area of the sphere ($4\pi a^2$) times the change in radius (equation 2), or

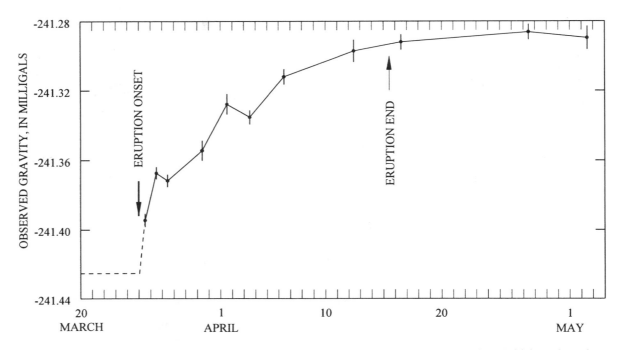

Fig. 3. Time plots of gravity at ML1 differenced with SAS during the 1984 Mauna Loa eruption. Initial gravity value from survey 6 weeks prior to eruption onset.

$$\Delta V_r = \frac{\pi a^3 \Delta P}{\mu} = \frac{3 V_r \Delta P}{4\mu} \qquad (3)$$

[*Johnson*, 1987, with volume relation $3V_r/4 = \pi a^3$ substituted]. Integration of equation (1) over the surface of the body gives

$$\Delta V_e = \frac{3 \Delta P V_r (1-\nu)}{2\mu} \qquad (4)$$

which is the volume change of the body due to displacement, Δh, of the free surface. Division of equation (4) by (3) gives

$$\Delta V_e = 2(1-\nu)\Delta V_r \qquad (5)$$

which is the volume change of the body as a function of volume change of the imbedded spherical source. Notice that for a Poisson's ratio of 0.25, typical of crustal rock, equation (5) predicts dilation, or expansion, of the crust equal to 50% of the volume change of the source. The volume of surface uplift would equal the volume change of the source only if the Poisson's ratio were 0.5; media with such a Poisson's ratio include rubber and fluids.

To the gravity modeler, the significance of crustal dilation predicted by equation (5) is the implied density change, which has a direct effect on the measured gravity field. The variation in density $\Delta\rho$ at a point located within the body at depth D below the surface is

$$\Delta\rho = \frac{(1-2\nu)}{\mu} \cdot \frac{3\rho V_r \Delta P}{4\pi} \cdot \frac{X^2 - 2(D+Z)^2}{(X^2+(D+Z)^2)^{5/2}} \qquad (6)$$

where Z is the depth of burial of the spherical source, and X is the horizontal distance between source and observation point [modified from *Hagiwara*, 1977]. Notice, again, that equation (6) predicts a changing density distribution within the body, except in the special case that $\nu = 0.5$.

To the volcanologist seeking to estimate the magma budget of a volcano by monitoring surface uplift, the significance of equations (5) and (6) is that a portion of the volume of expansion or contraction of a volcanic edifice is the consequence of crustal density change, not magma accumulation. Analyses that have assumed that one unit volume of uplift is equivalent to one unit volume of reservoir magma accumulation (of which there are many) are thus in error.

Model for Gravity Change from Reservoir Volume Change

The problem of modeling gravity change associated with an altered density distribution of the crust has been treated by *Hagiwara* [1977], *Rundle* [1978], *Walsh and Rice* [1979], and *Savage* [1984] for the case of a spherical source. The consensus is that deformation of the crust caused solely by the volume change of a spherical source does not produce a net gravity change. Previous gravity studies of Kilauea Volcano [*Jachens and Eaton*, 1980; *Dzurisin et al.*, 1980; *Johnson*, 1987; *Johnson*, 1992] have made the explicit assumption that crustal deformation yields no gravity change. (Some gravity change, however, is expected due to related vertical movement of the observation site with respect to the mass of the Earth plus any change in the mass of magma contained within the source.)

It is useful to review some details of the *Hagiwara* [1977] study of deformation-induced gravity change for a spherical source model. *Hagiwara* [1977] separated the deformation into three components and solved the gravity change for each separately: (1) the volume change of the spherical source cavity, (2) the surface uplift, and (3) the crustal density change. I have modified the original equations to reduce the number of elastic constants to only the Poisson's ratio v and to use a source volume change term ΔV_e rather than a pressure change.

The component of gravity change due to the volume change ΔV_r of the source is

$$\delta g_0 = \gamma(\rho_0 - \rho_c)\Delta V_r \cdot \frac{Z}{(X^2 + Z^2)^{3/2}} \tag{7}$$

where ρ_c is the crustal density. The value of γ is 6.67×10^{-11} Nm^{2}kg^{-2}. Included is an allowance for mass, such as magma, of density ρ_0 which replaces displaced crustal material. If no matter moves into or out of the source to balance ΔV_r, then $\rho_0 = 0$. The magnitude of (7) is essentially the gravitational attraction of a spherical shell that represents the mass gained or lost by a change in diameter of the source. For example, expansion of an empty reservoir would result in a spherical shell where empty space of density $\rho_0 = 0$ has displaced crustal rock of density ρ_c. This would give a negative gravity change component.

The component of gravity change due to surface uplift is

$$\delta g_1 = 2\gamma\rho_c(1 - v)\Delta V_r \cdot \frac{Z}{(X^2 + Z^2)^{3/2}} \tag{8}$$

To illustrate this component, consider a gravimeter fixed in space above the Earth's surface. An increase in V_r results in surface uplift - uplift moves mass upward and displace air with a thin layer of crustal material. This layer has an area equal to the uplift anomaly, and a variable thickness depending on the local uplift. The gravimeter located above the uplifted area would record a gravity increase due

to this component of deformation.

Finally, an outcome of deformation is variation in the density of the crust within the deformed region. The gravity change component due to density variation is

$$\delta g_2 = -\gamma\rho_c(1 - 2v)\Delta V_r \cdot \frac{Z}{(X^2 + Z^2)^{3/2}} \tag{9}$$

Notice that only in the unrealistic case of a Poisson's ratio of 0.5 does the density change term vanish. Otherwise, for a typical Poisson's ratio of 0.25, the effect of this term for inflation, as an example, is a gravity decrease corresponding to the net density decrease of the deformed crust.

Considering only the deformation-induced components of the gravity change (by setting $\rho_0 = 0$ to reflect no inflow or outflow of matter from the source sphere), the uplift component (δg_1) is exactly offset by the sum of the source volume change and crustal density change components ($\delta g_0 + \delta g_2$). In other words, the net gravity change due to deformation caused by a point source is nil as stated by *Hagiwara* [1977], *Rundle* [1978], *Walsh and Rice* [1979], and *Savage* [1984].

The residual gravity change, $\Delta g'$ using the notation of *Johnson* [1992], is defined as the sum of the above components, including the mass of material flowing into or out of the source, but not including the free-air change. Summing (7), (8), and (9) above gives

$$\Delta g' = \gamma\rho_0\Delta V_r \cdot \frac{Z}{(X^2 + Z^2)^{3/2}} \tag{10}$$

Division of (10) by (1) and substitution of (3) gives

$$\frac{\Delta g'}{\Delta h} = \frac{\pi\gamma\rho_0}{(1 - v)} \tag{11}$$

Equation (11) is of greater use if we modify ρ_0 to incorporate the concept of "effective density" which is used in previous work on Kilauea Volcano by *Jachens and Eaton* [1980]. To do this, recall that ρ_0 is, in effect, the mass change within the spherical source divided by the volume change of the source, or $\Delta M / \Delta V_r$. Equation (11) is further generalized if all mass change within the source volume is included in ΔM, regardless of whether the mass is accommodated specifically within ΔV_r or, as more likely the case, distributed within the entirety of the reservoir cavity V_r. Equation (11) is then rewritten as

$$\frac{\Delta g'}{\Delta h} = \frac{\pi\gamma}{(1 - v)} \cdot \frac{\Delta M}{\Delta V_r} \tag{12}$$

Equation (12) is made directly equivalent to the *Jachens and Eaton* [1980] effective density concept as well as previous equations given by *Johnson* [1987, equation 5; 1992, equation 3] by substitution of equation (5) giving

$$\frac{\Delta g'}{\Delta h} = 2\pi\gamma \cdot \frac{\Delta M}{\Delta V_e} \qquad (13)$$

The difference is use of the edifice volume change ΔV_e rather than the source volume change ΔV_r.

Gravity benchmarks ride on the Earth's surface, and so deformation also changes the height of the benchmarks with respect to the Earth's center of mass. Gravity changes observed at the surface include the free-air term which accounts for vertical reposition of the gravity site with respect to the Earth's gravity field. The Earth's gravity field decreases with height at a gradient of -0.3086 μGal/mm on average and at sea level. The complete expression for the gravity change (Δg) observed by a gravimeter resting on the surface is

$$\Delta g = \left\{ 2\pi\gamma \cdot \frac{\Delta M}{\Delta V_e} - 0.3086 \right\} \cdot \Delta h \qquad (14)$$

using units of μGal and mm.

Model for Deformation and Gravity Change from Dike Intrusion

Yang and Davis [1986] give analytic expressions for displacements and strain for an elastic half-space subject to opening of a rectangular crack. This model, published as FORTRAN subroutine DEFOR is adopted here to simulate the March 25 Mauna Loa dike-forming intrusion and the resulting deformation. The eight parameters of the dike are latitude and longitude of the dike center, dike length, azimuth and dip, depth to dike top and bottom, and dike thickness. Surface displacements are output directly, and density changes throughout the body are calculated from the output strains.

Savage [1984], *Sasai* [1986], and *Okubo* [1992] give expressions for the gravity changes caused by deformation associated with the widening of a vertical crack. In contrast to the spherical source example, the ratio of gravity change to uplift is not simply linear for the vertical crack.

Application of Models to Mauna Loa

A critical drawback of using (14) to analyze gravity changes on Mauna Loa is that the derivation of the equation is based on a model having a free surface which is flat. Mauna Loa, in reality, like all volcanoes has a free surface which is irregular and sloped. The topography of Mauna Loa affects the modeled gravity and deformation changes. In terms of gravity, the changes predicted by the density change component of the model involve parts of the crust that do not in reality exist and they involve a surface uplift component which is based on a model with a mislocated surface. As an example, station ML1 is located at a topographically high spot on the rim of Mokuaweoweo caldera; the surface of Mauna Loa adjacent to ML1 is generally lower in altitude. The model is based on surface uplift at a similar horizon as ML1, however, in reality the uplifted areas are significantly below the level of ML1. In terms of deformation, the unique geometry of Mauna Loa will make actual deformation differ from the model predictions.

Procedures used in the following analysis section take into consideration the effect of topography on the gravity change. To allow for the vertical relief of Mauna Loa in the gravity change calculations, a numerical model is used that has a grid of surface elevations at 100 meter spacings to correctly position the mass elements of Mauna Loa's edifice with respect to the observation points in three-dimensional space. Testing of the model using a flat surface grid gives identical results to (13) above. A correction for the effect of topography on deformation is not implemented in this analysis. Here deformation is calculated using a model which has a flat free surface at an elevation of 3970 meters, representative of the general elevation of the summit area of Mauna Loa.

ANALYSIS

Determining Subsidence of ML8

A problem that has to be dealt with first is that the leveling survey conducted prior to the eruption did not include ML8. I made pre-eruption gravity measurements at ML8 because, as I understood at the time, it was at the southeast terminus of the level line. Unfortunately, it was only after the eruption that I learned that, while ML8 was considered to be a "leveling benchmark", actual leveling surveys had never reached ML8 - previous leveling had extended only as far as ML7.

The height change of ML8 was reconstructed by fitting a least-squares "Mogi" solution to the leveling data, plotting measured height changes as a function of radial distance from the source, and extrapolating the model height change curve to predict the height change of ML8. Figure 4 shows the results of this procedure; ML8 is estimated to have moved up 27 mm relative to the zero datum at ML7. Note that the latitude and longitude of the source are among the estimated parameters, so the X position of the data on the plot are not fixed in the inversion.

The zero change at ML7, of course, is not absolute: increasing values of the model curve at radial distances further out than ML8 suggest that the entire leveling line subsided relative to land beyond the periphery of the surveyed area. Simply extrapolating the model curve shown in Figure 4 to where it flattens out gives a total of 90 mm absolute subsidence of the reference site ML7.

Determining Height Change of "Floating" Level Line

It is apparent from the subsidence curve plotted in Figure 4 that the distant stations ML7 and ML8 experienced subsidence relative to points further out. How much the level network "floated" relative to stable ground surrounding the subsidence region is important to the analysis that follows. There are no geodetic measurements that tie subsidence measurements to a stable datum. However, an estimate of the subsidence can be made from the gravity data since gravity change and uplift are related by the free-air gradient.

A positive gravity change of +13.5±13.3 µGal measured at ML8 (Table 1, relative to SAS) is evidence of slight subsidence. The classic interpretation of this change would apply a $\Delta g'/\Delta h$ ratio of around -0.2 µGal/mm (simply the Bouguer adjusted free-air gradient) to predict an elevation change of minus 67 mm. The problem with this approach is that actual $\Delta g'/\Delta h$ ratios are difficult to predict and vary greatly from volcano to volcano and over time at the same volcano [*Rymer*, 1994]. In fact, rather than applying a generic ratio to gravity data as another way to measure uplift, current volcano research using gravity methods is now focused on modeling the details that make observed ratios differ from the previously expected ones [e.g. *Johnson*, 1992].

So, given that there is really no universal conversion factor for gravity and height changes on volcanoes, we should use only one that is locally determined for the specific time period. Here a representative factor was picked using relative changes between ML1, ML3, and ML8 (site C1 was excluded because of it's proximity to the eruptive fissure). A linear fit to these data, shown in Figure 5, gives a ratio of -0.3096 µGal/mm. Thus, the 13.5 µGal increase at ML8 relative to SAS corresponds to 43 mm of subsidence at this site, assuming that the observed correlation is constant over the region of subsidence.

We now have an estimate of the subsidence at the reference benchmark ML7; this offset may be applied to all the level data to correct for the vertical "float" of the level line.

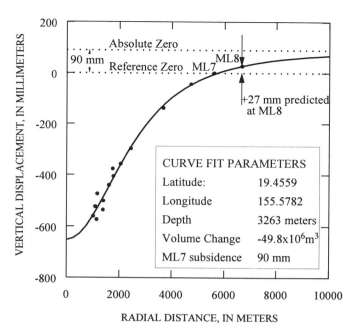

Fig. 4. Comparison of observed elevation changes with best-fit Mogi model solution. Only data from leveling benchmarks south of HVO133 are considered because an eruptive fissure crossed the northern portion of the leveling line and locally affected the data. Parameters of the point-source model used to fit the data are listed in inset. Extension of model subsidence curve to radial distance of ML8 gives a predicted change of +27 mm relative to reference zero at ML7. Extension of curve to an infinite radial distance gives 90 mm.

The ML7 subsidence value of -70 mm is a sum of -27 mm change relative to ML8, estimated from point-source model, plus -43 mm change of ML8 relative to SAS, estimated by extrapolating the correlation between height and gravity change. Height change values corrected for this network float are listed in Table 2. These values are presumed here to be absolute with respect to initial positions.

The network float estimate of -90 mm obtained earlier from a least-squares fit of a point-source model to the leveling data is similar, giving us some confidence that the values are accurate. The method used to obtain this value, while useful for extrapolating height changes over the short 1-km distance between ML7 and ML8, probably should not be extended as a predictor of network float because the model does not include the effect of the dike and topography.

Inversion of Geodetic Data for Dike and Reservoir Parameters

Displacement of Mauna Loa's surface resulted from a composite of the injection of magma into a vertical dike,

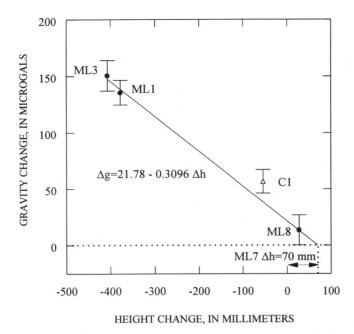

Fig. 5. Changes in gravity differenced with SAS plotted against changes in elevation differenced with ML7 for sites in the summit area of Mauna Loa during the 1984 eruption. Gravity measurements made February 13-14, 1984 and May 2, 1984; leveling data from surveys June 27, 1983 and May 7-8, 1984. Extrapolation of $\Delta g'/\Delta h$ correlation suggest that leveling reference station ML7 may have subsided by about 70 mm relative to gravity reference site SAS.

feeding the surface effusion of lava, and contraction of the summit reservoir as some previously stored magma exited into the dike and to the surface. Since the dike and reservoir superimposed displacements on each other, the inversion used here attempts to model both sources together. The goal here is to estimate the position and volume change of the dike and summit reservoir.

Equations used to describe deformation due to a dike are from *Yang and Davis* [1986]. Eight parameters of the dike are latitude and longitude of the dike center, dike length, azimuth and dip, depth to dike top and bottom, and dike thickness. In this case, visual observations of the eruptive fissure at the surface can be used to fill in some of the parameters and, thus, reduce the number of unknown parameters in the inversion. Figure 2 shows a plan view of the model dike which corresponds to the surface trace of the curtain of fire. It is centered at 19.4753°N, 155.5940°W, strikes 38.6°, and is 10 km in length. In reality the fissure extended further to the northeast and southwest, but increasing the model dike length was found to not affect the results of the inversion. This is because the ends of the model dike are well outside the survey area

and contribute little to the total change observed at each site. The depth to the dike top is set to zero and the dike is assumed to be vertical. This leaves two remaining unknowns: depth to the dike bottom and dike width.

Equations used in modeling deformation due to volume changes of a spherical source are given by *Hagiwara* [1977]. Unknowns are latitude, longitude, and depth of the source, and the volume change of the source. The magma reservoir is situated a few kilometers below the surface so there are no visual clues to the model parameters as there are for the dike - all reservoir parameters will come from the inversion.

Data used in the inversion consisted of all tilt and height changes shown in Figure 2. Based on the determination described above, height changes were adjusted by -70 mm to account for the network "float" and a change of -43 mm was assigned to ML8. The routine used to find the best-fit dike and summit reservoir parameters consisted of 3000 sets of forward dike model calculations and point-source model inversions. The 3000 sets cover all combinations of dike width between 400 and 1000 mm in 10 mm steps and depth to dike bottom between 2000 and 7000 meters in 100 meter steps. For each set, first a forward model was run which corrected the tilt and level observations for the effect of the dike intrusion. Then the adjusted data were inverted using a least-squares technique [*Dvorak et al.*, 1983] to determine the parameters for a best-fit elastic point source model. Finally, the list of 3000 resultant sum of squared residuals was searched for the minimum value. The parameters that corresponded with the best least-squares fit are listed in Table 2. The horizontal location of the reser-

TABLE 2. Results from Geodetic Data Inversion

Parameter	Value	Error	Unit
DIKE:			
length	10000	fixed	meters
azmuth	38.6	fixed	degrees
latitude of center	19.4753	fixed	°N
longitude of center	155.5940	fixed	°W
dip	90	fixed	degrees
depth to top	0	fixed	meters
depth to bottom	5000	±1000	meters
thickness	0.73	±0.1	meters
RESERVOIR:			
latitude	19.4583	±0.0014	°N
longitude	155.5854	±0.0008	°W
depth	3630	±150	meters
ΔV_e	-66.9x10^6	±4.6	cubic meters

voir source is mapped on Figure 2.

Results of the inversion are sensitive to the estimate of level line "float", or subsidence of the reference benchmark. If the level line as a whole is assumed to have subsided more than that predicted here, then the inversion yields a larger subsidence volume and a deeper Mogi source depth. An attempt was made to run the geodetic inversion with the network float as an additional unknown parameter. This endeavor did not yield a stable solution, however, probably because of the lack of far-field geodetic data. Without far-field data, the parameters of the Mogi source depth and the dike bottom depth tend to couple with, and also drift with, the level line float parameter. This problem is attacked here by using the far-field gravity tie to assess the vertical movement of the leveling reference benchmark; hence the accuracy of the gravity tie and assumptions used in the assessment have some bearing on the realiablity of the inversion results.

Determining the error bounds for the model solution is problematic because of the mixture of forward and inverse modeling techniques used. Adjustment of the dike parameters by up to ±0.1 meters in width and ±1000 meters depth of dike bottom can be made with only slight degradation of the model fit. The model fit is substantially poorer with parameters beyond the bounds given above, so we may be fairly confident that the true values lie within the interval. Error bounds for the three dimensional position of the spherical source is estimated at ±200 meters, and for the volume change ±5x10^{6}m^3. The error bounds given here indicate only the uncertainty of the model parameters as determined from the available data. Additional data would have helped to better resolve the parameters and, more importantly, to constrain the selection of appropriate models.

Adjustments to Gravity Data

The observed gravity changes are a composite of the changes due to mass movement and density change throughout Mauna Loa's edifice, plus the change due to variation of the vertical position of the gravity stations with respect to the Earth's mass. In this section, we adjust the gravity measurements for known changes resulting from the dike intrusion, contraction of the reservoir, and height change of the gravity sites. The gravity effect of the mass of the magma that flowed out of the reservoir during the deflation is left as the single remaining unknown. This will be estimated later from the residual gravity changes.

Rather than use the theoretical relations of $\Delta g'/\Delta h$ in this study (e.g. equations 13 and 15), I have chosen to make specific calculations of $\Delta g'$ and Δh using the basic equa-

tions for deformation and explicit source parameters. The main improvement results from the ability to specify in three-dimensions the location of mass components that are being displaced and dilated. The gravity change at a specified position (corresponding to a gravity station location) is calculated individually for all components and the values are summed to give the composite change.

I use a digital elevation model (DEM) for the Mauna Loa vicinity (modified by NOAA from USGS 7.5' quadrangle data to give 100 meter grid spacing) to insure that only mass components that actually exist are included in the summation. The DEM grid used here covers a 40 km by 40 km area of Mauna Loa, giving elevations at 100-meter-spaced grid nodes to the nearest meter. Three-dimensional calculations are based on subdivision of vertical columns below each DEM grid node into 100-m cubes Each column of cubes has an origin at the surface and extends 20 km below the surface.

Table 3 summarizes the adjustments made to the observed gravity changes and gives the residuals.

Free-air gradient. The free-air gradients used here to correct the gravity data for height change are up to 16 percent steeper than the -0.3086 µGal/mm value considered average for the Earth's surface. *Hammer* [1970] points out that the actual gradient may differ from theory in areas of local gravity anomalies due to density contrasts in the Earth's crust. At Kilauea Volcano, field measurements by *Johnson* [1992] gave a free-air gradient 6 percent steeper than normal; a positive Bouguer anomaly [*Kinoshita et al.*, 1963] centered over Kilauea may contribute to the steepness of the gradient. While the summit area of Mauna Loa also has a positive Bouguer anomaly indicating a dense intrusive core, similar to Kilauea, an even greater concern is the topographic effect of Mauna Loa's high profile on the gradient. There is a huge density contrast between the rock of Mauna Loa's cone and the air which surrounds its flanks - many times the density contrast between the intrusive core and edifice. A very steep free-air gravity gradient on Mauna Loa would be expected from this.

The free-air gradients are calculated individually for each station and are listed in Table 3. These values are derived from the theoretical gradient formula given by *Hammer* [1970, equation 1] for specified site elevation and latitude. Then a correction was made to the gradient for terrain above sea-level. This involved use of a DEM of the Island of Hawaii with elevations at 100 meter grid spacing as input to a program that calculated the gravitational attraction of the island's mass at a specific location. A density of 2300 kg/m^3 [Kinoshita et al., 1963] was used and only mass above sea level was included in the model. Two calculations were made for each station: one at the surface

TABLE 3. Adjusted Gravity and Height Changes: February 13-14 to May 2, 1984

Station	Observed Δg (μGals)	Estimated Absolute Δh (mm)	Free-air Gradient (μGal/mm)	Free-air Correction (μGal)	Mogi Source Correction (μGal)	Dike Source Correction (μGal)	Residual $\Delta g'$ (μGals)
C1	56.6	-124	-0.3451	-42.8	2.0	-72.4	-57
ML1	135.7	-448	-0.3572	-160.0	20.2	-53.2	-57
ML3	150.7	-477	-0.3495	-166.7	2.4	-32.7	-46
ML7	-	-70	-	-	-	-	-
ML8	13.5	-43	-0.3178	-13.7	-3.1	-9.3	-13
SAS	reference 0	reference 0	-0.3200	-	-	-	-

elevation and location of the station and another at an elevation one meter above the surface. The difference in gravity at the two heights is then added to the per-meter theoretical gradient as a terrain correction.

Field measurements of the free-air gradient were made at site MLO (Figure. 1) located on the north flank of Mauna Loa at an elevation of 3375 m. A wooden structure at MLO served as a convenient platform on which to make precise measurements of the gravity difference (to ±2.5 μGal) between two points separated vertically by 1.8 meters. The observed -0.3387 μGal/mm gradient compares favorably with a gradient of -0.327 μGal for MLO predicted by the computational procedure above. Agreement between the calculated and observed gradients at MLO gives confidence in the calculated values in Table 2 which have not been field checked.

The free-air correction is simply the calculated gradient times the height change and is subtracted from the observed gravity change.

Spherical reservoir source. The position and deflation volume of a spherical reservoir were determined in the inversion of geodetic data described above (Table 2). The modeled surface collapse volume ΔV_e of 66.9×10^6m^3 corresponds to a reservoir contraction ΔV_r of 44.6×10^6m^3 according to equation (5). Magma mass transfer from or into the reservoir is left as an unknown by setting $\rho_0=0$. The gravity change due to inlay of a spherical shell of crustal material, $\rho=2300$ kg/m^3, equal in volume to the reservoir volume loss ΔV_r above is calculated using equation (7).

Next, the gravity change due to crustal density change associated with contraction of the reservoir is estimated numerically. To do this, Mauna Loa's edifice is represented as a three dimensional grid of 100-meter cubes with density $\rho=2300$ kg/m^3. Equations (4) and (6) are used to find the density change for each cubic cell, the gravity change resulting from the density increment is calculated for each cube [*Parasnis*, 1986, equation 3.25], and the

values summed for all the cells in the model.

Finally, vertical displacements of the surface are calculated from equations (1) and (4) at horizontal positions defined by the DEM grid center point and vertical position defined by the elevation given in the DEM. The gravitational attraction of adding or subtracting square plates of crust (horizontal dimensions of 100 m, thickness given by Δh, density $\rho=2300$ kg/m^3) positioned according to the DEM grid node location and elevation is calculated. A summation is made of calculated gravity changes for all grid nodes in the model.

Resulting gravity corrections for contraction of the spherical source are listed in Table 3. Recall that the theoretical model (equation 10) which is based on elastic crust with a flat free surface gives $\Delta g'=0$, assuming, as we did in making the numerical calculation above, that magma mass within the reservoir remains constant (i.e. $\rho_0=0$). Numerical results for all gravity stations are within a few μGal of zero, with the exception of ML1 which is 20 μGal. The large change at ML1 compared to the theoretical value is probably a consequence of the placement of ML1 on a high point of the Mauna Loa's edifice, in proximity to a sudden drop off into Mokuaweoweo caldera.

Dike intrusion source. The gravity effect of the dike which formed during the early stages of the 1984 Mauna Loa eruption is estimated using a numerical procedure similar to that used to model the reservoir source. The specific position of the dike and opening are from the inversion of geodetic data described earlier (Table 2).

The gravity effect of the dike itself is calculated for each gravity station. [*Parasnis*, 1986, equation 3.25]. Since the process of dike intrusion essentially creates a crack, or void, in the crust and then immediately fills the crack with magma, the appropriate density to use in the calculation is the density difference between the extracted crust (ρ_c) and the magma (ρ_m) that replaced it. A density contrast of (ρ_m-ρ_c=2600 kg/m^3-2300 kg/m^3) 300 kg/m^3 between intruded magma and displaced crust is assumed here. Crustal density change and surface uplift is determined using the *Yang and Davis* [1986] DEFOR program for all grid cells in the three-dimensional model of Mauna Loa; the resultant gravity changes are calculated and summed.

Gravity changes specific to the dike intrusion and related deformation are listed in Table 3. The dike-related gravity change correction is substantial - ranging up to -72 µGal at C1 - and is strongly a function of the density change component.

Composite gravity adjustment. Residual gravity changes, after adjustment of the observed changes for free-air change and changes related to the dike injection and reservoir contraction described above, range up to a value of -57 µGal at C1 and ML1. Residual values for all gravity sites are listed in Table 3. The negative sign of the residual changes indicates mass (magma) loss from the reservoir which, so far, has not been accounted for by the adjustments.

Estimate of Reservoir Mass Loss

An estimate of the mass of magma removed from the summit reservoir may be made from the residual gravity changes. We assume that the center of mass of the removed magma is coincident with the location and depth of the ΔV_r source which we solved earlier from the geodetic data. This spot, to review, is positioned at a depth of 3630 m (below a 3970 m datum) within the edifice and is below a location mapped on Figure. 2 with a circle symbol. Calculated as a point mass [*Parasnis*, 1986], a loss from the reservoir of 133×10^9 kg is indicated by the ML1 gravity residual and a loss of 140×10^9 kg by the ML3 data. The mass loss indicated by the residual gravity change at C1 is much greater, however values for C1 are suspect because of proximity of the eruptive fissure and greater distance from the reservoir. Taking the average of ΔM estimates from ML1 and ML3 data only gives a reservoir mass loss figure of 136×10^9 kg.

I estimate that this mass loss value may have an uncertainty of $\pm 50 \times 10^9$ kg. While the precision of the gravity change data is adequate, the lack of redundant observations is limiting. Additional gravity data at sites distributed over the Mauna Loa summit area would help considerably.

DISCUSSION

Analysis of gravity and geodetic changes recorded during the 1984 Mauna Loa eruption has determined that $136(\pm 50) \times 10^9$ kg of mass, presumably magma, was removed from a reservoir situated at a depth of 3630 ± 200 m. Concurrent collapse of Mauna Loa's surface, attributed

to reservoir deflation, totaled $66.9(\pm 5) \times 10^6$ m^3. The "effective density" of the deflation, calculated by dividing the mass change by the edifice volume change, is $\Delta M/\Delta V_e = 2033$ kg/m^3. For comparison, two major deflations of Kilauea Volcano in August 1981 and January 1983 showed slightly higher $\Delta M/\Delta V_e$ ratios of 3050 kg/m^3 and 2625 kg/m^3, respectively [*Johnson*, 1992]. *Jachens and Eaton* [1980] found a $\Delta M/\Delta V_e$ ratio of 3290 kg/m^3 from analysis of gravity and elevation changes during a major summit collapse of Kilauea in November 1975; however, the effect of significant earthquake-related horizontal displacements and dike intrusion were not considered in the analysis. Larger ratios, averaging 7165 kg/m^3, accompanied brief episodes of eruption at the Pu'u O'o vent of Kilauea in 1985-1986 [*Johnson*, 1992]. Considering the Mauna Loa $\Delta M/\Delta V_e$ ratios are at the low end of the range observed at Kilauea, it appears that some aspects of the structure and mechanical properties may be dissimilar. The reader is referred to Johnson [1992] for a discussion of the mechanical properties of Kilauea's edifice and magmatic system that govern $\Delta M/\Delta V_e$ during deflation.

Elastic Properties of Edifice and Reservoir Content

The relationship between the mass of magma withdrawn from a reservoir and resultant edifice collapse is dependent on the mechanical properties of the edifice and magma as shown by *Johnson* [1992, equation 10]:

$$\frac{\Delta M}{\Delta V_e} = \frac{\rho_m}{2(1-\nu)} + \frac{2\rho_m\mu}{3(1-\nu)}\left[\frac{1}{K} + \frac{\rho_m NRT}{P^2\omega}\right] \qquad (15)$$

Edifice properties incorporated in (15) are Poisson's ratio ν, assumed to be 0.25, and shear modulus μ. Magma property terms in (15) are enclosed by brackets and are the gas-free magma compressibility ($1/K$) and the CO_2 gas phase compressibility; the sum of the bracketed items comprise the aggregate compressibility of the magma. The magma bulk modulus is K, and N is the CO_2 content of the magma expressed as total mass fraction. The mass of 1 mol of CO_2, ω, is 0.044 kg. R is the gas constant, equal to 8.314 m^{3}Pa/mol°K. *Russell* [1987] estimates a temperature T of 1423K for magma contained within Mauna Loa's reservoir. Pressure P within the reservoir, approximated here by the lithostatic pressure at the 3630 m reservoir depth, is 82 MPa. A value N less than n, given by *Harris* [1981] as

$$n = 5 \times 10^{-6} + bP \qquad (16)$$

where b is the solubility constant equal to 5.9×10^{-12}, indicates that the magma is undersaturated in CO_2 and will not exsolve a separate gas phase. In this case, only the $1/K$ term appears within the brackets in equation (15).

The first term on the right-hand side of (15) reflects the volume reduction of the reservoir by magma withdrawal. The second term reflects additional magma mass removed from the reservoir *without* further decrease in the reservoir volume. Here decompressional expansion of magma and CO_2, and CO_2 exsolution fills reservoir space formerly occupied by the magma. If the reservoir magma has a high compressibility (i.e. low K) and high CO_2 content, volumetric expansion of the reservoir contents will tend to limit reservoir contraction and surface subsidence. On the other hand, if the crustal shear modulus μ value is low, suggesting a relatively weak edifice, the amount of reservoir contraction and crustal subsidence ΔV_e will be enhanced. The Poisson's ratio appears in (15) as part of the conversion between ΔV_e and ΔV_r, accounting for crustal dilation associated with deformation (see equation 5).

Johnson [1992] applied values of $\mu=3$ GPa [based on *Davis et al.*, 1973, 1974; *Davis*, 1986; and *Rubin and Pollard*, 1987] and $K=11.5$ GPa [from *Murase et al.*, 1977] to evaluate the Kilauea $\Delta M/\Delta V_e$ data and obtained magmatic CO_2 concentrations of 0.0009 (August 1981 deflation) and 0.0005 (January 1983 deflation) by weight. These CO_2 concentrations are low compared to estimates of 0.0065 [*Gerlach and Graeber*, 1985] and 0.0032 [*Greenland et al.*, 1985] for the initial load of CO_2 contained in mantle-supplied magma at Kilauea. Apparently exsolved CO_2 in Kilauea's reservoir tends to escape, keeping the retained amount low. A case may be made that CO_2 gas migration and escape is so complete as to remove *all* exsolved CO_2. This is supported by the similarity between values reported by *Johnson* [1992] and the predicted saturation level of roughly 0.0005 CO_2 by weight as indicated by (16) for pressures equivalent to lithostatic at source depths active in the 1981 and 1983 events.

Returning to Mauna Loa, the $\Delta M/\Delta V_e$ value of 2033 kg/m^3 is interpreted to indicate a low CO_2 gas content for the reservoir. Parameters $\mu=3$ GPa and $K=11.5$ GPa, previously used in the Kilauea analysis [*Johnson*, 1992], are adopted here. Assuming a lithostatic 82 MPa pressure within the 3630 m-deep reservoir, the solubility relationship (16) indicates that up to a weight fraction of 0.0005 CO_2 may be dissolved in the magma. Evaluation of equation (15) using the above concentration of CO_2, and assuming presence of a free CO_2 gas phase in the melt, gives $\Delta M/\Delta V_e=2672$ kg/m^3. If, on the other hand, the concentration of CO_2 was slightly less than the saturation level and no free gas were present, then (15) is properly evaluated with the terms in brackets reduced to only $1/K$,

giving $\Delta M/\Delta V_e=2336$ kg/m^3. The measured $\Delta M/\Delta V_e$ value is near the unsaturated estimate. The details may not be significant, given the uncertainties in chosen values for elastic parameters and the process of deriving $\Delta M/\Delta V_e$ from the gravity and geodetic data. The fundamental conclusion is that the concentration of CO_2 resident within the reservoir is low, and I suggest that migration and escape of exsolved gas keeps CO_2 content trimmed to very near levels defined by the saturation limit.

Comparison of Mass of Lava Erupted to Reservoir Supply

Examination of the 1984 Mauna Loa eruption with the added perspective provided by the gravity change data has enabled us to estimate the mass change of the magma reservoir. This estimate is a more direct measure of magma supplied by the reservoir than previous estimates based on surface subsidence alone. The 136x10^9 kg mass loss so determined is comparable to a 52x10^6 m^3 volume of magma figured at fixed density of 2600 kg/m^3. This value is only 30 percent of the amount of erupted lava [*Lipman and Banks*, 1987, adjusted to 2600 kg/m^3 density equivalent]. Since additional magma filled a 22-km long dike that was intruded during the early stages of the eruption, the actual contribution of the summit reservoir to the eruption may be lower. A reservoir contribution in the 20 to 25 percent range is calculated for plausible dike dimensions of 0.75 m width, 5000 m height, and 10 km to 22 km in length.

Where then did the extra magma come from? An additional, undetected source is required to account for erupted lava and dike filling in excess of the amount estimated here to have been supplied by the monitored reservoir. The idea that additional reservoirs supplemented the eruption has been previously suggested by *Dvorak et al.* [1985]. I agree, and suggest that the extra magma was delivered from a source substantially deeper than the known reservoir.

If the 1984 Mauna Loa eruption were supplied by a deeper secondary source, that source is not apparent in the available data. Two factors work against us in being able to resolve a deep source. One is that the deformation and gravity change anomalies for a deep source are spread out over a very large area and have a low maximum amplitude. The other is that large changes caused by the shallow source plus dike intrusion overshadow the subtle changes that would indicate a deep source.

Although we cannot pin down a specific source depth, lack of a significant far-field gravity change increases the likelihood that the source was relatively deep. A gravity change of only +13.5±13.3 μGal at ML8 relative to SAS, which is largely explained as far-field subsidence from the known shallow source, is not large enough to indicate a secondary source of volumetric contraction (of the order of 136×10^6 m^3) in the intermediate depth range between 5 and 15 km. A deep secondary source below 15 km, on the other hand, would produce a much smaller change to the ML8-SAS gravity tie which would be within the error bounds. So, while the gravity tie does not directly indicate source deeper than 15 km, the data cannot disprove the hypothesis either.

A gravity tie between the local base station SAS and a location on the lower slope of Mauna Kea, 30 km from the subsidence maximum, showed a change of only $+1.6 \pm 11.2$ μGal between surveys spanning the eruption. Unfortunately, at the 16 km radial distance of SAS to the source, expected changes due to even a relatively large volume contraction are too small to resolve with enough certainty to model a source depth.

Model for Mauna Loa Magma Storage and Supply.

The lavas from the 1984 Mauna Loa eruption are remarkably uniform in composition, suggesting a common source [*Rhodes*, 1988]. Furthermore, *Russell* [1987] and *Rhodes* [1988] show that the lava geochemistry has characteristics that suggest a period of storage within a shallow magma reservoir. These observations do not support the idea of direct feeding of the eruption from a secondary, deep source.

Both geochemical and the present geophysical constraints may be satisfied with a model for the eruption that includes simultaneous magma discharge and resupply of the shallow reservoir. In other words, as previously stored magma flowed upward out of the reservoir to supply the eruption, new magma from a deep source flowed into the reservoir at the bottom. The apparent volume change of the reservoir during the eruption is simply the net difference between the inflow and outflow.

Rhodes [1988] explains that the long-term homogeneity of Mauna Loa's lavas is evidence of a steady state magmatic system where composition changes of stored magma due to crystallization are balanced by the influx of new magma of parental composition. A common view of Hawaiian volcanoes is that magma resupply to the shallow reservoir is a gradual process that accompanies periods of repose [e.g. *Decker*, 1987]. What I propose here is that rapid, major magmatic resupply to Mauna Loa reservoir may be a part of the eruption process. During eruption, new magma initially enters the reservoir from below, but does not mix with resident magma above rapidly enough to be incorporated into the surface eruption. With time, and

before a subsequent eruption, the new and resident magmas mix.

One possible indication of magma resupply during the 1984 Mauna Loa eruption is an increase in the rate of CO_2 venting from the summit observed during the later portion of the eruption [*Greenland*, 1987]. While some CO_2 may have been released from the resident magma, following depressurization and vessiculation, the analysis here suggests that the resident magma is fairly low in CO_2 concentration. A preferable source for the emission increase is newly delivered, CO_2-rich magma from the deeper source.

A record of CO_2 outgassing by *Ryan and Chin* [1992] and *Ryan* [this volume] shows peak rates of CO_2 emission immediately following eruptions in 1950, 1975, and 1984; after each peak the rate decayed exponentially. If CO_2 rich, parental magma were delivered to the reservoir at a more or less constant rate, one would expect the CO_2 emission rate to remain steady as well. The exponential decay in the degassing rate, in contrast, suggests progressive degassing of a single batch of gas-rich magma.

I propose that magma supply to Mauna Loa's 3-4 km deep reservoir is episodic and that a large portion of the influx of magma from depth occurs concurrently with eruptive events. Whether the cause of, or the effect of eruption, a pulse of magma resupply would involve a drastic increase in the rate of flow between a deep (>15 km) source and the shallow reservoir. Between eruptions, a low rate of resupply of CO_2-rich parental magma along with the inevitable reservoir degassing process allows reservoir CO_2 levels to fall with time. A pulse of resupply during eruption satisfies the constraint from the gravity analysis that the net mass reduction of the shallow reservoir was only 20-25 percent of the mass erupted.

NOTATION

D position below surface, m.
K magma bulk modulus, 11.5 GPa [*Murase et al.*,1977].
ΔM reservoir magma mass change, kg.
N total CO_2 in magma, weight fraction.
P pressure, MPa.
ΔP pressure change, MPa
R gas constant, 8.314 m^3 Pa/mol °K.
T reservoir magma temperature, K, Kelvins.
V_r volume of reservoir or spherical source, m^3.
ΔV_r reservoir or source volume change, m^3.
ΔV_e edifice volume change, m^3.
X horizontal distance to source, m.
Z depth to source, m.
b CO_2 solubility constant, 5.9×10^{-12} Pa^{-1} [*Harris*, 1981].

δg_0 gravity change component due to source volume change, μGal.

δg_1 gravity change component due to surface uplift, μGal.

δg_2 gravity change component due to density variation, μGal.

Δg observed gravity change, μGal.

$\Delta g'$ residual (after free-air correction) gravity change, μGal.

Δh height change, mm.

n limit of disolved CO_2 in magma, weight fraction.

γ universal gravitational constant, 6.67×10^{-11} Nm^2kg^{-2}.

μ crustal shear modulus, GPa.

ν edifice Poisson's ratio, dimensionless.

ω mass of 1 mol of CO_2, 0.044 kg/mol.

ρ_c crustal density, 2300 kg/m^3 [*Kinoshita, et al.,* 1963].

ρ_m magma density, 2600 kg/m^3 [*Fujii and Kushiro,* 1977].

ρ_0 exchange density, value arbitrary.

Acknowledgments. I thank Bob Decker and the staff of the Hawaiian Volcano Observatory for their help and generosity during my stay at HVO as a student/volunteer in 1984. That they woke me early on the morning of the eruption, putting me on a helicopter at dawn to do a gravity survey, shows the kind of "let's try it" attitude that makes HVO a leader at testing ideas, in innovation. I thank John Dvorak, Ron Hanatani, Arnold Okamura, Maurice Sako, and Ken Yamashita for their skillful job of collecting geodetic data in the thin air of Mauna Loa's summit. I gratefully acknowledge the constructive comments of Mike Ryan and Hazel Rymer.

REFERENCES

Davis, P. M., Surface deformation due to inflation of an arbitrarily oriented triaxial ellipsoidal cavity in an elastic half-space, with reference to Kilauea Volcano, Hawaii, *J. Geophys. Res., 91,* 7429-7438, 1986.

Davis, P. M., D. B. Jackson, J. Field, and F. D. Stacey, Dilauea Volcano, Hawaii: A search for the volcanomagnetic effect, *Science, 180,* 73-74, 1973.

Davis, P. M., L. M. Hastie, and F. D. Stacey, Stresses within an active volcano-with particular reference to Kilauea, *Tectonophysics, 22,* 355-362, 1974.

Decker, R. W., Dynamics of Hawaiian volcanoes: An overview: *U. S. Geol. Surv. Prof. Pap., 1350,* 997-1018, 1987.

Decker, R. W., R. Y. Koyanagi, J. J. Dvorak, J. P. Lockwood, A. T. Okamura, K. M. Yamashita, and W. R. Tanigawa, Seismicity and surface deformation of Mauna Loa Volcano, Hawaii: *Eos Trans. AGU, 64,* 545-547, 1983.

Dvorak, J., A. Okamura, and J. H. Dieterich, Analysis of surface deformation data, Kilauea Volcano, Hawaii, October 1966 to September 1970: *J. Geophys. Res., 88,* 9295-9304, 1983.

Dvorak, J. J., and A. T. Okamura, A hydraulic model to explain variations in summit tilt rate at Kilauea and Mauna Loa Volca-

noes: *U. S. Geol. Surv. Prof. Pap., 1350,* 1281-1296, 1987.

Dvorak, J. J., A. T. Okamura, and M. K. Sako, Summit magma reservoir at Mauna Loa Volcano, Hawaii: *Eos Trans. AGU, 66,* 851, 1985.

Dzurisin, D., L. A. Anderson, G. P. Eaton, R. Y. Koyanagi, P. W. Lipman, J. P. Lockwood, R. T. Okamura, G. S. Puniwai, M. K. Sako, K. E. and Yamashita, Geophysical observations of Kilauea Volcano, Hawaii, 2, Constraints on the magma supply during November 1975-September 1977, *J. Volcanol. Geotherm. Res., 7,* 241-269, 1980.

Dzurisin, D., R. Y. Koyanagi, and T. T. English, Magma supply and storage at Kilauea Volcano, Hawaii, 1956-1983, *J. Volcanol. Geotherm. Res., 21,* 177-206, 1984.

Fujii, T., and I. Kushiro, Density, viscosity, and compressibility of basaltic liquid at high pressures, in *Annual Report of the Director 1976-1977,* pp. 419-424, Geophysical Laboratory, Carnegie Institution, Washington, D.C., 1977.

Gerlach, T. M., and E. J. Graeber, Volatile budget of Kilauea Volcano, *Nature, 313,* 273-277, 1985.

Greenland, L. P., Composition of gasses from the 1984 eruption of Mauna Loa Volcano: *U. S. Geol. Surv. Prof. Pap., 1350,* 781-790, 1987.

Greenland, L. P., W. I. Rose, and J. B. Stokes, An estimate of gas emissions and magmatic gas content from Kilauea Volcano, *Geochim. Cosmochim. Acta, 49,* 125-129, 1985.

Hagiwara, Y., The Mogi model as a possible cause of the crustal uplift in the eastern part of Izu Peninsula and the related gravity change: *Bull. Earthquake Res. Inst. Univ. Tokyo, 52,* 301-309 (in Japanese), 1977.

Hammer, S., The anomalous vertical gradient of gravity: *Geophysics, 35,* 153-157, 1970.

Harris, D. M., The concentration of CO_2 in submarine tholeiitic basalts, *J. Geol., 89,* 589-701, 1981.

Jachens, R. C., and G. P. Eaton, Geophysical observations of Kilauea Volcano, Hawaii, 1, Temporal gravity variations related to the 29 November 1975, M=7.2 earthquake and associated summit collapse, *J. Volcanol. Geotherm. Res., 7,* 225-240, 1980.

Johnson, D. J., Elastic and inelastic magma storage at Kilauea Volcano: *U. S. Geol. Surv. Prof. Pap., 1350,* 1297-1306, 1987.

Johnson, D. J., Dynamics of magma storage in the summit reservoir of Kilauea Volcano, Hawaii: *J. Geophys. Res., 97,* 1807-1820, 1992.

Kinoshita, W. T., H. L. Krivoy, D. R. Mabey, and R. R. MacDonald, Gravity survey of the Island of Hawaii: *U. S. Geol. Surv. Prof. Pap., 475-C,* C114-C116, 1963.

Lipman, P. W., and N. G. Banks, Aa flow dynamics, Mauna Loa 1984: *U. S. Geol. Surv. Prof. Pap., 1350,* 1527-1567, 1987.

Lockwood, J. P., N. G. Banks, T. T. English, L. P. Greenland, D. B. Jackson, D. J. Johnson, R. Y. Koyanagi, K. A. McGee, A. T. Okamuara, and J. M. Rhodes, The 1984 eruption of Mauna Loa Volcano, Hawaii: *Eos Trans. AGU, 66,* 169-171, 1985.

Longman, I. M., Formulas for the tidal acceleration of gravity: *J. Geophys. Res., 64,* 2351-2355, 1959.

Mogi, K., Relations of eruptions of various volcanoes and the deformation of the ground surfaces around them: *Bull. Earth-*

quake Res. Inst. Univ. Tokyo, 39, 99-134, 1958.

Murase, T., I. Kushiro, and T. Fujii, Compressional wave velocity in partially molten peridotite, in *Annual Report of the Director 1976-1977*, pp. 414-416, Geophysical Laboratory, Carnegie Institution, 414-416, Washington, D.C., 1977.

Okamura, A. T., J. J. Dvorak, M. K. Sako, R. Y. Hanatani, D. J. Johnson, and K. M. Yamashita, Surface deformation associated with the March 25-April 15, 1984 Mauna Loa eruption: *Eos Trans. AGU, 65*, 1137, 1984.

Okubo, S., Gravity and potential change due to shear and tensile faults in a half-space: *J. Geophys. Res., 97*, 7137-7144, 1992.

Parasnis, D. S., *Principles of Applied Geophysics*, Chapman and Hill, 402 pp., 1986.

Rhodes, J. M., Geochemistry of the 1984 Mauna Loa eruption: Implications for magma storage and supply: *J. Geophys. Res., 93*, 4453-4466, 1988.

Rubin, A. M., and D. D. Pollard, Origins of blade-like dikes in volcanic rift zones, *U. S. Geol. Surv. Prof. Pap., 1350*, 1149-1470, 1987.

Rundle, J. B., Gravity changes and the Palmdale uplift: *Geophys. Res. Lett., 5*, 41-44, 1978.

Russell, J. K., Crystallization and vesiculation of the 1984 eruption of Mauna Loa: *J. Geophys. Res., 92*, 13731-13743, 1987.

Ryan, S., and J. F. S. Chin, Quiescent CO_2 outgassing of Mauna Loa Volcano: *Eos Trans. AGU, 73*, 628, 1992.

Ryan, S., Quiescent outgassing of Mauna Loa Volcano 1958-1994, This Volume.

Rymer, H., Microgravity change as a precursor to volcanic activity: *J. Volcanol. Geotherm. Res., 61*, 311-328, 1994.

Sasai, Y., Multiple tension-crack model for dilatancy: Surface displacement, gravity and magnetic change: *Bull. Earthquake Res. Inst. Univ. Tokyo, 61*, 429-473, 1986.

Savage, J. C., Local gravity anomalies produced by dislocation sources: *J. Geophys. Res., 89*, 1945-1952, 1984.

Walsh, J. B., and J. R. Rice, Local changes in gravity resulting from deformation: *J. Geophys. Res., 84*, 165-170, 1979.

Yang, X., and P. M. Davis, Deformation due to a rectangular tension crack in an elastic half-space: *Bull. Seism. Soc. Am., 76*, 865-881, 1986.

Daniel J. Johnson, 3616 NE 97th St., Seattle, WA 98115

Remote Sensing of Mauna Loa

Anne B. Kahle[1], Michael J. Abrams[1], Elsa A. Abbott[1], Peter J. Mouginis-Mark[2],

and Vincent J. Realmuto[1]

[1]*Jet Propulsion Laboratory, California Institute of Technology, Pasadena, California*

[2]*University of Hawaii, Honolulu, Hawaii*

The application of remote sensing to the study of volcanoes has advanced considerably in recent years. The availability of digital multispectral data, acquired from satellite or airborne instruments, has allowed us to infer many properties of the surface material. Each wavelength region offers specific information relating to the chemical and physical state of eruptive products. The reflected visible through thermal part of the spectrum reveals information relating to iron oxidation state, vegetation cover, development of silica coatings, and presence of SO_2, among others. The microwave part of the spectrum reveals information relating to surface roughness and dielectric constant variations. The use of these data to study and map volcanoes can save time and effort for mapping; and can greatly reduce the risk associated with doing fieldwork on active volcanoes. Satellite instruments due to be launched in the next five years promise to provide far better data than are currently available from space.

1. INTRODUCTION

Remote sensing is proving to be a valuable tool for studying volcanoes. Large areas can be observed rapidly, safely, and, if using satellites, frequently. In addition to providing images (sometimes stereoscopic) comparable to aerial photographs and amenable to the traditional photographic interpretation, remote sensing instrumentation generally provides multispectral data in digital format. From these data, we can infer many properties of the surface material. In this chapter, we will discuss remote sensing at visible, infrared and microwave wavelengths. There have been some satellite observations of Mauna Loa and the island of Hawaii with the Advanced Very High Resolution Radiometer (AVHRR), Japanese Earth Remote Sensing (JERS-1), Landsat, and French Societe Propri d'Observation de la Terre (SPOT) satellites and with Shuttle Imaging Radar (SIR-B and SIR-C) (see, for example, Plate 1 depicting the Island of Hawaii using SPOT satellite data combined with digital elevation; *Chadwick, et al.*, personal communication). However, the most detailed studies have

utilized NASA airborne sensors. We will emphasize these latter studies here, as they best demonstrate the potential of the techniques.

In section 2, we discuss the spectral characteristics of materials on the surface of Mauna Loa. In section 3, we show examples of multispectral remotely sensed data in the visible through infrared wavelength regions; in section 4, we discuss remote sensing of SO_2 plumes using thermal infrared (TIR) data; in section 5 we discuss the use of multispectral data for temperature determination; section 6 deals with radar remote sensing; and finally, in section 7, we discuss future remote sensing of Mauna Loa. Figure 1 is an index map showing the locations of the images that will be discussed in this paper.

2. EVOLUTION OF SPECTRAL FEATURES

2.1. *Introduction: Spectral basis for mapping from remote sensing data*

The surface weathering of Mauna Loa basalts is evidenced by distinct chemical and mechanical changes, beginning during initial cooling. These changes have been used to estimate the relative ages of individual flows [*Kahle, et al.*, 1988; *Abrams, et al.*, 1991] from remote sensing data in both the visible to short wavelength infrared (SWIR)

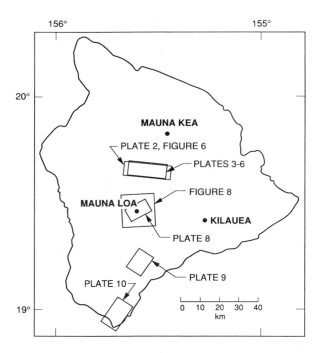

Figure 1. Index map of the island of Hawaii showing location of images discussed in this paper.

region (0.5—2.4 μm) and part of the thermal infrared (TIR) region (8—12 μm).

On the windward side of the island of Hawaii and in other humid areas, vegetation growth is so rapid that lava surfaces are obscured from aerial view in a very few years. In contrast, lava flows in arid regions on the island are nearly unvegetated and may remain barren for long periods of time. During this interval, the change most obvious to the eye is the oxidation of iron, so that the surface, originally black or dark brown, becomes reddish or tan [*Lockwood and Lipman*, 1987]. Other less obvious changes also occur, including the accretion of silica-rich veneers or coatings (~80 wt % SiO$_2$) derived largely from windblown soil [*Curtiss, et al.*, 1985] or tephra [*Farr and Adams*, 1984] and the devitrification of the thin (~50 μm) glassy crusts or chill coats so common on fresh pahoehoe. The coats may also spall to reveal a more vesicular crystalline substrate. While the flows are still active and during cooling, acid aerosols in the vicinity of gases escaping from the melt can cause chemical changes to the already emplaced rock, leaching out iron and other cations, and leaving a more silica-rich rock [*Realmuto, et al.,*, 1992]. In addition, mechanical weathering of the flows encourages colonization by vegetation; very old flows in kipukas, for example, tend to be heavily vegetated. The effects of these and other processes on the spectral characteristics of weathering lava flows will be discussed in more detail in the next sections. We will also show how it has been

possible to capitalize on these spectral changes to determine relative age relations of lava flows on Mauna Loa.

2.2. *Visible to short wavelength infrared region*

The spectrum of basalt in the visible to short wavelength infrared (0.4—2.4 μm) region is dominated by the presence of iron in its various oxidation states and by the presence of vegetation. Laboratory measurements of flows of different ages illustrate the effects of iron oxidation on the spectral behavior.

Fresh aa and pahoehoe flows are both nearly flat spectrally and have very low albedo throughout the reflectance part of the spectrum (Figure 2). Pahoehoe can be very smooth and flat and can exhibit strong specular reflection. Laboratory studies were done on samples of basalts collected from an area on the north slope of Mauna Loa [*Abrams, et al.*, 1991]. With increasing age, the overall reflectance increases (from 5-7% to about 12%) for flows up to about 4000 years old. At the same time, there develops a fall-off in reflectance at the blue end (~0.4 μm) of the spectrum that becomes progressively more pronounced with age. They attributed both of these effects to the conversion of ferrous iron to ferric iron by oxidation [*Hunt*, 1977]. Chemical analyses of the samples were done to measure the ferric iron content of the rocks. There is a systematic increase with age of ferric iron, from 3.2% to 4.6%, which correlates with the observed spectral characteristics. The slope of the reflectance curves between 0.8 and 0.4 μm is therefore a good measure of the increase in ferric iron content, and hence is correlated to increasing age.

Vegetation cover on Mauna Loa basalts varies as a function of age, rainfall and elevation. Historic flows are essentially unvegetated at higher elevation due to the lack of soil development. Lower down, in areas of heavy rainfall, lichen can begin to grow, especially on the rougher aa, within 2-3 years. Lichen grows early on these flows and after time is replaced by other types of vegetation. At an elevation level of about 2100 meters, flows whose C-14 ages are between 200 and 1500 years have only a few percent cover of shrubs and grasses; flows with C-14 ages between 1500 and 4000 have 5—15% cover; flows older than 4000 years old become heavily vegetated and extensively weathered. The spectrum of healthy vegetation (Figure 2) shows the presence of chlorophyll absorption bands near 0.65 μm and 0.45 μm, a relative high in reflectance near 0.55 μm (the reason vegetation is green to the eye), and very high reflectance starting at 0.76 μm, gradually decreasing towards longer wavelength due to the effects of water absorption bands.

The distinct, and systematic effects of both oxidation and increasing vegetation cover with age on the spectral reflectance characteristics have been exploited to map some of the older lava flows at middle elevations on Mauna Loa,

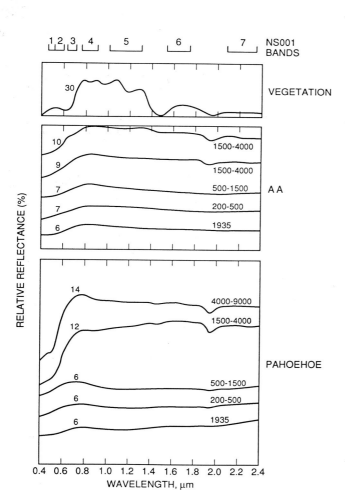

Figure 2. Laboratory reflectance spectra of green vegetation and Mauna Loa aa and pahoehoe flows dating from 1935, 1899, 200-500 years BP, 500-1500 years BP, 1500-4000 years BP and 4000-9000 years BP. Reflectance values at 0.8 μm are indicated above each curve. Spectra have been offset vertically for clarity. Positions of NS001 channels are indicated above the graphs. Modified from *Abrams, et al.* [1991].

and to determine the relative ages of the older flows. The technique has been shown to be in excellent agreement with standard field mapping techniques [*Abrams, et al.*, 1991].

2.3. Thermal infrared region

The TIR spectral characteristics of basalts are controlled by (1) macroscopic properties, such as the amount and size of vesiculation, (2) differences related to the Si-O bonding which are in turn related to the silica content, and (3) chemical alteration of the surfaces from acidic gases. Each of these effects will be discussed with regard to our ability to use these spectral differences to map

Mauna Loa flows in the TIR wavelength region.

The thermal infrared radiance emitted from a surface is a function of the temperature and spectral emittance, where the emittance is the parameter related to the composition and surface roughness of the material. For a perfect black-body, the emittance is unity, and the radiance is strictly a function of the temperature and wavelength, as described by Planck's Law. Most geologic materials have non-unity emittance at some wavelengths in the TIR; it is this deviation from unity that allows separation and identification of mineral composition. Compared to the visible/near-infrared spectra, the thermal infrared reflectance spectra of the younger basalts show prominent spectral differences. These spectral features are related to Si-O bonding, and facilitate the distinction of different lava flows.

Crisp et al. [1990] investigated the relationship between changes in spectral emittance features and mineralogy of Hawaiian basalts. They collected samples of molten pahoehoe lava from active effusions of Puu Oo in May 1989. The spectra of these fresh samples (after cooling into a glass) exhibit a broad feature peaked between 10.3 and 10.5 μm, indicative of a strong degree of disorder (Figure 3). Disordered glass is made up of silicate units with a wide variety of bond angles, bond strengths, and bonding arrangements that vibrate at different frequencies resulting in broad spectral features in the infrared [e.g., *Simon and McMahon*, 1953; *Brawer and White*, 1975; *Dowty*, 1987]. The spectra of the fresh Hawaiian basalts are almost identical to the spectra of samples from older (prehistoric, 1972, 1880, 1899, and 1984) Mauna Loa basaltic lavas that were fused in an oven and quenched [e.g., *Kahle, et al.*, 1988, Figure 3b]. They also closely match the spectra of the glassy interior of older basalts. The persistence of this broad spectral feature in the glassy interior of rocks, even in rocks over 4000 years old, shows how well the interior glass can be protected from weathering processes and can retain its original composition and degree of disorder.

However, as the Hawaiian rocks age and weather, the spectral character of their outer surfaces changes. The first change evident in the spectra of the Hawaiian rocks is that the single broad feature splits into two features that *Kahle et al.* [1988] and *Crisp et al.* [1990] call "B" at 9.2—9.5 μm and "C" at 10.5—10.8 μm (Figure 4). After just a few days or weeks of exposure to the elements (rain, atmosphere, and acidic volcanic fumes) the B and C features were evident in some, but not all, of the exposed surfaces. Features B and C were commonly found on the rapidly cooled top surface of flows, but also appeared in the interiors when rocks were broken after emplacement and cooling and these interiors were then exposed to the Hawaiian environment for a sufficient time. In samples that they examined that were only a week old, the C feature was always stronger than the B feature.

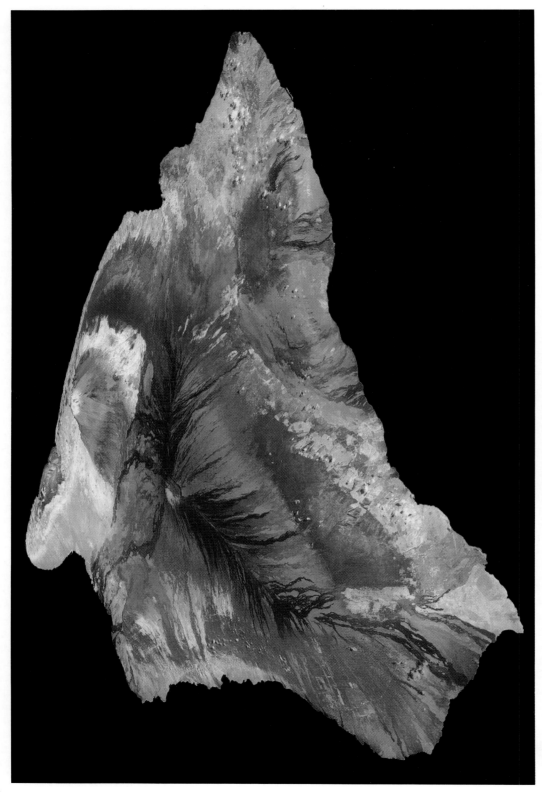

Plate 1. Simulated perspective view of the Island of Hawaii viewed from the southeast. This image was created by combining digital elevation data from the United States Geological Survey's 7-1/2' database, with SPOT multispectral satellite data. SPOT bands 1, 2, and 3 are dispalyed in green, blue, and red respectively. The vertical exaggeration is 2X. Vegetation appears in shades of red, lava flows in shades of blue and green. Image courtesy of Oliver Chadwick and Steven Adams of the Jet Propulsion Laboratory.

Plate 2. A decorrelation stretched TIMS image of part of the north flank of Mauna Loa. Bands 1, 3, and 5 are displayed in blue, green and red, respectively. Age and vegetation cover information are given in Figure 6. After *Kahle, et al.* [1988].

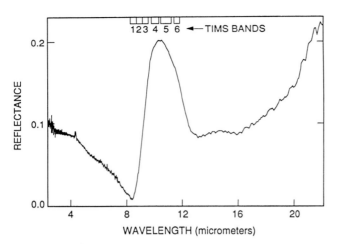

Figure 3. Laboratory spectrum of a cooled sample collected from an active Puu Oo flow in May, 1989. TIMS channel locations are shown on upper axis. The laboratory spectrometer measures reflectance, which is equal to (1-emissivity) according to Kirckhoff's law. After *Crisp et al.* [1990].

With increasing age the overall trend is for the B feature to strengthen relative to the C feature. A simple explanation for the evolution of these two emission peaks in the top surfaces of Hawaiian rocks is that the structure of the metastable glass is becoming more ordered with time. Immediately after its rapid quenching in air, the glass is strongly disordered. With time this unstable configuration breaks down as the silica tetrahedra become organized into silica tetrahedral sheetlike (B feature) and chainlike units (C feature) [e.g., *Brawer and White*, 1975]. With more time the sheetlike units become the preferred mode. This could be an isochemical process in which the glass structure becomes more ordered by chain units progressively converting to sheet units, or it could be a

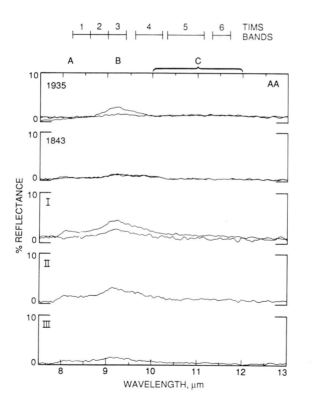

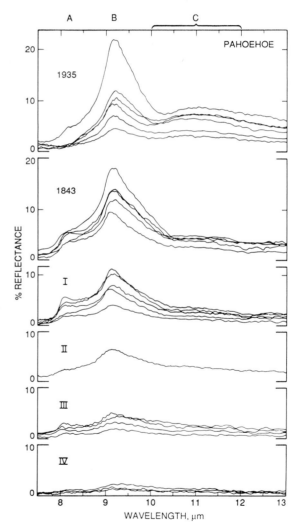

Figure 4. Laboratory reflectance spectra of Mauna Loa lavas collected along the Hilo-Kona jeep trail (Figure 6 and Plate 2). Letters A, B, and C refer to spectral features discussed in the text. (a) Aa (b) Pahoehoe. After *Kahle, et al.* [1988].

structural change accompanying a compositional change at the surface.

With continued aging, the height of the C feature decreases until it is undetectable. It is likely that the C feature disappears as the glass becomes more ordered, without any chemical change in the glass. Soon after the B feature overtakes the C feature in height, a new feature they named "A" can appear at about 8.1 or 8.2 μm. The A feature is commonly a shoulder on the side of the B feature. It is very common for samples greater than 50 years old to show an A feature. The A feature has not been found in the week-old samples from Puu Oo but does appear in some samples of the 1984 Mauna Loa flow collected three years after emplacement but only at locations that receive more than 250 cm rainfall per year and in Puu Oo flows 1-5 years after emplacement. *Kahle et al.* [1988] proposed that the A feature results from the addition of a silica-rich coating. *Farr and Adams* [1984] describe the development of this coating as the accumulation and leaching of windblown tephra and dust. A scanning electron microscope (SEM) and thin section investigation [*Crisp et al.*, 1990] of Mauna Loa 1984 samples confirmed that the A feature is associated with a silica-rich rind that appears to be the results of addition of material to the basalt surface rather than leaching of the substrate basalt. Previously, the shortest time documented for the formation of a silica coating was 13 years in an area southwest of Kilauea where the rainfall rate is about 150 cm/year [*Farr and Adams*, 1984]. As the lava ages further, the spectral contrast decreases slowly until, after a few hundred years, most of the units appear spectrally flat in the infrared due to spalling of glass or development of iron oxide coatings (Figure 4). However, as noted in section 2.2, at this time the spectral contrast in the VIS and SWIR becomes more evident.

The spectral character of aa in the thermal infrared is partially controlled by the texture of the material. Radiation emitted by the aa is often partially trapped by the roughness of the surface, which tends to act like a large number of blackbody cavities. However, the spectral signature, while reduced, has features very much like those of similar age pahoehoe. Figure 5 shows lab spectra of aa and pahoehoe of the same age, which have been normalized to the same scale.

Vegetation spectra are essentially equal to blackbody spectra, so the addition of vegetation to the surface serves to reduce the spectral contrast of the materials.

Another influence on the TIR spectral characteristics of basalts is alteration associated with fumarolic activity. Near active venting areas and areas undergoing degassing, hydrous, silica-rich areas form, accompanied by the deposition of sulfate salts [*Realmuto, et al.*, 1992]. This is thought to be caused by acid-leaching of cations by aerosols, and resulting enrichment of silica. The spectral characteristics of this material resemble those produced by the secondary silica coatings described by *Crisp et al.* [1990].

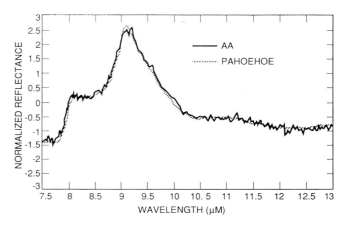

Figure 5. Laboratory reflectance spectra of aa and pahoehoe of the same age (0.2—0.5 ka) which have been normalized.

3. MAPPING USING REMOTELY SENSED DATA

NASA's Thermal Infrared Multispectral Scanner (TIMS) and the NS001 scanner (a Thematic Mapper simulator) were flown aboard a NASA C-130 aircraft over the island of Hawaii during November, 1985 at about 1100 local time. TIMS acquires digital radiance data in image format. There are six spectral channels of data between 8 and 12 μm (Table 1) [*Palluconi and Meeks*, 1985]. The sensitivity is 0.1—0.5K. Images are acquired using a mirror that scans an arc of +38° to -38° about nadir, with an angular resolution of 2.5 mrad. The NS001 acquires eight channels of data in the visible, reflected infrared, and thermal infrared parts of the spectrum (Table 1). Images are acquired with a scan angle of 100°, and an angular resolution of 2.5 mrad. The C-130 operates at altitudes up to 7.7 km above sea level; thus, the scanners acquire data with a nadir ground pixel size of 25 m or less. For these data flights, the aircraft operated at an altitude of 3.5 km above terrain, producing 8 m pixels.

Digital data from TIMS bands 1, 3 and 5 were processed using a decorrelation stretch algorithm [*Gillespie, et al.*, 1986]. This procedure exaggerates subtle color differences in image data by increasing the saturation and intensity while generally preserving the hue information. Thus, the resulting images can be related back to the spectral information of the components used to create the color triplet image. Because the decorrelation stretch depends upon the frequency distribution of radiance values within the particular image being stretched, the displayed colors of the same materials generally vary slightly from image to image.

Plate 2 shows a decorrelation-stretched image of basalt flows on the north flank of Mauna Loa, where TIMS bands 1, 3, and 5 are displayed in blue, green, and red respectively. (The road running diagonally across the image is the Hilo-Kona jeep trail.) Despite the chemical

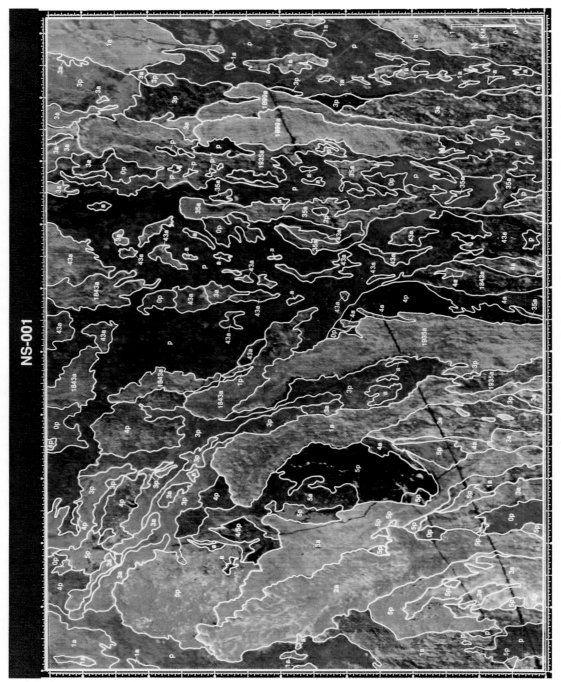

Plate 3. NS001 principal components image. Bands 1 through 7 were used for the analysis. The loadings for component 2, displayed in red, produced a difference between the IR and VIS bands; component 3, in green, was heavily loaded as the difference between bands 3 and 1; component 4, in blue, was heavily loaded as the difference between bands 4 and 7. The overlay shows interpreted flow contacts, with age labels as in Plate 2; p = pahoehoe; a = aa; Op = undifferentiable prehistoric pahoehoe; 43a = 1843aa; 35a = 1935aa. After *Abrams, et al.* [1991].

TIMS

Plate 4. TIMS decorrelation stretched image of the Mauna Loa test area. Bands, 1, 3, and 5 are displayed in blue, green, and red, respectively. The overlay shows interpreted flow contacts, with age labels as in Plate 2; p = pahoehoe; a = aa; units labeled "a" alone = undifferentiable aa; 35p = 1935p; 43p = 1843p.   After *Abrams, et al.* [1991].

TABLE 1. Band Passes of Scanners

NS001		TIMS	
CHANNEL	WAVELENGTH, μm	CHANNEL	WAVELENGTH, μm
1	0.45—0.52	1	8.2—8.6
2	0.52—0.60	2	8.6—9.0
3	0.63—0.69	3	9.0—9.5
4	0.76—0.90	4	9.6—10.2
5	1.0—1.30	5	10.2—11.2
6	1.55—1.75	6	11.2—11.7
7	2.08—2.36		
8	10.4—12.5		

and petrologic similarity of the unweathered basalts, Plate 2 shows a wide range of different thermal IR "colors", both within and among the numerous individual lava flows.

Field checking of these and other Hawaiian TIMS images and comparison with geologic maps [*Holcomb*, 1987; *Lockwood, et al.*, 1988; *Lockwood*, unpublished data] reveal systematic relationships between the TIMS colors and the type of basalt (pahoehoe or aa) and degree of weathering and, hence, age. Pahoehoe and aa flows are consistently separable in the images where there is little or no vegetation. Single basalt flows of either type may show some image color differences soon after eruption [*Realmuto, et al.*, 1992]; however, the greatest color differences appear to be related to age.

Figure 6 is a field-derived geologic map of the same area covered by Plate 2 for comparison. The contacts of some flows are more accurately portrayed in the images. In a few cases, geologic relations that were difficult to map in the field can easily be seen in the images. One example of this is the boundary between the 1843 (red) and 1935 (blue) pahoehoe flows, which is difficult to distinguish in the field, but is very distinct in the TIMS image.

Freshly broken and unweathered basaltic cinders and crushed basalt exposed in quarries consistently appear cyan or light blue-green in the false-color pictures. In contrast, young aa flows appear dark blue-green, and this color shifts to dark brown or orange with increasing age. The oldest flows are heavily vegetated and appear dark green. In the false-color pictures these flows are not always easy to distinguish from young, largely unvegetated aa. However, they may be separated by temperature. In Plate 2, lightly vegetated aa flows ranged in surface temperature from 35° to 43°C; heavily vegetated flows were ~29°C.

Unvegetated pahoehoe was warmer than aa. Surface temperatures for pahoehoe in Plate 2 ranged from 43° to 54°C.

There is a pronounced and systematic color change with increasing age of pahoehoe. The TIMS color shifts from dominantly blue to purple and magenta (compare, for instance, the 1935 and 1880 flows in Plate 2). This range of colors mimics the range for the different initial states. Increasingly older flows show colors not observed for young flows: red (1843) and orange (0.2—1.5 ka), mixed orange and green (1.5—4 ka), and ultimately light green (4—8 ka). The oldest lavas (>8 ka) are forested and appear dark green. Heavily vegetated aa and pahoehoe are probably indistinguishable from each other.

Because of the known color assignments of the TIMS wavelength bands, and by comparison with spectra of samples collected in the field, we can relate the colors in the images to the spectral features observed in Figure 4. Spectral feature C is associated with the blue color in the image and A with red. As spectral feature A appears and increases and feature C diminishes, the colors move progressively from blue to magenta to red. Finally, as the spectral contrast diminishes, the units on the images become green — which here corresponds to spectrally flat.

For aa, young rough flows are associated with dark blue-green TIMS colors and very weak spectral features, including subdued B and even some C. Because aa and pahoehoe of the same flow have essentially identical compositions, we attribute the subdued spectrum of the aa to multiple scattering in the rough surface (cavity or blackbody radiation). The dark colors in the images are due to low temperatures in the shadowed portions of the rough surface, not yet warmed by the sun at the time (~1100 LT) of data acquisition.

The shift to brown image color with increasing age of aa is directly caused by the silica-rich rinds. The increased color saturation of these flows is attributed to filling of small cavities with the thermally opaque silica, thereby reducing the amount of blackbody radiation. There is no very old unvegetated aa in the study area. The color of the oldest aa flows in Plate 2 is controlled by the admixture of vegetation, which consistently appears dark green in the false-color images.

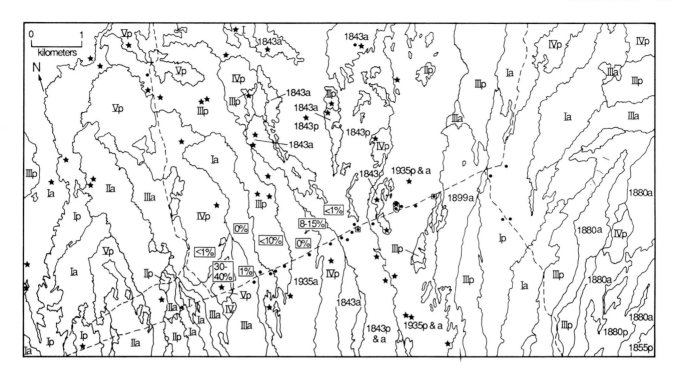

Figure 6. Index map for Mauna Loa TIMS image (Plate 2) showing flow outlines and ages (dates for historical flows; radiocarbon age groups for prehistoric flows I = 0.2—0.5 ka; II = 0.5—1.5 ka; III = 1.5—4 ka; IV = 4—8 ka; V = >8 ka); "a" is aa; "p" is pahoehoe. Circles mark location for samples of Figure 3. Numbers in boxes give the vegetation cover measured in the field for selected flows [*J. B. Adams*, personal communication, 1988]. After *Kahle, et al.* [1988].

Abrams et al. [1991] analyzed the NS001 visible and SWIR data for roughly the same area as Plate 2. The NS001 data were processed by using Karhunan-Loeve or principal components transformations. This is a dimension reduction procedure that forms linear combinations of the original data based on the variance-covariance matrix [*Soha and Schwartz*, 1978; *Gillespie*, 1980]. The resulting image (Plate 3) depicts the flows in various colors; the overlain interpretation map delineates flow boundaries recognized from the image. On the image, the young pahoehoe flows (1935, 1843, 200-500 years old and 500-1500 years old) are all displayed as indistinguishable blue colored units, in contrast with the TIMS image (Plate 4) which allowed easy separation of these units. They are labeled p (pahoehoe) and Op (older pahoehoe) to indicate that they were not differentiable. These flows are still very dark and fresh-looking in the field, and they have not as yet developed significant oxidation of mafic minerals. Spectrally, they are all dark and show no distinguishing features. Older pahoehoe flows become greener on the image, and the oldest, with vegetation cover, are magenta. The aa flows progress from reddish-brown (1935) to brown (1899) to light brown (1843) to blue-brown (200-500 years old) to light blue-green (500-1500 years old) to green-yellow (1500-4000 years old) to dark green (>4000 years old). The relations between the image colors and the increase in iron oxidation and change in reflectance spectra are consistent.

When the NS001 and TIMS data were combined by principal components analysis, it was shown that emittance and reflectance spectral properties of the rocks were correlated to some degree, as would be expected. As shown in Plates 5 and 6, the flows are differentiated by the color changes, which are systematic with relative age.

These results show the utility of remote sensing in the relative dating of similar basalts in an arid environment, using combined data from the reflectance and emittance parts of the spectrum. The existence of weathering systematics implies that it may prove possible to estimate flow age from remotely sensed data, provided that progression of colors has already been calibrated for a given region. Important influences on the thermal spectra appear to be surface texture, physical and chemical degradation of glassy crusts (spalling, devitrification, alteration), accretion of silica-rich coats, and vegetation growth. Important influences on the reflectance spectra appear to be development of iron oxide minerals from weathering, vegetation growth, and surface texture. At least for the north flank of Mauna Loa, these effects occur in characteristic sequences

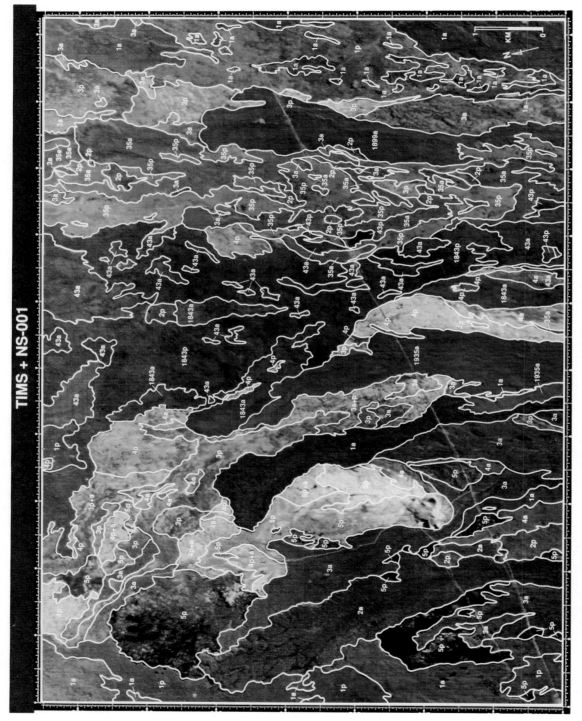

Plate 5. Combined NS001 and TIMS data processed using a principal components transformation. NS001 bands 1 through 7, and TIMS bands 1 through 6 were used for the analysis. PC1, PC5, PC6 displayed in red, green and blue, respectively. The overlay shows interpreted flow contacts, with age labels as in Plate 2; p = pahoehoe; a = aa; 35a(p) = 1935a(p); 43a(p) = 1843a(p). *After Abrams, et al.* [1991].

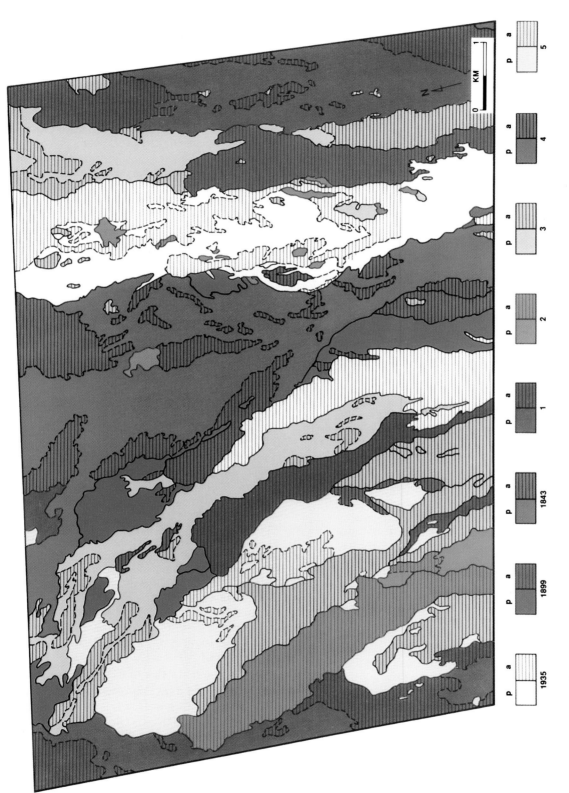

Plate 6. Geologic map of the Mauna Loa test site. Spectral sample sites are indicated by numbers 1-12; p = pahoehoe; a = aa; 1 = unit 1 (0.2—0.5 ka); 2 = unit II (0.5—1.5 ka); 3 = unit III (1.5—4 ka); 4 = unit IV (4—8 ka); 5 = unit V (>8 ka). Modified from *J. Lockwood* [unpublished data, 1988]. After *Kahle, et al.* [1988].

and at different rates, so that color pictures created from visible to thermal infrared data depict flows at different stages of their development in different colors. The combined use of data in the two wavelength regions provides more information than the use of either separately. These relationships should prove useful in reconnaissance geologic mapping in volcanic fields in other arid or semiarid areas.

4. REMOTE SENSING OF VOLCANIC SO_2 PLUMES

4.1. *Measurement technique*

The absorption spectrum of the sulfur dioxide (SO_2) molecule exhibits ultraviolet (UV), infrared (IR), and microwave bands that are amenable to remote sensing techniques. All three of these absorption features have been used to measure SO_2 with passive ground-based, airborne, and spaceborne instruments.

Dispersive correlation spectroscopy became the first remote sensing technique to be widely applied to the study of volcanic plumes [cf *Stoiber et al.*, 1983] using the portable correlation spectrometer, or COSPEC [*Moffat and Millán*, 1971]. COSPEC instruments convert measurements of the UV radiation transmitted through a volcanic plume into estimates of its SO_2 burden [*Hamilton et al.*, 1978; *Hoff and Millán*, 1981]. The collection of COSPEC measurements at volcanoes throughout the world has allowed volcanologists to estimate the annual volcanic contribution to the global atmospheric SO_2 budget [cf *Berresheim and Jaeschke*, 1983; *Stoiber et al.*, 1987] and recognize that a rapid change in the SO_2 flux from a volcano can signal an eruption [i.e., *Casadevall et al.*, 1981; *Chartier et al.*, 1988; *Malinconinco*, 1979, 1987; *Caltabiano et al.*, 1994].

A correlation spectrometer can measure only a small portion of a plume; operators must either move the instrument in traverses beneath plumes or occupy locations near vents and pan the field of view horizontally across plumes. Wind speed must be measured or known from other sources. The estimation of either SO_2 flux or spatial and temporal variations in the SO_2 contents of volcanic plumes can therefore involve considerable labor, logistical planning, and in some cases, personal risk.

Satellite-based remote sensing techniques, employing the instruments such as the Total Ozone Mapping Spectrometer (TOMS) or the Microwave Limb Sounder (MLS), are largely free of the limitations imposed by field logistics and hazards. A major advance in the remote sensing of volcanogenic SO_2 occurred in 1983, when *Krueger* [1983] demonstrated that the SO_2 clouds from the eruption of El Chichón could be imaged, or mapped at a 50 km spatial resolution at nadir by TOMS. TOMS data have since been used to map the SO_2 clouds produced by the recent eruptions of Mauna Loa [*Casadevall et al.*, 1984], Nevado del Ruiz [*Krueger et al.*, 1990], Cerro Hudson [*Doiron et*

al., 1991], Mount Pinatubo [*Bluth et al.*, 1992a], and Mount Spurr [*Bluth et al.*, 1992b]. In addition, the TOMS data archive (which dates back to 1979) has been used to estimate the contribution of explosive eruptions to the global SO_2 budget [*Bluth et al.*, 1993].

Read et al. [1993] showed that MLS data, acquired from the Upper Atmosphere Research Satellite (UARS), could be used to recover altitude profiles of the SO_2 clouds produced by the Pinatubo eruption. The MLS procedure is sensitive to SO_2 concentrations as low as 3 ppb(v).

Realmuto et al. [1994] recently demonstrated that thermal IR image data could be used to map the SO_2 content of volcanic plumes. This technique was developed using TIMS data acquired at Mount Etna, Sicily on 26 July 1986, when the estimated column abundance of SO_2 within the plume ranged from 1 to 17 g m^{-2} and the overall flux was estimated at 78 kg/s. The TIMS-based estimation technique is based on the detection of a broad feature between 8 and 9.5 μm in the absorption spectrum of SO_2. The maximum absorption is located at 8.5 μm, and falls between the first and second channels of TIMS. The estimation procedure is based on radiative transfer modeling, which is used to fit the radiance observed by TIMS as it views the ground through an intervening SO_2 plume. Based on the data collected at Mount Etna in 1986, the TIMS-based procedure has an estimation error of approximately 15%.

4.2. *Applications to Mauna Loa eruption plumes*

The 1984 eruption of Mauna Loa began on 25 March and ended on 14 April [*Lockwood et al.*, 1985; *Smithsonian Institution*, 1989; *Casadevall et al.*, 1984]. On the first day of the eruption, the plume reached an altitude of approximately 11 km. The plume did not reach the stratosphere, since the tropopause was at an altitude of 18 km. Between 27 March and 14 April 1984 the maximum altitude of the plume was less than 4.6 km. During this period the plume travelled westward at a rate of 800 km/d, reducing visibility at airports up to 5000 km away from Mauna Loa.

The flux rates of SO_2 and CO_2 were determined with a series of airborne COSPEC and infrared analyzer surveys of the plume [*Casadevall et al.*, 1984]. Between 2 April and 14 April the flux of SO_2 ranged between 2200 and 6600 x 10^3 kg per day, while the flux of CO_2 ranged between 240 and 1400 x 10^3 kg per day. Estimates of the SO_2 comumn abundance with TOMS indicated a maximum of approximately 2 g m^{-2} on 27 March. TOMS data were also used to estimate that the Mauna Loa plume contained 130,000 x 10^3 kg of SO_2 on 26 March and 190,000 x 10^3 kg on 27 March.

SO_2 abundances in the 26 July 1986 Mount Etna plume were estimated to be as much as eight times higher

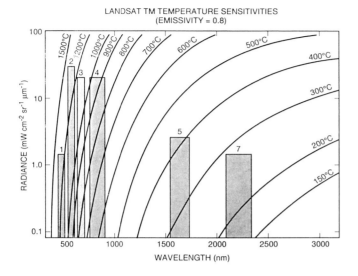

LANDSAT TM TEMPERATURE SENSITIVITIES
(EMISSIVITY = 0.8)

Figure 7. Wavelength dependence of thermal radiance, according to Planck's formula, for surfaces with a uniform emissivity of 0.8, plotted for a range of temperatures. The boxes indicate the band passes and operational range of radiance for the visible and short wavelength infrared Landsat Thematic Mapper (TM) sensors. After *Francis and Rothery* [1987].

[*Realmuto et al.*, 1994] than those reported for the 1984 Mauna Loa plume, yet the Etna plume was not detected by TOMS. This lack of detection is probably due to narrow width (1-5 km) and low altitude (<5 km MSL) of the Etna plume relative to the Mauna Loa plume. The coarse spatial resolution of TOMS (50 km at nadir) is a limiting factor on the size of plume that can be detected.

5. REMOTE SENSING OF TEMPERATURE

Multispectral remote sensing can be used for temperature determinations in volcanic regions [*Francis and Rothery*, 1987; *Rothery, et al.*, 1988; *Realmuto, et al.*, 1992; *Mouginis-Mark, et al.*, 1994a; *Flynn, et al.*, 1994]. While the techniques described here have not been used on Mauna Loa, their use in the future seems likely. Because of the wide range of temperatures encountered, it is desirable to acquire multispectral data from the visible and the short-wave infrared as well as the thermal infrared. While one commonly thinks of using the thermal infrared to measure temperature, in volcanic regions the radiation from very hot sources will saturate the majority of thermal IR sensors. Planck's Law indicates that the intensity of emitted radiation increases as a function of temperature, but also that the wavelength of the maximum emission shifts to progressively shorter wavelengths as the temperature increases.

Figure 7, after *Francis and Rothery* [1987] shows a plot of radiance as a function of wavelength for a variety of different temperatures. The plot also includes the location and saturation level of the Landsat Thematic Mapper channels. When measuring very hot volcanic targets, one needs to select that wavelength region where the emitted radiation dominates over reflected radiation, but at the same time does not saturate the sensor. *Rothery et al.* [1988] and *Oppenheimer et al.* [1993] have also shown how using multiple wavelengths allows for the derivation of more than one temperature within a pixel — often the case where there is a small very hot area within a pixel while the majority of the pixel is near ambient temperature.

Ground-based spectroradiometric measurements, a study using the Advanced Very High Resolution Radiometer (AVHRR) and Landsat Thematic Mapper (TM) data, and work done using the NS001 and TIMS aircraft multispectral thermal data, have illustrated some of the applications of multispectral measurements to temperature observations of volcanic features on Kilauea, Hawaii, though not over Mauna Loa.

Using a spectroradiometer in the 0.4-2.5 m region, *Flynn and Mouginis-Mark* [1992] and *Flynn et al.* [1993] made nighttime observations of active flows, and observations of the active Kupaianaha lava lake; both areas are on the East Rift Zone of Kilauea. They found that the crust of a newly emplaced lava flow cooled by about 350° in less than an hour. The agreement between their observations and theoretical models of flow emplacement help to constrain such models. Over the Kupaianaha lava lake, they found that the temporal variability of the thermal output occurred on time scales of seconds to minutes, implying that satellite-based observations would, at best, provide only a snapshot of a rapidly changing phenomenon.

In a study examining the effects of viewing geometry on detection of thermal features in AVHRR weather satellite compared with Landsat Thematic Mapper, *Mouginis-Mark et al.* [1994a] found that with nadir-looking, 1 km spatial resolution data, overturning lava lakes, surface flows, and lava tubes with skylights could be detected thermally. However, in off-nadir looking data, where the pixel size is as large as 14 km, some of the smaller of these phenomena could not be separated from solar heated, older lava flows. Using TM data, with 30 m spatial resolution, all of the phenomena could be detected thermally.

Realmuto et al. [1992] used TIMS thermal data acquired from the NASA C-130 to map the temperature distribution related to underground transport and storage of lava in the Kupaianaha flow field, Kilauea. Active lava tubes were clearly mapped. The authors' temperature maps also defined the boundaries of hydrothermal plumes which resulted from entry of lava into the ocean. The multispectral data also allowed determination of emissivity variations attributed to acid-induced leaching near fumaroles (described earlier).

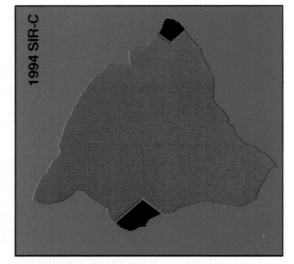

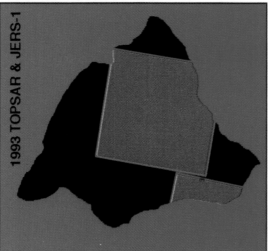

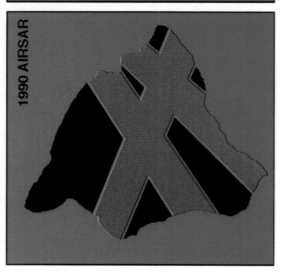

Plate 7. Maps showing the coverage of Mauna Loa with imaging radar systems. Areas in red show places where data were collected in 1990 (left), 1993 (center) and 1994 (right). The AIRSAR data collected in 1990 were obtained with multiple parallel flight lines, so that there is some overlapping coverage at different incidence angles along the center of each of the three broad bands shown here. The large square in the 1993 map shows the location of the JERS-1 data, and the smaller area on the southwestern tip of the island marks the location of the TOPSAR data. The map for the 1994 SIR-C data shows the coverage that was obtained during the two flights in April and October, 1994. SRL-1 Data Take 52.1 provides the coverage that extends from South Point to Hilo, SRL-1 Data Take 122.2 the coverage from Kohala and Hualalai to Kilauea, and SRL-2 Data Take 36.1 the coverage of western Mauna Loa. Additional data were also collected by SRL-2 in October 1994 at higher spatial resolution (12.5 m/pixel) for subareas within this broader coverage.

Plate 8. AIRSAR aircraft radar image of Mokuaweoweo caldera, obtained August 6, 1990. The geometry of the caldera is distorted due to the oblique viewing geometry of the radar. This image is a combination of P-band HV displayed in red, L-band HV displayed in green, and C-band HV displayed in blue. The viewing direction is from the east, and the incidence angle varies from 35° in the near-range to 62° in the far-range. This combination of radar wavelengths and polarizations enhances the differences between the historical aa flows (in green) and the prehistoric pahoehoe flows (in purple). Other features can also be identified: (a) the 1975 and 1984 flows South Pit; (b) the ash deposit associated with the 1949 cone; (c) individual flow lobes in the latest prehistoric flows [*Macdonald*, 1971]; (d) segments of the 1984 lava flows on the floor of North Pit; and (e) the 1940 cone Also visible (arrowed) is the 1984 eruptive vent.

TABLE 2. Characteristics of Radar Systems That Have Imaged [ ]a Loa

Sensor	Wavelength[a]	P1 Pixel size	Angle	[ ]zation
AIRSAR	C, L, P	12.5 m/pixel	20 - 60°	HH, [ ]VV
TOPSAR	L, P	12.5 m/pixel	20 - 60°	
	C	12.5 m/pixel	20 - 60°	VV
JERS-1	L	18 m/pixel	35°	HH
SIR-C[b]	X, C, L	25 m/pixel	56° and 60°	X-VV, C-VV, [ ]
ERS-2[c]	C	25 m/pixel	23°	VV
RADARSAT[3]	C	25, 50, 100 m/p	20 - 50°	HH

[a] Radar wavelengths are as follows: X-band is 3 cm, C-band is 5.6 cm, L-band is 24 cm, and P-band is 68 cm.
[b] SIR-C/X-SAR operated in many different wavelength/polarization modes [*Evans, et al.*, 1993]. Only those th[ ] were used to image Hawaii are given here.
[c] Future radar system if value for Mauna Loa studies.

6. RADAR REMOTE SENSING OF MAUNA LOA

6.1. *Data coverage*

In a comparable manner to visible and infrared imaging the Earth's surface, radar systems provide the capability [ob]serve the ground at several different wavelengths, [polar]izations, and viewing geometries. An attribute of all [s]ystems is that they can image the surface irrespective [of] cover or time of day, thus complete coverage of [ ] can be obtained. The most important parameter [fo]r investigation is the wavelength, which is typi-[cally 3] cm (X-band), 5.6 cm (C-band), 24 cm (L-[band, 68] cm (P-band). Other parameters that can be [p]olarization (horizontally-transmit and receive [as "]HH"; horizontally-transmit and vertically [receive] and vertically transmit and receive as [ang]le (called the "incidence angle") at which [ ] the surface (typically in the range 20-[ ]. B]ecause radars provide their own illu-[mination] also possible to control the "look [ ] [dir]ent, so that structural features in a [b]e enhanced. The specific charac-[teristics of] systems that have been used to [summa]rized in Table 2.

[radar] (SIR-B) data were col-[lected in O]ctober 1984 [*Gaddis, et al.*, [were] collected over Mauna Loa [by the N]ASA/JPL aircraft radar [of] the Big Island (Plate [ ] the eastern half of the [ ] by the spaceborne [ ] radar returned to [ti]me to perform [souther]rn flank of the

volcano using the TOPSAR radar system [*Zebker, et al.*, 1992]. TOPSAR can produce digital elevation models (DEMs) of landscapes at a spatial resolution of 10 m/pixel with a vertical accuracy of ~2-5 m depending on the roughness of the terrain. These TOPSAR data for Mauna Loa have not yet been processed, but it is expected that the data will be of value for the measurement of lava flow thicknesses. TOPSAR data were collected across the 1950 lava flows in the hope that thickness variations associated with the local slope could be measured; good observations of the volumetric flow rate of these flows [*Finch and Macdonald*, 1950; *Macdonald and Finch*, 1950] should enable numerical models for the longitudinal variation in lava flow rheology [e.g., *Fink and Zimbelman*, 1989] to be tested.

The two flights of the Space Shuttle SIR-C/X-SAR [*Evans, et al.*, 1993] radar also produced almost complete coverage of Mauna Loa and Kilauea volcanoes (Plate 7). The first SIR-C/X-SAR flight (in April 1994) was primarily devoted to using Kilauea as a calibration target for subsequent comparison with data obtained for other basaltic shield volcanoes around the world [*Mouginis-Mark*, 1994b]. Nevertheless, most of the eastern side of Mauna Loa was also imaged (Data Take 52.1 at 52.3°). Data were collected at three wavelengths (X-, C-, and L-band), various polarizations (HH, HV, and VV), multiple incidence angles (18-59°) on both ascending and, in one instance, a descending orbit. In most cases, only the eastern lower flank of Mauna Loa was imaged during these studies of Kilauea, except on the descending orbit when data for all of the northern half of Mauna Loa were obtained (Plate 7). During the second SIR-C/X-SAR mission (in October 1994), almost all of the western half of Mauna Loa was covered in a single data take (Data Take 36.10) at 32.1° incidence angle, complementing the April 1994 coverage of

Figure 8. SPOT panchromatic image of Mokuaweoweo caldera, showing the area included in the AIRSAR image (Plate 8). Notice that at optical wavelengths the shelly, "fountain-fed", pahoehoe flows appear darker than older, dense, "tube-fed", pahoehoe, and aa appears darker than pahoehoe of equivalent or younger ages. North is towards the top.

Plate 9. AIRSAR aircraft radar image of part of the Ninole Hills on the eastern flank of Mauna Loa, obtained August 4th, 1990. This image is a combination of P-band HV displayed in red, L-band HV displayed in green, and C-band HV displayed in blue. The viewing direction is from the west, and the incidence angle varies from 35° in the near-range to 62° in the far-range. The P-band data enables the 1950 Kahuku flow ("A") to be identified, but the prehistoric Pohina flow ("B") has radar backscatter characteristics that are comparable to the rain forest and so cannot be discriminated. Notice that no other flow margins can be identified in this scene, so that this radar data set cannot be used to extend existing geologic maps of the area. The radar does, however, clearly delineate the Makaalia ridge ("C"), and subtle breaks in slope (arrows point downslope). Two jeep tracks ("D") can also be seen. Image width is ~12 km, the viewing direction is from the left.

Plate 10. AIRSAR aircraft radar image of the South Point area on the distal end of the Southwest Rift Zone of Mauna Loa, obtained August 4th, 1990. This image is a combination of P-band HV displayed in red, L-band HV displayed in green, and P-band VV displayed in blue. The viewing direction is from the northwest (top of image), and the incidence angle varies from 35° in the near-range to 60° in the far-range. At "A", several lobes of the 7.2 Ka Puu Poo Pueo Flow (which is covered by 20-40 cm of ash) can be identified. The 1868 flow is at "B". By virtue of the look-direction, the sea cliff and Pali o Mamalu (black arrows) is clearly seen. Dunes are located at "C". The combination of two P-band polarizations is particularly good at showing the morphology of flows that lie beneath a surficial layer of the Pahala Ash, which is 2—8 m thick at this locality [*Stearns and Macdonald*, 1946] and is predominately dark green in this image. This penetration enables the structure of some near-surface prehistoric flows to be seen at "D" and "E". Interestingly, the barren 1868 flow ("B") has the same radar chracteristics as flow "D", which is deeply covered in ash. No fieldwork to investigate these radar properties has, however, been done at this time. The light at South Point is at "F". The white arrows point to several of the numerous wind streaks that have formed in the loose surficial materials in response to the prevailing trade winds. Image width is 13 km.

the eastern side of the volcano. Additional data for Mauna Loa were also obtained during the last four days of the mission when exact-repeat orbits were selected for radar interferometry experiments. Data were obtained at high resolution (12.5 m/pixel) over the 1989, 1942, and 1984 lava flows between the 1300-2000 m elevation levels. These data, which were obtained at C- and L-band VV-polarization) will be processed to produced a high resolution DEM as part of a broader investigation to study day-to-day variations in the distribution and thickness of the lava flow field of Kilauea.

6.2. *Examples of radar observations*

To date, the radar data for Mauna Loa have been largely ignored in favor of the comparable information for Kilauea volcano, due to the easier field logistics for data validation and the strong interest in the ongoing eruption at the Puu Oo vent for topographic change studies. For Kilauea, the primary use of the aircraft data has been the quantitative analysis of the radar backscatter properties of the individual lava flows [*Campbell and Campbell*, 1992], and their subsequent comparison to lava flows on Venus. Almost no investigations of the TOPSAR or SIR-C/X-SAR data for Hawaii have yet been published, but comparable analyses of other volcanoes [e.g., *Mouginis-Mark and Garbeil*, 1993; *Mouginis-Mark*, 1994b] suggest some of the studies that can be accomplished with the available data. In general, these studies focus on lithologic mapping, analyses of the structure of the volcano, the search for buried lava flows beneath ash deposits, and topographic mapping. All but the last of these applications can be demonstrated with Mauna Loa data collected in August 1990 by the AIRSAR system, and will now be briefly reviewed.

The use of multiwavelength radar data to study differences in the surface roughness of lava flows has been well known for more than a decade. Using AIRSAR data, *Campbell and Campbell* [1992] showed that aa and pahoehoe lava flows have different radar backscatter characteristics at L-band (24 cm). In the case of the summit of Mauna Loa (Plate 8), the multiwavelength radar enables different flow units to be identified, as well as subtle details of the structure of Mokuaweoweo caldera. Few of these phenomena have not been recognized from detailed field mapping [*Macdonald*, 1971; *Lockwood, et al.*, 1987]. The radar data show many flow contacts within the Mokuaweoweo overflow units more clearly than does conventional air photography, but the significance of some of these contacts has not yet been established. The difference in the radar and optical properties of the lava flows around Mokuaweoweo caldera can be seen from a comparison of the AIRSAR data and the SPOT satellite panchromatic (10 m/pixel) data presented in Figure 8.

Details of the structure of a volcano can often be enhanced using radar data; the careful selection of radar inci-

dence angle and viewing direction can reveal subtle topographic features that may be difficult to identify due to the vegetation cover. Such an example is shown in Plate 9, which is part of the Ninole Hills on the southeastern flank of Mauna Loa. This area, which is part of the Kau Forest Reserve, has proven to be difficult to map on the ground due to the dense vegetation [e.g., *Lipman and Swenson*, 1984; *Lockwood, et al.*, 1988] and yet offers an insight into the earlier eruptive activity of the volcano. In particular, the P-band wavelength cross-polarized (HV) radar data provide some capabilities to image surface flows irrespective of the vegetation cover, and to infer subtle topographic features (arrowed in Plate 9) by virtue of tonal variations in the vegetation canopy. This break in slope is also evident on the JERS-2 and SIR-C radar images of the eastern flank of Mauna Loa, and so this appears to be a real feature. Parts of the Kahuku flow can be delineated in the radar image, although the Pohina flow is indistinguishable from the high background radar backscatter from the forests.

The ability of radar to penetrate dry, unconsolidated materials was first demonstrated through analysis of Space Shuttle (SIR-A) radar data for the Eastern Sahara [*McCauley, et al.*, 1982], where buried drainage channels were identified beneath the desert sand. In the case of Mauna Loa, the South Point area at the distal end of the Southwest Rift Zone provides a similar opportunity for radar penetration studies, since there is 2-8 m of the Pahala ash in this area [*Sterns and Macdonald*, 1946, p. 73]. Particularly in the case of the longest radar wavelength (P-band, 68 cm), the possibility of delineating the edges of lava flows beneath the ash is high. Plate 10 shows one of the AIRSAR images of this area. A comparison between these radar data, air photographs and field observations shows that, in addition to distinguishing individual lobes on the Puu Poo Pueo Flow mapped by *J. Lockwood* [pers. comm., 1995], the radar can detect braided flow boundaries that are reminiscent of lava channels beneath the veneer of soil and wind-blown ash (e.g., pts. "D" and "E" in Plate 10). Other morphological features are also easily seen in the radar data, including wind streaks and dunes.

7. FUTURE REMOTE SENSING OF MAUNA LOA

Additional orbital radars will be used to study Hawaii. The European Space Agency (ESA) has been operating the ERS-1 radar since 1991, and this has been used to study many of the volcanoes in Alaska and the Aleutians [*Rowland, et al.*, 1994]. However, the ERS-1 spacecraft, and its successor ERS-2, cannot currently be used to study any part of Hawaii due to the fact that the spacecraft do not carry onboard tape recorders. This requires that data are transmitted in real-time to ground stations that are in the line of sight of the spacecraft. In order to collect ERS-2 and JERS-1 radar data, the University of Hawaii at Manoa (on the island of Oahu) is building a ground station specifically to study Mauna Loa and Kilauea using data

TABLE 3. ASTER Channels and Wavelength (µm)

	VNIR		SWIR		TIR
1	0.52-0.60	4	1.600-1.700	10	8.125-8.475
2	0.63-0.69	5	2.145-2.185	11	8.475-8.825
3	0.76-0.86	6	2.185-2.225	12	8.925-9.275
		7	2.235-2.285	13	10.25-10.95
		8	2.295-2.365	14	10.95-11.65
		9	2.360-2.430		

from these satellites. It is expected that the Hawaii ground station will be operational by June 1995.

The ground reception capability for radar data of Hawaii creates the opportunity for extensive topographic and geodetic investigations of Hawaiian volcanoes. Using a technique called radar interferometry, which coherently compares the phase information contained within two radar images of the same scene, the value of the ERS-1 radar data in the analysis of ground movements associated with earthquakes was demonstrated for the 1992 Landers Earthquake in California [*Massonnet, et al.*, 1993, 1994; *Zebker, et al.*, 1994]. These radar observations enable ground movement at a fraction of a radar wavelength (typically ~3 mm in line-of-sight movement) to be identified over the entire image for dry targets in arid environments. Provided that the phase data from the radar can be correlated (the moist atmosphere in Hawaii may prove to be a problem at low elevations), such observations are predicted to be of great value for the study of ground deformation prior to a new eruption, or for the mapping of the planimetric shape of new lava flows [*Mouginis-Mark*, 1994c]. Current plans are for ERS-1 and ERS-2 to be flown in tandem so that exact-repeat coverage will be obtained once during a 24 hour interval sometime within a 35 day observation period. Following the expected demise of ERS-1 in late 1995, exact-repeat coverage will only be achieved from successive orbits of the ERS-2 spacecraft, which will have a 35-day repeat interval. JERS-1 radar data will be collected every 24 days. Thus it should be possible to generate a deformation map of Mauna Loa approximately once per month.

The Canadian RADARSAT spacecraft [*Raney, et al.*, 1991] will also image Hawaii, except in this instance the data will be recorded on board. Data from RADARSAT will be particularly useful for frequent large-area coverage of the Big Island, due to the variable swath width, multiple incidence angles, and variable spatial resolution capabilities of the sensor. However, because of the absence of onboard GPS equipment, it is unlikely that RADARSAT data will be useful for interferometry experiments because the baseline separation of the orbits will not be known to an adequate accuracy. RADARSAT is planned for launch in early 1996.

A major improvement to optical remote sensing from space will occur in 1998 with the launch of the NASA's Earth Observing System AM-1 platform. This satellite will include the Advanced Spaceborne Thermal Emission and Reflection Radiometer (ASTER) and the Moderate-Resolution Imaging Spectrometer (MODIS). ASTER is a 14-channel imaging instrument, with bands in the VNIR, SWIR and TIR (Table 3). The pixel size is 15, 30 or 90 m, respectively for the three wavelength regions, with a repeat cycle of 16 days or less, depending on latitude; the swath width is 60 km [*Kahle, et al.*, 1991]. MODIS will provide images in 36 bands between 0.4 and 14.5 µm with a pixel size of 250 m and 1 km. The 1 km resolution images will be available every one or two days. MODIS will thus allow monitoring on a 1-2 day basis while ASTER can be used for detailed process studies.

ASTER's VNIR bands are similar to the Landsat TM bands and the TMS bands in this wavelength region, allowing detection of changes in iron oxidation and vegetation development. The SWIR bands will improve the capability to detect and map the development of clay minerals, associated with soil development and weathering in more humid climates. The spectral resolution of the ASTER TIR system is very similar to TIMS; therefore, ASTER should allow scientists to map both the change in surface properties of lava flows, and to map the SO_2 in eruption plumes from Mauna Loa.

The improvement in pixel size of the thermal bands compared to AVHRR (90 m vs 1 km) will allow mapping of lava flows, lava tubes, and lava ponds, instead of just detecting them. It will also be possible to determine temperatures for volcanic features whose hottest components are smaller than a full pixel (generally the case for the flows, tubes and ponds). These new capabilities will improve our ability to map Mauna Loa, as well as other volcanoes worldwide.

Acknowledgements. We would like to acknowledge the thoughtful advice of manuscript reviewers David Rothery and Ronald Greeley. Portions of this work were performed at the

Jet Propulsion Laboratory, California Institute of Technology, under contract to the National Aeronautics and Space Administration.

REFERENCES

Abrams, M., E. Abbott, and A. Kahle, Combined use of visible, reflected and thermal infrared images for mapping Hawaiian lava flows, J. Geophys. Res., 96, 475-484, 1991.

Berresheim, H., and W. Jaeschke, The contribution of volcanoes to the global atmospheric sulfur budget, J. Geophys. Res., 88, 3732-3740, 1983.

Bluth, G. J. S., S. D. Doiron, C. J. Scott, C. C. Schnetzler, A. J. Kreuger, and L. S. Walter, Global tracking of the SO_2 clouds from the June 1992 Mount Pinatubo eruptions, Geophys. Res. Lett., 19, 151-154, 1992a.

Bluth, G. J. S., C. J. Scott, M. Schoeberl, C. C. Schnetzler, A. J. Kreuger, and L. S. Walter, SO_2 cloud tracking from the June and August eruptions of Mount Spurr, Alaska, Trans. Amer. Geophys. Union, 74, 614, 1992b.

Bluth, G. J. S., C. C. Schnetzler, A. J. Krueger, and L. S. Walter, The contribution of explosive volcanism to global atmospheric sulfur dioxide concentrations, Nature, 366, 327-329, 1993.

Brawer, S. A., and W. B. White, Raman spectroscopic investigation of the structure of silicate glasses, I, the binary alkiline silicates, J. Chem. Phys., 63, 2421-2432, 1975.

Caltabiano, T., R. Romano, and G. Budetta, SO_2 flux measurements at Mount Etna (Sicily), J. Geophys. Res., 99, 12809-12819, 1994.

Campbell, B. A., and D. B. Campbell, Analysis of volcanic surface morphology on Venus from comparison of Arecibo, Magellan, and terrestrial radar data, J. Geophys. Res., 97, 16293-16314, 1992.

Casadevall, T. J., D. A. Johnston, D. M. Harris, W. I. Rose, L. L. Malinconico, R. E. Stoiber, T. J. Bornhorst, S. N. Williams, L. Woodruff, and J. M. Thompson, SO_2 emission rates at Mount St. Helens from March 29 through December 1980, in USGS Prof. Paper 1250, edited by P. Lipman and D. R. Mullineaux, pp. 193-200, 1981.

Casadevall, T. J., A. J. Kruegar, and B. Stokes, The volcanic plume from the 1984 eruption of Mauna Loa, Hawaii, Trans. Amer. Geophys. Union, 65, 1133, 1984.

Chartier, T. A., W. I. Rose, and J. B. Stokes, Detailed record of SO_2 emissions from Puu Oo between episodes 33 and 34 of the 1983-1986 ERZ eruption, Kilauea, Hawaii, Bull. Volcanol., 50, 215-228, 1988.

Crisp, J., A. Kahle, and E. Abbott, Thermal infrared spectral character of Hawaiian basaltic glasses, J. Geophys. Res., 95, 21657-21669, 1990.

Curtiss, B., J. B. Adams, and M. S. Ghiorso, Origin, development, and chemistry of silica-alumina rock coatings from the semiarid regions of the island of Hawaii, Geochim. et Cosmochim. Acta, 49, 49-56, 1985.

Doiron, S. D., G. J. S. Bluth, C. C. Schnetzler, A. J. Krueger, and L. S. Walter, Transport of Cerro Hudson SO_2 clouds, Trans. Amer. Geophys. Union, 72, 489-498, 1991.

Dowty, E., Vibrational interactions of tetrahedra in silicate glasses and crystals, III, Calculations on simple sodium and lithium silicates, thorveitite and rankinite, Phys. Chem. Miner., 14, 542-552, 1987.

Evans D. L., C. Elachi, E. R. Stofan, B. Holt, J. B. Way, M. Kobrick, H. Ottl, P. Pampaloni, M. Vogt, S. Wall, J. van Zyl, and M. Schier, The Shuttle Imaging Radar-C and X-SAR mission, Eos, 74, pp. 145, 157, 158, 1993.

Farr, T., and J. Adams, Rock coatings in Hawaii, Geo. Soc. Amer. Bull., 95, 1077-1083, 1984.

Finch, R. Y., and G. A. Macdonald, The June 1950 eruption of Mauna Loa, Part 1, Volcano Letter, 508, 12 pp., 1950.

Fink, J., and J. Zimbelman, Longitudinal variations in rheological properties of lavas: Puu Oo basalt flows, Kilauea volcano, Hawaii, in Lava Flows and Domes, IAVCEI Proc. Volc, vol. 2, eidted by J. Fink, pp. 15-173, Springer-Verlag, New York, 1989.

Flynn, L., and P. Mouginis-Mark, Cooling rate of an active lava flow from nighttime spectroradiometer measurements, Geophys. Res. Let., 19, 1783-1786, 1992.

Flynn, L., P. Mouginis-Mark, J. Gradie, and P. Lucey, Radiative temperature measurements at Kupaianaha Lava Lake, Kilauea Volcano, Hawaii, J. Geophys. Res., 98, 6461-6476, 1993.

Flynn, L. P., P. J. Mouginis-Mark, and K. A. Horton, Distribution of thermal areas on an active lava flow field: Landsat observations of Kilauea, Hawaii, July 1991, Bull. Volc., 56, 284-296, 1994.

Francis, P. W., and D. A. Rothery, Using the Landsat Thematic Mapper to detect and monitor active volcanoes: An example from Lascar volcano, northern Chile, Geology, 15, 614-617, 1987.

Gaddis, L. R., P. J. Mouginis-Mark, R. B. Singer, and V. H. Kaupp, Geologic analyses of Shuttle Imaging Radar (SIR-B) data of Kilauea Volcano, Hawaii, Geol. Soc. Amer. Bull., 101, 317-332, 1989.

Gaddis, L. R., P. J. Mouginis-Mark, and J. Hayashi, Lava flow surface textures: SIR-B radar image texture, field observations, and terrain measurements, Photogram. Eng. Rem. Sen., 56, 211-224, 1990.

Gillespie, A. R., Digital techniques of image enhancement, in Remote Sensing in Geology, edited by B. Siegal and A. Gillespie, pp. 139-226, John Wiley, New York, 1980.

Gillespie, A. R., A. B. Kahle, and R. E. Walker, Color enhancement of highly correlated images, I, Decorrelation and HSI contrast stretches, Remote Sens. Environ., 20, 209-235, 1986.

Hamilton, P. M., R. H. Varey, and M. M. Millán, Remote sensing of sulfur dioxide, Atmos. Environ., 12, 127-133, 1978.

Hoff, R. M., and M. M. Millán, Remote SO_2 mass flux measurements using COSPEC, J. Air Pollut. Control Assoc., 31, 381-384, 1981.

Holcomb, R. T., Eruptive history and long-term behavior of Kilauea volcano, in Volcanism in Hawaii, Volume I, U.S. Geol. Surv. Prof. Pap. 1350, edited by R. W. Decker, T. L. Wright, and P. H. Stauffer, pp. 261-350, 1987.

Hunt, G., Spectral signatures of particulate minerals in the visible and near infrared, Geophys., 42, 501-513, 1977.

Kahle, A. B., A. R. Gillespie, E. A. Abbott, M. J. Abrams, R. E. Walker, G. Hoover, and J. P. Lockwood, Relative dating of Hawaiian lava flows using multispectral thermal infrared images: A new tool for geologic mapping of young volcanic terranes, J. Geophys. Res., 93, 15239-15251, 1988.

Kahle, A. B., F. D. Palluconi, S. J. Hook, V. J. Realmuto, and G. Bothwell, The Advanced Spaceborne Thermal Emission and Reflectance Radiometer (ASTER), Intl. J. Imaging Sys. Tech., 3, 144-156, 1991.

Krueger, A. J., Sighting of El Chichón sulfur dioxide clouds with the Numbus 7 Total Ozone Mapping Spectrometer, Science, 220, 1377-1379, 1983.

Krueger, A. J., L. S. Walter, C. C. Schnetzler, and S. D. Doiron, TOMS measurement of the sulfur dioxide emitted during the 1985 Nevado del Ruiz eruptions, J. Volcanol. Geotherm. Res., 41, 7-15, 1990.

Lipman, P. W., and A. Swenson, Generalized geologic map of the Southwest Rift Zone of Mauna Loa, Hawaii, U.S. Geological Survey Miscellaneous Investigation Series Map I-1323, 1984.

Lockwood, J. P., N. Banks, T. English, L. Greenland, D. Jackson, D. Johnson, R. Koyanagi, K. McGee, A. Okamura, and J. Rhodes, The 1984 eruption of Mauna Loa volcano, Hawaii, EOS Trans. Amer. Geophys. Union, 66, 169-171, 1985.

Lockwood, J. P., J. J. Dvorak, T. T. English, R. Y. Koyanagi, A. T. Okamura, M. L. Summers, and W. R. Tanigawa, Mauna Loa 1974—1984: A decade of intrusive and extrusive activity, in Volcanism in Hawaii, U.S. Geological Survey Prof. Paper 1350, pp. 537-570, 1987.

Lockwood, J., and P. Lipman, Holocene eruptive history of Mauna Loa, in Volcanism in Hawaii, Volume I, U.S. Geol. Surv. Prof. Paper 1350, edited by R. Decker, and P. Stauffer, 1987.

Lockwood, J. P., P. W. Lipman, L. D. Petersen, and F. R. Warshauer, Generalized ages of surface lava flows of Mauna Loa volcano, Hawaii, U.S. Geol. Surv. Misc. Inv. Map, I-1908, 1988.

Macdonald, G. A., Geologic map of the Mauna Loa Quadrangle, Hawaii, U.S. Geological Survey Geologic Map, Mauna Loa Quadrangle, Hawaii, GQ-897, 1971.

Macdonald, G., and R. Y. Finch, The June 1950 eruption of Mauna Loa, Part 2, Volcano Letters, 509, 1-6, 1950.

Malinconico, L. L., Fluctuations in SO_2 emission during recent eruptions of Etna, Nature, 278, 43-45, 1979.

Malinconico, L. L., On the variation of SO_2 emissions from volcanoes, J. Volcanol. Geotherm. Res., 33, 231-237, 1987.

Massonnet, D., K. Feigl, M. Rossi, and F. Adragna, Radar interferometric mapping of deformation in the year after the Landers earthquake, Nature, 369, 227-230, 1994.

Massonnet, D., M. Rossi, C. Carmona, F. Adragna, G. Peltzer, K. Feigl, and T. Rabaute, The displacement field of the Landers earthquake mapped by radar interfero-metry, Nature, 364, 13-142, 1993.

McCauley, J. F., G. G. Schaber, C. S. Breed, M. J. Grolier, C. V. Haynes, B. Issawi, C. Elachi, and R. Blom, Subsurface valleys and geoarcheology of the eastern Sahara revealed by Shuttle radar, Science, 218, 1004-1020, 1982.

Moffat, A. J., and M. M. Millán, The applications of optical correlation techniques to the remote sensing of SO_2 plumes using sky light, Atmos. Environ., 5, 677-690, 1971.

Mouginis-Mark, P. J., and H. Garbeil, Digital topography of volcanoes from radar interferometry: An example from Mt. Vesuvius, Italy, Bull. Volcan., 55, 566-570, 1993.

Mouginis-Mark, P., H. Garbeil, and P. Flament, Effects of viewing geometry on AVHRR observations of volcanic thermal anomalies, Rem. Sens. Environ., 48, 51-60, 1994a.

Mouginis-Mark, P. J., Preliminary observations of volca-noes with the SIR-C/X-SAR radar. Submitted to IEEE Trans. Geosci. Rem. Sen., 1994b.

Mouginis-Mark, P. J., Mitigating volcanic hazards through radar interferometry, Geotimes, July, 11-13, 1994c.

Oppenheimer, C., P. Francis, D. Rothery, R. Carlton, and L. Glaze, Infrared image analysis of volcanic thermal features: Lascar volcano, Chile, 1984-1992, J. Gephys. Res., 98, 4269-4286, 1993.

Palluconi, F. D., and G. R. Meeks, Thermal infrared mul-tispectral scanner (TIMS): An investigator's guide to TIMS data, JPL Publ. 85-32, 1985.

Raney, R. K., A. P. Luscombe, E. J. Langham, and S. Ahmed, RADARSAT, Proc. IEEE, 79, 839-849, 1991.

Read, W. G., L. Froidevaux, and J. W. Waters, Microwave Limb Sounder (MSL) measurements of SO_2 from Mt. Pinatubo volcano, Geophys. Res. Let., 20, 1299-1302, 1993.

Realmuto, V. J., M. J. Abrams, M. F. Buongiorno, and D. C. Pieri, The use of multispectral thermal infrared image data to estimate the sulfur dioxide flux from volcanoes: A case study from Mount Etna, Sicily, July 29, 1986, J. Geophys. Res., 99, 481-488, 1994.

Realmuto, V. J., K. Hon, A. B. Kahle, E. A. Abbott, and D. C. Pieri, Multispectral thermal infrared mapping of the 1 October 1988 Kupaianaha flow field, Kilauea volcano, Hawaii, Bull. Volc., 55, 33-44, 1992.

Rothery, D. A., P. W. Francis, and C. A. Wood, Volcanic monitoring using short wavelength infrared data from satellites, J. Geophys. Res., 93, 7993-8008, 1988.

Rowland, S. K., G. A. Smith, and P. J. Mouginis-Mark, Preliminary ERS-1 observations of Alaskan and Aleutian volcanoes, Remote Sens. Environ., 48, 358-369, 1994.

Simon, I., and H. O. McMahon, Study of some binary silicate glasses by means of reflection in infrared, J. Am. Ceram. Soc., 36, 160-164, 1953.

Smithsonian Institution, Global Volcanism 1975—1985, Prentice Hall, New Jersey, and American Geophysical Union, Washington, DC, 657 pp., 1989.

Soha, J.M. and A.A. Schwartz, A multispectral histogram normalization contrast enhancement, in Proc. of Fifth Canadian Symp. on Remote Sensing, Victoria, BC, Canada, pp. 86-93, 1978.

Stearns, H. T., and G. A. Macdonald, Geology and groundwater resources of the Island of Hawaii, Hawaii Divn. Hydrography, Bull. 9, 363 pp., 1946.

Stoiber, R. E., L. L. Malinconico, and S. N. Williams, Use of the correlation spectrometer at volcanoes, in Forecasting

Volcanic Events, edited by H. Tazieff and J. C. Sabroux, Elsevier, pp. 425-444, 1983.

Stoiber, R. E., S. N. Williams, and B. Huebert, Annual contribution of sulfur dioxide to the atmosphere by volcanoes, J. Volcanol. Geotherm. Res., 33, 1-8, 1987.

Zebker, H. A., S. N. Madsen, J. Martin, K. B. Wheeler, T. Miller, Y. Lou, G. Alberti, S. Vetrella, and A. Cucci, 1992, The TOPSAR interferometric radar topographic mapping instrument, IEEE Trans. Geosci. Rem. Sen., 30, 933-940, 1992.

Zebker, H. A., P. Rosen, R. M. Goldstein, A. Gabriel, and C. L. Werner, On the derivation of coseismic displacement fields using differential radar interferometry: The Landers earthquake, J. Geophys. Res., 99, 19617-19634, 1994.

Continuous Monitoring of Volcanoes
With Borehole Strainmeters

Alan T. Linde and I. Selwyn Sacks

Department of Terrestrial Magnetism, Carnegie Institution of Washington, Washington, DC

Monitoring of volcanoes using various physical techniques has the potential to provide important information about the shape, size and location of the underlying magma bodies. Volcanoes erupt when the pressure in a magma chamber some kilometers below the surface overcomes the strength of the intervening rock, resulting in detectable deformations of the surrounding crust. Seismic activity may accompany and precede eruptions and, from the patterns of earthquake locations, inferences may be made about the location of magma and its movement. Ground deformation near volcanoes provides more direct evidence on these, but continuous monitoring of such deformation is necessary for all the important aspects of an eruption to be recorded. Sacks-Evertson borehole strainmeters have recorded strain changes associated with eruptions of Hekla, Iceland and Izu-Oshima, Japan. Those data have made possible well-constrained models of the geometry of the magma reservoirs and of the changes in their geometry during the eruption. The Hekla eruption produced clear changes in strain at the nearest instrument (15 km from the volcano) starting about 30 minutes before the surface breakout. The borehole instrument on Oshima showed an unequivocal increase in the amplitude of the solid earth tides beginning some years before the eruption. Deformational changes, detected by a borehole strainmeter and a very long baseline tiltmeter, and corresponding to the remote triggered seismicity at Long Valley, California in the several days immediately following the Landers earthquake are indicative of pressure changes in the magma body under Long Valley, raising the question of whether such transients are of more general importance in the eruption process. We extrapolate the experience with borehole strainmeters to estimate what could be learned from an installation of a small network of such instruments on Mauna Loa. Since the process of conduit formation from the magma sources in Mauna Loa and other volcanic regions should be observable, continuous high sensitivity strain monitoring of volcanoes provides the potential to give short time warnings of impending eruptions. Current technology allows transmission and processing of rapidly sampled borehole strain data in real-time. Such monitoring of potentially dangerous volcanoes on a global scale would provide not only a wealth of scientific information but also significant social benefit, including the capability of diverting nearby in-flight aircraft.

1. INTRODUCTION

Monitoring of volcanoes with a variety of techniques has a long history [e.g., *Tilling and Dvorak*, 1993]. Seismic and chemical monitoring allow inferences about

volcanic unrest, but deformation measurements [*Dvorak and Dzurisin*, 1995] give the most direct indications of changes in the status of the magma underlying the volcano. Few of the world's active volcanoes have deformation monitoring and only a very small number have reasonable areal and temporal coverage.

Here we concentrate on the results obtained for several cases of volcanic activity using data from Sacks-Evertson borehole strainmeters [*Sacks et al.*, 1971], and attempt to determine what we would be able to learn about the volcanic process if suitable arrays of these instruments were deployed on Mauna Loa and other active volcanoes. These instruments are recognized as having provided a significant improvement in our capability to monitor deformations in the earth's crust reliably by means of short baseline measurements. Power and maintenance requirements of these strainmeters are both low, so that field deployment and operation are readily achievable in remote areas. They have very high sensitivity (10^{-12} in strain), broad frequency response (zero frequency to about 20 Hz) and high dynamic range (~ 130 dB). The typical installation of the instrument is at a depth of about 200 m. The instrument follows deformation in the surrounding rock as a result of being bonded in with a grout which expands as it cures. The response characteristics of the instruments are such that they also serve as extremely broad band seismometers (effectively constant velocity response from 20 Hz down to zero frequency). The down-hole low ground noise is the limit to the detection capability of these instruments which have dynamic range greater than that of modern seismometers. Our discussion here will not address this high frequency capability, but we remind readers that an array of these instruments will also allow detection and location of earthquakes.

Although our work, with colleagues in a number of collaborating institutions, has concentrated on investigations of the seismogenic processes and of earthquakes, some instruments have been installed in volcanic areas. In south east Honshu, Japan, and in southern Iceland episodes of volcanic unrest have resulted in recordings of very clear strain changes as magma forced its way from a deep reservoir to the surface. Also, a deformation transient at Long Valley, California, which followed the 1991 Landers earthquake raises the possibility that such episodes may be of more general significance in initiation of eruptions.

2. VALUE OF BOREHOLE STRAIN DATA FOR VOLCANIC STUDIES

We discuss below some studies in which the advantages of borehole strain data have been useful in adding to our knowledge of volcanic processes. The value of geodetic techniques, including particularly GPS measurements, is widely appreciated [e.g., see *Dvorak and Dzurisin*, 1995]. Provided stable benchmarks can be installed, these techniques currently provide the best estimates of deformational changes over long periods and therefore comprise a necessary component in the monitoring of active volcanoes. The use of borehole strainmeters in volcanic situations is not yet widespread, and consequently the advantages they provide for monitoring shorter period changes (shorter than about a year) is not commonly realized.

Published work comparing the capabilities of GPS measurements with those of strainmeters [*Tralli*, 1991] focus on a comparison in the frequency domain and on the problems of detecting large scale tectonic deformation. Also important is to evaluate the relative capability of the techniques for detecting localized sources of deformation. We examine here in a more general manner why continuous strain data provide a valuable addition to our tools for investigating volcanoes. This can be appreciated from some simple model calculations. We use buried spherically symmetric pressure sources (Mogi model) as illustrations since they have been used to explain most of the observed surface deformations associated with volcanic activity. Similar conclusions are, however, applicable whether we use models of dikes, sills or shear faults.

We calculate the surface displacements and dilatations produced by deflating Mogi sources [*Mogi*, 1957] at centroid depths of 5 km and 10 km, since this range is typical of depths estimated for various magma chambers under volcanoes. Figure 1 shows the vertical and radial displacements and the dilatational strains as a function of radial distance from the surface point vertically above the source. The source at 10 km depth has a pressure decrease of 40 MPa, radius of 1 km; the 5 km deep source has 10 MPa pressure change and the same radius. The deeper source was given greater strength simply for convenience in plotting the displacements; those parameters result in the same vertical displacement for the point immediately above the sources. We do not mean to imply that these sources are representative of any particular eruption; the point here is to compare the sensitivities of displacement monitoring and of strain monitoring.

Since GPS measurements may become more common on volcanoes, we evaluate displacement detection capability using a limit of 3 mm for horizontal displacements and about 1 cm for vertical changes [see e.g., *Dixon*, 1991 for a discussion of GPS errors]. From Figure 1 we see that, for the test sources, the horizontal signals would be just above the resolution level at a distance range of about 2 km to 6 km for a 5 km deep source and from about 4 km to 12 km for the 10 km deep source. Vertical displacements are barely resolvable within a few km of the point above the sources only. The test sources used for illustration thus represent the smallest changes that could take place at those depths and be detected by GPS measurements.

Also shown in Figure 1 are the dilatational strain changes produced over the same distance range by the same strength sources as for the displacement changes. The threshold limit of detection for a borehole strainmeter is a function of the duration of the signal; as the time dur-

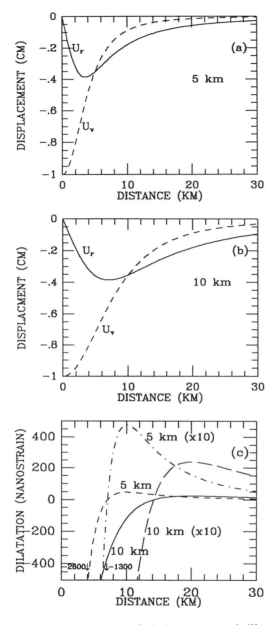

Fig. 1. Calculated variaton of displacements and dilatational strain with distance from the point above a buried spherical pressure source (Mogi model): (a) shows the changes in radial position (U_r) and in vertical position (U_v) for a source at a depth of 5 km; (b) for a depth of 10 km; (c) shows the changes in dilatation for both sources. Note the change in sign for dilatation at a distance approximately 1.4 times the source depth. The curves labeled with the (x10) notation have been magnified 10 times. The 5 km deep source has a radius of 1 km and a pressure decrease of 10 MPa, the 10 km deep has 1 km and 40 Mpa. These test sources would be just detectable with GPS monitoring but, for short durations, produce strain changes much greater than the resolution capability of the strainmeters.

ing which the change takes place increases, so does the earth noise level. For time intervals of hours and days before and during an eruption, a conservative estimate of the threshold level is about 10 nanostrain, comparable to the amplitude of the solid earth tides. From Figure 1, we see that sources which are at the limit of detection by GPS techniques will result in signals that are as much as 100 times the resolution threshold at distances within a few km of the source, and are 5 times threshold at distances out to more than 10 km from the volcano. Thus, strain monitoring provides a very significant increase in our sensitivity for detecting changes in magma chambers. Because the Sacks-Evertson instruments have very high frequency response, analysis of the time history of even very rapid changes is possible. Additionally, the variation of strain amplitude with distance is a strong function of depth because of the depth dependent change in sign with distance. With strain monitoring at only a few distances, we are able to constrain the source depth much more effectively than with comparably spaced displacement measurements.

Clear displacement changes have, of course, been measured during eruptions, but the above example shows that the advantage of strain monitoring is not only that the "before-after" state change can be determined if we have access to borehole strain data, but also that we will be able to determine with much greater capability the time history of those changes. This capacity of borehole strain data to illuminate the details of how and when the magma propagates from depth to the surface and of variations in the pressure changes in the supply reservoir presents the opportunity for new insight into the processes of volcanic eruptions. We next discuss an example to illustrate this: detailed modeling of the course of the eruption of Hekla in 1991, which was possible even though the closest instrument was about 15 km from the volcano and the others more than 35 km away.

3. THE ERUPTION OF HEKLA, ICELAND IN 1991

Iceland is crossed by the mid-Atlantic ridge and thus provides an opportunity for land-based studies of the spreading process. The transform zone just to the north of Iceland failed with a magnitude 7 earthquake during an episode of spreading in northern Iceland during 1975-76. This spreading event included a significant eruption of Krafla. It is now widely thought that a comparable tectonic episode may soon occur in southern Iceland with failure of the on-land transform zone (magnitude 6.5+ earthquake) and perhaps associated volcanic activity. The Nordic countries have supported the development and installation of a real-time seismic network [*Stefansson et al.*, 1993]. In a cooperative experiment between Carnegie Institution of Washington's Department of Terrestrial Magnetism and Vedurstofa, the Iceland Meteorological Office, a small network of Sacks-Evertson borehole strainmeters was installed in southern Iceland in late 1979 (Figure 2).

Currently, data from the strainmeter network are transmitted in real time by radio frequency to Vedurstofa in Reykjavik. The signals are sampled at 50 per second with 20 bit dynamic range. The data acquisition system is a 486 PC using the UNIX operating system. Relatively little data processing is required in order to remove the solid earth tidal components and the strains produced by changes in atmospheric pressure (also monitored at each site), and

to convert from digital count to strain. This processing is performed routinely and the strain data are displayed graphically (on any of the workstations in the local network) with an update interval of 1 minute. The archived continuous data stream retains low-pass filtered data sampled once per second while the original data (50/sec) are saved for detected events.

Hekla eruptions are characterized by fissuring, with a dominate strike direction of N60°E. The topography of Hekla exhibits an elongation in that direction; presumably the edifice has grown in the strike direction as a result of the dikes bringing magma from a deep reservoir to the surface. A continuing puzzle is that the orientation of Hekla is clearly rotated clockwise (about 15°) from the trend of tectonically-controlled topography in the area. The 1991

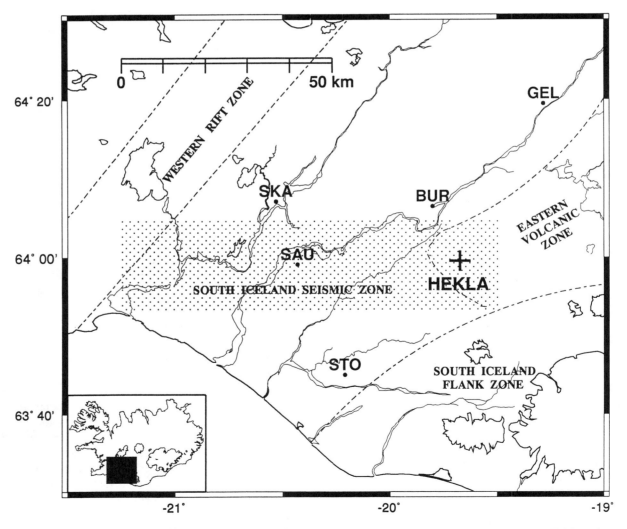

Fig. 2. Map of southwest Iceland showing the location of the borehole strainmeters (3 letter codes) and Hekla volcano (+).

eruption was well recorded by the strainmeter network, with the closest (BUR) at 14.6 km from the summit [*Linde et al.*, 1993]. The others lie at distances between 37 and 45 km and cover a wide azimuthal range. Previous work [*Kjartsson and Gronvold*, 1983; *Sigmundsson et al.*, 1992] suggests that the magma chamber beneath Hekla has a centroid depth of about 7 - 9 km; the data available for those studies allowed modeling of only a spherical pressure source at depth. The continuous strain data (at that time 16 bit resolution with sampling interval 1.4 sec) provide the additional capability of determining parameters of the conduit forced open to allow magma to escape to the surface, as well as an estimate of the speed of magma ascent. The modeling of this eruption was described in *Linde et al.* [1993]. Here we review that analysis since similar characteristics may be expected for an eruption of Mauna Loa if monitored by a network of borehole strainmeters.

Figure 3 shows data for the month of January; for all traces, strain changes due to the eruption are clearly visible. (The data shown in all plots have been low-pass filtered and resampled at 1/min. Also, earth tides and pressure effects have been removed.) Figure 4 shows data for 5 days covering the eruption. The overall nature of the strain changes, with most of the deformation occurring during the first several hours of the eruption, is consistent with observations of the rate of lava production [*Gudmundsson et al.*, 1992]. The strong coherence be-

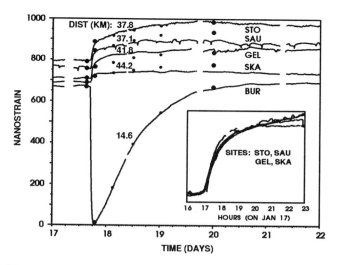

Fig. 4. Five days of data from the borehole instruments. Insert (covering 7 hours, and with amplitudes normalized) shows that the more distant sites undergo expansion in a remarkably coherent manner. Large solid circles show the strain changes calculated from a 2 stage model: deflation of a deep reservoir together with dike formation followed solely by continued deflation of the deep source. Smaller circles show the results of calculations for intermediate times during stage 2. [After *Linde et al.*, 1993.]

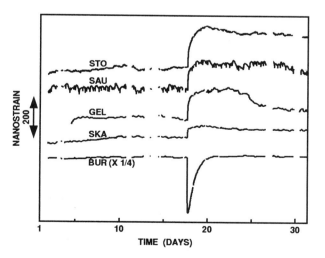

Fig. 3. Data from the borehole array for the month of January, 1991. The original data were low-pass filtered and resampled at 1 minute intervals. Expansion is positive. Earth tides and atmospheric pressure-induced strain changes have been removed. For some traces, a linear trend was removed. Changes due to the eruption of Hekla on January 17 are clearly visible on all traces.

tween the distant stations (see insert, Figure 4) indicates that simple models should be capable of satisfying the data, while the change of sign in the rate at BUR requires that the model include both a deep deflating source and growth of a dike to the surface. From the strain rate at BUR (Figure 5), *Linde et al.* [1993] showed that the dike broke through to the surface at 17:01 hours on January 17, consistent with other available observations which set the eruption time between 17:00 and 17:10. Figure 5 clearly shows that deformations due to dike growth were observable starting about 30 minutes before the eruption.

Linde et al. [1993] used elastic halfspace expressions to calculate surface deformations due to the source model as a function of time. Figure 6 shows the first few hours of the eruption, with the geometry of the source model shown as an insert. Solid circles on the plots indicate results of calculations as dike growth proceeds and the deep reservoir undergoes loss of magma. The dike orientation was taken as that of the dominant fissuring of Hekla in this and previous eruptions, while the deep reservoir was taken to be directly beneath the summit; these were chosen as being the simplest geometries consistent with the eruptive behaviour of Hekla. During the later stages of the eruption (see Figure 4), *Linde et al.* [1993] satisfy

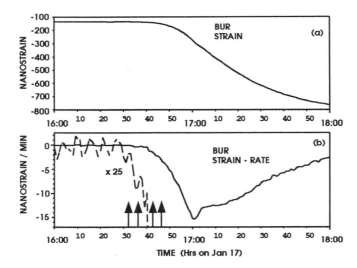

Fig. 5. Two hours of (a) strain data and (b) strain-rate data from BUR. The dashed strain-rate curve is magnified 25 times. Minimum in the strain-rate gives the inferred time of surface breakout. Arrows indicate the times of the first located earthquakes associated with the eruption. [After *Linde et al.*, 1993.]

the data, as indicated by solid circles, by taking the dike growth to be complete and having only continued loss of magma from the deep source.

The parameters that *Linde et al.* [1993] derived from modeling are consistent with other information. The estimated dike thickness was 85 cm, which is similar to that of mapped dikes [*Gudmundsson*, 1990]. Calculation of the volume loss from the deep reservoir is in very good agreement with the estimates of the erupted volume of lava. Finally, *Linde et al.* [1993] used parameters of their model to calculate horizontal displacements (of up to several cm) at sites for which GPS observations gave measurements of displacements over about 2 years spanning the eruption. Those calculated values provided a good match to the data, about as good as that from the model of *Sigmundsson et al.* [1992], who found a best fit single deflation source.

This study by *Linde et al.* [1993] illustrates the potential of data from Sacks-Evertson borehole strainmeters. Even though the nearest instrument was about 15 km from the volcano, the recorded signals had very high signal/noise ratio and the rapid data sampling allowed modeling of the time history. In contrast, data from GPS campaigns provided signals only a few times the error estimates (although for future events those errors would be lower because of improved satellite coverage) and allowed modeling only of the "before-after" changes. From the *Linde et al.* [1993] work it is clear that with a borehole

strainmeter experiment designed to monitor a volcano we can expect to record and hence analyse the progression of an eruption in much greater detail.

4. IZU-OSHIMA (1986-87) AND OFF-ITO (1989) ERUPTIONS IN JAPAN

The Japanese geophysical community expects that a great earthquake may occur in the subduction zone between Japan and the Philippine Sea plate to the south west of Tokyo in Suruga Bay (the west side of the Izu Peninsula). A large observation program for earthquake

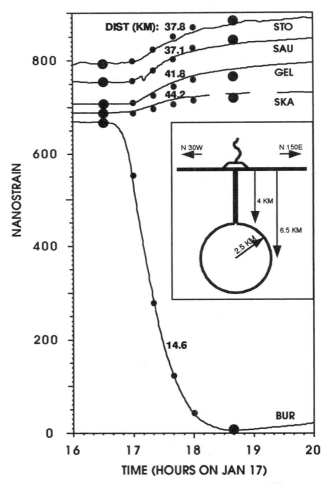

Fig. 6. Four hours of strain data. The insert is a sketch of the model used for calculation, with cross-section perpendicular to the strike direction (N60°E) of Hekla and its main fissures. Large circles indicate the pre-eruption levels and the model strain changes calculated at the end of stage 1 (dike formation complete). Smaller circles show the results of intermediate calculations. [After *Linde et al.*, 1993.]

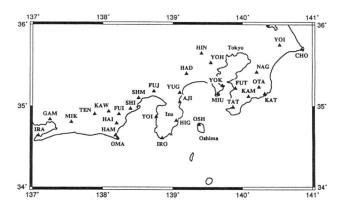

Fig. 7. Map of south east Honshu, Japan showing the locations of 31 Sacks-Evertson borehole strainmeters installed and operated by the Japanese Meteorological Agency.

prediction research has been established in that area, including a network of 31 Sacks-Evertson borehole strainmeters (Figure 7). That array was installed by the Japanese Meteorological Agency (JMA), who continue to operate the array. Two episodes of volcanic unrest have occurred in that area since the first 10 instruments became operational.

The Izu-Oshima eruption of 1986 was well recorded by many of the strainmeters; the one on the island (just a few km from the summit of Miharayama) experienced strain changes of about 10^{-4} (Figure 8), which is more than 100 times greater than the largest signal recorded during the eruption of Hekla. Studies of this eruption [*Kanjo et al.*, 1988; *Yamamoto and Kumagai*, 1988; *Hashimoto and Tada*, 1990] have used strain and other deformation data to estimate the geometry of the magma sources under the volcano. The most complete study, by *Hashimoto and Tada* [1990], uses all the available deformation data to determine a model which satisfies the cumulative changes during the eruptive sequence in 1986. The resultant model has broad similarity to the model for Hekla (above) in that both a deep deflation source and a tensile crack (dike) to the surface are needed. Much of the data used in that study derives from geodetic surveys and thus allows modeling only of the total change during the interval between surveys. Continuous borehole strain and tilt data were recorded and allow calculations of the time history of the eruption. *Yamamoto and Kumagai* [1988] and *Kanjo et al.* [1988] show that, starting about 2 hours before the eruption on November 21, clear anomalous deformational changes were recorded, consistent with a dike propagating to the surface. *Kanjo et al.* [1988] also reported a significant change in the strain-rate recorded by the borehole strainmeter on Oshima beginning several days before the eruption.

Clear deformational changes were also recorded for the 1989 off-Ito under-water eruption. *Shimada et al.* [1990] reported on the changes in line lengths as determined from GPS measurements. *Okada and Yamamoto* [1991] used a variety of deformation data, including tilt, strain and line length changes, to determine the geometry of the magma sources together with a shear fault. In this study, the modeling indicates 3 stages (each covering several days) in the development of the eruption. The sparsity of temporal sampling of most of the data precluded more detailed modeling.

Another potentially important effect relating to pre-eruption volcanic activity is that the amplitudes of solid earth tidal components may increase measurably. To our knowledge, the only clear case reported of tidal amplitude changes is based on data from the Oshima borehole strainmeter (Figure 9). That study [*Sawada et al.*, 1984]

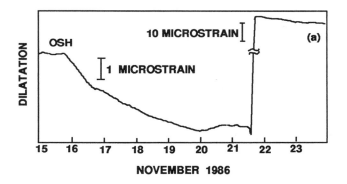

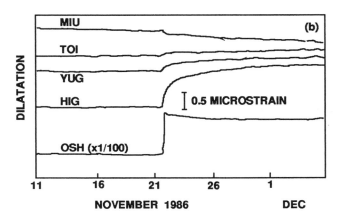

Fig. 8. (a) Record of strain changes recorded by the borehole strainmeter on Oshima during the eruption of Miharayama in November 1986. (b) Strain records from other nearby strainmeters showing changes due to the eruption on November 21. [After the Reports of the Coordinating Committee for Earthquake Prediction.]

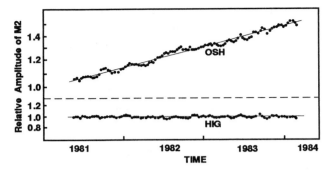

Fig. 9. Changes in amplitude of the M2 component of the solid earth tides at the Oshima strainmeter for several years preceding the Oshima eruption. The upper panel shows that the M2 amplitude at Oshima changed during the interval shown by about 50%, while there were no detectable changes at Higashi-izu on the east coast of the Izu Peninsula. [After *Sawada et al.*, 1984.]

showed changes in the M2 tidal amplitude of 50% over several years. The work was done before the Izu-Oshima eruption, and the authors attempted to relate the changes to an eruption of Miyakejima, south of Oshima. It now seems clear that the tidal amplitude changes were related to changes in the magma chamber under Oshima as it prepared for eruption, the large changes in tidal amplitudes being indicative of changes in the elastic constants of the underlying material. This potential needs to be explored.

5. REMOTELY TRIGGERED SEISMICITY AT LONG VALLEY

The 1992 Landers earthquake in southern California had some unusual characteristics including the triggering of seismicity at remote locations [*Hill et al.*, 1993]. These triggered earthquakes came as a surprise because aftershock locations are usually indicators of the dimensions of rupture zone of the main shock. A variety of explanations have been proposed to explain this apparently anomalous effect, but deformation data recorded during the episode of enhanced seismicity in Long Valley seem to hold the key to an understanding of this activity. Figure 10 is a map of the Long Valley caldera area showing the location of a borehole strainmeter (POP), a long baseline tiltmeter (LBT) and lines (radiating from site CASA) whose lengths are measured repeatedly using a two-color laser ranging system. These measurements showed no clear change in any of the line lengths during several days following the Landers earthquake (see Figure 11). In Figure 12, however, we see that there is a strong correspondence between the increase in seismic activity and the changes in dilata-

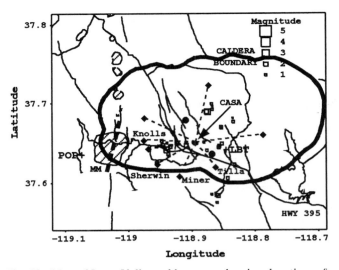

Fig. 10. Map of Long Valley caldera area showing location of Sacks-Evertson borehole strainmeter (POP) and long baseline tiltmeter (LBT). Lines radiating from CASA have lengths monitored with a two color laser system. Earthquakes shown are those occurring in the five day interval preceding the Landers earthquake.

tion at POP (plotted with contraction positive for comparison purposes) just outside the caldera and with tilt, inside the caldera, at LBT. *Linde et al.* [1994] have proposed that, since the triggered seismicity and the background activity are similar in all respects except rate of occurrence, the stress in the area due to ongoing pressurization of a magma body was increased by the rising in the

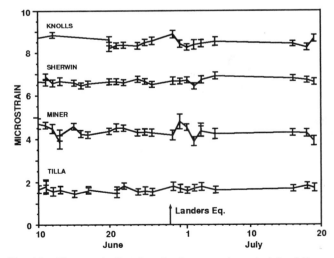

Fig. 11. Changes in line lengths (expressed as strain) of lines in Figure 10. Note that there are no clear changes in the days following the Landers earthquake. [After *Johnston et al.*, 1995.]

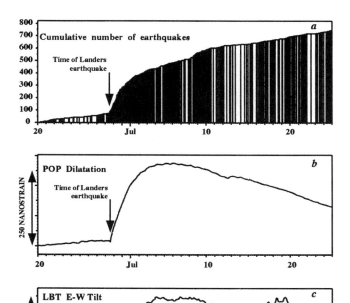

Fig. 12. Data from the Long Valley caldera area following the Landers earthquake of 1992. (a) shows the cumulative number of earthquakes with a clear rapid increase following Landers. (b) is the variation in dilatational strain recorded at a borehole strainmeter just outside the caldera. Tides, atmospheric pressure effects and a seasonal variation have been removed. Contraction is shown positive for comparison purposes. (c) variation in east-west tilt (tides removed) recorded on a long baseline tiltmeter inside the caldera. Strain and tilt excursions have rise times and durations remarkably similar to those of the triggered seismicity. [After *Linde et al.*, 1994.]

magma of bubbles liberated from the bottom by the shaking of long period seismic waves from Landers. Other mechanisms, including rupturing of overpressured fluid chambers, have been discussed by *Johnston et al.* [1995] and *Hill et al.* [1995].

For our present purposes, two aspects of this are relevant, irrespective of which model (if any) is correct. First, the observed deformations in strain and tilt are large in amplitude and highly correlated with each other (as well as with the changes in seismic activity); the instruments are well separated and clearly are providing information about significant local (to the caldera area) deformation. Other geodetic monitoring in the area, including line

length changes measured daily with a two color laser ranging system [see *Langbein et al.*, 1993], did not detect this transient. Since the line length measurements have resolution of about 1 mm, it appears that none of the position measuring techniques (including GPS) has the capability of detecting such a transient even though it was observed as quite a large signal by instruments sensitive to the gradients of the displacements (strain, tilt). The second point is that, as noted in *Johnston et al.* [1995], more extensive volcanic activity, including eruptions, might be triggered as a result of a deformation transient; *Yokoyama* [1971], *Nakamura* [1975] and *Williams* [1995] have shown that a number of eruptions may have been triggered by earthquakes. Other Long Valley recent history [*Langbein et al.*, 1993] shows that an increase in seismicity (after 1989) lagged a marked change in deformation rate. Thus it would seem advantageous, in any program of volcano monitoring, to include an array of borehole strainmeters. Long baseline tiltmeters may be more difficult logistically, particularly since monitoring of radial tilt, presumably the most important tilt component, faces the difficulty of large changes in elevation.

6. AN EXPERIMENT ON MAUNA LOA

The studies discussed above, for Hekla in Iceland, for eruptions near Izu in Japan and for the transient at Long Valley in California, all demonstrate the value of continuous borehole strain monitoring of active volcanoes. Here we extrapolate from those experiences to estimate what constraints could be placed on an eruption of Mauna Loa by using data from a small network of Sacks-Evertson instruments installed close to the volcano. Deformation surveys of Mauna Loa show that the volcano is inflating, presumably as a result of increased pressure in the underlying magma chamber [*Okamura et al.*, 1992], leading to an expectation that the volcano could erupt within the next several years. Thus mounting such a strain monitoring experiment would provide significant scientific return within a reasonable time span.

Figure 13 is a map of the Mauna Loa area showing five possible sites for installation of borehole instruments. The exact locations are not critical in terms of the experiment design: what is desired is to have a number of sites which cover comparable distance and azimuth ranges from the summit. Logistical details would be a factor in selecting actual locations. The array as shown in Figure 13 is designed on the basis of (1) current estimates of the depth to the magma reservoir, (2) knowledge of areas of lava flow in previous eruptions (to be avoided), (3) derived estimates of the geometry of the conduit formed as the

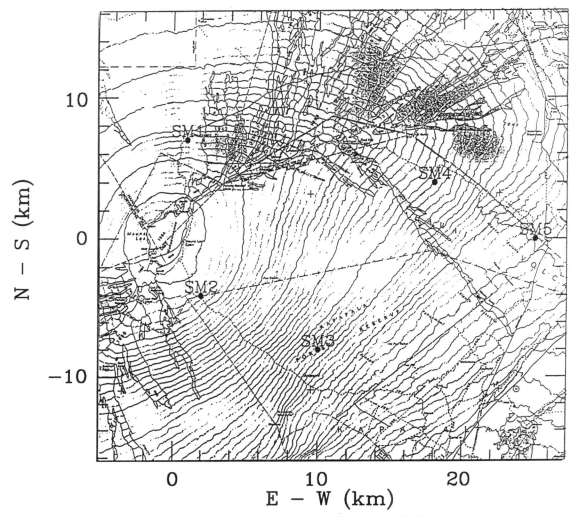

Fig. 13. Map of Mauna Loa area, with Kilauea on the right edge. SM1 through SM5 shows possible sites for borehole strainmeters. These sites are used in model calculations of strain changes due to pressure sources and dike under Mauna Loa (Figures 14, 15, 16).

magma forces its way to the surface, and (4) the consideration that Kilauea is active and the experiment could also serve to add to our understanding of eruptions of Kilauea. This would also allow discrimination between strain changes generated by Mauna Loa activity and Kilauea activity because of the differing changes in source distance across the array.

The recent eruptive history of Mauna Loa includes rift zone eruptions to the north east and to the south west [*Lockwood*, this volume]. This suggests that transport of magma to the surface takes place as dikes with strikes corresponding to those of the rift zones. Previous work indicates that the magma reservoir which supplies the erupted material is at a centroid depth of about 5 km. We use this

geometry together with estimates of the strength of previous eruptions [*Decker et al.*, 1983] to make model calculations for the strain signals at the sites shown in Figure 11. We also vary the depth of the deep reservoir to indicate the resolution such data would have in determining the source geometry.

First, we look at the variation of dilatational strain with radial distance from a buried pressure source in which pressure is decreasing (Figure 14). Here we consider only a simple spherical pressure source (Mogi model) as representing the deep reservoir, although the site locations would have the potential to resolve azimuthal variations in the deep source. Because of free surface effects, there is a change in sign of the dilatation with distance, with the

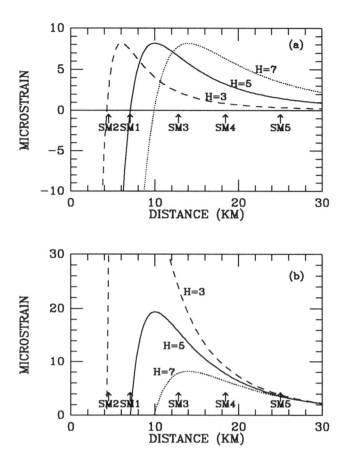

Fig. 14. Dilatational strain changes due to pressure sources at depths of 3, 5 and 7 km under Mauna Loa. (a) Plots normalized to have the same maximum amplitude. (b) Plots normalized to have the same amplitude at a distance of 30 km. Because of the free surface effects, there is a change in sign of the strain with distance, with the reversal distance depending directly on the depth of the source. The result is that the variation of amplitude with distance is a strong function of depth. SM1 through SM5 show the positions of sites in Figure 13. The strain amplitude ratios among these sites allows very strong control on the depth of the source.

nodal circle having radius 1.4 times the source depth. This sign reversal means that measurements of dilatation at the range of distances for the proposed array provides an extremely powerful constraint on the depth of the reservoir. We show calculations of dilatational strain for depths of 3 km, 5 km and 7 km (bracketing the range of depths estimated previously). Clearly, the ratio of signal strengths at the different sites shown in Figure 13 would allow depth control to a fraction of a km.

The recorded strain changes are of course a combination of the effects from deflation of the deep reservoir and from growth of the dike to the surface. In Figure 15 we show contour plots of dilatational strain for dikes (using the strike direction of surface fissuring), spherical pressure sources at one of the chosen depths, and the resultant strain due to the combined sources at an intermediate stage, when dike growth is complete, and at completion of the eruption. Suggested site locations are also shown. Little more than cursory inspection of these plots is required to realise that there is nothing critical about the site locations which impacts on the capability of the network to resolve the geometric and strength parameters of the magma sources. Provided we are able to ensure a reasonable azimuthal and distance coverage, almost any set of sites will allow robust estimates of the source parameters.

During the pre-eruptive interval, after magma begins to force open a conduit to the surface and before surface rupturing, we can apply conservation of mass requirements to balance the dike growth and loss of pressure in the deep chamber. We note in passing that such a calculation allows an excellent match to the 30 minutes of strain changes recorded at BUR, the nearest instrument to Hekla, immediately prior to that surface breakout. In the absence of better constraints, we assume that the dike growth for Mauna Loa is at rates similar to those for Hekla; however, all of the time scales used in these model calculations are arbitrary and none of our inferences about resolving the geometry of the magma system under Mauna Loa depend at all on the time scales. For an actual experiment, of course, the time scales for the various stages of the eruption have the potential to provide additional information about the eruptive process. Calculated time histories of strain changes are shown in Figure 16. Again, for the purposes of these model calculations we assume dike growth is complete within 2 hr. of the initiation of the eruption process and that continued deflation of the deep source continues for some time. We take that time to be several days and assume that the pressure decrease follows an exponential decay. This is a reasonable approximation to the behaviour at Hekla, but the actual variation is not relevant to our comments about parameter estimation. What is clear from these time series is that we would expect to record the strain changes at extremely high signal/noise ratio. Shown at the bottom of each panel is an estimate of the solid earth tidal signal (which is very well recorded by these instruments). On the amplitude scales for the eruption generated signals, the tides are barely discernible. These time series also show, as noted above from the spatial variation of the signals, that such an array would allow well-constrained estimates of the geometry of the magma sources. While we make no attempt to show the initial detail for this synthetic eruption, the growth of

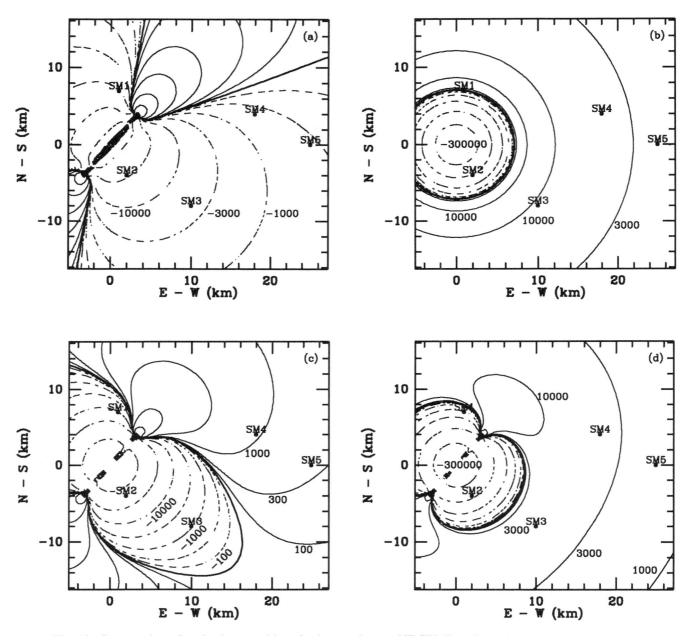

Fig. 15. Contour plots of strain changes. (a) strain changes due to a NE-SW dike. (b) strain changes due to a deflating source at 5 km depth. (c) changes due to a dike and partial deflation of the 5 km source, representing a stage of an eruption after a few hours. (d) final changes due to the combined sources. Plots for the deflating source at 3 km and 7 km, although similar, allow robust source depth determination.

the dike to the surface immediately preceding the surface breakout must generate signals similar to that recorded at BUR (Figure 5) for the eruption of Hekla. In the experiment we are modeling here, with five instruments close to the volcano, we would obtain high quality redundant information about this initial stage of the eruption. We would be able to determine the rates at which the dike grows, perhaps with the capacity to estimate both vertical and horizontal speeds of propagation. If the eruption is accompanied by deeper movement of magma (e.g., rising from depths of several tens of km), the strain data may allow constraints on that process.

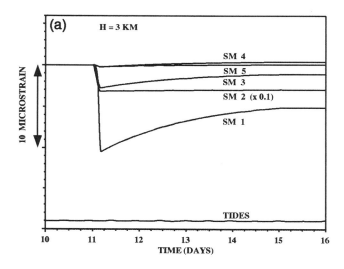

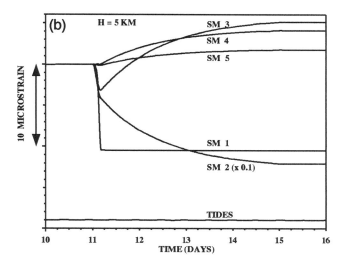

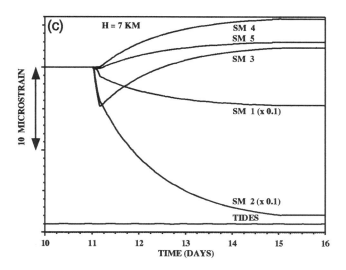

The continuous high sensitivity data from the instrument array shown in Figure 11 would also allow detection of any transients in deformation which may precede an eruption. The deformational event at Long Valley following Landers may be relevant to the triggering of eruptions on other volcanoes, including Mauna Loa [*Yokoyama* 1971, *Nakamura* 1975, *Williams* 1995]. The experience at Long Valley is consistent with the model calculations in this work in that it indicates that such strain measurements may provide the only feasible means of detecting such transients. In the several days preceding the initial eruption of Izu-Oshima in November 1986, there was a change in strain-rate recorded by the borehole strainmeter on the island [*Kanjo et al.*, 1988]. Had a small array of instruments been in place, it would have been possible to determine if this was in fact due to precursory changes in the magma supply system, and to calculate estimated parameters for the locations, depths and magnitudes of those changes.

7. FEASIBILITY OF AN ALERT SYSTEM

The borehole strain data sets recorded in the vicinity of volcanic eruptions, particularly in the case of Hekla, have provided an unprecedented opportunity to determine the mechanism of a volcanic eruption: *Linde et al.* [1993] were able to monitor the movement of magma from the deep reservoir and determine the source characteristics and changes in them during the eruption. In addition, since the magma source was at depths below about 4 - 5 km, we know that the ascent velocity of the Hekla magma is about 10 km/hr.

Despite differences in detailed characteristics of volcanoes, we may be optimistic that this type of observation would provide valuable immediate pre-eruptive signals for Mauna Loa and other volcanoes since eruptions must be preceded and accompanied by magma forcing a path from a reservoir, at some depth, to the surface. This movement, whether through dikes or otherwise, deforms the surround-

Fig. 16. Calculated strain time histories for a synthetic eruption in which a deflating deep reservoir supplies magma through a dike which is forced open during the first hours of the eruption. SM1 through SM5 are the sites shown in Figure 13. Three depths for the deep source are illustrated: (a) 3 km; (b) 5 km; (c) 7 km. These plots have quite different characteristics depending on the depth of the supply reservoir, and such actual data thus provide very strong constraints on the eruption process. The lowest trace in each panel is of the calculated solid earth tides; it serves to indicate that the eruption-generated signals will be very large.

ing rock and necessarily generates strain changes detectable by near instruments. The time interval during which we would be able to observe deformation before the eruption would depend directly on the depth to the reservoir as well as the speed of magma ascent.

In addition to the scientific value of such monitoring, social benefits would accrue. It is feasible to operate the network as a real-time alert system. All of the technical requirements are in place for such a program. This may not be of high value in the case of Mauna Loa but we mention it briefly here because it is relevant more generally. Our collaborators in the USGS use the GOES satellite to retrieve in real time data sampled at 10 minute intervals at a large number of sites in California [*Silverman et al.*, 1989]. In our Iceland collaborative program, we use RF telemetry for the real time acquisition of 50 sps 20 bit data, with on-line processing routinely giving graphical display of calibrated, detided, pressure-corrected data updated every minute. Updates could easily be performed more rapidly (e.g., every second) since only two multiplications and two subtractions are required for each datum. Analog to digital converters with 24 bit resolution and adequate sample rates are now inexpensive and would allow the recording of the complete instrumental dynamic range in a single channel. The central recording site for such a system could also function in a remote area if necessary, since a system with adequate computational features and data storage capacity can be operated without requiring power from a utility company. Operation of strainmeters in relatively close proximity to a volcano would result in such large signals during and immediately preceding an eruption (see the model calculations above) that even very simple algorithms applied to the array data would allow automatic determination of alert or eruption status. Use of slightly more sophisticated processing such as suggested by *Voight* [1988] could result in a more robust procedure for a warning state. Operation of continuous GPS receivers at the same sites would provide the best long term information about deformational changes. Other sensors which provide valuable additional information could be added for small incremental costs.

Communication of an alert status to the user community then requires only the transmission of a small amount of information; this is a relatively simple task no matter how remote the data acquisition site. Such a short term warning of an imminent eruption may, depending on the interval, be of significant value in reducing disaster for the surrounding populations. More certainly, such a short term warning has large value if communicated to nearby in-flight aircraft. There have, in the last several years, been a number of instances of commercial aircraft flying through volcanic ash clouds [e.g., *Casadevall*, 1994] with near catastrophic and highly expensive (~ $80 M) results. Reliable continuous monitoring of dangerous volcanoes on a global scale is feasible and would result in safer flight conditions.

8. CONCLUSION

Compared with other deformational measurements (geodetic, GPS), there is relatively little borehole strainmeter data available which pertains to episodes of volcanic unrest. Those that do exist, especially when considered in the light of model calculations, demonstrate the potential for such data sets to provide otherwise unobtainable new and important information about the character of volcanic systems during eruptions and during short intervals preceding eruptions. An experiment on Mauna Loa combining the sensitivity and high frequency capacity of borehole strainmeter data with the long term stability of GPS monitoring would certainly add significantly to our understanding of the nature of the volcano.

Acknowledgments. We thank Tim Dixon, Mike Rhodes, and anonymous reviewer for constructive comments which led to improvements in this paper.

REFERENCES

Casadevall, T. J., The 1989-1990 eruption of Redoubt volcano, Alaska: impacts on aircraft operations, *J. Volcan. Geothermal Res.*, 62 (1-4), 301-316, 1994.

Decker, R. W., R. Y. Koyanagi, J. J. Dvorak, J. P. Lockwood, A. T. Okamura, K. M. Yamashita, and W. R. Tanigawa, Seismicity and surface deformation of Mauna Loa volcano, Hawaii, *Trans. Am. Geophys. Un.*, 64, 545-547, 1983.

Dixon, T. M., An introduction to the Global Positioning System and some geological applications, *Rev. Geophys.*, 29, 249-276, 1991.

Dvorak, J. J., and D. Dzurisin, How volcanoes move, *Reviews of Geophys.*, in press, 1995.

Gudmundsson, A., Emplacement of dikes, sills and crustal magma chambers at divergent plate boundaries, *Tectonophys.*, 176, 257 - 275, 1990.

Gudmundsson, A., N. Oskarsson, K. Gronvold, K. Saemundsson, O. Sigurdsson, R. Stefansson, S. R. Gislason, P. Einarsson, B. Brandsdottir, G. Larsen, H. Johannesson, and T. Thordarson, The 1991 eruption of Hekla, Iceland, *Bull. Volcanol.*, 54, 238-246, 1992.

Hashimoto, M., and T. Tada, Crustal deformations associated with the 1986 fissure eruption of Izu-Oshima volcano, Japan, and their tectonic significance, *Phys. Earth Planet. Inter.*, 60, 324-338, 1990.

Hill, D. P., et al, Seismicity remotely triggered by the magnitude 7.3 Landers, California earthquake, *Science, 260*, 1617-1623, 1993.

Hill, D. P., M. J. S. Johnston, J. O. Langbein, and R. Bilham, Response of the Long Valley caldera to the $M_w = 7.3$ Landers, California, earthquake, *J. Geophys. Res.*, in press, 1995.

Kanjo, K., K. Sato, and O. Kamagaichi, The 1986 eruption of Izu-Oshima volcano, Japan, from volumetric strainmeasurements, Proc. Kagoshima International Conf. on Volcanoes, 300-303, 1988.

Johnston, M. J. S., D. P. Hill, A. T. Linde, J. Langbein, and R. Bilham, Transient deformation during triggered seismicity from the June 28, 1992, $M_w = 7.3$ Landers earthquake at Long Valley volcanic caldera, *Bull. Seismol. Soc. Am., 85*, 787-795, 1995.

Kjartansson, E. and K. Gronvold, Location of a magma chamber reservoir beneath Hekla volcano, Iceland, *Nature, 310*, 139-141 (1983).

Langbein, J. O., D. P. Hill, T. N. Parker, and S. K. Wilkinson, An episode of reinflation of the Long Valley caldera, eastern California; 1989-1991, *J. Geophys. Res., 98*, 15,851-15,870, 1993.

Linde, A. T., K. Agustsson, I. S. Sacks, and R. Stefansson, Mechanism of the 1991 eruption of Hekla from continuous borehole strain monitoring, *Nature, 365*, 737-740, 1993.

Linde, A. T., I. S. Sacks, M. J. S. Johnston, D. P. Hill, and R. Bilham, Increased pressure from rising bubbles as a mechanism for remotely triggered seismicity, *Nature, 371*, 408-410, 1994.

Lockwood, J. P., Mauna Loa eruptive history - the radiocarbon record, this volume.

Mogi, K., On the relations between the eruptions of Sakurazima volcano and the crustal movements in its neighbourhood, *Bull. Volcanological Soc. Japan, 2*, 9-18, 1957.

Nakamura, K., Volcano structure and possible mechanical correlation between volcanic eruptions and earthquakes, *Bull. Volc. Soc. Japan, 20*, 229-240, 1975.

Okada, Y., and E. Yamamoto, A model for the 1989 seismovolcanic activity off Ito, central Japan, derived from crustal movement data, *J. Phys. Earth, 39*, 177-195, 1991.

Okamura, A., Miklius, A., Okubo, P., and Sako, M. K., Forecasting eruptive activity in Mauna Loa volcano, Hawaii, *Trans. Amer. Geophys. Un., 73*, 343, 1992.

Sacks, I. S., S. Suyehiro, D. W. Evertson, and Y. Yamagishi, Sacks-Evertson strainmeter, its installation in Japan and some preliminary results concerning strain steps, *Papers Meteorol. Geophys. 22*, 195-208, 1971.

Sawada, Y., K. Fukui, K. Sato, S. Nihei, and A. Fukudome, Notable phenomena registered by the borehole type volume strainmeter at Izu-Oshima before and after the 1983 Miyakejima eruption, *Bull. Volcanological Soc. Japan, 29*, S141-S152, 1984.

Shimada, S., Y. Fujinawa, S. Sekiguchi, S. Ohmi, T. Eguchi and Y. Okada, Detection of a volcanic fracture opening in Japan using Global PositioningSystem measurements, *Nature, 343*, 631-633, 1990.

Silverman, S., C. Mortensen, and M. Johnston, A satellite-based digital data system for low frequency geophysical data, *Bull. Seismol. Soc. Am., 79*, 189-198, 1989.

Sigmundsson, F., P. Einarsson, and R. Bilham, Magma chamber deflation recorded by the Global Positioning System; the Hekla 1991 eruption, *Geophys. Res. Lett., 19*, 1483-1486, 1992.

Stefansson, R., R. Boovarsson, R. Slunga, P. Einarsson, S. Jakobsdottir, H. Bungum, S. Gregerson, J. Havskov, J. Hjelme, and H. Korhonen, Earthquake prediction research in the South Iceland seismic zone and the SIL project, *Bull. Seismol. Soc. Am., 83*, 696-716, 1993.

Tilling, R. I., and J. J. Dvorak, Anatomy of a basaltic volcano, *Nature, 363*, 125-133, 1993.

Tralli, D. M., Spectral comparison of continuous Global Positioning System and strainmeter measurements of crustal deformation, *Geophys. Res. Lett., 18*, 1285-1288, 1991.

Voight, B., A method for prediction of volcanic eruptions, *Nature, 332*, 125-130, 1988.

Williams, S. N., Erupting neighbors - at last, *Science, 267*, 340-341, 1995.

Yamamoto, E., and T. Kumagai, Precursory tilt changes of the 1986-1987 volcanic eruption of the Izu-Oshima volcano obtained by continuous crustal tilt observations, *Proc. Kagoshima International Conf. on Volcanoes*, 308-311, 1988.

Yokoyama, I., Volcanic eruptions triggered by tectonic earthquakes, *Geophys. Bull. Hokkaido Univ., 25*, 129-139, 1971.

A. T. Linde and I. S. Sacks, Department of Terrestrial Magnetism, Carnegie Institution of Washington, 5241 Broad Branch Road, N.W., Washington, DC 20015.

A Seismological Framework for Mauna Loa Volcano, Hawaii

Paul G. Okubo

U.S. Geological Survey, Hawaiian Volcano Observatory
Hawaii National Park, Hawaii

The Mauna Loa eruptions of 1975 and 1984 occurred after much of the current seismographic monitoring practice had been established at the Hawaiian Volcano Observatory. Observations made during these eruption sequences have provided valuable insights into some of the behavior of Mauna Loa Volcano, and they contain rather clear seismic precursors to these eruptions. Although the earlier instrumental record offers considerably less seismological detail, similar precursory regional seismicity had been noted by earlier investigators. Large earthquakes beneath the flanks of Mauna Loa have preceded a number of eruptions. Flank movement may facilitate magma migration such that these earthquakes could be considered among key eruption precursors. The distribution of earthquake hypocenters indicates a possible magma storage complex beginning at approximately 4 km below the south-southeastern part of Mokuaweoweo Caldera. This position is consistent with interpretations of geodetic monitoring data that show a pattern of steady reinflation since the 1984 eruption. Earthquake counts and seismicity rates since 1984 remain low and suggest that, if the seismicity patterns preceding the 1975 and 1984 sequences will be repeated, Mauna Loa's next eruption is not imminent.

INTRODUCTION

Seismographic monitoring of the Hawaiian volcanoes began in 1912, when the first Omori and Bosch-Omori instruments were installed in the Whitney Laboratory of the Hawaiian Volcano Observatory (HVO) [*Apple*, 1987; J. P. Eaton, unpublished manuscript, 1986]. These earliest instruments were mechanical systems, offering limited magnification of seismic ground motions at the recording site [J. P. Eaton, unpublished manuscript, 1986; *Klein and Koyanagi*, 1980]. Nevertheless, in late September 1914, after a 6-week-long period of repair and improvement, registration of what was interpreted as a rather clear seismic prelude to the November 25, 1914, eruption of Mauna Loa began [*Wood*, 1915]. The first eruption recorded by HVO, then, occurred on Mauna Loa Volcano.

Since that time, as instrumentation on Hawaii has both improved and increased in number, HVO has recorded and reported on numerous eruptions of both Kilauea and Mauna Loa Volcanoes. Earlier in the history of HVO, Mauna Loa was the more active of the two volcanoes [e.g., *Klein*, 1982, Table 2]. In the recent past, however, since the 1950 eruption of Mauna Loa, Kilauea has erupted more frequently than Mauna Loa. Thus, HVO has obtained more detailed instrumental observations of Kilauea derived from the more frequent eruptions occurring on Kilauea, allowing rather detailed interpretations and models of magma transport and other aspects of Kilauea. By comparison, Mauna Loa has been less studied.

The Mauna Loa eruptions of 1975 and 1984 both occurred during the relatively modern seismographic era at HVO. Geologic, seismic, and geodetic studies centered about times of eruption have provided insights into the signatures of magma movement and dike emplacement at Mauna Loa as at Kilauea [*Lockwood et al.*, 1987]. Continuing geodetic and seismic monitoring has helped establish a mechanical framework within which the volcanic processes occur.

Since the 1984 eruption, Mauna Loa Volcano has been quiescent, and eruptive activity in Hawaii has been restricted to the present Kilauea east-rift-zone eruption that started in January 1983. This report mostly reviews earlier seismographic observations of the seismic activity preceding the 1975 and 1984 eruptions. I will describe present HVO seismologic monitoring practice and then discuss the

Mauna Loa Revealed: Structure,
Composition, History, and Hazards
Geophysical Monograph 92
Published in 1995 by the American Geophysical Union

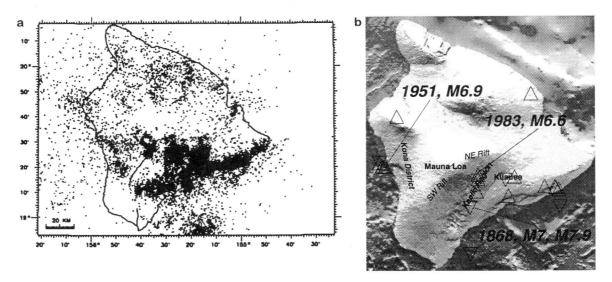

Figure 1. Schematic map of the Island of Hawaii, showing earthquake epicenters, plotted without regard to magnitude or focal depth, determined by the Hawaiian Volcano Observatory for the years 1971 through 1994. Epicenters of Hawaiian earthquakes M ≥ 6 are shown on shaded relief map.

1975 and 1984 pre-eruption seismicity. Finally, I will review the present seismicity beneath Mauna Loa and suggest elements for future research and monitoring of Mauna Loa Volcano.

PRESENT MONITORING PRACTICE

From its days of operating the early mechanical seismographs at isolated installations on the Island of Hawaii, HVO has expanded and improved upon the capabilities of its seismographic-network operations as developments in seismographic instrumentation have allowed [*Klein and Koyanagi*, 1980; *Klein et al.*, 1987]. The HVO seismographic network now comprises 53 stations covering the entire island, producing a total of 86 channels of data centrally recorded and processed at HVO. Particular monitoring focus is on Kilauea, and most of the seismographic stations are concentrated about its summit and rift zones and the adjacent volcanic flanks. A smaller number of stations are concentrated about the Mauna Loa summit, and additional stations are relatively sparsely distributed about the rest of the volcano.

The basis for seismologic monitoring of the active Hawaiian volcanoes is the precise estimation of earthquake hypocentral locations and magnitudes. To accomplish this goal, the network consists of high magnification, short-period seismometers and a computer-based recording system that allows clear registration and precise timing of the onsets of the first-arrivals of seismic waves from even very small earthquakes. Each year, HVO publishes a summary containing the locations earthquake hypocenters and other seismic activity, now distributed as U.S. Geological

Survey Open-File Reports [e.g., *Nakata et al.*, 1994]. These summaries include details of the HVO seismographic network and seismic data-processing procedures.

Present seismic processing routinely affords formal uncertainties of less than 1 km in both the epicentral and focal-depth estimates for earthquakes of M ≥ 1 [e.g., *Klein et al.*, 1987; *Nakata et al.*, 1994]. Final preparation of the catalog of earthquakes located for the seismic summaries includes recalculation of hypocenters, using the computer program HYPOINVERSE [*Klein*, 1978; 1989]. The earthquake-location procedure uses a depth-dependent, laterally homogeneous velocity model, featuring linearly increasing seismic wave speeds within two crustal layers overlying a homogeneous mantle [*Klein*, 1981]. Even those earthquakes recorded before the current location practice was implemented have been relocated with these same procedures. Thus, HVO has built as consistent a catalog of located earthquakes as possible, dating back to 1959.

Seismologic observations, principally earthquake locations, are used to pinpoint magma movement and indicate changes within the volcanoes. The number and distribution of seismographic stations on Kilauea and the greater frequency of eruptions from and intrusions into Kilauea detected with modern monitoring capabilities in place enabled *Klein et al.* [1987] to compile an authoritative description of the seismicity of Kilauea. In addition to defining parts of the volcano's magma storage and transport systems, their study of earthquake distributions in space and time have helped establish elements of a physical or dynamic understanding of the process by which magma moves into and through the volcano. Different earthquake types beneath Kilauea Caldera have also been identified, distinguishable according to spectral content of the seismic waveform [e.g., *Koyanagi et al.*, 1976; *Aki and Koyanagi*, 1981]. These earthquakes occur within a sufficiently dense part of the seismographic network to allow instrumental locations to be estimated. This capability has also contributed substantially to building a model of the summit magma-storage complex and understanding the mechanisms of magma transport [e.g., *Koyanagi et al.*, 1976, 1987; *Aki and Koyanagi*, 1981].

Figure 1 shows the locations of earthquake epicenters, drawn from the HVO seismicity catalog for 1971 through 1994, on the entire Island of Hawaii. Also indicated in this figure are the epicenters of large earthquakes (M≥6) beneath the flanks of Mauna Loa. For earthquakes occurring before 1962, because of limited or no local instrumental recording, the locations of earthquake epicenters are rough estimates which are strongly influenced by historical macroseismic, or felt earthquake, accounts [*Wyss and Koyanagi*, 1992].

Figure 1 clearly shows the highly seismically active regions associated with the summit, east and southwest rift zones, and south flank of Kilauea Volcano. Also appearing as a highly seismically active area is the Kaoiki region between the summits of Kilauea and Mauna Loa. The Mauna Loa summit and northeast rift zone are associated with clusters of earthquakes. West of the Mauna Loa summit, earthquakes are less numerous, although their distribution clusters noticeably just to the west-northwest of the summit and a trend exists in the epicentral distribution that corresponds to the Kealakekua Fault in the Kona district of western Hawaii.

RECENT ERUPTIVE ACTIVITY

Mauna Loa Summit Seismicity

Included in the earthquake catalogs derived from the modern seismographic-network configuration and practice is the seismicity associated with the 1975 and 1984 eruptions of Mauna Loa. Other investigators have reported on the seismicity associated with these eruptions, as well as the behavior observed during the 1975-84 intereruption period [*Decker et al.*, 1983; *Lockwood et al.*, 1985; *Lockwood et al.*, 1987]. Other aspects of the seismic activity related to forecasting eruptions of Mauna Loa were discussed by *Klein* [1982] and *Decker et al.* [this volume]. In this section, I discuss some of the preeruptive seismicity beneath the Mauna Loa summit region reported on earlier and present some differing aspects of the seismicity recorded before these two eruptions.

Pre-1975-eruption seismicity. The 1975 eruption of Mauna Loa followed 25 years of repose after the large southwest-rift-zone eruption of June 1950. In the months leading up to the July 1975 eruption, HVO seismologists recognized a marked increase in activity beneath the vol-

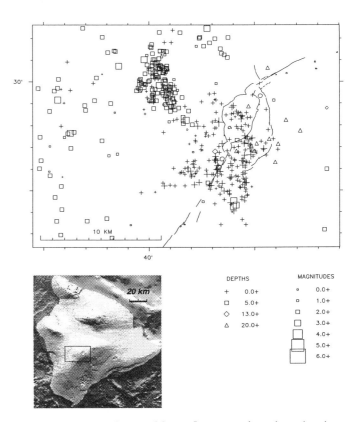

Figure 2. Schematic map Mauna Loa summit region, showing locations of earthquakes during 2-year period before the July 5, 1975 eruption (July 5, 1973 - July 10, 1975), using data from the Hawaiian Volcano Observatory seismicity catalog. Earthquake symbols are coded according to magnitude and focal depth. Shaded relief inset shows the location on the Island of Hawaii.

cano. The sequence and increases in seismicity were described by *Koyanagi et al.* [1975], who concluded that the data suggested that Mauna Loa was recovering from its post-1950 quiescent state. Although their report was drafted before the eruption, the final version was not submitted until after the volcano had erupted.

In April 1974, as noted by *Koyanagi et al.* [1975], both counted and located earthquakes showed definite increases above previous levels of seismic activity. Previously, HVO seismologists had noted an increase in the number of counted long-period earthquakes beneath Mauna Loa, beginning with a sharp increase in mid-February. Then, as described by *Decker et al.* [1983] and *Lockwood et al.* [1987], Mauna Loa seismicity continued to increase, with notable earthquake swarms in August and December 1974. After a brief delay that followed the December swarm, strong seismic activity at Mauna Loa returned in February 1975 and persisted throughout the months before the July 5 eruption.

Figure 2 shows the locations of epicenters of earthquakes

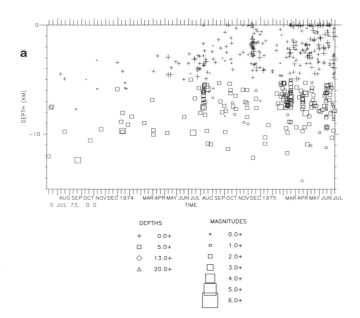

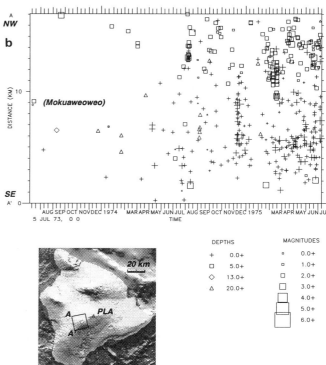

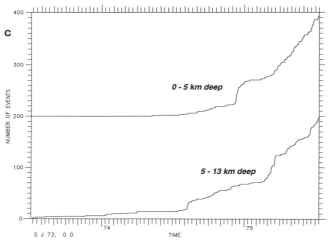

Figure 3. Summary plots of Mauna Loa seismicity during 2-year period before 1975 eruption. (a) Earthquake focal depth versus time. (b) Distance versus time for epicenters contained within box and projected along line A-A' shown in inset. (c) Cumulative number of located earthquakes in two depth ranges: upper curve: 0 to 5 km ; lower curve: 5 to 13 km.

beneath the summit region of Mauna Loa, during a 735-day period beginning July 5, 1973, 2 years before the onset of the 1975 eruption. During this period, two groupings of earthquakes are visible: a shallow group centered about the summit caldera and upper part of the southwest rift zone, and a slightly deeper group centered northwest of the caldera.

Figure 3 provides additional details of the seismicity beneath the Mauna Loa summit during this period. Increased seismic activity leading to the eruption began with a relatively deep earthquake swarm to the northwest of Mokuaweoweo Caldera at the Mauna Loa summit. The August 1974 swarm consisted primarily of earthquakes occurring at focal depths of 5 to 8 km; the December 1974 swarm was generally shallower, with focal depths of 0 to 5 km, and closer to the caldera. As the rate of shallower

seismic activity leveled off, the deeper earthquakes reemerged in late February 1975. Then, the seismicity rates accelerated until the time of the eruption, with sharp increases in both depth categories recorded in June 1975.

Lockwood et al. [1987] described the 1975 eruption and subsequent activity. The eruptive vents were restricted to the summit region, and erupting lava was no longer visible after July 6. Continuing intense seismic activity, including an M = 4.5 earthquake on July 9 near Pu'u Ulaula (station PLA; see Figure 3b), along the northeast rift zone of the volcano, was interpreted to indicate dike intrusion into the northeast rift zone and potential downrift eruptions, but by July 12 seismic activity had subsided sufficiently to end the eruption alert.

Pre-1984-eruption seismicity. Because of the historical pattern of Mauna Loa eruptions -- that is, relatively short lived eruptive activity localized about the Mauna Loa summit followed within several years by sustained eruption from a downrift vent -- and because of continued inflation of the volcano, *Lockwood et al.* [1976] suggested that a

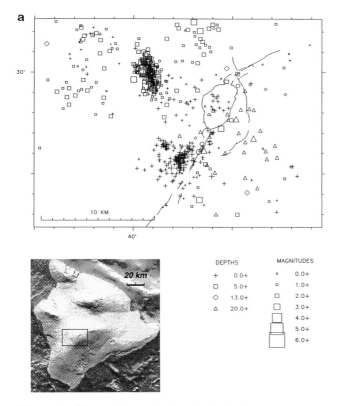

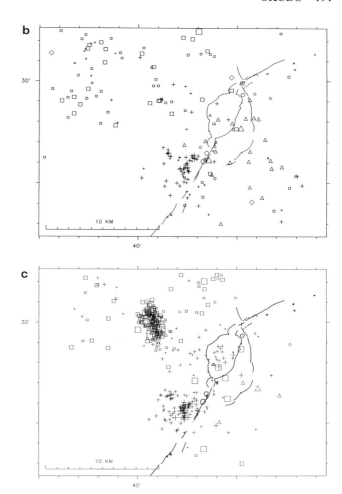

Figure 4. Mauna Loa summit seismicity between 1975 and 1984 and several days after onset of 1984 eruption. (a) August 5, 1975, through March 30, 1984.   (b) August 5,1975, through March 25, 1982. (c) March 25, 1982, through March 30, 1984. Shaded relief inset shows location on the Island of Hawaii.

larger northeast-rift-zone eruption would occur within 3 years of the 1975 eruption. Although no eruption occurred between 1975 and 1978, continuing seismic monitoring of Mauna Loa indicated a slight increase seismicity beginning in 1978. Based on continuing geodetic and seismic monitoring, *Decker et al.* [1983] identified an increase in eruption probability for the 2-year period beginning in mid-1983. Their forecast could be regarded as accurate with respect to the 1984 northeast-rift-zone eruption.

Decker et al., [1983] and *Lockwood et al.*, [1987] noted the similarities between the pre-1975 and pre-1984 precursory seismicity patterns. However, while the same regions to the northwest and southwest of Mokuaweoweo caldera were again seismically active, the sequence of seismicity leading to the 1984 eruption differs from that recorded before the 1975 eruption. Figures 4 through 6 show the locations of earthquake epicenters, cumulative numbers in the 0- to 5-km and 5- to 13-km depth ranges, and distance versus time, respectively, for this period.

During the intereruption period, seismicity clearly continued. Figure 4a shows the locations of epicenters be-

neath the Mauna Loa summit region, as in Figure 2 during the intereruption period, beginning 1 month after the 1975 eruption and continuing through the onset of the 1984 eruption. This seismicity is divided into two groups, August 5, 1975, to March 25, 1982 (Figure 4b), and March 25, 1982, to March 30, 1984 (Figure 4c), to show the increase in seismicity during the 2 years before the eruption.

The intereruption cumulative numbers and the temporal distribution of earthquakes differ from the patterns of pre-1975 seismicity. As mentioned earlier, a slight increase in seismicity was noted in 1978. This activity ended the very quiet period immediately after the 1975 eruption. Throughout the rest of the intereruption period, the rate of shallow seismicity steadily increased and showed a marked increase beginning in March 1984, several weeks before the eruption (Figure 5). Seismicity in the 5- to 13-km-depth range occurred at a relatively more uniform rate, except for an earthquake swarm in September 1983 that activated the same region to the northwest of the caldera as had the pre-

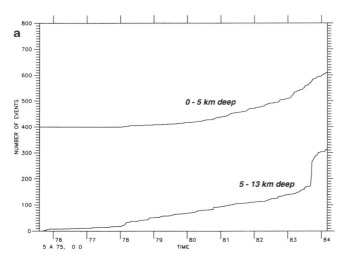

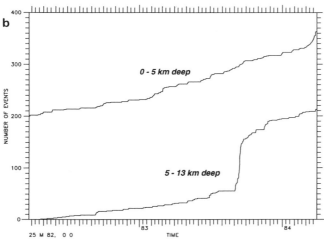

Figure 5. Cumulative number of located earthquakes versus time for periods August 5, 1975, through March 30, 1984 (a) and March 25, 1982, through March 30, 1984 (b).

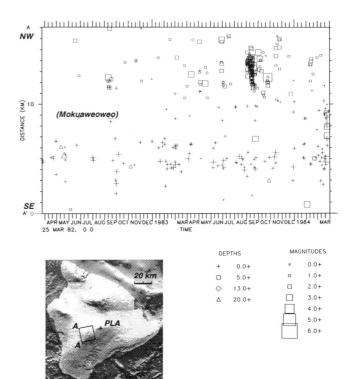

Figure 6. Distance versus time distribution of earthquakes, projected along line A-A', for same 2-year period as in Figures 4c and 5b.

1975 eruption sequence (Figure 6). After the November 16, 1983 Kaoiki earthquake, the number of deeper earthquakes beneath the Mauna Loa summit increased slightly, but subsequently no sustained increase in seismicity rate between 5- and 13-km depth occurred before the 1984 eruption as before the 1975 eruption.

Unlike the 1975 eruption, which was heralded by months of increasing summit seismicity, possibly the strongest indicator of the elevated levels of unrest before the 1984 eruption was the September 1983 earthquake swarm to the northwest of Mokuaweoweo Caldera. This swarm included about 100 earthquakes recorded well enough to locate and 2 earthquakes of M ≥ 4.0. A subset of seven of these events with M ≥ 3.0 were included in a study of earthquake focal mechanisms and regional stress [*Gillard et al.*, 1992]. *Gillard et al.* found a homogeneous set of focal

mechanisms for this small subset of the swarm, consisting of strike-slip faulting with a normal faulting component. Even before the work of *Gillard et al.*, these earthquakes had been understood to reflect tectonic adjustments to strains resulting from volcanic loading [*Lockwood et al.*, 1987, based on communication from E.T. Endo].

Mauna Loa Flank Seismicity

On November 16, 1983, a large (M = 6.6), damaging earthquake occurred beneath the Kaoiki region, on the flank of Mauna Loa between the summits of Kilauea and Mauna Loa Volcanoes. The location of the earthquake epicenter is indicated in Figure 1. Several investigators have studied aspects of this earthquake and its associated effects [e.g., *Endo*, 1985; *Chen and Nabelek*, 1990; *Jackson et al.*, 1992]. Seismologic studies of this earthquake indicate that significant, if not the predominant, faulting motions were associated with seaward sliding along a subhorizontal rupture plane at about 12-km-depth [*Endo*, 1985; *Chen and Nabelek*, 1990].

Owing to the size, location, and timing of this earthquake, it is justifiably included in discussions of precursory phenomena associated with the 1984 eruption [e.g., *Lockwood et al.*, 1985; 1987]. I mentioned earlier that the 1983 Kaoiki earthquake did not appear to affect seismicity rates beneath the Mauna Loa summit. The aftershock sequence, however, extended over a large area beneath the southeast flank of Mauna Loa, merging with the northeast rift zone of the volcano. Because of the extent of the aftershock activity and the manner in which the 1975 eruption ended after a strong intrusion of magma into the northeast rift zone, expectations of a rift-zone eruption were probably only heightened by the 1983 Kaoiki earthquake. The seaward motions of the volcanic flank associated with such large earthquakes may facilitate magma movement through the volcano.

The Kaoiki region is one of the most active seismic zones on the Island of Hawaii. Moderate earthquakes, including the November 1983 event, occur there repeatedly. *Wyss* [1986] has suggested a regularity of moderate earthquake occurrence in this region and projected an anticipated window between 1992 and 1996 for the next earthquake in this series. Since 1962, a total of eight Kaoiki earthquakes M ≥ 4.5 have occurred, three of which were of M ≥ 5: June 27, 1962 (M6.1), November 30, 1974 (M5.4), and November 16, 1983 (M6.6). The two most recent earthquakes in this series have been followed by eruptions of Mauna Loa.

Besides these two modern eruptions, the early instrumental record reinforces the notion that large earthquakes beneath the volcanic flanks are related to Mauna Loa eruptions. The eruptions of 1919, 1935, 1942, and 1950 all occurred after significant flank earthquakes [*Wyss and Koyanagi*, 1992]. For these eruptions, the discussion of the April 1942 northeast-rift-zone eruption includes mention of a precursory flank earthquake [*Finch*, 1943], the M ~ 6 Kaoiki earthquake of September 25, 1941 [*Wyss and Koyanagi*, 1992].

The June 1950 eruption of Mauna Loa occurred along the southwest rift zone, 18 months after the January 1949 eruption and only days after the May 29, 1950 earthquake in Kona. The 1949 eruption was preceded by a small increase in seismicity which was regarded as neither sufficiently definitive nor sufficiently large to allow an eruption prediction [*Macdonald*, 1954]. Macdonald's narrative of the 1950 eruption includes mention of an increase in the number of Mauna Loa earthquakes in May 1950, including the May 29 earthquake. Because of the limited number and range of instruments in operation at the time, details of microseismicity and earthquake mechanisms during this sequence have not been established.

Macdonald's [1954] narrative placed the May 29, 1950 earthquake beneath the uppermost part of the southwest rift zone of Mauna Loa. However, from the macroseismic accounts and limited instrumental recordings of this earthquake [e.g., *Wyss and Koyanagi*, 1992], it is thought to be of at least M = 6, beneath central Kona, and similar in focal mechanism to the August 21, 1951, M = 6.9 Kona earthquake. *Beisser et al.* [1994], who analyzed teleseismic

recordings of the 1951 earthquake, inferred that the mechanism of the 1951 earthquake, similar to other large Hawaiian earthquakes [e.g., *Chen and Nabelek*, 1990], was characterized by seaward sliding along a subhorizontal rupture plane or decollement at about 13-km depth.

Thus, a Kona analog exists to the pattern of deformation occurring in the Kaoiki region, defined on the scales of large earthquakes [*Beisser et al.*, 1994] and microearthquakes [*Gillard et al.*, 1992]. The regional, seaward deformations along a decollement associated with the Kaoiki and Kona earthquakes could facilitate magma migration through the shallower parts of Mauna Loa. The large earthquakes are also part of the mechanical response of the volcanic flanks to intrusions and eruptions. For the Kaoiki earthquakes, the associated region is the northeast rift zone, whereas the Kona earthquakes are related to activity along the southwest rift zone.

The largest earthquake ever recorded in Hawaii occurred in April 1868, centered beneath the southeast flank of the volcano [*Wyss*, 1988]. Without seismographic records, estimation of a precise epicenter is difficult, but the historical accounts reviewed by Wyss suggest that the earthquake occurred beneath the flank and was part of a series of events which included a southwest-rift-zone eruption of Mauna Loa. However, the relation of volcanic flank seismicity to eruptions and magma intrusion for events in the poorly instrumented past is difficult to establish.

CURRENT SEISMICITY AND DISCUSSION

The 1975 and 1984 eruptive sequences provided critical opportunities to observe and analyze many aspects of Mauna Loa seismic activity. The observations briefly described above clearly suggest the existence, if not a strictly repeating pattern, of seismic precursors to these eruptions in zones to the northwest of Mokuaweoweo Caldera and beneath the southwest rift zone. The pre-1975 seismicity buildup was more clearly recognizable than that before the 1984 eruption, possibly due to the longer repose period since the preceding eruption in 1950.

Since the 1984 northeast-rift-zone eruption, Mauna Loa has been in repose. Continued geodetic monitoring indicates continuing inflation in the summit region [e. g., *Okamura et al.*, 1991, 1992; *Miklius et al.*, 1993, and this volume]. Figures 7 and 8 plot seismicity for the Mauna Loa summit region, as in Figures 2 and 3c, for the period 1985-94. The seismicity rate beneath the Mauna Loa summit fluctuates. Throughout early 1992, earthquakes at 5- to 13-km depth were more frequent than those at 0- to 5-km depth. The rate of deeper events then decreased, and the rate of shallower events increased slightly, in 1993, although the cumulative numbers of located earthquakes remained low. This observation suggests that Mauna Loa remains some time away from renewed eruptive activity.

The earthquake locations and the relative positions of clusters or swarms suggest some key details in the eventual

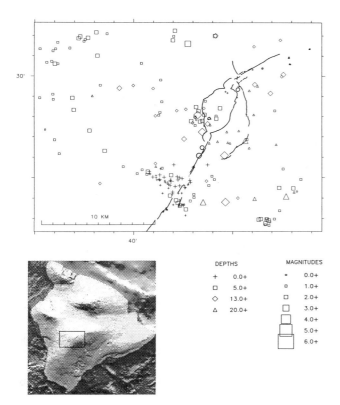

Figure 7. Schematic map of Mauna Loa summit region, showing locations of earthquakes during period January 1, 1984- December 31, 1994.   Shaded relief inset shows locations on Island of Hawaii.

reactivation of Mauna Loa. Figure 9 shows two cross sections of hypocenters through Mauna Loa. The large concentrations of earthquakes in the eastern parts of the cross sections are Kaoiki earthquakes. The earthquake swarms are included in the other clusters of earthquakes in the vicinity of Mokuaweoweo Caldera. Although the seismicity beneath the west flank of Mauna Loa is less intense than in the Kaoiki region, a decollement is suggested at the boundary separating the shallow, more seismically active volume from the relatively aseismic, lower western section.

Two types of earthquake activity have been linked with Mauna Loa eruptions, and increases in the numbers of either type would indicate a possible impending eruption. For both of the recent eruptions of Mauna Loa, earthquake clusters, deeper than 5 km, northwest of the Mauna Loa summit caldera were clearly identified and recognized as abnormal. Farther from the summit, large earthquakes with mechanisms tied to the decollement surfaces underlying the west and southeast flanks of Mauna Loa would raise our level of anticipation. Seaward motions of the flanks would presumably facilitate the upward migration of magma.

Following the suggestion of *Koyanagi et al.* [1975] that, similar to the magma-storage complex beneath the Kilauea summit outlined by the relative locations of different earthquake families [*Koyanagi et al.*, 1976; *Klein et al.*, 1987], the aseismic region below the shallow earthquakes in the Mokuaweoweo Caldera region feasibly represents a magma source. To the extent allowed by the hypocentral detail, this region lies beneath the caldera at greater than about 4-km depth, corresponding to the area indicated in Figure 9. Modeling of geodetic data compiled from several different geodetic monitors has led to the proposal of a magma source at about the same position and depth [*Decker et al.*, 1983; *Lockwood et al.*, 1987]. This geodetic modeling is consistent with continued inflation beneath the Mauna Loa summit in about the same position [*Okamura et al.*, 1991, 1992; *Miklius et al.*, 1993 and this volume].

With the inferred inflation occurring approximately beneath the summit caldera, Mauna Loa is evidently building toward its next eruption. Large earthquakes beneath the flanks of Mauna Loa have been linked with eruptions on Mauna Loa, principally those occurring along the rift zones. Although not all of the recorded eruptions of Mauna Loa clearly have seismic precursors, the large flank earthquakes mentioned earlier are probably clear precursors to only a subset of Mauna Loa eruptions or only under certain conditions.

The above discussion is based on the earthquake catalogs routinely maintained at HVO, using a relatively simple, regional seismic crustal-velocity model. Work is under way to improve the relative precision of earthquake locations beneath Kilauea [*Got et al.*, 1994; *Gillard et al.*, 1994] Similar studies should be attempted to improve

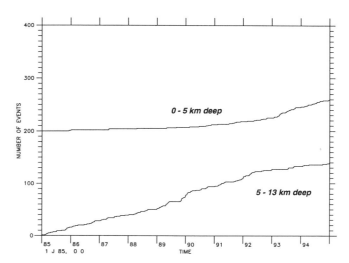

Figure 8. Cumulative number of located earthquakes beneath Mauna Loa summit, during period January 1, 1984 - December 31, 1994.

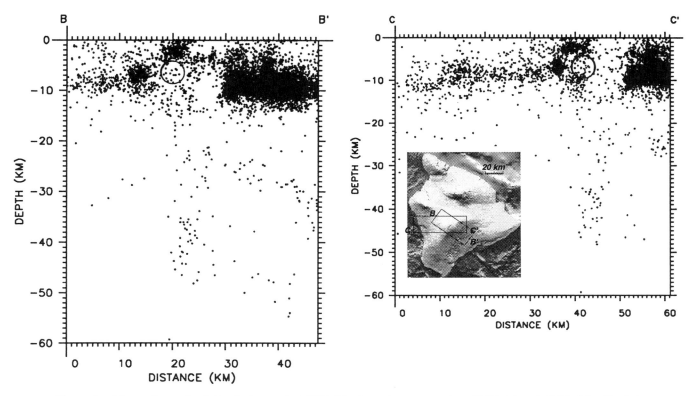

Figure 9.   Mauna Loa seismicity during period 1971-94, through cross sections B-B' (a) and C-C' (b).   Shaded relief inset shows location on Island of Hawaii.

structural detail for Mauna Loa. These efforts would enhance estimation of more detailed, three-dimensional velocity models based on earthquake arrival times [e. g. , *Thurber*, 1984, 1987] and a more confident estimation of earthquake focal mechanisms [*Karpin and Thurber*, 1987; *Got et al.*, 1994; *Gillard et al.*, 1994].

Recent work on Kilauea is aimed at reexamining the relations between geodetic and seismic data over longer time scales and larger regional scales than those typically examined for specific eruption-related detail. These studies have led to interesting implications for the relation between volcanic and tectonic processes [e.g., *Borgia et al.*, 1990; *Borgia*, 1994; *Clague and Denlinger*, 1994; *Denlinger and Okubo*, 1992 and 1995]. Because of the sizes of the active structures [*Okubo et al.*, 1992; *Denlinger and Okubo*, 1992, 1995], an important mechanical interaction may well exist between Kilauea and Mauna Loa that at present is still incompletely understood.

CONCLUSION

Mauna Loa Volcano continues to show signs of activity. Present levels of seismic activity, relative to that which preceded the 1984 and 1975 eruptions, indicate that the volcano is some months, if not longer, from its next eruption. A large earthquake beneath the flanks of Mauna Loa could signal the start of a period of buildup toward an eruption, as in 1942, 1950, 1975, and 1984.

Because the Kaoiki earthquake source region lies between Kilauea and Mauna Loa, the seismic activity at Kilauea, especially along its southwest rift zone, probably affects both the Kaoiki region and Mauna Loa. Future activity beneath Mauna Loa might even require a substantial change in the ongoing 12-year-long eruption occurring in the east rift zone of Kilauea. More comprehensive regional geodetic and seismologic monitoring is necessary to identify any reasonable relations.

Acknowledgments. This report was made possible only through the dedicated efforts of the HVO staff directed at seismographic monitoring of the Hawaiian volcanoes. Many people have contributed to the building of the network operations resulting in the long HVO seismological record; in particular, I would like to acknowledge the contributions by Robert Koyanagi, Fred Klein, Jennifer Nakata, Wilfred Tanigawa, and Alvin Tomori. I would like to thank David Clague, David Harlow, Laura Kong, and Arnold Okamura for helpful reviews. I would also like to thank Jack Lockwood for his encouragement during the preparation of this report. The bathymetry shown in shaded relief maps was compiled by J. Smith and T. Duennebier of the University of Hawaii at Manoa.

REFERENCES

Aki, K. and R. Y. Koyanagi, Deep volcanic tremor and magma ascent mechanism under Kilauea, Hawaii, *J. Geophys. Res.*, *86*, 7095-7109, 1981.

Apple, R. A., Thomas A. Jaggar, Jr., and the Hawaiian Volcano Observatory, in *Volcanism in Hawaii*, 2 v., *U.S. Geol. Surv. Prof. Paper 1350*, edited by R. W. Decker, T. L. Wright, and P. H. Stauffer, *2*, chap. 61, 1619 - 1644, 1987.

Beisser, M., D. Gillard, and M. Wyss, Inversion for source parameters from sparse data sets: test of the method and application to the 1951 [M=6.9] Kona, Hawaii, earthquake, *J. Geophys. Res.*, *99*, 19661-19678, 1994.

Borgia, A., Dynamic basis of volcanic spreading, *J. Geophys. Res.*, *99*, 17791-17804, 1994

Borgia, A., J. Burr, L. D. Montero, W. Morales, and G. E. Alvarado, Fault propagation folds induced by gravitational failure and slumping of the Central Costa Rica Volcanic Range: Implications for large terrestrial and Martian volcanic edifices, *J. Geophys. Res.*, *95*, 14357 - 14382, 1990.

Chen, W. -P., and J. Nabelek, Source parameters of the June 26, 1989 Hawaiian earthquake [abs.], *Eos Trans. AGU supp.*, *71*, 562, 1990.

Clague, D.A., and R. P. Denlinger, Role of olivine cumulates in destabilizing the flanks of Hawaiian volcanoes, *Bull. Volcanol.*, *56*, 425-434, 1994.

Decker, R.W., R. Y. Koyanagi, J. J. Dvorak, J. P. Lockwood, A. T. Okamura, K. M. Yamashita, and W. R. Tanigawa, Seismicity and surface deformation of Mauna Loa Volcano, Hawaii, *Eos Trans. AGU*, *64*, 545-547, 1983.

Denlinger, R.P., and P. G. Okubo, A mechanical model for the south flank [abs.], *Eos Trans. AGU supp.*, *73*, 505, 1992.

Denlinger, R. P., and P. G. Okubo, Structure of the mobile south flank of Kilauea volcano, Hawaii, in press, *J. Geophys. Res.*, *100*, 1995.

Endo, E. T., Seismotectonic framework for the southeast flank of Mauna Loa volcano, Hawaii, Ph. D. thesis, University of Washington, Seattle, 1985.

Finch, R.H., The seismic prelude to the 1942 eruption of Mauna Loa, *Bull. Seismol. Soc. Am.*, *33*, 237-241, 1943.

Gillard, D., M. Wyss, and J. S. Nakata, A seismotectonic model for western Hawaii based on stress tensor inversion from fault plane solutions, *J. Geophys. Res.*, *97*, 6629-6641, 1992.

Gillard, D., A. M. Rubin and J.- L. Got, Precise relocations and focal mechanisms of earthquakes accompanying dike intrusions in Hawaii [abs.], *Eos Trans. AGU supp.*, *75*, 715, 1994.

Got, J.-L., J. Frechet, and F. W. Klein, Deep fault plane geometry inferred from multiplet relative relocation beneath the south flank of Kilauea, *J. Geophys. Res.*, *99*, 15375-15386, 1994.

Jackson, M. D., E. T. Endo, P. T. Delaney, T. Arnadottir, and A. M. Rubin, Ground ruptures of the 1974 and 1983 Kaoiki earthquakes, Mauna Loa Volcano, Hawaii, *J. Geophys. Res.*, *97*, 8775-8796, 1992.

Karpin, T.L., and C. H. Thurber, The relationship between earthquake swarms and magma transport: Kilauea Volcano, Hawaii, *Pure Appl. Geophys.*, *125*, 971-991, 1987.

Klein, F.W., Hypocenter location program HYPOINVERSE, part 1: users' guide to versions, 1, 2, 3, and 4, *U.S. Geological Survey Open-File Report 78-694*, 1-113, 1978.

Klein, F.W., A linear gradient crustal model for south Hawaii, *Bull. Seismol. Soc. Am.*, *71*, 1503-1510, 1981.

Klein, F.W., Patterns of historical eruptions at Hawaiian volcanoes, *J. Volcanol. Geotherm. Res.*, *12*, 1-35, 1982.

Klein, F. W., User's guide to HYPOINVERSE, a program for VAX computers to solve for earthquake locations and magnitudes, *U.S. Geological Survey Open-File Report 89-314*, 58 pp., 1989.

Klein, F.W., and R. Y. Koyanagi, Hawaiian Volcano Observatory seismic network history 1950-1979, *U.S. Geological Survey Open-File Report 80-302*, 84 pp., 1980.

Klein, F.W., R. Y. Koyanagi, J. S. Nakata, and W. R. Tanigawa, The seismicity of Kilauea's magma system, in Decker, R.W., Wright, T.L., and Stauffer, P.H., in *Volcanism in Hawaii*, 2 v., *U.S. Geol. Surv. Prof. Paper 1350*, edited by R. W. Decker, T. L. Wright, and P. H. Stauffer, *2*, chap. 43, 1019-1185, 1987.

Koyanagi, R.Y., E. T. Endo, and J. S. Ebisu, Reawakening of Mauna Loa Volcano, Hawaii: a preliminary evaluation of seismic evidence, *Geophys. Res. Lett.*, *2*, 9, 405-408, 1975.

Koyanagi, R. Y., J. D. Unger, E. T. Endo, and A. T. Okamura, Shallow earthquakes associated with inflation episodes at the summit of Kilauea Volcano, Hawaii, in *Proceedings of the Symposium on Andean and Antarctic Volcanology and Chemistry of the Earth's Interior*, special series, 11 p., 1976

Koyanagi, R. Y., B. Chouet and K. Aki, Origin of volcanic tremor in Hawaii, in Decker, R.W., Wright, T.L., and Stauffer, P.H., in *Volcanism in Hawaii*, 2 v., *U.S. Geol. Surv. Prof. Paper 1350*, edited by R. W. Decker, T. L. Wright, and P. H. Stauffer, *2*, chap. 45, 1221-1257 ,1987.

Lockwood, J. P., R. Y. Koyanagi, R. I. Tilling, R. T. Holcomb, and D. W. Peterson, Mauna Loa threatening, *Geotimes*, *21*, 12-15, 1976.

Lockwood, J.P., N. G. Banks, T. T. English, L. P. Greenland, D. B. Jackson, D. J. Johnson, R. Y. Koyanagi, K. A. McGee, A. T. Okamura, and J. M. Rhodes, The 1984 eruption of Mauna Loa Volcano, Hawaii, *Eos Trans. AGU*, *66*, 169-171, 1985.

Lockwood, J.P., J. J. Dvorak, T. T. English, R. Y. Koyanagi, A. T. Okamura, M. L. Summers, and W. R. Tanigawa, Mauna Loa 1974-1984: a decade of intrusive and extrusive activity, in *Volcanism in Hawaii*, 2 v., *U.S. Geol. Surv. Prof. Paper 1350*, edited by R. W. Decker, T. L. Wright, and P. H. Stauffer, *1*, chap. 19, 537-570, 1987.

Macdonald, G. A., Activity of Hawaiian volcanoes during the years 1940-1950, *Bulletin Volcanologique*, *2*, 119-179, 1954

Miklius, A., A. T. Okamura, M. K. Sako, J. Nakata, and J. Dvorak, Current status of geodetic monitoring of Mauna Loa Volcano [abs.], *Eos Trans. AGU* , *74*, 629, 1993.

Nakata, J.S., A. H. Tomori, J. P. Tokuuke, R. Y. Koyanagi, W. R. Tanigawa, and P. G. Okubo, Hawaiian Volcano Observatory summary 88, part 1. Seismic data, January to December 1988 [Chronological summary by T.L. Wright], *U.S. Geological Survey Open-File Report 94-169*, 78 pp., 1994.

Okamura, A.T., A. Miklius, M. K. Sako, and J. Tokuuke, Evidence for renewed inflation of Mauna Loa Volcano,

Hawaii [abs.], *Eos Trans. AGU, 72,* 566, 1991.

Okamura, A.T., A. Miklius, P. Okubo, and M. K. Sako, Forecasting eruptive activity of Mauna Loa Volcano, Hawaii [abs.], *Eos Trans. AGU, 73,* no. 43, 343, 1992.

Okubo, P., J. Nakata, A. Tomori, and W. Tanigawa, Delineation of large-scale seismotectonic structures in Hawaii [abs.], *Eos Trans. AGU, 73,* no. 43, 512, 1992.

Thurber, C.H., Seismic detection of the summit magma complex of Kilauea Volcano, Hawaii, *Science, 223,* 165-167, 1984.

Thurber, C.H., Seismic structure and tectonics of Kilauea Volcano, in *Volcanism in Hawaii,* 2 v., *U.S. Geol. Surv. Prof. Paper 1350,* edited by R. W. Decker, T. L. Wright, and P. H. Stauffer, *2,* chap. 38, 919-934, 1987.

Wood, H.O., The seismic prelude to the 1914 eruption of Mauna Loa, *Bull. Seismol. Soc. Am., 5,* 39-51, 1915.

Wyss, M., Regular intervals between Hawaiian earthquakes; implications for predicting the next event, *Science, 234,* 726-728, 1986.

Wyss, M., A proposed source model for the great Ka'u, Hawaii, earthquake of 1868, *Bull. Seismol. Soc. Am., 78,* 1450-1462, 1988.

Wyss, M., and R. Y. Koyanagi, Isoseismal maps, macroseismic epicenters, and estimated magnitudes of historical earthquakes in the Hawaiian Islands, *U.S. Geological Survey Bulletin 2006,* 93 pp.; addendum, 1 p., 1992.

_____

Paul G. Okubo, U. S. Geological Survey, Hawaiian Volcano Observatory, P. O. Box 51, Hawaii National Park, Hawaii 96718.

Recent Inflation and Flank Movement of Mauna Loa Volcano

Asta Miklius[1], Michael Lisowski[1], Paul T. Delaney[2], Roger P. Denlinger[1],

John J. Dvorak[3], Arnold T. Okamura[1], and Maurice K. Sako[1]

Geodetic measurements on the summit of Mauna Loa reveal that since the last eruption in 1984, the shallow summit magma chamber has inflated approximately the same amount as between the 1975 and 1984 eruptions. However, it does not appear to have recovered the entire volume withdrawn during the 1984 eruption. Together with the lack of increased shallow earthquake activity, this observation suggests that, as of June 1995, the next eruption of Mauna Loa is not yet imminent. Global Positioning System measurements in 1993 and 1994 show southeastward movement of the southeast flank of over 4 cm/year, comparable to displacements measured on adjacent Kilauea Volcano's south flank over the same interval. The upper west flank appears to be stable, producing a strong asymmetry of motion about the summit. Gradients of motion on the southeast flank result in about one microstrain/year of compression and shear across the Kaoiki seismic zone, an area of persistent seismicity that has produced large historic earthquakes. The flank motions observed between 1993 and 1994 could be caused by the combined effects of slip along the basal Kaoiki decollement and inflation of a deep source.

INTRODUCTION

The hazards posed by Mauna Loa Volcano include both lava flows that rapidly endanger inhabited areas and large earthquakes that produce severe ground shaking. The most recent eruption, from the northeast rift zone in 1984, sent lava flows within 6.5 km of the city of Hilo (Figure 1); eruptions from the southwest rift zone are potentially even more hazardous [see *Trusdell*, this volume]. Mauna Loa has generated damaging earthquakes in historical time, on the Hilea and Kaoiki fault systems on the east flank and on the Kealakekua fault system on the

[1]U.S. Geological Survey, Hawaiian Volcano Observatory, Hawaii National Park, Hawaii
[2]U.S. Geological Survey, Flagstaff, Arizona
[3]Hilo, Hawaii

Mauna Loa Revealed: Structure,
Composition, History, and Hazards
Geophysical Monograph 92
Published in 1995 by the American Geophysical Union

lower west flank [see *Okubo*, this volume].

Geodetic monitoring on the summit of Mauna Loa, begun in the mid-1960s, spans the inflationary period between the brief summit eruption of 1975 and the three-week-long summit and northeast rift zone eruption of 1984. This record gives us a basis for comparison with the current period of inflation. Estimation of flank motion has been difficult in the past because of limitations of terrestrial surveying methods. Results of Global Positioning System (GPS) campaigns of 1993 and 1994 offer the first glimpse of these motions.

In this paper, we present a brief overview of what we have learned about the current status of Mauna Loa through monitoring of ground-surface deformation and pose a few questions to guide future studies.

DEFORMATION OF THE SUMMIT

The simplest deformation monitor on the summit of Mauna Loa, and the most frequently occupied, is a set of three electro-optical distance measurement (EDM) baselines from a single station (HVO93 in Fig-

Fig. 1. Map of the island of Hawaii showing geomorphic features, the Kaoiki, Hilea, and Kealakekua seismic zones, the area of Figure 2, and the current Mauna Loa GPS network (open circles). Bathymetry compilation provided by John Smith (unpublished data).

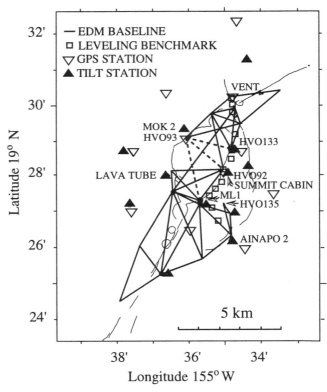

Fig. 2. Deformation monitoring network on the summit of Mauna Loa. EDM baselines to permanently mounted reflectors are dashed. Thin solid lines show the summit caldera, outer caldera and rift zone faults.

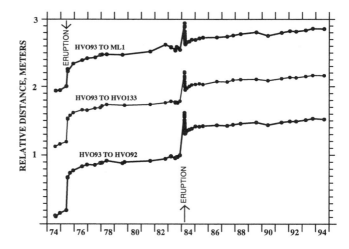

Fig. 3. Time-series of line-length changes measured with EDM from HVO93 to permanently mounted reflectors located across the caldera (see Figure 1). The large extensions at the start of the 1975 and 1984 eruptions were caused by the formation of dikes that fed the eruptive fissures. The size of the data symbols.corresponds to approximately two sigma of expected error.

ure 2) to permanently-mounted reflectors across the caldera. It has also proven to be a reliable indicator of the status of the shallow summit magma system, in that epochs over which we measure extension across the caldera correlate with periods when we measure significant uplift, tilt and horizontal displacement of stations around the summit. At these times, both vertical displacement gradients and horizontal displacements radiate away from an area just east of the main caldera, consistent with inflation of a shallow magma body. Since 1984, the rate of extension has varied somewhat, with higher rates during the year after the eruption, and at least two periods, from 1989 to 1990 and from 1993 to 1994, when line lengths across the caldera decreased or did not change (Figure 3). Since the 1984 eruption, lines across the summit have extended an average of about 20 cm, approximately equal to the amount of extension observed between the 1975 and 1984 eruptions. Magnitudes of horizontal displacement estimated from trilateration of the summit EDM network from

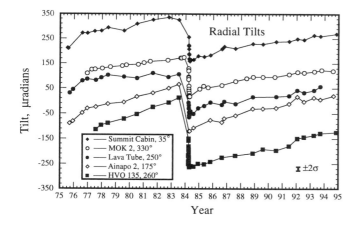

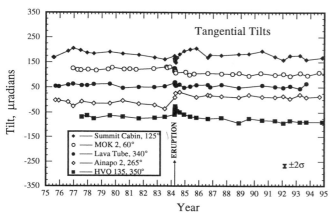

Fig. 4. Radial (a) and tangential (b) components of tilt at representative stations near the summit of Mauna Loa (see Figure 2 for locations). Directions of tilt components (given in inset) are chosen such that tangential tilts are minimized. A positive increase in tilt corresponds to a downward tilt in the specified direction.

1984 to the time of the last occupation of that network, in 1991, also equal or exceed those measured from 1975 to 1984. Horizontal displacements of stations near the summit from 1993 to 1994, measured with GPS, are less than the approximately 1 cm uncertainty of the measurements.

A 7-km level line along the east side of the summit caldera (Figure 2) was last occupied in 1991. The maximum uplift, relative to station Vent (Figure 2), recorded on this line from 1984 to 1991 is 32 +/- 2 cm, approximately the same as that measured from 1975 to 1983. However, the vertical change associated with deflation during the 1984 eruption was more than 60 cm.

Ground-surface tilts have been measured with spirit-level methods at stations on the summit since

late 1975. These stations generally tilt in directions that are radial from an area just east of the main caldera. Accordingly, tilt signals at representative stations are shown by their best-fit radial and tangential components (Figure 4). Consistent with the EDM data, these tilt data show that Mauna Loa started reinflating immediately after the 1984 eruption, and has reinflated at least as much since 1984 as during the inflation interval before the 1984 eruption (Figures 4 and 5). The directions of inflationary tilting largely oppose those of deflationary tilting during the 1984 eruption (see *Lockwood et al.*, 1987). The magnitude of inflationary tilts since 1984 is, however, substantially less than the magnitude of co-eruptive deflationary tilts.

Line-length changes measured with EDM and vertical changes measured by leveling from 1984 to 1991

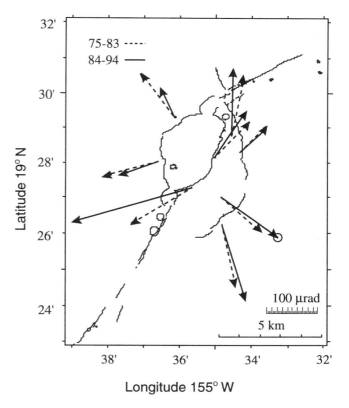

Fig. 5. Comparison of spirit-level tilt data from the inter-eruptive period from 1975 to 1983 and the period after the 1984 eruption to the most recent survey in 1994. Note the similarity in both directions and magnitudes. Errors expected from this type of measurement are on the order of 16 microradians for the difference between two surveys (*Delaney et al.*, 1993). The error circle is representative of expected error on all vectors.

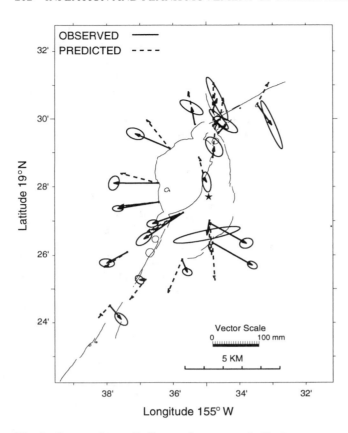

Fig. 6. Comparison of trilateration-network displacements from 1984 to 1991 and the displacements resulting from a point-source model of inflation. Both EDM and leveling data, weighted by their expected uncertainty, were used in the inversion. The epicenter of inflation is indicated by the star and is buried about 3.7 km below the surface. Indeterminate components of the displacement field (rigid body and rotaion and translation) are fixed by the model coordinate method of *Segall and Matthews* (1988), in which the misfit between the computed and predicted displacements is minimized. Error ellipses represent the expected 95% confidence interval.

are fitted to an elastic model for the expansion of a spherical source. The modeled source is located just east of the main caldera (Figure 6), at about 3.7 km depth. Using a Poisson's ratio of .25, the estimated magma-accumulation volume is 25 million m^3. The location and depth of this source are very close to those modeled for the loci of the 1975-1984 inflation and the 1984 deflation [*Lockwood et al.*, 1987]. The model accounts for about 90% of the measured signal, although model residuals are too great to be explained entirely by anticipated measurement errors. Displacements of stations east of the caldera are not

fit to the model as well as those on the west side of the caldera (Figure 6). The level data are everywhere well accounted for by the model (Figure 7).

Only the spirit-level tilt data span the current inter-eruptive period, from 1984 to 1994. Fitting just these data to the elastic point-source model yields similar source parameters, with a total change in volume from the 1984 eruption to 1994 of about 35 million m^3. The model fits the data reasonably well, but with greater parameter uncertainties.

FLANK MOVEMENT

A GPS network capable of detecting flank motion on Mauna Loa was established and occupied in September 1993. The first reoccupation of a part of that network started in late July 1994, and continued in September 1994 and in January 1995. In order to account for the differences in reoccupation time, we present the results as velocities, but emphasize that there is no data to assess whether the observed motions occur at a constant rate. Data were processed with NASA's Jet Propulsion Lab's (JPL) GIPSY software [*Lichten*, 1990], using improved orbits from JPL.

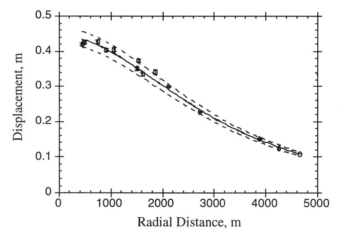

Fig. 7. Comparison of relative vertical displacements (open circles) measured by leveling from 1984 to 1991 with displacements resulting from the best-fitting point-source model, shown by the solid line. Dashed lines represent the range of model values allowed by the inversion at 1 standard deviation. The uplift of the benchmark farthest from the center of inflation (Vent, see Figure 2) is set to the amount predicted by the model (10.8 cm), and, although height differences between benchmarks are input into the inversion, the data is shown here as cumulative change from Vent. Error bars on the data points show expected error assuming the elevation of Vent is known.

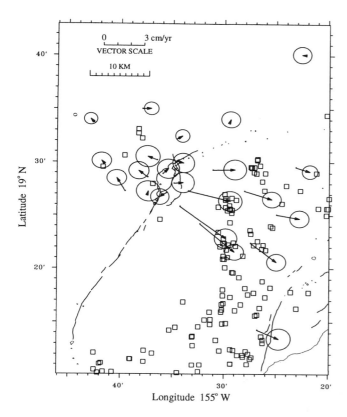

Fig. 8. Horizontal velocities measured by GPS between 1993 and 1994-1995. 95% confidence ellipses are scaled by the residuals of the network adjustment. Also plotted are the epicenters of earthquakes of M>1.5 over the same time period.

Five global tracking sites were included in the solutions. An average of 85% of integer phase cycle ambiguities were resolved; most of the remaining ambiguities were associated with either the 500 km-long baseline to the Kokee tracking site or very short observation periods.

Results of these data show southeastward movement of the upper southeast flank of more than 4 cm/yr (Figure 8) relative to a station on the summit of Mauna Kea (Figure 1). The most rapidly moving stations are located east of the faults of the outer summit caldera, while stations on the east rim of the main caldera, only 2-3 km away, did not move significantly. The upper west flank of Mauna Loa appears to be stable.

The largest horizontal displacements occur northwest of the Kaoiki seismic zone (Figure 1) and decrease gradually across the zone to about 2 cm on the lower flank. No correlation appears to exist between

the sharp western boundary of Kaoiki seismicity (Figure 8) and the magnitudes of displacement. There is, however, substantial contraction across this zone, about 1 microstrain/year. Magnitudes of displacement also decrease eastward along the flank, resulting in left lateral shear, also about 1 microstrain/year. Kaoiki seismicity is dominated by low-angle thrust faulting between 10 and 12 km depth, but right-lateral strike-slip faulting on nearly vertical planes is also a significant component of seismicity [*Endo*, 1985; *Wyss*, 1992]. The locations of decollement-type events in the Kaoiki seismic zone appear to be consistent with a low velocity layer at about 11.5 km depth below the southeast flank of Mauna Loa, interpreted to be the interface between the oceanic crust and the volcanic pile [*Thurber et al.*, 1989].

DISCUSSION

Monitoring of geodetic networks near the summit of Mauna Loa provides a picture of inflation of a shallow chamber with a magma-accumulation rate of about 3.5 million m³/year. The total volume change from 1984 to 1994 is approximately equal to that measured during the inter-eruptive period from 1975 to 1984. By this measure, it would seem that an eruption of Mauna Loa is likely in the near future. However, the magma-withdrawal volume during the 1984 eruption (~65 million m³, based on *Lockwood et al.'s* [1987] reported volume of surface subsidence) is substantially larger than the accumulation volume since the eruption. Moreover, the volume of lava erupted in 1984, about 220 million m³ [*Lipman and Banks*, 1987], greatly exceeds the estimated accumulation and withdrawal volumes, even taking into account that errors in calculation of magma-equivalent eruption volumes may be as much as 50%. Finally, the heightened rate of shallow seismicity beneath the summit in the year preceding both the 1975 and 1984 eruptions has not yet been observed [*Okubo*, this volume]. Thus, the next eruption of Mauna Loa does not appear to be imminent, as of June, 1995.

The discrepancy between the volume of lava erupted in 1984 and volumes estimated from surface displacement led *Dvorak et al.* [1985] to suggest that the 1984 eruption tapped sources of magma that were deeper than our summit networks could detect. The current GPS network addresses the need for far-field measurements, but results from the 1993-1994 surveys showing strong asymmetry of motion in the far-field about the summit preclude uniform inflation of

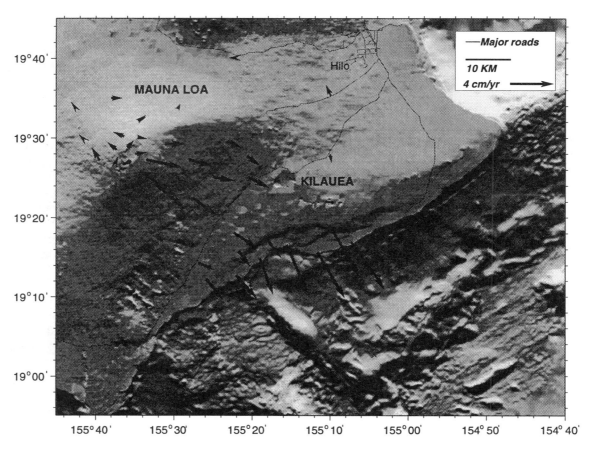

Fig. 9. Displacement rates, relative to Mauna Kea, of stations on Mauna Loa from 1993 to 1994-1995, and of selected stations on Kilauea over approximately the same time period.

a point source at any depth as the primary source of the flank deformation. Modeling of all the 1993-1994 Mauna Loa data (summit tilt, summit EDM, and summit and flank GPS) does not fit a point-source model, but more complex magma reservoir geometries have not yet been tested. Inflation of a deep source could provide at least part of the driving force for motion of the flank, as has been suggested for the motion of the south flank of Kilauea [e.g., *Owen et al.*, 1995].

Although previously unconfirmed by geodetic data, motion of the southeast flank of Mauna Loa has been implied by various analyses of low-angle thrust faulting in the Kaoiki seismic zone [e.g., *Endo*, 1985, *Bryan and Johnson*, 1991, *Wyss* et al.,1992]. These studies have suggested that Mauna Loa's southeast flank slips seaward on the basal Kaoiki decollement at about 12 km depth. *Bryan and Johnson's* [1991] focal mechanism analysis of basal slip closely predicts the azimuth of motion observed

on both Mauna Loa and Kilauea's flanks (Figure 9). Westward movement of the west flank of Mauna Loa is also suggested by their analysis, and that of *Gillard et al.* [1992]. Although the upper west flank does not appear to have moved significantly from 1993-1994, GPS data from 1987 to 1990 show westerly movement of a station on the west coast of up to 8 cm [*Dvorak et al.*, 1994]. The Kealakekua fault system on Mauna Loa's lower west flank has been the site of large historic earthquakes [*Okubo*, this volume], as has the Kaoiki area on the southeast flank, and the south flank of Kilauea. Although the relationship between steady or episodic ground motion and the generation of large earthquakes on volcano flanks remains enigmatic, defining such motion will doubtless increase our understanding of the processes which lead to flank instability.

An important question for earthquake risk assessment in the Kaoiki seismic zone is whether the velocities measured from 1993 to 1994-95 are typical of the rates of deformation on the southeast flank. Displacements near the summit from 1984 to 1991 appear to fit a point-source model of inflation fairly well, but the movements of stations east of the summit caldera are more easterly than the model predicts. This misfit may be an indication that southeasterly flank motion had occurred during that interval. This possibilty also raises the question of the nature of the interaction between the inflation of the shallow magma chamber and the movement of the flank. Measurements in the summit area appear to show a decrease in inflation rate from 1993 to 1994, but an alternative explanation is that movement of the flank results in a lessened rate of pressure increase around the chamber. In this case, periods such as 1993-1994 could be times of increased rate of flank movement, rather than decreased rate of magma accumulation.

On the other hand, the velocities of Kilauea's south flank from 1993 to 1994 are comparable to those on Mauna Loa's southeast flank (Figure 9) and are less than the rates measured from 1990 to 1993 [Owen et al., 1995]. If the movement of Mauna Loa's flank is related to the movement of the adjacent south flank of Kilauea, then it is conceivable that the rate of movement of Mauna Loa's flank was greater in the past.

Continued deformation monitoring is needed to provide answers to the numerous questions raised by the recent GPS data from Mauna Loa's southeast flank regarding the interaction between the movements of Kilauea and Mauna Loa, and the relationship of these movements with the magma systems and seismicity of these volcanoes.

Acknowledgements. We thank the former and current staff, volunteers and associates of the Hawaiian Volcano Observatory who helped to collect the data presented herein. Thanks to Laura Kong for help with graphics.

REFERENCES

Bryan, C. and C. Johnson, Block tectonics of the island of Hawaii from an analysis of basal slip, *Bull. Seismol. Soc. Am., 81*, 491-507, 1991.

Chen, W.-P. J. Nabelek, and M.A. Glennon, Source parameters of the June 26, 1989 Hawaiian earthquake, *Eos, Trans. AGU, 77*, 562, 1990.

Delaney, P.T., A. Miklius, T. Arnadottir, A.T. Okamura, and M.K. Sako, Motion of Kilauea Volcano during sustained eruption from the Puu Oo and Kupaianaha vents, 1983-1991, *J. Geophys. Res., 98*, 17,801-17,820, 1993.

Dvorak, J.J., A.T. Okamura, M.K. and Sako, Summit magma reservoir at Mauna Loa Volcano, Hawaii, *Eos, Trans. AGU, 66*, 851, 1985.

Dvorak, J.J., A.T. Okamura, M. Lisowski, W. Prescott, and J. Svarc, GPS Measurements on the island of Hawaii: 1987 to 1990, *U.S.G.S. Bull. 2092*, 1994.

Endo, E.T., Seismotectonic framework for the southeast flank of Mauna Loa Volcano, Hawaii, Ph.D. thesis, University of Washington, Seattle, 1985.

Gillard, D., M. Wyss, and J.S. Nakata, A seismotectonic model for western Hawaii based on stress tensor inversion from fault plane solutions, *J. Geophys. Res. 97*, 6629-6641, 1992.

Lichten, S.M., Estimation and filtering for high precision Global Positiong System applications, *Manuscripta Geodetica, 15*, 159-176, 1990.

Lipman, P.W., and N.G. Banks, Aa flow dynamics, Mauna Loa 1984, A decade of intrusive and extrusive activity, *U.S.G.S. Prof. Pap. 1350*, 1527-1568, 1987.

Lockwood, J.P., J.J. Dvorak, T.T. English, R.Y. Koyanagi, A.T. Okamura, M.L. Summers, and W.R. Tanigawa, Mauna Loa 1974-1984: A decade of intrusive and extrusive activity, *U.S.G.S. Prof. Pap. 1350*, 537-570, 1987.

Okubo, P.G., A seismological framework for Mauna Loa Volcano, Hawaii, *J. Geophys. Res.*, this volume.

Owen, S., P. Segall, J. Freymuller, A. Miklius, R. Denlinger, T. Arnadottir, M.K. Sako, and R. Burgmann, Rapid deformation of the south flank of Kilauea Volcano, Hawaii, *Science, 267*, 1328-1332, 1995.

Segall, P. and M.V. Matthews, Displacement calculatons from geodetic data and the testing of geophysical deformation models, *J. Geophys. Res., 93*, 14,954-14,966, 1988.

Thurber, C.F., Y. Li, and C. Johnson, Seismic detection of a low-velocity layer beneath the south flank of Mauna Loa, Hawaii, *Geophys. Res. Let., 16*, 649-652, 1989.

Trusdell, F.A., Lava flow hazards and risk assessment on Mauna Loa Volcano, Hawaii, *J. Geophys. Res.*, this volume.

Wyss, M., B. Liang, W.R. Tanigawa, and S. Wu, Comparisons of orientations of stress and strain tensors based on fault plane solutions in Kaoiki, Hawaii, *J. Geophys. Res., 97*, 4769-4790, 1992.

An Empirical Glass-Composition-Based Geothermometer
for Mauna Loa Lavas

Charlene Montierth, A. Dana Johnston, and Katharine V. Cashman

Department of Geological Sciences, University of Oregon, Eugene, Oregon

An empirical glass-composition-based geothermometer has been calibrated at 1 atmosphere for Mauna Loa lavas. A linear relationship exists between the weight percent of MgO in the glass and the temperature from which the samples were quenched, and is described by the equation: $T(°C) = 23.0(MgO) + 1012 (±10°C)$. This equation is valid across the crystallization interval for Mauna Loa lavas and may be applied to any glassy Mauna Loa sample that contains some combination of olivine ± plagioclase or pigeonite + clinopyroxene + plagioclase. It is significantly different from a similar geothermometer calibrated for Kilauean lavas [*Helz and Thornber*, 1987], which is described by the equation: $T(°C) = 20.1(MgO) + 1014 (±10°C)$. The glass geothermometer is useful as both a monitoring tool for current eruptions and particularly as a device for studying past eruptions [e.g., *Cashman et al.*, 1994; *Helz et al.*, in review]. For example, vent lavas from the 1984 eruption at Mauna Loa were reported to have increased in crystallinity from 0 vol% to nearly 30 vol%, with little apparent change in temperature [*Lipman et al.*, 1985]. This increase was inferred to result from degassing-induced undercooling of the lava. Application of the geothermometer to vent spatter samples supports this hypothesis.

INTRODUCTION

Mauna Loa, rising nearly 9 km from the ocean floor, is the largest shield volcano on Earth and is one of the five volcanoes that make up the island of Hawaii. It has been remarkably active in historic time, erupting an average of once every 3.9 years since 1832 [*Klein*, 1982]. The most recent eruption of Mauna Loa occurred in 1984 and lasted 22 days. Lavas were erupted from the summit and along the northeast rift zone, and threatened the westernmost edge of the coastal community of Hilo. Monitoring at Mauna Loa continues currently and activity at the volcano suggests that future eruptions are inevitable [e.g., *Miklius*, 1993].

Monitoring and study of Mauna Loa, as well as of other volcanoes of the Hawaiian islands over the past several decades, have been accomplished employing a variety of disciplines ranging from petrology to hydrology to geophysics. A fundamental piece of information about a volcano is the temperature of its erupted lavas, and how temperature changes as a function of time and location, both during a single eruption and between eruptive episodes. The temperatures of erupted lavas reflect pre-eruption processes such as magma mixing, fractional crystallization and cumulate formation, cooling, and degassing, factors which are intimately related to the physical conditions of magma storage and transport [*Lipman et al.*, 1985]. Changes in lava temperature will also occur during and after eruption as the result of degassing at the vent and cooling on transport, which in turn cause crystallization and affect lava rheology, volume flux and flow morphology [*Sparks and Pinkerton*, 1978; *Lipman and Banks*, 1987; *Rowland and Walker*, 1990]. Thus, determination of eruption temperatures is fundamental to understanding pre-eruptive magmatic processes and the evolution of erupting lavas through time. Moreover, temperature changes with surface transport are crucial for predicting flow behavior and aiding in volcano hazard analysis.

Active fire fountain and flow temperatures are commonly monitored by *in situ* thermocouple and optical pyrometer measurements [e.g., *Lockwood et al.*, 1987; *Lipman and Banks*, 1987]. Thermocouple measurements, however, are sometimes difficult or even dangerous to collect, optical pyrometer measurements are affected both by the surface conditions of the lava and by atmospheric variables

Mauna Loa Revealed: Structure,
Composition, History, and Hazards
Geophysical Monograph 92

Table 1. Bulk Compositions (wt%) of Starting Materials and Comparisons

	TLW67-61[a]		ML-206[b]		1984 ML ave[c]		K175-1-143.8[d]	K167-3-83.8[e]
SiO_2	49.68	(0.13)	51.71	(0.20)	51.54	(0.24)	48.46	49.93
TiO_2	1.75	(0.19)	2.25	(0.35)	2.08	(0.02)	2.34	2.99
Al_2O_3	11.02	(0.06)	13.92	(0.08)	13.65	(0.07)	12.33	13.61
FeO	11.35	(0.18)	10.54	(0.29)	-	-	9.02	9.18
Fe_2O_3	-	-	-	-	12.07	(0.06)	1.61	1.92
MnO	0.18	(0.05)	0.17	(0.03)	0.18	(0.01)	0.16	0.16
MgO	14.66	(0.10)	6.75	(0.08)	6.73	(0.08)	12.15	7.54
CaO	8.53	(0.16)	10.70	(0.16)	10.50	(0.06)	10.76	10.8
Na_2O	1.70	(0.07)	2.35	(0.05)	2.54	(0.11)	1.98	2.46
K_2O	0.37	(0.04)	0.40	(0.03)	0.39	(0.01)	0.49	0.67
P_2O_5	m.d.l.	(0.01)	0.24	(0.07)	0.24	(0.01)	0.28	0.36
Total	99.25		99.05		99.92		99.58	99.62

1-σ standard deviations given in parentheses
n.d., not determined; m.d.l., below minimum detection level
a. Picrite starting material (microprobe analysis of melted sample)
b. Basalt starting material (microprobe analysis of melted sample)
c. Average 1984 Mauna Loa basalt [*Lipman and Banks*, 1987
d and e. Starting materials used by *Helz and Thornber* [1987]

[*Lipman and Banks*, 1987], and both can only be obtained for observed eruptions. Post-eruption geothermometry based on co-existing oxide phases [e.g., *Helz and Thornber*, 1987] has been used to arrive at quenching temperatures of appropriate samples. The use of co-existing oxide pairs, however, is limited to the rather narrow temperature and compositional range over which both phases coexist [*Helz and Thornber*, 1987]. An alternative means of determining the temperature of erupted lavas, particularly one that would not require the direct measurement of erupted lava and thus could be more widely applied in the study of older glass-bearing flows, would be useful.

Helz and Thornber [1987] presented an experimentally-calibrated empirical geothermometer for lavas erupted from Kilauea volcano. They found that the variation of the MgO (and CaO) content of glasses produced experimentally from basaltic and picritic Kilauean starting compositions varied linearly with temperature. The MgO geothermometer is calibrated over a temperature range of 1060°C to 1260°C and is applicable to glassy samples from Kilauea that contain olivine. The uncertainty associated with temperatures calculated using the geothermometer is ± 10°C [*Helz and Thornber*, 1987]. This geothermometer has been used to monitor changes in lava temperature both at the vent [*Helz et al.*, 1991] and during transport within the lava tube system [*Helz*, 1993; *Cashman et al.*, 1994] during the current (1983 - present) Kilauea eruption, and to analyze the thermal histories of other recent eruptions at Kilauea [*Helz et al.*, in press].

Mauna Loa lavas have more SiO_2 and less TiO_2, FeO, K_2O, and P_2O_5 than Kilauean lavas for comparable MgO contents (Table 1) and are therefore more refractory. As a result, they have somewhat higher liquidus temperatures and lower glass MgO contents at a given temperature [*Helz*

et al., 1991]. Consequently, application of the *Helz and Thornber* [1987] geothermometer to Mauna Loa lavas does not yield accurate temperatures. This study reports on an experimentally determined calibration of a new geothermometer similar to that of *Helz and Thornber* [1987], for lavas of Mauna Loa compositions.

EXPERIMENTAL METHOD

Two samples of glass-bearing Mauna Loa lava were chosen as starting compositions for 1 atmosphere melting experiments designed to calibrate the glass geothermometer. The samples were chosen to cover the range in composition of Mauna Loa lavas. The first, ML206, is a basalt spatter sample from the 1984 eruption. The second, TLW67-61, is a picrite that was erupted in 1852. The bulk compositions, obtained from microprobe analysis of completely melted samples, are provided in Table 1.

To evaluate the degree to which TLW67-61 represents a liquid composition, the ferrous iron and magnesium distribution coefficient between the most Fo-rich olivine phenocryst composition (Fo_{88}; Table 2a) and the bulk rock composition (Table 1) were calculated. Ferrous/ferric iron ratios were calculated for the melt using the technique of *Kilinc et al.* [1983]. The distribution coefficient (K_D), is expressed by the relation:

$$K_D = (X_{FeO}^{Ol}/X_{FeO}^{Liq})(X_{MgO}^{Liq}/X_{MgO}^{Ol}) \qquad (1)$$

where X is the mole fraction of the oxide in the phase noted. *Roeder and Emslie* [1970] showed that for basalts in equilibrium with olivine at one atmosphere this value is 0.30 ± 0.03. We arrived at a calculated value of 0.31, slightly higher than the generally accepted value. This indicates that TLW67-61 has accumulated some olivine, as also concluded by *Rhodes* [this volume]. Thus, TLW67-61 does deviate from a true liquid composition by having 14.7 weight percent MgO, compared to the value corrected for accumulation by *Rhodes* [this volume] of 13%. Although this would result in higher liquidus temperatures for our sample than a true primitive Mauna Loa liquid, we do not consider it to be a problem for our purposes because it simply results in the extension of the same olivine-control line to higher temperatures.

The samples were ground into fine powders ($\approx 10 - 50$ μm grain size) and mixed with polyvinyl alcohol to produce a slurry. A small bead (approximately 15-25 mg) of the resulting slurry was suspended from a 0.05 mm Pt-10Rh wire, allowed to dry, and loaded into the hot spot of a Deltech MoSi$_2$ vertical quench furnace. Temperature was monitored using a Pt-Pt10Rh thermocouple that was suspended adjacent to the sample in the furnace hot spot. The thermocouple was calibrated against the melting point of gold. Reported run temperatures are accurate to ±5°C or better.

The oxygen fugacity within the furnace was held at nickel-nickel oxide (NNO) buffer conditions [*Huebner and Sato*, 1970] with a downward flowing gas mixture of H$_2$ and CO$_2$, and was monitored using an SiRO$_2$ solid electrolyte fO_2 sensor using air as a reference gas. The oxygen sensor was calibrated against the iron-wustite buffer [*Myers and Eugster*, 1983] at appropriate temperatures.

Run times varied from 15-142 hours and were long enough to closely approach equilibrium, as discussed below. The run products were quenched by dropping them out of the furnace into water. The beads were then mounted in epoxy and cut and polished for microprobe analysis.

Microprobe analyses of the samples were performed on the CAMECA SX-50 electron probe microanalyzer (EPMA) at the University of Oregon. Glass analyses were performed using a 15 keV accelerating voltage and a 10 nA beam current, with beam diameters ranging from 10-20 μm, depending upon the size of available glassy regions in the samples. Analyses of all major crystalline phases were performed as well, using a 15 keV accelerating voltage and a 20 nA beam current, with a beam diameter of 1 μm.

RESULTS

The temperature, run time, and averaged composition of all major phases for each experimental run used to calibrate the geothermometer are provided in Table 2a. Also given are the 1-σ standard deviations from the means of the individual analyses. The mean counting errors associated with each oxide are given in Table 2b. The standard deviations on the oxides for the experimentally produced phases are close to or less than the mean counting errors, and thus, with the exception of the pyroxenes in the basalt samples,

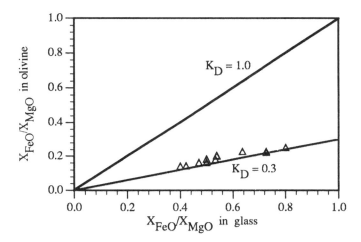

Fig. 1. Fe/Mg distribution coefficients (K$_D$'s) between olivine and melt for olivine-bearing experimental samples.

the phases are homogeneous within the counting statistic errors. The calculated modes and associated errors of major phases in each run product are provided in Table 3. Modes were calculated from averaged microprobe analyses using a least-squares mass-balance technique; analytical uncertainties were propagated using the algorithms developed by *Albarede and Provost* [1977]. All calculations were performed using SIMPLEX3, a FORTRAN program written by M.B.Baker.

Attainment of Equilibrium

As a test for equilibrium in the olivine-bearing experimental run products, distribution coefficients relating the distribution of ferrous iron and magnesium between olivine and melt were calculated using the technique described above. The calculated values for our olivine-bearing run products, illustrated in Figure 1, vary from 0.31 to 0.34, slightly but consistently on the high side of the generally accepted value of 0.30 ± 0.03 [*Roeder and Emslie*, 1970] for basalts in equilibrium with olivine. The slightly elevated values may reflect minor loss of iron from the melt to the Pt-10Rh wires used in the experiments or minor error in the ferrous/ferric iron correction for the melt.

As a further check on attainment of equilibrium, two reversals were run as follows. Basalt samples that were previously completely melted were crystallized at a target temperature. Unmelted samples were simultaneously mounted in the furnace so that both melting and crystallization could occur concurrently. The samples were run at the target temperature for approximately 48 hours in order to ensure that equilibrium was closely approached. Although the textures of the two sets of simultaneously run samples were markedly different (Figures 2a, 2b), the compositions of the glasses are identical within the analytical error (Table

Table 2a. Electron Microprobe Analyses of Experimental Run Products (wt%)

sample	run T (°C)	t(hr)	phase	SiO_2	TiO_2	Al_2O_3	FeO^a	MnO
TLW-18	1310	45	gl	50.0 (0.32)	1.77 (0.18)	11.8 (0.10)	11.4 (0.37)	0.17 (0.03)
			ol	40.1 (0.18)	0.04 (0.03)	n.d.	11.8 (0.02)	0.15 (0.03)
TLW-17	1300	70	gl	50.2 (0.33)	1.92 (0.21)	12.0 (0.15)	11.2 (0.36)	0.17 (0.03)
			ol	39.9 (0.32)	0.02 (0.03)	n.d.	12.2 (0.56)	0.14 (0.03)
TLW-4	1260	47	gl	51.2 (0.47)	1.94 (0.04)	12.9 (0.19)	10.8 (0.20)	0.13 (0.03)
			ol	40.0 (0.22)	0.12 (0.08)	n.d.	13.3 (0.14)	0.17 (0.02)
TLW-6	1240	76	gl	51.0 (0.31)	2.05 (0.21)	13.1 (0.15)	10.4 (0.42)	0.20 (0.08)
			ol	39.6 (0.30)	0.05 (0.04)	n.d.	13.8 (0.12)	0.18 (0.02)
TLW-7	1230	88	gl	52.6 (0.28)	1.89 (0.13)	13.6 (0.08)	9.3 (0.19)	0.16 (0.02)
			ol	40.1 (0.13)	0.01 (0.02)	n.d.	13.6 (0.11)	0.21 (0.02)
TLW-8	1220	95	gl	51.6 (0.26)	2.02 (0.14)	13.4 (0.11)	10.0 (0.19)	0.17 (0.06)
			ol	39.6 (0.18)	0.03 (0.02)	n.d.	14.4 (0.21)	0.18 (0.01)
TLW-9	1210	46	gl	51.6 (0.25)	2.17 (0.23)	13.4 (0.22)	9.9 (0.23)	0.14 (0.03)
			ol	39.2 (0.22)	0.05 (0.04)	n.d.	15.3 (0.20)	0.20 (0.02)
TLW-11	1200	141	gl	51.9 (0.35)	2.16 (0.24)	13.7 (0.03)	9.9 (0.06)	0.20 (0.01)
			ol	39.5 (0.21)	0.05 (0.04)	n.d.	15.4 (0.38)	0.18 (0.02)
TLW-12	1190	43	gl	51.6 (0.51)	2.01 (0.16)	14.0 (0.22)	10.2 (0.31)	0.18 (0.02)
			ol	39.3 (0.21)	0.07 (0.04)	n.d.	16.8 (0.24)	0.20 (0.04)
TLW-13	1180	74	gl	51.6 (0.21)	2.39 (0.39)	13.9 (0.11)	10.4 (0.11)	0.22 (0.01)
			ol	39.2 (0.13)	0.08 (0.05)	n.d.	16.4 (0.30)	0.20 (0.02)
TLW-14	1170	53	gl	52.2 (0.14)	2.37 (0.25)	13.9 (0.18)	10.4 (0.14)	0.19 (0.03)
			ol	38.7 (0.20)	m.d.l.	n.d.	18.2 (0.50)	0.20 (0.02)
			pl	51.6 (0.41)	n.d.	29.6 (0.59)	1.0 (0.26)	n.d.
TLW-16	1160	95	gl	52.9 (0.64)	2.14 (0.16)	14.0 (0.23)	9.5 (0.35)	0.17 (0.03)
			ol	39.2 (0.10)	m.d.l.	n.d.	17.0 (0.19)	0.24 (0.02)
			pl	51.7 (0.13)	n.d.	30.3 (0.03)	0.8 (0.00)	n.d.
ML-28	1150	54	gl	52.3 (0.50)	2.53 (0.27)	13.3 (0.18)	11.4 (0.48)	0.19 (0.05)
			cpx	52.2 (0.53)	0.73 (0.12)	2.5 (0.47)	8.7 (0.49)	0.22 (0.04)
			pl	51.7 (0.26)	n.d.	29.5 (0.58)	1.1 (0.30)	n.d.
ML-16	1140	21	gl	51.6	2.54	12.5	13.5	0.22
			cpx	52.2 (0.64)	0.84 (0.17)	3.11 (0.27)	9.4 (1.28)	0.21 (0.06)
			pig	54.9 (0.46)	0.39 (0.08)	1.4 (0.57)	12.3 (0.77)	0.28 (0.01)
			pl	52.4 (0.47)	n.d.	30.0 (0.21)	1.1 (0.11)	n.d.
ML-29	1130	48	gl	51.6 (0.19)	3.01 (0.22)	12.5 (0.15)	13.2 (0.11)	0.20 (0.03)
			cpx	52.0 (0.90)	0.78 (0.23)	2.5 (0.52)	9.8 (1.00)	0.21 (0.04)
			pig	54.7 (0.51)	0.32 (0.14)	1.5 (0.65)	11.4 (0.72)	0.22 (0.06)
			pl	52.0 (0.25)	n.d.	29.4 (0.37)	1.1 (0.19)	n.d.
ML-30	1120	141	gl	51.3 (0.80)	3.50 (0.44)	11.9 (0.84)	14.6 (1.52)	0.22 (0.04)
			cpx	52.0 (0.40)	0.82 (0.12)	2.7 (0.67)	9.7 (1.62)	0.19 (0.03)
			pig	53.3	0.44	1.0	15.4	0.33
			pl	52.4 (0.46)	n.d.	29.0 (0.45)	1.1 (0.18)	n.d.
ML-31	1110	48	gl	51.4 (0.52)	3.36 (0.07)	12.2 (0.28)	14.4 (0.14)	0.18 (0.08)
			cpx	52.5 (0.26)	0.85 (0.17)	2.4 (0.48)	10.1 (1.24)	0.24 (0.01)
			pig	54.5 (0.51)	0.39 (0.17)	1.5 (0.43)	12.0 (0.44)	0.26 (0.06)
			pl	51.7 (0.43)	n.d.	29.8 (0.30)	1.0 (0.07)	n.d.

Table 2a. (continued)

MgO	CaO	Na$_2$O	K$_2$O	P$_2$O$_5$	NiO	total	n
13.1 (0.09)	9.1 (0.21)	1.74 (0.02)	0.34 (0.04)	0.21 (0.05)	n.d.	99.61	5
48.0 (0.19)	0.25 (0.02)	n.d.	n.d.	n.d.	0.34 (0.04)	100.61	4
12.5 (0.36)	9.3 (0.15)	1.60 (0.18)	0.35 (0.05)	0.20 (0.05)	n.d.	99.37	12
47.3 (0.68)	0.25 (0.05)	n.d.	n.d.	n.d.	0.33 (0.06)	100.04	7
10.9 (0.08)	9.8 (0.16)	1.88 (0.08)	0.38 (0.02)	0.22 (0.09)	n.d.	100.16	6
46.3 (0.28)	0.37 (0.03)	n.d.	n.d.	n.d.	0.23 (0.04)	100.48	4
9.9 (0.14)	9.9 (0.18)	1.81 (0.07)	0.36 (0.03)	0.17 (0.07)	n.d.	98.86	4
45.5 (0.48)	0.33 (0.04)	n.d.	n.d.	n.d.	0.21 (0.07)	99.60	4
9.6 (0.15)	10.4 (0.06)	1.43 (0.07)	0.31 (0.03)	0.15 (0.04)	n.d.	99.42	7
46.9 (0.18)	0.26 (0.02)	n.d.	n.d.	n.d.	0.24 (0.01)	101.30	4
8.9 (0.08)	10.4 (0.23)	1.81 (0.04)	0.38 (0.03)	0.20 (0.06)	n.d.	98.84	6
45.5 (0.14)	0.28 (0.02)	n.d.	n.d.	n.d.	0.27 (0.05)	100.21	3
8.5 (0.06)	10.4 (0.10)	2.09 (0.13)	0.41 (0.01)	0.25 (0.02)	n.d.	98.88	2
44.2 (0.33)	0.24 (0.02)	n.d.	n.d.	n.d.	0.24 (0.03)	99.34	10
8.3 (0.05)	10.5 (0.07)	2.21 (0.05)	0.45 (0.02)	0.26 (0.02)	n.d.	99.49	3
44.3 (0.54)	0.32 (0.04)	n.d.	n.d.	n.d.	0.24 (0.03)	100.00	10
7.6 (0.24)	10.5 (0.15)	2.38 (0.04)	0.46 (0.04)	0.21 (0.07)	n.d.	99.09	3
43.1 (0.24)	0.33 (0.03)	n.d.	n.d.	n.d.	0.30 (0.03)	100.06	5
7.1 (0.03)	10.5 (0.09)	2.36 (0.02)	0.46 (0.02)	0.23 (0.04)	n.d.	99.12	3
43.7 (0.16)	0.33 (0.02)	n.d.	n.d.	n.d.	0.32 (0.03)	100.11	6
6.4 (0.32)	10.5 (0.14)	2.34 (0.08)	0.49 (0.05)	0.23 (0.05)	n.d.	99.04	3
41.7 (0.48)	0.39 (0.04)	n.d.	n.d.	n.d.	0.29 (0.05)	99.46	7
n.d.	13.9 (0.20)	3.23 (0.09)	0.11 (0.02)	n.d.	n.d.	99.53	7
6.8 (0.16)	10.6 (0.20)	2.32 (0.09)	0.49 (0.02)	0.25 (0.10)	n.d.	99.12	5
43.4 (0.20)	0.35 (0.02)	n.d.	n.d.	n.d.	0.24 (0.03)	100.38	5
n.d.	14.1 (0.08)	3.38 (0.08)	0.01 (0.03)	n.d.	n.d.	100.17	2
6.0 (0.14)	10.3 (0.22)	2.35 (0.09)	0.47 (0.04)	0.26 (0.07)	n.d.	99.03	10
18.2 (0.64)	16.7 (1.05)	0.20 (0.02)	0.01 (0.01)	n.d.	n.d.	99.47	9
n.d.	13.9 (0.24)	3.42 (0.20)	0.10 (0.02)	n.d.	n.d.	99.76	7
5.4	9.9	2.37	0.49	0.31	n.d.	98.73	1
19.2 (3.27)	14.9 (5.12)	0.20 (0.06)	0.02 (0.00)	n.d.	n.d.	100.10	2
26.1 (1.12)	5.6 (1.50)	0.07 (0.02)	0.01 (0.01)	n.d.	n.d.	100.96	3
n.d.	13.7 (0.22)	3.50 (0.12)	0.14 (0.03)	n.d.	n.d.	100.81	5
5.3 (0.06)	9.6 (0.08)	2.43 (0.10)	0.57 (0.04)	0.34 (0.08)	n.d.	98.76	6
18.9 (1.27)	14.9 (1.58)	0.17 (0.01)	0.01 (0.01)	n.d.	n.d.	99.26	9
27.3 (1.40)	4.0 (1.81)	0.07 (0.01)	0.10 (0.01)	n.d.	n.d.	99.62	3
n.d.	13.5 (0.25)	3.51 (0.14)	0.10 (0.02)	n.d.	n.d.	99.64	6
4.8 (0.37)	9.23 (0.20)	2.42 (0.04)	0.64 (0.02)	0.37 (0.05)	n.d.	99.11	4
18.3 (0.97)	15.4 (1.31)	0.20 (0.03)	0.01 (0.01)	n.d.	n.d.	99.38	6
23.1	5.24	0.07	0.02	n.d.	n.d.	98.90	1
n.d.	13.3 (0.20)	3.61 (0.13)	0.14 (0.02)	n.d.	n.d.	99.51	7
4.6 (0.07)	9.4 (0.28)	2.38 (0.01)	0.65 (1.20)	0.47 (3.11)	n.d.	99.03	2
17.9 (1.31)	15.4 (0.73)	0.22 (0.08)	0.03 (0.02)	n.d.	n.d.	99.70	5
26.8 (0.75)	3.8 (1.23)	0.09 (0.03)	0.01 (0.00)	n.d.	n.d.	99.39	3
n.d.	13.7 (0.23)	3.44 (0.13)	0.10 (0.02)	n.d.	n.d.	99.75	8

[a]All Fe calculated as FeO; 1-σ standard deviations given in parentheses.
abbreviations: n.d., not determined; m.d.l.,below minimum detection limit; gl, glass; ol, olivine; pl, plagioclase; cpx, clinopyroxene; pig, pigeonite.

Table 2b. Mean Relative % Uncertainty of Microprobe Analyses From Counting Statistics

	SiO$_2$	TiO$_2$	Al$_2$O$_3$	FeO	MnO	MgO	CaO	Na$_2$O	K$_2$O	P$_2$O$_5$	NiO
gl	0.62	9.73	1.41	2.73	24.53	1.49	1.86	3.60	6.95	26.06	n.a.
ol	0.65	25.05	n.a.	1.68	17.52	0.70	7.28	n.a.	n.a.	n.a.	15.41
pl	0.65	0.60	n.a.	1.21	4.68	n.a.	n.a.	n.a.	2.42	12.00	n.a.
pyx	0.28	7.32	1.25	1.09	8.40	0.57	0.71	8.85	31.65	n.a.	n.a.

abbreviations: n.a., not analyzed; gl, glass; ol; olivine; pl, plagioclase; pyx, pyroxene

4), supporting a close approach to equilibrium in our experiments.

Geothermometry

The MgO contents of the experimental glasses are plotted versus temperature in Figure 3, with the least-squares linear regression. The line representing the calibration of the Kilauea geothermometer [*Helz and Thornber*, 1987] is also shown. The weight percent of MgO in the glass varies virtually linearly with temperature over the entire temperature range investigated (1310°-1110°C; 0-40 wt% crystals) and thus provides a useful empirical geothermometer for Mauna Loa lavas. This linear relationship is best fit by the equation (r^2 = 0.99):

$$T(°C) = 23.0(MgO) + 1012 \qquad (2)$$

where MgO is the weight percent MgO in the glass. This is significantly different from the Kilauea MgO geothermometer of *Helz and Thornber* [1987], which is given by the equation T(°C) = 20.1(MgO) + 1014.

The MgO contents of the Mauna Loa glasses vary quite strongly with temperature so the uncertainties in temperature estimates generated by the counting errors inherent in microprobe analysis are small: ±1.5% of the amount present. This yields an uncertainty in the calculated temperature of ±2°- 4°C. In combination with the uncertainty in run temperature, this yields a total uncertainty in calculated temperature of ±7°- 9°C. We therefore assign a conservative error of ±10°C to the geothermometer.

Also included in Figure 3 are the major crystalline phases present over the temperature range investigated. It is important to note that even though olivine was absent in the lowest temperature runs (<1160°C), the MgO-temperature correlation continues to be linear through that range. This is in contrast with both the Kilauea geothermometer, which is linear because of the continued presence of olivine in the low temperature runs, and the data presented by *Helz and Thornber* [1987] for MORBs, which only extend to 1120°C and are linear for all olivine bearing assemblages. The key difference in the crystal assemblage of these low temperature Mauna Loa run products is that they contain two pyroxenes; apparently the relatively MgO-rich pigeonite crystallizes in proportions that enable it to proxy for olivine in controlling the MgO concentration in the lower temperature glasses.

To help visualize the phase relations of our samples we have plotted them in the projection of *Grove* [1993] with

Table 3. Calculated Modes (wt%) of Experimental Run Products

Run	T(°C)	glass	olivine	plagioclase	clinopyroxene	pigeonite
TLW-18	1310	95.2 (0.4)	4.8 (0.3)			
TLW-17	1300	93.2 (0.6)	6.8 (0.6)			
TLW-4	1260	89.2 (0.6)	10.8 (0.3)			
TLW-6	1240	86.3 (0.4)	13.7 (0.3)			
TLW-7	1230	84.5 (0.3)	15.4 (0.3)			
TLW-8	1220	83.9 (0.3)	16.1 (0.2)			
TLW-9	1210	82.6 (0.4)	17.4 (0.2)			
TLW-11	1200	81.2 (0.3)	18.8 (0.2)			
TLW-12	1190	79.8 (0.5)	20.2 (0.4)			
TLW-13	1180	79.5 (0.3)	20.5 (0.2)			
TLW-14	1170	76.1 (1.1)	22.2 (0.6)	1.7 (0.1)		
TLW-16	1160	77.7 (1.6)	21.9 (0.4)	0.4 (1.0)		
ML-28	1150	82.4 (2.4)		8.8 (1.1)	8.9 (1.2)	
ML-16	1140	70.7 (2.2)		9.7 (5.9)	3.5 (5.2)	16.1 (1.1)
ML-29	1130	67.0 (1.6)		11.5 (1.8)	3.8 (1.3)	17.8 (0.9)
ML-30	1120	56.8 (2.7)		11.7 (2.2)	7.3 (2.1)	24.2 (1.8)
ML-31	1110	59.7 (1.6)		11.7 (1.7)	6.9 (1.3)	21.7 (0.9)

1-σ standard deviation in parentheses

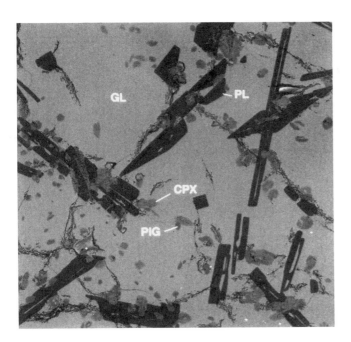

Fig. 2a. Back-scattered electron image of sample ML206-c1a. Sample was fused at 1180°C (the liquidus temperature for this bulk composition is 1170±10°C) for 24 hours, quenched, and then re-run at 1140°C for 48 hours.

Crystallinity

The calculated crystal contents of each of the experimental runs (Table 3) are plotted as a function of temperature in Figure 5 (closed symbols). The error bars on the modes calculated for the basalt samples are large because of the difficulty in obtaining representative analyses of the extremely heterogeneous pyroxenes present in those samples. For comparison, results obtained from the MELTS program [*Ghiorso and Sack*, 1993] for each of the bulk compositions are also plotted in Figure 5 (open symbols). The break in slope results from the change from crystallizing a single phase (olivine) in the picrite at higher temperatures to crystallizing a multi-phase assemblage in the basalt at lower temperatures. The release of large quantities of latent heat during the multiphase crystallization serves to effectively buffer the temperature, resulting in rapid crystallization over a short temperature interval [*Lange et al.*, 1994]. The linear variation of crystallinity with changing temperature provides a quick method for obtaining temperatures from quenched samples, and is particularly sensitive over the temperature range for the basalts. The success of the MELTS program in matching our experimental data suggests that it may be used to derive modifications to our geothermometer for other bulk compositions as long as they do not deviate from the compositions we studied too much. In the event that they do deviate substantially, addi-

the reaction curves of *Grove et al.* [1983] in Figure 4. The correspondence between the projected glasses and the reaction curves is remarkably good considering that the reaction curves were determined using calc-alkaline compositions. For example, most of the glasses produced from the picrite plot in the olivine field and trend away from olivine. The basalt glasses, by contrast, plot parallel and close to the pigeonite-augite-plagioclase cotectic, just beyond the point where olivine reacts out by the reaction olivine + liquid = augite + pigeonite + plagioclase [*Grove et al.*, 1983]. Thus, the geothermometer is applicable to Mauna Loa glasses from either side of the reaction point coexisting with olivine (± minor plagioclase) or clinopyroxene, pigeonite, and plagioclase. We note that some of the glasses coexisting with olivine only project below the olivine-control line in Figure 4. We have traced the problem to slight calibration errors for SiO_2 resulting in low abundancy of SiO_2 (by weight percent) in the analyses, and low quartz values in the projection scheme. Although regrettable, this does not significantly affect the temperature-MgO calibration shown in Figure 3. Also shown is the field occupied by projected Kilauean glass compositions [*Helz et al.*, in press], which trend away from olivine and then turn and follow a path parallel and close to the olivine-augite-plagioclase cotectic, in agreement with the assemblages observed in these samples.

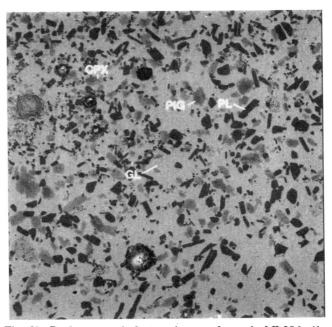

Fig. 2b. Back-scattered electron image of sample ML206-c1b, run at 1140°C for 48 hours, simultaneously with the sample shown in Figure 2a. Phases present are plagioclase (PL), clinopyroxene (CPX), pigeonite (PIG), and glass (GL); horizontal field of view is 0.9 mm.

Table 4. Electron Microprobe Analyses (wt%) of Reversal Glasses

type	ML-c1a crys.	ML-c1b melt.	ML-c2a crys.	ML-c2b melt.
n	5	3	3	3
SiO_2	52.55 (0.20)	51.90 (0.29)	52.98 (0.11)	51.96 (0.14)
TiO_2	3.00 (0.18)	2.73 (0.10)	3.09 (0.06)	2.97 (0.29)
Al_2O_3	12.79 (0.10)	12.62 (0.12)	12.53 (0.11)	12.47 (0.00)
FeO	12.12 (0.33)	12.30 (0.21)	11.64 (0.07)	12.32 (0.51)
MnO	0.18 (0.04)	0.20 (0.07)	0.23 (0.04)	0.24 (0.08)
MgO	5.76 (0.08)	5.68 (0.07)	5.35 (0.03)	5.38 (0.11)
CaO	10.03 (0.09)	10.06 (0.04)	9.95 (0.08)	9.69 (0.11)
Na_2O	2.45 (0.10)	2.39 (0.01)	2.11 (0.05)	2.45 (0.03)
K_2O	0.52 (0.04)	0.52 (0.04)	0.60 (0.07)	0.57 (0.01)
P_2O_5	0.29 (0.05)	0.29 (0.04)	0.37 (0.07)	0.36 (0.12)
Total	99.67	98.69	98.84	98.41
in. melting T(°C)	1180	---	1190	---
final run T(°C)	1140	1140	1130	1130

n: number of points analysed; 1-σ standard deviation given in parentheses
crys: crystallization experiment; melt: melting experiment
Crystallization experiments were fused at the initial melting T, quenched, and re-run at the final run T simultaneously with a second sample of the same starting material that was not previously fused.

tional experimental checks on the MELTS results would be desirable.

DISCUSSION

1984 Eruption of Mauna Loa.

The 1984 eruption of Mauna Loa was one of one of the best documented long-lasting basaltic eruptions in history. The northeast-rift eruption occurred over about three weeks, from March 25 - April 14, and supplied a'a lavas that extended up to 27 km from the source vent within a few days of the start of the eruption [Lipman and Banks, 1987].

Temperature monitoring of the erupted lavas began with the early fountaining events and continued throughout the eruption. Optical pyrometers were used to measure the temperature of early fountains and, in some cases, incandescent standing waves in vigorous flow channels. Most of the flow measurements, however, were made with thermocouples [Lipman and Banks, 1987]. Measured temperatures at the vent were remarkably consistent throughout the eruption, remaining within a range of 1,140 ± 3°C [Lipman et al., 1985]. Downstream cooling was minimal as well; most downstream flow measurements were within the range of 1,135 ± 5°C [Lipman and Banks, 1987].

The crystallinity of the erupted lavas, in marked contrast, varied quite dramatically over the course of the eruption. Three texturally-distinct populations of crystals were identified by Lipman et al. [1985]: (1) sparse resorbed phenocrysts of magnesian olivine up to 5 mm in diameter; (2)

microphenocrysts of augite, plagioclase, and olivine typically 0.05 -1.0 mm in length; and (3) dusty microlites less than 0.01 mm in length. These were inferred to have formed well before the eruption began, in the vent immediately before and during the eruption, and during surface flow, respectively. Olivine phenocrysts comprised < 0.5 vol% of the lava everywhere and the microlite population was absent in vent samples but increased downstream. In contrast, the abundance of microphenocrysts at the vent increased from <0.5 vol% to 30 vol% over the three week course of the eruption [see also Crisp et al., 1994]. This is remarkable in view of the nearly constant measured vent temperatures.

The bulk composition of the erupted lava remained essentially constant throughout the eruption (Table 1). The anhydrous liquidus temperature for this bulk composition was calculated by Lipman, et al. [1985] to be 1,186 ± 15 °C using the technique of Nielsen and Dungan [1983]. Calculations performed with the MELTS program [Ghiorso and Sack, 1993], using the bulk composition of the basalt sample ML-206 (Table 1), placed the liquidus temperature at 1170 ± 10°C ; this liquidus temperature was confirmed experimentally. The early, nearly crystal-free lavas erupted in 1984 therefore appear to have been 20 - 30°C undercooled [Lipman et al., 1985; Lipman and Banks, 1987]. They inferred this undercooling to have been the result of the separation and release of volatiles as the magma migrated to the north-east rift zone eruption site. This volatile loss shifted the relevant liquidus to the anhydrous one, some 30°C above the temperature of the erupted lava.

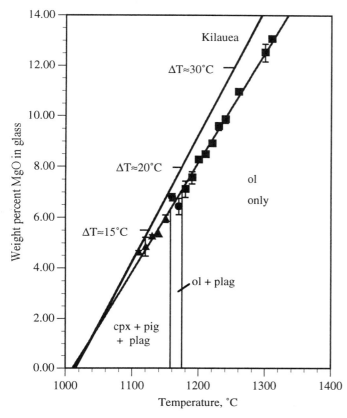

Fig. 3. Weight percent MgO in experimental glasses plotted as a function of temperature. Symbols distinguish crystalline phases present in samples: olivine (squares); olivine + plagioclase (circles); and clinopyroxene + orthopyroxene + plagioclase (triangles). Olivine-bearing samples are picrite starting composition, pyroxene-bearing samples are basalt. The linear least-squares regression through the data given by the equation: T(°C) = 23.0(MgO) + 1012. The Kilauea line [*Helz and Thornber*, 1987] is provided for comparison. Uncertainties are smaller than the symbol unless indicated otherwise.

Crystallization occurred in response to this undercooling and caused the marked increase in the microphenocryst content of the erupted lavas, unaccompanied by any measured temperature change.

The textures of the microphenocrysts are indicative of the rapid crystallization that would accompany undercooling of the magnitude described above [*Lipman et al.*, 1985; *Crisp et al.*, 1994]. Most of the microphenocrysts occur as skeletal, intergrown, and radiating blades and plates and many are elongate, hollow or swallow-tailed . Many of the pyroxenes are actually composed of intergrown clinopyroxene and pigeonite, indicative of disequilibrium growth . These textures predominate, particularly in the early eruptive samples; later samples have textures indicative of equilibrium crystallization.

Application of the Geothermometer

We have applied the glass geothermometer presented above to analyze the temperature evolution of a suite of samples collected during the 1984 Mauna Loa eruption. Field temperature measurements for this eruption [*Lipman and Banks* ;1987] are compared to temperatures calculated using the MgO content of the interstitial glass of the samples (Figure 6). Most of the field measurements were flow temperatures made with thermocouples; a few early fountaining event temperatures were obtained using a Hot Shot optical pyrometer. Many of the microprobe analyses used to calculate the geothermometer temperatures were provided by R. T. Helz [*Helz et al.*, in press] ; the rest were performed at the University of Oregon as described above.

Virtually all of the calculated temperatures using the Mauna Loa geothermometer are higher than the measured field temperatures. Since the systematic variables that affect field temperature measurements largely act to decrease the measured temperatures, this general pattern of divergence is not unreasonable [*Helz et al.*, in press]. However, even though the differences between the field temperatures and the calculated temperatures are small (within the error assigned the geothermometer in most cases) and in the ex-

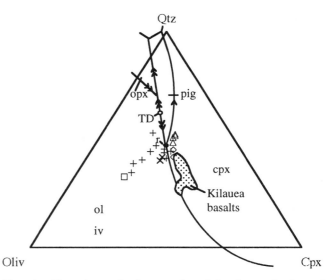

Fig. 4. Experimental glasses plotted in the projection of *Grove* [1993] with reaction curves from *Grove et al.* [1982]. Glasses span a considerable range in composition and pass through the reaction point (r) where olivine is consumed. Symbols distinguish the phase assemblages present in the run products: bulk picrite (squares); olivine-bearing picrite (crosses); olivine+plagioclase-bearing picrite (x's); bulk basalt (diamonds); plagioclase+pigeonite+clinopyroxene-bearing basalt (triangles). The field labeled "Kilauea basalts" was determined by plotting natural Kilauean glasses [*Helz et al.*, in review] in the same projection described above.

pected direction, the pattern of the divergence is not random and requires another explanation.

The compositional geothermometer is based upon cooling-induced crystallization and calibrates the change in the composition of residual melt with the changing temperature of a lava assuming equilibrium cooling and crystallization at 1 atmosphere. If, however, a lava erupts that is significantly undercooled and disequilibrium crystallization occurs with no temperature change, the geothermometer will record a much greater variation in temperature than actually occurs with time. The divergence of calculated and field temperatures should be greatest initially (the field temperatures should be much lower than the calculated temperatures) and should converge as the lava approaches equilibrium at atmospheric pressure. This trend (labeled "expected devolatilization trend") is schematically represented in Figure 6 [*Helz et al.,* in press]. The Mauna Loa data fall along this hypothetical trend. The earliest measurements are the Hot Shot points that were measured during the initial fountaining events that marked the onset of the eruption. These points should fall at the upper end of the trend illustrated in Figure 6, with the greatest divergence of field and calculated temperature, which they do. The lowest temperature measurements are from a'a flows, which are the most completely devolatilized and the most crystalline. The calculated temperatures of these samples converge with the measured field temperatures. Nearly all of the other samples fall between the a'a samples and the fountain samples, with the exception of one pahoehoe flow sample that had an anomalously low field measurement. In sum, this analysis supports the assertion made by *Lipman et al.*

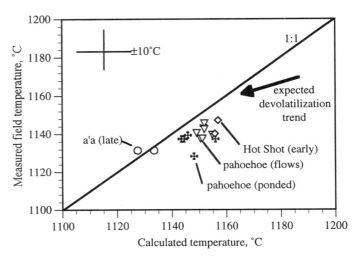

Fig. 6. Measured field temperature [*Lipman and Banks,* 1987] versus temperature calculated using the glass geothermometer for samples from the 1984 eruption of Mauna Loa. All of the field measurements except those noted in the diagram were thermocouple measurements. See text for discussion of the significance of the divergence from 1:1 correspondence.

[1985] that the crystallization of the 1984 Mauna Loa lavas occurred as a result of undercooling of the lava produced by the separation of volatiles during the eruption.

CONCLUSION

The glass geothermometer for Mauna Loa presented here is useful on several counts. It is significantly different from a similar glass geothermometer for Kilauea [*Helz and Thornber,* 1987]. It may be used to obtain the quench temperature of any glass-bearing sample from Mauna Loa, and is of particular value in obtaining quench temperatures for older glass-bearing lavas. It may be used to obtain vent and flow temperatures for eruptions that pre-date modern monitoring and thus to contribute to understanding the thermal history of Mauna Loa. Additionally, the crystallinity of quenched samples may be used directly to rapidly estimate temperature during eruptive episodes. However, as the discussion of the 1984 eruption of Mauna Loa above highlights, care must be taken in interpreting the temperatures obtained using either of these techniques, since changes in crystallinity and therefore in the composition of the residual melt may be driven by devolatilization as well as by changing temperature.

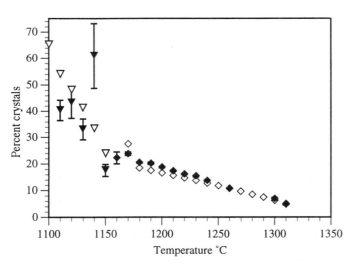

Fig. 5. Total crystallinity plotted as a function of temperature. Closed symbols are the summed calculated crystal abundances (associated errors are error bars) reported in Table 3; open symbols are the total percent crystals determined using the MELTS program [*Ghiorso and Sack,* 1993]. Diamonds are picrite samples and triangles are basalt.

Acknowledgments. This work was funded by support from the Environmental Careers Organization to C. M. and National Science Foundation grant EAR92-18908 to K. V. C. and A. D. J.. The electron beam instruments at the University of Oregon were funded by National Science Foundation grants EAR88-03960 and EAR92-04446 and matching grants from the W. M.

Keck Foundation. Starting materials for the experiments were graciously provided by R. T. Helz and J. P. Lockwood; the authors thank them both. Thanks as well go to R. A. Lange for providing the results of calculations and to M. Mangan and R. T. Helz for suggesting the project and for their continued interest in our results. Finally, this manuscript benefited greatly from the careful, thoughtful reviews and editing provided by R. T. Helz, R. J. Kinzler, and J. M. Rhodes.

REFERENCES

Albarede, F. and A. Provost, Petrological and geochemical mass-balance equations: an   algorithm for least-square fitting and general error analysis, *Comput. Geosci., 3:*, 309-326, 1977.

Cashman, K. V., M. Mangan, and S. Newman, Surface degassing and modifications to vesicle size distributions in active basalt flows, *J. Volcan. Geotherm. Res., 61*, 45-68, 1994.

Crisp, J., K. V. Cashman, J. A. Bonini, S. Hougen, and D. Pieri, Crystallization history of the 1984 Mauna Loa flow, *J. Geophys. Res., 99*, 7177-7198, 1994.

Ghiorso, M. S., MELTS: Software for the thermodynamic analysis of phase equilibria in magmatic systems, *Abstracts With Programs: Geol. Soc. Am. Annual Mtg.*, A-96, 1993.

Grove, T. L., Corrections to expressions for calculating mineral components in "Origin of calc-alkaline series lavas at Medicine Lake volcano by fractionation, assimilation, and mixing" and "Experimental petrology of normal MORB near the Kane Fracture Zone: 22°-25° N, mid-Atlantic ridge", *Contrib. Mineral. Petrol., 114*, 422-424, 1993.

Grove, T. L., D. C. Gerlach, T. W. Sando and M. B. Baker, Origin of calc-alkaline series lavas at Medicine Lake volcano by fractionation, assimilation, and mixing: corrections and clarifications, *Contrib. Mineral. Petrol., 82*, 407-408, 1983.

Helz, R. T., Thermal efficiency of lava tubes at Kilauea volcano, Hawaii, in *Ancient Volcanism and Modern Analogues, IAVCEI abstracts, Canenberra* , pg. 47, 1993.

Helz, R. T. and C. R. Thornber, Geothermometry of Kilauea Iki lava lake, *Bull. Volcanol., 49*, 651-658, 1987.

Helz, R. T., C. Heliker, M. Mangan, K. Hon, C. A. Neal, and L. Simmons, Thermal history of the current Kilauean east rift eruption, *Program and Abstracts: Eos, Transac., Am. Geophys. Union, supp., 72*, 44, 557-558, 1991.

Helz, R. T., N. G. Banks, C. Heliker, C. A. Neal, and E. W. Wolfe, Comparative geothermometry and thermal history of recent Hawaiian eruptions, *J. Geophys. Res.*, in press, 1995.

Huebner, J. S. and M. Sato, The oxygen fugacity-temperature

relationships of manganese oxide and nickel oxide buffers, *Am. Mineral., 55*, 934-952, 1970.

Kilinc, A., I. S. E. Carmichael, M. L. Rivers, and R. O. Sack, The ferric-ferrous ratio of natural silicate liquids equilibrated with air, *Contrib. Mineral. Petrol., 83*, 136-140, 1983.

Klein, F. W., Patterns of historical eruptions at Hawaiian volcanoes, *J. Volcan. Geotherm. Res., 12*, 1-35, 1982.

Lange, R. A., K. V. Cashman, and A. Navrotsky, Direct measurements of latent heat during crystallization of a ugandite and an olivine basalt, *Contrib. Mineral. Petrol., 118*, 169-181, 1994.

Lipman, P. W., N. G. Banks, and J. M. Rhodes, Degassing-induced crystallization of basaltic magma and effects on lava rheology, *Nature ,317*, 604-607, 1985.

Lipman, P. W. and N. G. Banks, Aa flow dynamics, Mauna Loa 1984, edited by R.W.Decker, T.L.Wright, and P.H. Stauffer, *U. S. Geol. Surv. Prof. Pap., 1350*, 1527-1567, 1987.

Lockwood, J. P., J. J. Dvorak, T. T. English, R. Y. Koyangi, A. T. Okamura, M. L. Summers, and W. R. Tanigawa, Mauna Loa 1974-1984: A decade of intrusive and extrusive history, *U. S. Geol. Surv. Prof. Pap., 1350*, 537-570, 1987.

Miklius, A., A. T. Okumura, M. K. Sato, J. Nakata, and J. Dvorak, Current status of geodetic monitoring of Mauna Loa volcano, *Program and Abstracts: Eos, Transac., Am. Geophys. Union supp., 74*, 43, 629, 1993.

Myers J. and H. P. Eugster, The system Fe-Si-O: oxygen buffer calibrations to 1,500 K, *Contrib. Mineral. Petrol., 82*, 75-90, 1983.

Nielsen, R. L. and M. A. Dungan, Low pressure mineral-melt equilibria in natural anhydrous mafic systems, *Contrb. Mineral. Petrol., 84*, 310-326, 1983.

Rhodes, J. M., The 1852 and 1868 Mauna Loa picrite eruptions: clues to parental magma compositions and the magmatic plumbing system, *this volume*.

Roeder, P. L., and R. F. Emslie, Olivine liquid equilibrium, *Contrib. Mineral. Petrol., 29*, 275-289, 1970.

Rowland S. K., and G. P. L. Walker, Pahoehoe and aa in Hawaii: volumetric flow rate controls on the lava structure, *Bull. Volcanol., 52*, 615-628, 1990.

Russell J. K., Crystallization and vesiculation of the 1984 eruption of Mauna Loa, *Jour. Geophys. Res., 92*, B13, 13731-13743, 1987.

Sparks R. S. J., and H. Pinkerton, The effect of degassing on the rheology of basaltic lava. *Nature, 276*, 385-386, 1978.

C. Montierth, A. D. Johnston, and K. V. Cashman, Department of Geological Sciences, University of Oregon, Eugene, OR 97403-1272

Olivine-Rich Submarine Basalts From the Southwest Rift Zone of Mauna Loa Volcano: Implications for Magmatic Processes and Geochemical Evolution

Michael O. Garcia, Thomas P. Hulsebosch

Department of Geology & Geophysics, University of Hawaii, Honolulu, Hawaii

J. Michael Rhodes

University of Massachusetts, Amherst, Massachusetts

The east Ka Lae landslide on the submarine south flank of Mauna Loa exposed a 1.3 km thick section into the interior of its southwest rift zone. We sampled this section in four dredge hauls and four submersible dives and made a multibeam survey of the rift zone. New magnetic data and our observations and bathymetric results indicate that the axis of the southwest rift is two to three kilometers west of the present topographic high. Our submersible observations of old beach deposits and the low sulfur content of pillow-rim glasses indicate that this portion of the southwest rift zone has subsided >400 m. Olivine-rich basalts are extremely abundant along the submarine portion of Mauna Loa's southwest rift zone but their abundance decreases significantly in the upper parts of the two sections examined. This change probably occurred ~ 60 ka when Mauna Loa's eruption rate slowed and was unable to keep up with its subsidence rate. The dense magmas for these olivine-rich basalts were probably intruded into the deeper portions of the rift zones and erupted from its distal regions during periods of high magma supply. The preferential eruption of olivine-rich lavas on the flanks of Mauna Loa and other Hawaiian volcanoes is a strong indication that a density filter operates within these volcanoes. These lavas contain abundant euhedral, undeformed olivine with high forsterite contents (typically 90%). Some of these olivines grew in magmas with 17.5 wt% MgO at temperatures of 1415°C, indicating that Hawaiian tholeiitic magmas are some of the most mafic and hottest magmas erupted during the Cenozoic. All of the submarine lavas have major element contents typical of Mauna Loa, but unlike its subaerial lavas, some of the submarine lavas have trace element and isotope ratios that overlap with those of Kilauea lavas. Thus, the source for Mauna Loa contained a Kilauea-like component that has been consumed during the last hundred thousand years, but the melt extraction conditions that have controlled the major elements in Mauna Loa lavas has remained relatively constant.

INTRODUCTION

Mauna Loa is the largest volcano on Earth. It stands ~8.5 km above the seafloor and has a keel under the summit of the volcano that extends about 5 km below the seafloor [*Hill and Zucca, 1987*]. The volume of Mauna Loa above the sea floor has been estimated at ~42,000 km³ [*Bargar and Jackson,*

Mauna Loa Revealed: Structure,
Composition, History, and Hazards
Geophysical Monograph 92
Copyright 1995 by the American Geophysical Union

1974]. *Lipman* [*this volume*] suggests that the Bargar and Jackson [*1974*] volume estimate is too low by 23,000 km³ because the size of the adjacent volcano, Kilauea, was overestimated. Our estimate for the keel volume is ~40,000 km³ based on limited seismic refraction data [*Hill and Zucca, 1987*]. This agrees with *Moore's* [*1987*] suggestion that ~50% of the volume of Hawaiian volcanoes is beneath the level of the sea floor. Thus, Mauna Loa may have a total volume of ~105,000 km³. *Lipman* [*this volume*] estimates the volcano's volume at ~80,000 km³. The differences in these volumes are probably within the uncertainties of our estimates for the locations of the volcano's base and it boundaries with its three neighbor volcanoes.

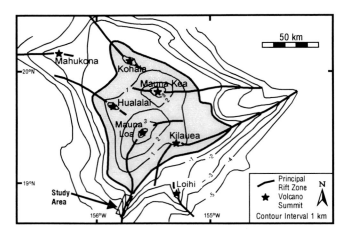

Fig. 1. Topographic map of the island of Hawaii showing the location of the summit and rift zones for the seven volcanoes that comprise the island [after *Lonsdale, 1989*].

The submarine portion of Mauna Loa Volcano has received limited geologic study because a detailed bathymetric map was not available until recently [*Chadwick et al., 1993; Moore and Chadwick, this volume*]. The previous studies of submarine Mauna Loa have focused on its shallow portions in order to examine the growth rate and subsidence history of the volcano [*Moore and Clague, 1987; Moore et al., 1990*]; features from the 1877 submarine eruption [*Fornari et al., 1980*]; and the morphology of the southwest rift zone [*Fornari et al., 1979*]. The giant landslides on the submarine flanks of Mauna Loa were examined by *Lipman et al.* [*1988*], *Moore et al.* [*1989*], and *Moore et al.* [*1995*]. Two reconnaissance dredging programs along the southwest rift zone recovered basalts that were included in broader studies of the submarine lavas of Hawaiian volcanoes [*Moore, 1966; Garcia et al., 1989*]. In contrast, the subaerial portion of the volcano is relatively well studied as a result of past and current geologic mapping by U.S. Geological Survey [*e.g., Lipman and Swenson, 1984; Lockwood et al., 1988*] and geochemical studies [*e.g., Wright, 1971; Rhodes, 1983; 1988; Kurz and Kammer, 1991; Rhodes and Hart, this volume*].

This study focused on the intermediate-depth portion of Mauna Loa's submarine south west rift zone (550-2000 m below sea level; mbsl). We present here the results of a new Seabeam survey along the rift zone and our petrographic, glass, and mineral chemistry studies of lavas collected from four submersible dives and four dredge hauls on the deeply dissected west flank of the rift zone. Our results show that Mauna Loa's submarine flanks are dominated by olivine-rich basalts. These basalts contain high forsterite olivine phenocrysts (up to 91.3% forsterite; Fo) which grew in magmas with up to 17.5 wt% MgO. The major element

contents of glasses from these basalts are identical to modern Mauna Loa lavas, indicating that there has been little or no change in the volcano's partial melting processes for perhaps a few hundred thousand years. Some of these rocks, however, have distinct trace element ratios that require a different source, one geochemically similar to that which is currently supplying Kilauea Volcano. This source component was apparently exhausted during the last hundred thousand years and may have had a lower melting temperature.

Geologic Setting

Mauna Loa has two prominent rift zones, one trending northeast and the other southwest (Figure 1). The southwest rift zone extends ~65 km subaerially and ~35 km below sea level to a depth of ~5000 m. It has >8 km of relief along which it erupts lava. There is a distinct bend in the rift about 2400 m above sea level where its trend becomes southward. Below this bend, the rift is marked by the Kahuku scarp. The subaerial portion of this scarp has up to 200 m of relief and is buried up rift by younger lavas. The submarine extension of the Kahuku scarp has up to 1800 m of relief. Unlike other subaerial scarps on Mauna Loa and its neighbor Kilauea, the Kahuku scarp is not draped by younger lavas [*Fornari et al., 1979; Moore et al., 1990*]. Thus, although Mauna Loa is still active (the most recent eruption was in 1984), apparently little Holocene volcanic activity has occurred along the submarine portion of the rift. This makes this portion of the Kahuku scarp a superb area to sample the deepest exposures of Mauna Loa's interior, and it provides the best available opportunity to evaluate the long-term magmatic history of the volcano.

The ages of the lavas in the submarine section of Mauna Loa's southwest rift are poorly known. The thickness of the section (~1.3 km) and its distance from the summit (where volcanic activity is more frequent; *Lockwood and Lipman, 1987*) indicate that the submarine section probably contains lavas older than the oldest exposed subaerial lavas on Mauna Loa (the 100 to 200 ka Ninole basalts; *Lipman et al., 1990*). Attempts to date the submarine lavas were unsuccessful because of the low radiogenic argon content of the lavas. The ages obtained range from 0.12 to 1.5 Ma with no correlation with stratigraphic position [*Lipman, this volume*]. Our best estimate for the age of the submarine lavas is 100 to 300 ka based on extrapolation of the subaerial lava ages. *Lipman's* [*this volume*] estimate is 200-350 ka.

Sampling and Geological Observations

Four dredge hauls were made on Mauna Loa (Figure 2) during a 1982 R/V Kana Keoki cruise to sample the submarine rifts of all the volcanoes that form the island of Hawaii.

These dredge hauls were made up the steep western face of the southwest rift zone at depths between 1650-2600 mbsl for hauls 1, 2, and 4 and 1000-2200 mbsl for dredge haul 3. A variety of petrographically distinct rocks was recovered from each dredge haul. A representative sample of each distinct rock type from each dredge haul was selected for petrographic and geochemical analysis. The petrography, glass, and volatile chemistry of these samples was presented by *Garcia et al.* [*1989*]. Whole-rock analyses for some of these lavas were reported by *Gurriet* [*1988*]. These samples are coded ML followed by the dredge haul and sample numbers.

Our 1991 submersible sampling program was designed to collect from two sections up the face of the Kahuku scarp and to sample along the topographic axis of the rift zone (Figure 2). We used the Hawaii Undersea Research Lab's Pisces V submersible which has a depth limit of 2000 m. Three dives up the scarp (numbered 182-184) produced 36 samples collected in situ from Mauna Loa's thickest stratigraphic section (~1.3 km; the thickest subaerial section is ~600 m; *Lipman et al., 1990*). Eleven samples were collected during dive 185 along the crest of the ridge between 1505 and 1825 mbsl. Samples were collected from depths about 20 to 30 m apart whenever possible. Each sample was treated as a separate flow unit. One dike was sampled.

The geologic features we observed from the submersible's portholes and from reviewing the videotapes taken during the dives are summarized in Figure 3. Dive 182 started on a flat bench just west of the cliff face at a depth of 1870 mbsl. The steep face started at ~1800 mbsl. The base of the cliff is mantled to a depth of 1430 mbsl with mostly columnar-jointed talus derived from dikes in the cliff face [*as noted by Fornari et al., 1979*]. Above 1430 mbsl, outcrops of pillow lavas are abundant in near-vertical faces with intervening areas of moderately sloping talus. The pillow lavas are cut locally by numerous, steeply dipping dikes, which appear to be buttressing the steep cliffs (45° to vertical). The dikes are 1 to 3 m wide and trend roughly north-south. Dive 183 continued the traverse up the steep face of the ridge starting at 1200 mbsl (where dive 182 ended) and ending at 40 mbsl. We observed more pillow lavas cut by dikes during this dive, except at 425 mbsl where boulders and carbonate sand were encountered, and at 165-190 mbsl where a dead coral reef was found (Figure 3). The shallowest pillow lavas were observed at 300 mbsl.

Dive 184 started about 3.5 km south of the area traversed during dives 182 and 183, beginning at a depth of 1790 mbsl and ascending to the ridge crest at 550 mbsl. On this traverse we found the same features noted in dives 182 and 183, except that 0.5 to 1 m thick tephra layers were found at 745 and 775 mbsl interbedded with pillow lavas, and no

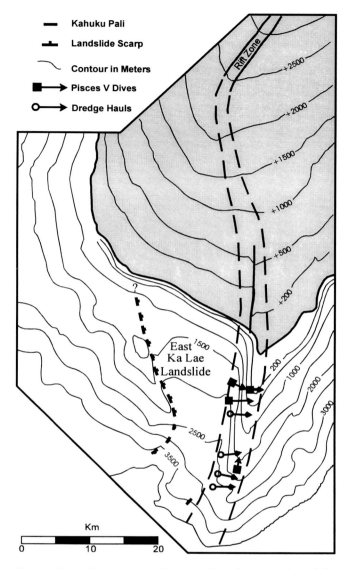

Fig. 2. Generalized topographic map of the lower portion of the southwest rift zone [after *Chadwick et al, 1993*] showing the location of dredge hauls, Pisces V dives, the Kahuku Fault, and the east Ka Lae landslide (boundaries shown by strike and dip symbols). Dashed parallel lines show the proposed location of Mauna Loa's southwest rift zone.

boulders or coral were observed. The top of the ridge at this location is a bench underlain by a ~ 30 m thick massive flow. Pillow lavas were collected during this dive from 1790 to 550 mbsl.

Dive 185 was along the crest of the ridge between 1825 and 1500 mbsl. The crest consisted of moderately dipping (10-20°) segments with rare pillow cones, separated by cliffs. No signs of recent volcanic or hydrothermal activity were observed.

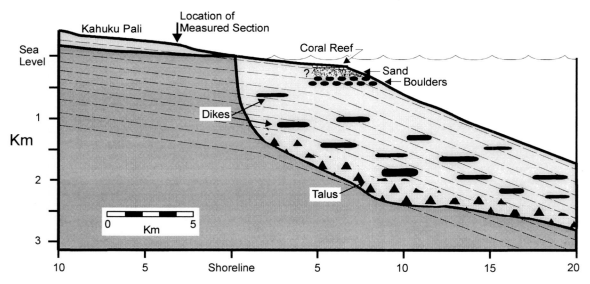

Fig. 3. Diagrammatic representation of the geology of the cliff face along the southwest rift zone (light gray area) based on observations from submersible dives. The dip of the lavas (dashed lines) increases just south of the dead coral reef, reflecting the steep submarine slope of Hawaiian volcanoes. Note the old shoreline deposits (sand and rounded boulders) below the reef. Dikes (black bands) are common to abundant in the cliff face; they trend nearly parallel to the cliff face and outcrop as walls. Vertical exaggeration is about 4x.

BATHYMETRY

The bathymetry of Mauna Loa's southwest rift was not well known prior to our study. Single-beam profiles had been made of the rift [*Fornari et al., 1979; Moore et al., 1990*] and a general (100 m contour) bathymetric map was constructed by *Moore et al.* [*1990*]. A more detailed map by *Moore and Clague* [*1992*] combined multibeam surveys west of the rift zone (longitude 155°45') with single-beam surveys across the rift. We had the opportunity in 1992 to make a multibeam survey of the rift zone with the U. S. Navy ship Laney Chouset. This new bathymetric map (Plate 1) combines our results with those of two previous NOAA multibeam surveys on either side of the rift. Our survey coverage is virtually continuous on the flanks of the rift and below 1200 mbsl. The coverage is 50-75% for depths between 500 -1200 mbsl and <50% above 500 mbsl along the rift.

The most striking feature of the new bathymetric map is the steep scarp along the west flank of the rift zone (Plate 1). This scarp extends onshore, where it is known as the Kahuku pali and only has ~10% of the relief of the submarine scarp. This scarp may have been created by the east Ka Lae landslide, rather than by faulting as proposed by *Lipman and Swenson* [*1984*] (Figure 2). The landslide is thought to have formed in one catastrophic event [*Moore et al., 1989*]. The smaller subaerial expression of the scarp may be a result of subsequent subaerial volcanism. The western detachment

surface of the landslide may be defined by a prominent ridge, which was interpreted by *Lipman* [*1980*] to be a down-dropped block similar to those on the south flank of Kilauea. In the new bathymetry, however, there is no obvious bathymetric expression of the block slumping. We interpret the ridge as an erosional remnant of the landslide. The base of the slide may have several steps at 1300-1550, 2050-2400, and 2480-2590 mbsl (Plate 1). Down slope from the deepest step there is a debris chute that turns to the south at 4200 mbsl.

Another significant feature on Plate 1 is the continuation of the subaerial slope of Mauna Loa on its south flank to a depth of ~160 mbsl. At this depth, there is a break in slope that is thought to mark the old subaerial/ submarine boundary when the volcano's growth rate was equal to or greater than its subsidence rate [*see Moore and Campbell, 1987*]. If this is an old shoreline, then Mauna Loa has subsided at least 160 m. *Moore and Clague* [*1992*] identified a flat crest on the rift down to at least 500 mbsl and perhaps as deep as 750 mbsl, based on single-beam profiles across the rift. They used these depths and the average subsidence rate of the volcano (2.6 mm/yr) to suggest that there had been little volcanic activity along this portion of the rift during the last 170 to 270 ky. The coastal portion of the rift, however, underwent extensive volcanism until probably just before deposition of the Pahala Ash, which caps the lower rift zone and is at least 31 ka [*Lipman and Swenson, 1984*]. We saw no soil horizons separating any of the flows in the ~150-m-thick section we measured up the Kahuku Pali, so there probably were no significant time breaks in volcanic activity prior to deposition of the ash. Furthermore, we observed pillow lavas as shallow as 300 mbsl during dive 183. If our revised depth for the slope break is used (160 m), then the age when volcanic activity may have slowed relative to subsidence would be ~60 ka. The absence of the old shoreline terrace on the southwest side of Mauna Loa and the ruggedness of the landslide scar indicate that the landslide is young and probably <60 ka.

One surprising feature of the map is the paucity of cones along the ridge that is considered the axis of Mauna Loa's southwest rift zone [*Lipman and Swenson, 1984*]. This impression was confirmed by examining a 10 m contour map of the ridge and the videotapes taken during dive 185 along its crest. In contrast, cones are common in a 3-km-wide zone along the submarine portion of Kilauea's east rift zone [*Lonsdale, 1989; Clague et al., in press*]. One explanation for the scarcity of cones along Mauna Loa's submarine ridge is that it is not the axis of the southwest rift zone but instead a remnant of the east flank of the rift zone left behind by the east Ka Lae landslide. This interpretation is supported by the abundance of dikes that were observed in the submarine cliff face during our submersible dives and those of *Fornari et al.* [*1979*] and new magnetic data for the rift zone, which show that the axis of the magnetic dipole is offset to the west of the ridge near the base of the cliff [*Smith et al., in press*]. Along Kilauea's submarine east rift zone, the dipole is centered along the topographic axis of the rift zone [*Smith et al., in press*]. Therefore, the center of Mauna Loa's southwest rift zone has been displaced two to three kilometers west of the present topographic high (Figure 2). The westward shift in the axis of the submarine portion of the southwest rift would make the bend in the rift at about 700 m above sea level less severe and would not require an eastward offset near the upper end of the Kahuku Pali as previously thought [*Lipman and Swenson, 1984*].

PETROGRAPHY

The lavas from the submarine southwest rift zone vary dramatically in mineralogy (Table 1) ranging from weakly phyric (<2 vol.% phenocrysts) to extremely olivine phyric (~47 vol.%). Most of the samples from the cliff face contain only olivine phenocrysts. The other ~40% contain <1 to 4 vol.% small phenocrysts of plagioclase and augite, in addition to variable amounts of olivine. In contrast, all of the dive 185 samples contain rare to common plagioclase and augite phenocrysts (Table 1). Only a few of the dredge haul lavas contain only olivine phenocrysts; most also have plagioclase and augite phenocrysts [*Garcia et al., 1989*]. Among our suite of submarine Mauna Loa lavas, about 65%

of these samples are olivine-rich (>10% olivine phenocrysts); 44% are picritic (>15 vol.% phenocrysts). The abundance of olivine in these submarine Mauna Loa lavas contrasts with previous studies of subaerial Mauna Loa lavas, which found that picrites constitute 7-11% of historical flows and ~15-20% of prehistoric flows [*Macdonald, 1949; Lockwood and Lipman, 1987*]. Hypersthene is common in subaerially erupted Mauna Loa lavas [*Macdonald, 1949*], but is rare in the submarine lavas.

About 90% of the olivine phenocrysts are undeformed euhedra. Some of these grains share crystal faces forming multicrystal aggregates, many with pockets of glass between crystals. Some euhedral grains with glass inclusions have weakly developed subgrain boundaries or kink bands. Traditionally, kink-banded olivines are thought to have been deformed in the solid state [*e.g., Raleigh, 1968*]. Glass inclusions, however, are unlikely to be preserved in such olivines. *Wilkinson and Hensel* [*1988*] suggested that olivine phenocrysts can be deformed in a magma during its ascent through narrow cracks. Some of the olivine-rich lavas contain <1 to 3 vol.% strongly deformed, anhedral grains and aggregates without glass inclusions. These olivines are probably xenocrysts or remnants of disaggregated xenoliths (see Table 1).

Spinel commonly occurs as inclusions in olivine, although it also is present as microphenocrysts in some of the more mafic samples. These more mafic samples contain both brown and opaque spinels. The other lavas contain only opaque spinel. Both types of spinel are always euhedral.

The plagioclase and pyroxene phenocrysts and microphenocrysts are generally subhedral to euhedral, although rare, round or ragged xenocrysts are present in some lavas. Many of the augites display hour-glass zoning; some have weak concentric zoning. The orthopyroxenes are unzoned optically and are somewhat elongate. The plagioclase crystals show no obvious petrographic signs of compositional zoning.

Lavas with obvious disequilibrium textures (e.g., resorbed grain boundaries, disrupted zoning, abundant inclusions) are identified as "mixed magmas" in Table 1. The olivine-rich lavas rarely show these features, but they are common among the samples collected along the ridge axis and from the dredge hauls.

GEOCHEMISTRY

Analytical Methods

Mineral and glass compositions were measured using the University of Hawaii, five-spectrometer Cameca SX-50 electron microprobe. Natural mineral and glass standards were used for calibration and a PAP- ZAF matrix correction

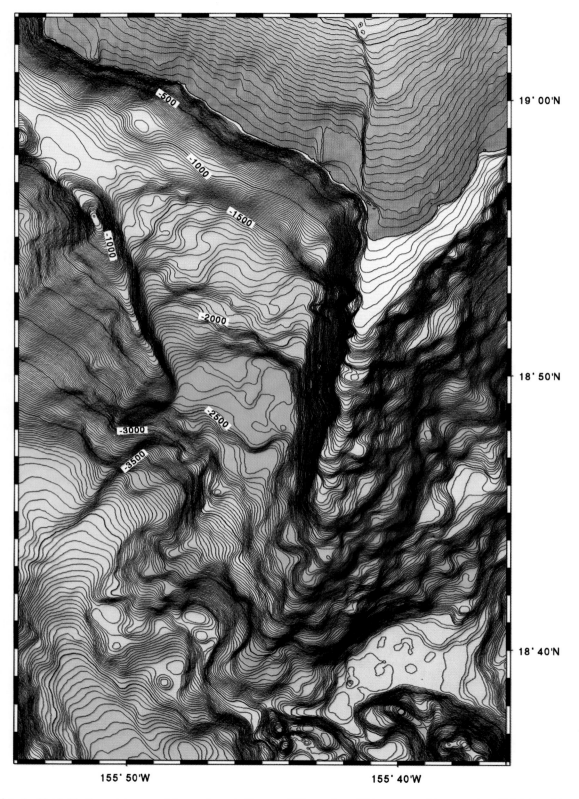

Plate 1. Shaded bathymetric map of the southwest rift zone of Mauna Loa. This map combines previous NOAA multibeam survey data from both sides of the rift with new multibeam data along the rift. The illumination angle is from the north and the contour interval is 20 m with every 500 m interval below sea level shown by a heavier line.

TABLE 1. Petrography[1] and Water Depth of Collection (mbsl) for Submersible Collected Lavas
from Mauna Loa's Southwest Rift Zone

Sample	Depth Collected	Olivine ph	mph	Plagioclase ph	mph	Cpx ph	mph	Opaques mph	Matrix	Glassy Margin	Mixed Magma; Comments
182-1	1420	33.4	4.2	<0.1	0.2	0.6	0.2	0.2	61.2	No	No; oliv xenos
182-2	1405	11.8	7.6	0.6	0.6	-	0.6	0.2	78.6	No	No
182-3	1395	11.0	8.4	0.2	1.2	<0.1	1.4	<0.1	77.8	Yes	Mixed; gabbro clots
182-4	1375	33.0	22.2	-	-	-	2.4	0.2	42.2	No	No; oliv xenos
182-5	1335	1.8	1.6	-	<0.1	-	0.4	<0.1	96.2	Yes	No; oliv xenos
182-6	1310	30.0	20.4	-	-	-	-	0.2	49.4	Yes	No; oliv xenos
182-7	1265	44.0	20.4	-	-	-	-	0.2	35.4	Yes	No
182-8	1235	29.0	10.8	-	-	-	-	0.2	60.0	Yes	No; oliv xenos
183-1	1200	47.2	12.8	-	-	-	-	0.4	39.6	No	No; oliv xenos
183-2	1135	30.2	8.4	-	1.4	-	-	<0.2	59.0	No	No; oliv xenos
183-3	1100	28.2	5.6	-	-	-	-	<0.1	66.2	Yes	No; oliv xenos
183-4	1075	-	<0.1	-	2.4	-	<0.1	-	97.6	No	No
183-5	1030	20.4	5.2	-	-	-	<0.1	0.2	74.2	No	No
183-6	1015	8.2	3.0	-	2.0	-	<0.1	<0.1	86.8	No	No
183-7	960	10.2	4.0	-	0.6	-	-	<0.1	85.2	Yes	No; oliv xenos
183-8	920	21.1	4.4	-	-	-	-	<0.1	74.2	No	No; oliv xenos
183-9	875	0.6	0.2	0.2	0.8	<0.1	0.6	-	97.6	No	No; gabbro clots
183-10	830	4.6	1.4	0.4	0.4	0.2	0.2	-	92.8	No	No?; oliv xenos
183-11	790	13.4	6.6	-	<0.1	-	<0.1	<0.1	80.0	Yes	No; oliv xenos
183-12	750	11.6	8.0	-	-	-	-	<0.1	80.4	Yes	No; oliv xenos
183-13	695	13.6	6.8	0.2	2.4	-	1.4	<0.1	75.6	No	Mixed; oliv xenos
183-14	650	11.8	6.8	1.4	3.0	0.4	2.0	<0.1	74.6	Yes	Mixed
183-15	630	15.8	6.6	-	-	-	<0.1	<0.1	77.6	Yes	No; oliv xenos
184-1	1790	9.6	2.0	2.4	1.6	1.6	1.0	-	81.8	No	Mixed
184-2	1735	12.6	1.6	-	-	-	-	-	85.8	Yes	No; oliv xenos
184-3	1680	32	2.8	-	-	-	-	0.2	65.0	No	No; oliv xenos
184-4*	1610	1.2	<0.1	1.2	12.6	2.0	0.2	-	82.8	No	No
184-5	1535	18.8	3.0	-	0.4	-	-	0.4	77.4	Yes	No; oliv xenos
184-6	1425	21.0	1.0	-	-	-	-	-	78.0	Yes	No
184-7	1240	21.0	8.2	-	-	-	0.2	<0.1	70.6	No	No
184-8	1020	0.8	<0.1	0.6	1.4	1.0	2.0	-	94.6	Yes	No; gabbro clots
184-9	775	19.6	2.8	0.4	1.8	0.2	1.2	<0.1	74.0	Yes	Mixed
184-11	745	5.4	0.4	1.7	9.4	2.2	2.0	-	79	Yes	Mixed; gabbro clots
184-12	575	15.4	1.4	1.6	3.2	1.4	2.4	0.2	74.0	Yes	Mixed
184-13	550	15.2	5.2	2.6	5.6	1.8	4.0	-	75.6	No	Mixed
185-1	1825	6.6	3.0	<0.1	1.8	0.4	2.6	-	85.6	Yes	No?; gabbro clots
185-2	1795	5.6	2.8	0.6	1.2	<0.1	1.2	<0.1	88.6	Yes	Mixed
185-3	1735	10.0	4.6	0.6	0.8	0.2	2.0	<0.1	81.8	Yes	Mixed
185-4	1700	5.8	0.6	3.4	2.2	1.8	2.4	<0.1	83.8	Yes	Mixed
185-5	1670	9.6	1.6	2.0	2.4	2.4	1.6	<0.1	81	No	Mixed
185-6	1655	7.8	1.0	1.6	1.8	3.2	1.8	<0.1	82.6	Yes	Mixed; gabbro clots
185-7	1620	1.6	1.2	1.8	1.4	3.4	2.0	<0.1	82.6	Yes	Mixed; gabbro clots
185-8	1560	9.0	0.4	5.4	2.6	5.4	3.2	<0.1	74.2	No	No
185-9A	1565	15.2	2.8	2.8	2.0	4.4	2.2	<0.1	70.6	No	Mixed; oliv xenos
185-9B	1565	5.0	0.4	6.0	3.4	8.0	2.4	<0.1	74.8	No	Mixed
185-10	1515	4.6	1.2	0.6	1.0	1.6	0.6	<0.1	90.4	Yes	No

[1]Modes are based on 500 points counted/sample.
*Dike.

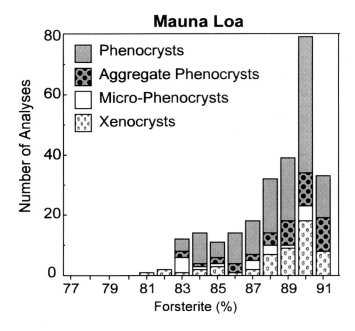

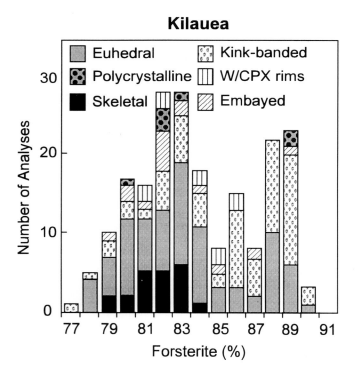

Fig. 4. Histogram of olivine core compositions (% forsterite) for submarine Mauna Loa lavas based on analysis of 255 crystals. Most of the grains contain high forsterite (89-91%). In contrast, Kilauea submarine lavas have lower average forsterite contents and more xenocrysts [*Clague and Denlinger, 1994*].

procedure was applied to all analyses. For olivines, spinels, and pyroxenes, a focused 20 nA beam was used; peak counting times were 40-80 s for major elements and 110 s for minor elements. A defocused (10 μm), 15 nA beam was used for plagioclase analyses. For glasses, a defocused (20 μm), 10 nA beam was used; counting times were 60 s, except for Na (40 s), S (300 s), K and P (100 s). Off-peak backgrounds were measured for half the peak counting times. The reported glass analyses are an average of 5 spot analyses; mineral analyses are an average of 3 spot analyses. Relative analytical error, based on repeated analysis of the Smithsonian standards A99 glass, hypersthene, Lake County plagioclase and San Carlos olivine, is <1% for major elements, <5% for minor elements and <2% for S in glass.

ICP-MS analyses of glass were made at Washington State University using methods described by *Garcia et al.* [*1993*]. About 100 mg of carefully hand-picked glass (i.e., free of minerals and alteration) were used for each analysis. For these analyses, accuracy and precision were <1 to 2 % for all elements based on repeat analyses of several samples including a Hawaiian basalt standard.

Mineral Chemistry

Olivine. Phenocrysts and microphenocrysts from representative lavas were analyzed to determine their compositions and zoning patterns (Table 2). They show a wide range of core compositions (82.0-91.3 Fo%) with no apparent correlation with crystal type (euhedral, undeformed

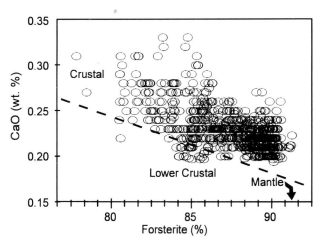

Fig. 5. CaO content (wt%) vs. percent forsterite in the cores of olivine phenocrysts. Note the modest increase in CaO content with decreasing Fo%, which is consistent with the experimental work of *Jurewicz and Watson* [*1988*]. All of the olivine is of crustal origin, based on their moderate CaO contents. The dashed line is an estimate of the boundary between olivines crystallized at upper and lower crustal depths. The highest forsterite crystals are of shallow origin.

vs. kinked; see Figure 4). Within individual lavas, the range may be small to large. Some of this variation may be an artifact of sectioning (see *Pearce, 1984*, for summary of this problem). Nevertheless, a histogram of the core compositions from the submarine lavas shows that they are forsterite-rich, with Fo 89-90 as the most common composition (Figure 4). In contrast, historically erupted Mauna Loa picrites have olivines with lower forsterite contents (87-89% Fo; *Wilkinson and Hensel, 1988; Rhodes, this volume*) and those from submarine Kilauea lavas have a bimodal variation with peaks at 82-83 and 88-89% Fo, with a strong preference for kink banding in the higher forsterite crystals (Figure 4).

The CaO contents of the Mauna Loa olivines are moderate and increase with decreasing forsterite content (Figure 5) as predicted from the experiments of *Jurewicz and Watson* [*1988*]. The CaO content in these olivines is consistent with equilibrium at shallow crustal pressures [*Stormer, 1973*].

The zoning in the olivines is normal in all analyzed grains (Figure 6) except for a few grains that have no apparent compositional zoning or modest reverse zoning (~1-2% Fo). The rare olivines with reverse zoning have lower forsterite contents (<84%). In contrast, about 20% of the olivine ana-

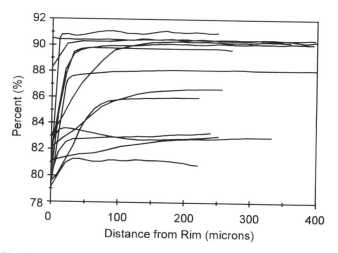

Fig. 6.   Core to rim compositional zoning profiles for olivine phenocrysts from Mauna Loa submarine basalts. Note that all but one of the profiles show normal zoning with a narrow, strongly zoned rim. The crystals with wider rim zoning may have partially re-equilibrated in the magma or they may not have been sectioned through their core. One grain has no zoning and another has reverse zoning.

TABLE 2. Representative Microprobe Analyses of Olivine from Mauna Loa Submarine Lavas

Sample	SiO_2	MgO	CaO	FeO	NiO	Total	Fo
183-7	39.53	43.19	0.23	16.38	0.24	99.80	82.4
182-2	39.10	43.48	0.24	16.09	0.26	99.34	82.8
182-3	39.45	44.46	0.25	15.47	0.24	100.10	83.6
182-8	39.92	45.34	0.25	14.32	0.26	100.03	84.9
182-4	40.12	45.31	0.24	13.65	0.31	99.82	85.5
182-8	40.34	45.84	0.24	13.47	0.31	100.08	85.8
182-6	40.03	46.71	0.22	12.51	0.34	99.99	86.9
183-1	40.57	46.87	0.22	12.13	0.31	100.09	87.3
182-2	40.14	47.02	0.22	11.46	0.30	99.32	88.0
183-2	40.69	47.50	0.13	10.86	0.37	99.74	88.6
182-4	40.23	48.58	0.25	10.40	0.31	99.93	89.3
182-4	40.48	48.75	0.21	10.13	0.37	100.06	89.6
182-6	40.51	49.61	0.26	9.21	0.35	100.06	89.6
182-8	41.08	49.44	0.20	9.01	0.48	99.87	90.6
182-7	40.89	49.95	0.21	8.52	0.45	100.03	90.7
ML1-10	39.58	45.77	0.24	14.32	0.27	100.18	91.3
ML1-11	40.63	48.81	0.21	10.14	0.40	100.19	85.1
ML1-11*	39.07	43.30	0.25	17.60	0.18	100.50	89.6
ML2-3*	39.13	43.47	0.24	17.26	0.19	100.26	81.4
ML4-44	40.82	48.38	0.20	9.83	0.45	99.68	81.8

*Reversely zoned.

TABLE 3. Microprobe Analyses of Spinels from Mauna Loa Submarine Lavas

Sample	TiO_2	Al_2O_3	Cr_2O_3	FeO	MgO	Total	Fe_2O_3*	FeO*	Total	Mg#	Cr/Cr+Al	Host[x]
183-5	1.16	3.45	50.90	19.50	14.30	99.31	6.93	13.26	100.00	65.77	71.7	OL-90.0
183-5	1.57	3.90	43.85	28.40	10.45	98.17	10.13	19.28	99.18	49.13	67.9	OL-87.0
183-5	1.31	13.45	48.15	23.10	12.75	98.76	8.32	15.62	99.59	59.26	70.6	Matrix
182-3	1.26	14.75	49.50	19.70	14.05	99.26	6.43	13.92	99.90	64.28	69.2	OL-89.7
182-3	1.12	12.70	46.65	28.00	10.20	98.67	9.75	19.23	99.65	48.6	71.1	OL-83.6
182-3	1.22	4.80	46.15	24.40	12.05	98.62	8.47	16.78	99.47	56.14	67.6	OL-87.3
182-3	1.77	14.05	42.75	30.40	10.10	99.07	11.19	20.33	100.19	46.96	67.1	Matrix
183-11	1.10	13.87	49.25	19.75	13.90	97.87	7.03	13.43	98.57	64.85	70.4	OL-89.5
183-11	1.85	14.40	42.00	28.35	11.15	97.75	11.03	18.42	98.85	51.89	66.2	OL-83.2
183-11	1.85	13.10	49.78	23.65	11.50	99.88	5.97	18.28	100.48	52.85	71.8	Matrix
182-2	2.17	16.15	39.40	28.11	12.10	97.93	11.68	17.60	99.10	55.06	62.1	OL-85.3
182-2	1.03	14.10	49.40	19.15	13.95	97.63	6.57	13.24	98.29	65.25	70.1	OL-90.2
182-2	3.92	13.25	33.65	38.75	9.00	98.57	16.66	23.75	100.24	40.3	63.0	Matrix

[x]Host olivine forsterite content.
*Fe_2O_3 calculated assuming stoichiometry.

TABLE 4. Microprobe Analyses of Pyroxenes in Mauna Loa Submarine Lavas

Sample		SiO_2	TiO_2	Al_2O_3	Cr_2O_3	FeO	MnO	MgO	CaO	Na_2O	Total
182-2	core	52.75	0.61	2.79	--	6.48	0.17	17.78	18.9	0.23	99.71
	rim	54.09	0.52	1.68	--	6.24	0.17	18.80	17.86	0.22	99.58
182-3	core	52.15	1.12	2.65	--	7.98	0.15	16.48	19.19	0.32	100.04
	rim	52.31	0.85	3.30	--	6.97	0.19	17.86	18.0	0.23	99.71
183-14	core*	52.10	0.69	2.61	0.61	6.78	0.19	17.75	18.7	0.23	99.66
	core*	53.2	0.54	1.89	0.24	7.55	0.18	18.61	17.55	0.21	99.97
184-1	core	53.05	0.59	1.98	--	7.14	0.17	18.65	17.55	0.21	99.34
	rim	52.75	0.61	2.23	--	6.90	0.18	18.13	18.4	0.22	99.42
	core*	54.92	0.34	1.92	--	10.79	0.22	29.1	2.32	0.03	99.64
184-11	core	51.92	1.01	2.42	0.32	8.70	0.19	16.68	18.65	0.28	100.17
	rim	53.1	0.65	2.15	0.70	6.70	0.19	18.25	18.20	0.22	100.16
	core	54.4	0.65	1.5	0.11	14.2	0.28	26.95	2.20	0.04	100.33
	rim	54.7	0.50	1.9	0.38	12.4	0.25	27.5	2.8	0.04	100.47
185-3	core*	52.05	0.71	2.62	0.75	7.72	0.20	17.45	18.05	0.27	99.82
	core*	54.61	0.48	1.64	0.07	13.08	0.26	27.75	2.53	0.04	100.46
	core*	54.18	0.55	1.26	0.12	14.75	0.28	26.71	2.17	0.04	100.06

*Analyses with only cores are normally zoned or unzoned.

TABLE 5. Microprobe Analyses of Plagioclases in Mauna Loa Submarine Lavas

Sample		SiO$_2$	Al$_2$O$_3$	FeO	CaO	Na$_2$O	K$_2$O	Total	An%
182-2	core*	51.55	30.25	0.65	13.96	3.19	0.08	99.65	70.4
182-3	core*	49.02	32.33	0.60	15.81	2.35	0.06	100.17	78.5
	core	52.25	29.95	0.61	13.6	3.47	0.10	99.98	68.0
	rim	51.84	30.35	0.65	14.0	3.30	0.09	100.23	69.7
183-14	core	50.08	30.95	0.58	14.84	2.82	0.07	99.34	74.1
	rim	49.9	31.4	0.70	15.10	2.63	0.07	99.80	75.7
184-1	core	49.31	31.75	0.57	15.5	2.4	0.04	99.57	77.9
	rim	48.0	32.8	0.64	16.5	1.8	0.03	99.77	83.4
	core*	50.0	31.0	0.60	14.95	2.69	0.06	99.31	75.2
184-11	core	50.45	31.08	0.47	14.75	2.83	0.06	99.63	74.0
	rim	49.7	31.7	0.45	15.24	2.53	0.05	99.67	76.7
	core	50.81	31.06	0.58	14.65	2.86	0.06	100.02	73.6
183-3	core*	50.90	30.7	0.65	14.45	2.96	0.08	99.74	72.6
	core	52.95	29.3	0.77	12.67	3.92	0.15	99.76	63.5
	rim	52.22	29.7	0.72	13.24	3.48	0.13	99.49	67.2
ML1-10	core*	52.77	29.15	0.86	12.66	4.15	0.14	99.73	62.2
ML1-11	core*	52.49	27.28	1.59	12.07	4.44	0.11	99.48	59.6
ML2-3	core*	50.28	31.08	0.61	14.74	3.04	0.03	99.75	72.7
ML4-44	core*	52.05	29.67	0.53	13.69	3.59	0.07	99.60	67.5

*Analyses with only cores reported are normally zoned or unzoned.

lyzed from Kilauea submarine lavas have reverse zoning [*Clague et al., in press*].

Spinel. Microprobe analyses were made of spinel inclusions and microphenocrysts in four representative samples (Table 3). The microphenocryst and inclusion spinels are compositionally identical except in sample 182-2; its microphenocrysts have higher TiO2 and FeO contents (Table 3). The TiO2 content of the spinels is <4.0 wt%, and except for a few matrix spinels, usually <2 wt%. This is characteristic of magnesiochromites which crystallize in picritic magmas [*Wilkinson and Hensel, 1988*]. The Mg# and Cr/Cr+Al ratios of these spinels range from 40-66 and 62-77, which are identical to values obtained for spinels from Mauna Loa subaerial picrites [*Wilkinson and Hensel, 1988*] and Kilauea submarine olivine-rich basalts [*Clague et al., in press*]. The Mg# of the spinel inclusions correlates well with the forsterite content of the host olivine (Table 3). Titanomagnetite was found in the matrix of some of the more slowly cooled basalts.

Pyroxene. The clinopyroxenes in the submarine Mauna Loa lavas are augites with moderate to low Al, Ti, Cr, and Na contents (Table 4). The orthopyroxene are bronzites. The Mg# of the pyroxene cores range from 76-83 for the bronzites and 79-83 for the augites. Many of these pyroxenes are unzoned or have weak normal zoning, although reverse zoning is common in pyroxenes from some of the submarine lavas (e.g., 184-1).

There are few published pyroxene analyses for Mauna Loa lavas (i.e., a few grains from five historical flows; *BVSP, 1981; Rhodes, this volume*) to compare the data for the submarine lavas with. Our new pyroxene analyses are within the compositional range reported for these subaerial lavas, which are thought to have formed in low MgO, summit reservoir magmas [*Rhodes, this volume*].

Plagioclase. This mineral was analyzed in representative lavas from each dive and dredge haul. The anorthite content (An) of plagioclase in these submarine lavas ranges from 60-83, although most grains have An contents of 67-78 An (Table 5). This range overlaps and extends to higher An contents the few reported analyses of Mauna Loa lavas [*BVSP, 1981; Rhodes, this volume*] and those reported for submarine Kilauea lavas (64-76; *Clague et al., in press*). Many of the larger plagioclase crystals have modest reverse zoning (1-2% An), although some have greater variations (3-4% An). These reversals start 40 to 100 μm from the edge of the crystal and end 10 to 20 μm from the edge.

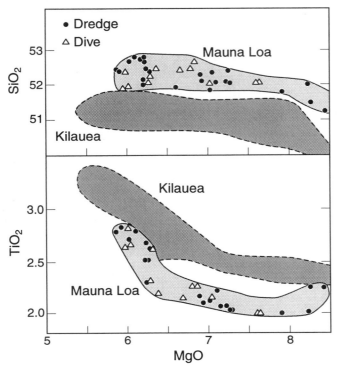

Fig. 7. MgO variation diagrams for TiO2 and SiO2 in glasses from Mauna Loa and Kilauea. The Mauna Loa glasses include dredged (Δ) and submersible-collected samples (●) The Kilauea field is based on submarine glasses [*Garcia et al., 1989*] and historically erupted subaerial glasses [*Garcia, unpublished*]. These elements and CaO are the best major element discriminants for separating lavas with MgO > 7 wt% from the two volcanoes [*Rhodes et al., 1989*].

Glass Chemistry

Major elements. Clear brown pillow-rim glasses were analyzed in this study to minimize the effects of mineral accumulation and thus obtain compositions more representative of Mauna Loa's magmas. Most of the dredge haul samples and about half of the dive samples have glassy margins [*Garcia et al., 1989* and Table 1]. These glasses were analyzed by microprobe for major and minor elements (including S). Two-thirds of the dive glasses were analyzed by ICP-MS for trace elements. The microprobe data for the dredge haul glasses were given in *Garcia et al.* [*1989*]. The microprobe data for the dive glasses are presented in Table 6. The submarine Mauna Loa glasses have a broad range in MgO content (5.8 to 8.3 wt%) and, like subaerial Mauna Loa lavas, have higher SiO2 and lower TiO2 contents at a given MgO content than glasses from its neighbor, Kilauea Volcano (Figure 7). The glasses from dive 185 have consistently lower MgO contents and more microlites of

plagioclase and pyroxene than glasses from the other dives. The dive 185 lavas were probably erupted at lower temperatures than the lavas from the other dives (10° to 50°C based on the Mauna Loa empirical glass thermometer; *Monierth et al., this volume*).

The S contents of the glasses range from <0.01 to 0.101 wt%. There is a crude correlation of S content with depth for the dive glasses except for one sample (Figure 8). The dredged glasses overlap in S content with the deeper dive glasses. The S content of Hawaiian submarine glasses decreases sharply with depth of eruption above 1000 mbsl as a result of degassing [*Moore and Fabbi, 1971; Killingley and Muenow, 1975*]. All but two of our Mauna Loa glasses have low S contents for their depth of recovery compared to Kilauea glasses (Figure 8). The most likely explanation for their low S content is subsidence of the lavas after eruption. If the Mauna Loa lavas had S contents similar to those of Kilauea, then the section has subsided >400 m, with the deeper samples showing the most apparent subsidence. The subsidence interpretation is consistent with the bathymetry data, which indicate at least 160 m of subsidence of the old shore line, and with the discoveries along the southwest rift of rounded boulders at depths of 425 mbsl during dive 183 and at 450 mbsl by *Moore et al.* [*1990*].

Trace elements. ICP-MS analyses were made to determine trace element abundances (REE, Ba, Th, Nb, Y, Ta, Hf, U, Rb, Cs) in the dive glasses (Table 7). The most incompatible elements vary by a factor of ~2, with Th showing the most variation (217%). Plots of highly incompatible trace

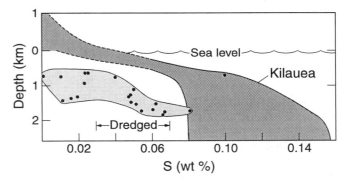

Fig. 8. S content vs. depth of sample collection (km below sea level) for glasses from submersible-collected and dredged lavas from Mauna Loa's southwest rift zone. The field for Kilauea glasses is shown for comparison [data from *Byers et al., 1985; Dixon et al., 1991; Muenow, unpubl.*]. At water depths <1 km, the S content of the Kilauea glasses decreases markedly because of degassing during eruption [*Moore and Fabbi, 1971; Killingley and Muenow, 1975*]. All but two of the Mauna Loa glasses have substantially lower S contents than predicted by the Kilauea field for their depth of collection. The low S content of the Mauna Loa glasses may be caused by subsidence of the rift zone.

TABLE 6. Microprobe Analyses of Glasses from Mauna Loa Southwest Rift Dives

Glass	SiO_2	TiO_2	Al_2O_3	FeO	MnO	MgO	CaO	Na_2O	K_2O	P_2O_5	S	Total	Mg#
182-3	52.25	2.18	14.04	10.56	0.17	6.88	11.15	2.27	0.32	0.18	0.016	100.02	53.7
182-4	52.15	2.05	14.13	9.90	0.15	7.25	11.10	2.33	0.30	0.19	0.020	99.57	56.6
182-6	51.50	2.17	13.45	10.35	0.16	8.25	10.75	2.21	0.31	0.19	0.049	99.39	58.7
182-7	51.26	2.17	13.41	10.53	0.16	8.42	10.70	2.16	0.31	0.19	0.049	99.36	58.8
183-3	52.03	2.01	13.45	10.30	0.20	8.24	10.46	2.10	0.31	0.19	0.051	99.34	58.8
183-7	52.08	2.11	13.94	10.20	0.17	7.25	10.93	2.21	0.33	0.19	0.023	99.43	55.9
183-11	52.35	2.09	14.07	10.04	0.16	7.08	10.95	2.30	0.30	0.18	0.015	99.54	55.7
183-12	52.45	2.04	14.16	9.95	0.18	7.29	10.88	2.22	0.31	0.14	0.001	99.62	56.6
183-14	51.95	2.48	13.80	10.69	0.17	6.65	10.82	2.37	0.38	0.20	0.026	99.54	52.6
183-15	52.05	2.24	13.98	9.90	0.18	7.13	11.10	2.20	0.37	0.20	0.025	99.38	56.2
184-2	51.81	2.01	13.73	10.49	0.16	7.90	10.90	2.18	0.30	0.18	0.082	99.74	57.3
184-5	52.06	2.11	14.01	10.20	0.16	6.95	11.22	2.29	0.32	0.20	0.063	99.58	54.8
184-8	52.01	2.54	13.60	11.78	0.18	6.26	10.30	2.23	0.40	0.23	0.012	99.54	48.7
184-9	51.91	2.12	13.99	10.42	0.15	7.01	11.11	2.21	0.31	0.20	0.040	99.47	54.5
184-11	52.19	2.32	13.66	10.68	0.18	6.24	10.63	2.43	0.38	0.18	0.101	98.99	51.0
185-1	52.67	2.71	13.57	11.46	0.17	6.20	10.03	2.49	0.44	0.27	0.066	100.08	49.1
185-2	52.40	2.85	13.55	11.53	0.19	5.91	9.88	2.54	0.50	0.28	0.067	99.70	47.7
185-3	52.45	2.80	13.62	11.50	0.19	5.84	9.71	2.52	0.50	0.34	0.054	99.52	47.5
185-4	52.35	2.70	13.88	11.25	0.17	6.23	10.22	2.39	0.51	0.29	0.061	100.05	49.7
185-6	52.43	2.61	13.89	10.81	0.15	6.29	9.95	2.41	0.47	0.32	--	99.33	50.9
185-7	52.50	2.51	13.71	11.46	0.18	6.22	9.89	2.47	0.38	0.22	0.063	99.60.	49.2
185-10	52.28	2.70	13.54	12.03	0.21	6.03	10.00	2.45	0.42	0.27	0.052	99.98	47.2
185-11	51.95	2.81	13.28	12.10	0.16	6.09	10.39	2.34	0.44	0.26	0.049	99.87	47.3

elements for the glasses from the dive-collected lavas and for whole-rock, XRF and INAA data for the dredge-collected lavas [*Gurriet, 1988*] define colinear trends (Figure 9). Plots of ratios of highly incompatible over moderately incompatible trace elements for the same glasses and lavas show broad linear and overlapping trends (Figure 10). Some of these ratios are greater than previously found for any subaerial Mauna Loa lava and overlap with ratios observed for Kilauea lavas (Figure 10).

DISCUSSION

Rock Type Variations: Spatial vs. Temporal Controls

Picritic basalts are rare in the volcanic record, especially among subalkaline lavas (<1%; *Wilkinson, 1986*). Geologic mapping of Hawaiian volcanoes has shown that picritic basalts are much more common on Hawaiian volcanoes (~5% for Kilauea; ~10-20% for Mauna Loa; *Macdonald, 1949*). Our current and previous studies of the submarine flanks of Hawaiian volcanoes [*Garcia et al., 1989*] indicate that picritic basalts are abundant on the flanks of these volcanoes (44-50%). The high temperature of the Hawaiian plume [*Sleep, 1990*] may be the cause of this greater abun-

dance. Temperatures of >1550°C have been estimated for the Hawaiian plume [e.g., *Watson and McKenzie, 1991*], which would generate magmas with high MgO content.

A section along the subaerial portion of the Kahuku Pali

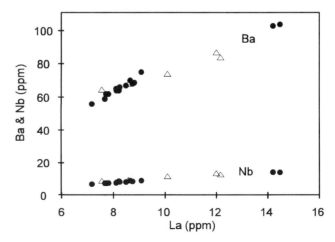

Fig. 9. Incompatible trace element variations for Mauna Loa submarine rift zone lavas. Symbols as in Figure 8. Data are from Table 5 for dive glasses and *Gurriet [1988]* for dredge whole-rocks. Note the collinear trend of the lavas and glasses.

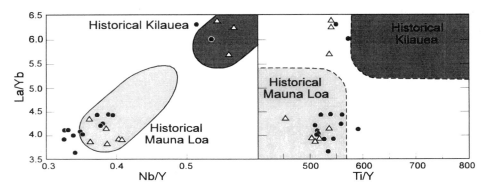

Fig. 10. Ratio-ratio plots for incompatible elements in Mauna Loa submarine rift zone lavas. The fields are for historical Mauna Loa [*Rhodes, unpublished*] and Kilauea lavas [*Garcia, unpublished*]. Although most of the submarine lavas plot in the Mauna Loa field for trace elements, five plot in the Kilauea field. These Kilauea-like lavas, however, have typical Mauna Loa Ti/Y ratios.

was measured to assist in our evaluation of the possible spatial and temporal controls on olivine abundance. This is one of the thickest subaerial sections on the volcano and the farthest from its summit (see Figure 3). We visually estimated the olivine content of the lavas and classified them into four broad categories: picritic (>15 vol.%), olivine-rich basalt (10-15 vol.%), olivine basalt (5-10 vol.%), and basalt (<5 vol.%). There is a clear temporal variation in this section. The upper third of the section is dominated by basalt (80%); the lower third of the section is mostly olivine-rich lavas (>95%; Figure 11). The same interpretation can be made for the submarine section of the rift. The samples from the ridge crest (dive 185) are younger than the lavas from the cliff sections. Only about one-third of the dive 185 lavas are olivine-rich compared to >80% for lavas from the other dives. Thus, there has been a substantial decrease in the abundance of olivine-rich lavas in the upper part of the two sections we examined on Mauna Loa's southwest rift. Nonetheless, the volcano is still occasionally erupting picritic basalts (e.g., 1852 and 1868 eruptions;). Unlike most of the submarine lavas, however, these subaerial lavas have lower forsterite content olivines (<89% Fo) and are interpreted to be cumulates flushed from shallow magma chambers [*Wilkinson and Hensel, 1988*].

The abundance of high forsterite olivines in the submarine lavas (Figure 4) indicates that they were derived from more mafic magmas, which avoided the buffering effects of evolved magma in the shallow summit magma chamber (3-7 km; *Decker et al., 1983; Rhodes, this volume*). The neutral buoyancy model of *Ryan [1987]* provides a good mechanism to allow the denser olivine-rich magmas to enter the rift zone without passing through the summit reservoir. The density of these olivine-rich magmas would have been much greater than expected for normal, subaerially erupted magmas (2.80 to 2.95 g/cm³ vs. 2.65-2.70 g/cm³; *Ryan, 1987*). The basics

of a "magma density stratification" model are summarized in Figure 12. This model proposes that the distal portions of the rift zone are primarily supplied with olivine-rich magmas by deep dikes (7-10 km) that intrude along density contrast zones. The existence of deep dikes within Hawaiian volcanoes has been supported by recent geophysical work on Kilauea [*Delaney et al., 1990*]. Lower density, less MgO-rich magmas ascend into the shallow (2-6 km) reservoir system of the volcano and are primarily erupted subaerially.

The ability of Mauna Loa to maintain a deep conduit within its rift zones may depend on its magma supply rate. The dead coral reef and drowned paleo-shorelines on the south flank of Mauna Loa indicate that the volcano's growth rate has slowed relative to its subsidence rate [*Moore et al., 1990*]. Our age estimate for this slowdown (~60 ka) is consistent with the change in abundance of olivine in the upper part of the subaerial Kahuku and submarine sections.

Origin of Olivine-Rich Mauna Loa Submarine Lavas

Many of the olivine-rich lavas that we recovered from the submarine flanks of Mauna Loa are cumulates. This can be inferred from a plot of olivine forsterite content vs. whole-rock Mg# (Figure 13). On this plot, the lavas that have accumulated olivine plot below the equilibrium field. These lavas are strongly olivine-phyric (21-47 vol.%) and have very high MgO contents (19.4 to 32.8 wt%). A few low Mg# lavas have olivines that are too forsteritic for their bulk rock Mg#; these olivines may be xenocrysts or are early formed olivines that were not separated from the magma. Another group of lavas have olivines that plot within or near the equilibrium field. These may be the lavas most representative of parental magmas. These samples contain 11-16 vol.% olivine, have Mg# of 71.9-75.6 and MgO contents of ~13.3-17.4 wt%.

Kahuku Pali Section

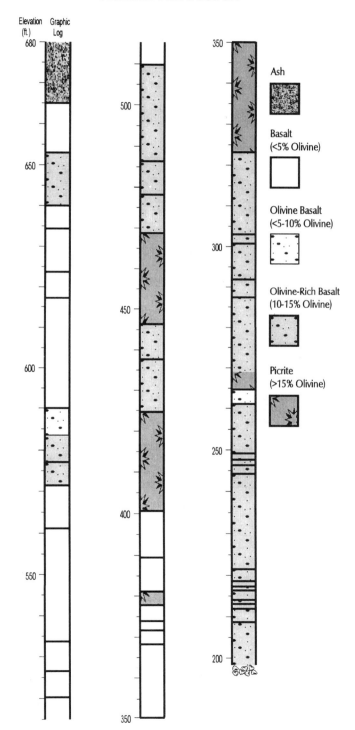

Some of the submarine lavas display petrographic (Table 1) and mineral compositional evidence (Table 6; Figure 6) of disequilibrium. Some of the minerals in these lavas are resorbed and have reverse zoning. The pyroxenes have low Mg# and must have grown in more evolved magmas than the coexisting olivines. If we assume a KD of 0.23 for Fe/Mg in augite/melt [*Grove and Bryan, 1983*], then the pyroxenes grew in magmas with Mg#s of 45 to 54; the olivine grew in magmas with Mg#s of 59 to 75 (Figure 13). These features are indicative of magma mixing involving olivine-rich and differentiated magmas. These differentiated magmas are similar to those erupted from the summit of Mauna Loa [e.g., *Rhodes, this volume*]. They may have been intruded from the summit reservoir through the shallower portions of the rift zone (2-4 km) and were mixed with the olivine-rich magmas from deeper dikes, or they may have been residual magmas from earlier intrusions that differentiated in the rift zone prior to mixing [c.f, *Wright and Fiske, 1971*]. Similar features were noted in, and explanations given for submarine lavas from Mauna Kea and Kilauea [*Yang et al., 1994; Clague et al., in press*]. Unlike lavas in those suites, our Mauna Loa suite includes lavas with no obvious signs of magma mixing (Table 1) and, therefore, they are more useful for inferring parental magma compositions.

Implications of Olivine-Rich Lavas for Mauna Loa Primary Magma Compositions

The MgO content of primary magmas for Hawaiian tholeiitic basalts has been a subject of considerable debate [see *Wilkinson and Hensel, 1988*, for a summary]. Estimates have ranged from 8 wt% [*Maaloe, 1979*] to 25 wt% [*Wright, 1984*]. A previous study of Mauna Loa subaerial lavas concluded that they were derived from parental magmas with 14-15 wt% MgO [*Wilkinson and Hensel, 1988*]. The discovery of glasses with up to 15 wt% MgO near the base of Kilauea [*Clague et al., 1991*] establishes a minimum parental magma MgO content and confirms that Hawaiian tholeiitic magmas are some of the most mafic magmas erupted during the Cenozoic.

The forsterite content of undeformed, euhedral olivine has frequently been used to infer the MgO content of the host magma [e.g., *Francis, 1985*]. This approach assumes equilibrium crystallization and that the early formed olivine is

Fig. 11. Stratigraphic section of the Kahuku Pali (lower, subaerial portion of Mauna Loa's southwest rift zone). Rock modes are based on hand specimen identification. The section is capped by the >31,000 year old Pahala Ash [*Rubin et al., 1987*]. The upper third of the section is dominated by basalts with variable amounts of small plagioclase phenocrysts. The lower third of the section consists of mostly olivine-rich basalts (>10 vol.% phenocrysts).

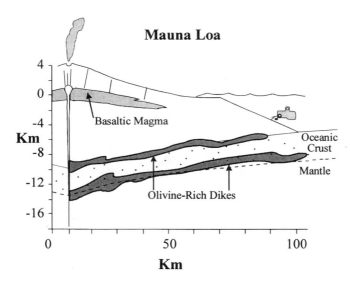

Fig. 12. Cartoon cross section of Mauna Loa illustrating the density stratification model for the volcano's magmatic plumbing system. The dense, olivine-rich magmas are intruded into the deeper portions of the rift zone at density contrast boundaries such as the Moho and are predominantly erupted from the distal portions of rift zones. Lower-density basaltic magmas rise into the upper part of the volcano's plumbing and are erupted preferentially on the subaerial parts of the volcano. Submersible not to scale.

carried with the ascending magma, a process that *Maaloe et al.* [*1988*] called "delayed fractionation". A correction for the effects of pressure on the olivine/melt partition coefficient [*Ulmer, 1989*] can be estimated from the olivine CaO content. Low CaO olivines form at higher pressure [*Stormer, 1973*], although the forsterite content of the olivine must also be considered [*Jurewicz and Watson, 1988*]. This approach generally yields host magmas with 12-15 wt% MgO because the olivines have <90% forsterite [*Wilkinson and Hensel, 1988; van Heerden and Le Roex, 1988*]. A recent study of Kilauea submarine olivine-rich basalts, however, found a Fo 90.7% olivine, which was related to a magma with 16.5 wt% MgO [*Clague et al., in press*].

Mauna Loa submarine basalts have olivines with the highest forsterite content ever reported for a Hawaiian tholeiitic basalt (91.3%). The moderate CaO contents of the olivines from these submarine lavas (0.18-0.31 wt%) indicate that an olivine/magma KD of 0.30 for FeO/MgO is appropriate. The calculated MgO content of the parental magma for the highest forsterite content olivine, assuming a ferrous iron content of 10% (based on measured values in Hawaiian lavas of 7 to 12%; e.g., *Byers et al., 1985*), would be ~17.5 wt% MgO. This Mauna Loa primary magma would have been most mafic and hottest (1415°C based on the empirical geothermometer for Mauna Loa tholeiites;

Montierth et al., this volume) magma erupted during the Cenozoic.

If Hawaiian primary magmas have >16 wt% MgO, it would eliminate an apparent paradox between experimental studies on Hawaiian tholeiites that assumed a parental magma with only 16 wt% MgO and found no garnet [e.g., *Eggins, 1992*], and trace element studies that require garnet in the source for Hawaiian magmas [e.g., *Hofmann et al., 1984*]. *Eggins* [*1992*] predicted that experiments using parental magmas with MgO contents of 17 wt% or more would contain garnet as predicted by the REE studies.

Temporal Geochemical Variation

Mauna Loa lavas are known to be compositionally distinct from other Hawaiian volcanoes [e.g., *Rhodes et al., 1989*]. How long has this distinction persisted? *Wright* [*1971*] observed no difference in the major elements between historical and prehistoric Mauna Loa lavas. An attempt to test this interpretation using the oldest exposed subaerial

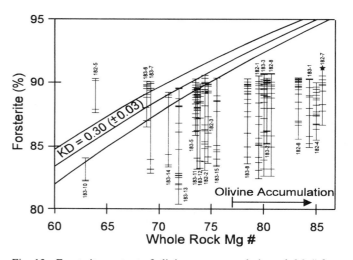

Fig. 13. Forsterite content of olivine cores vs. whole rock Mg# for 277 olivines from 21 submersible-collected lavas. The horizontal lines show the range of olivine core compositions for individual rocks (vertical lines). The equilibrium field is from *Roeder and Emslie* [*1970*] and *Ulmer* [*1989*] for lower pressure crystallization (<0.5 Gpa). Rocks with high Mg# (>76) have olivine forsterite contents too low to be in equilibrium with their whole rock Mg#; they have probably accumulated olivine. A few of the lavas have olivines with forsterite contents too high for their whole rock Mg#. They have either picked up xenocrysts or, more likely, not all of the early-formed olivine was separate from the magma prior to eruption. The lavas in the center of the figure with maximum olivine forsterite contents within the equilibrium field may be the most representative of near-primary compositions. The wide range in olivine composition within these lavas may be a consequence of delayed fractionation [c.f., *Maaloe et al., 1988*].

lavas from Mauna Loa (the Ninole basalts) yielded some-what ambiguous results because these older lavas are altered [*Lipman et al., 1990*]. The submarine southwest rift zone lavas are unaltered and some must be even older lavas than the Ninole basalts. The major element contents of these lavas are identical to historical Mauna Loa lavas (Figure 7). Thus, there is no discernible difference in major element content in Mauna Loa lavas over the last several hundred thousand years.

Some studies of trace elements and isotopes in Mauna Loa lavas have suggested temporal variations in these ratios. *Budahn and Schmitt* [*1985*] and *Tilling et al.* [*1987*] noted lower incompatible element abundances in historical vs. prehistoric lavas at the same MgO content. *Kurz and Kammer* [*1991*] measured lower $^3He/^4He$ ratios in historic lavas, which are consistent with the trace elements. *Lipman et al.* [*1990*] and *Rhodes and Hart* [*this volume*], however, argue that alteration-resistant trace element and isotope ratio variations over the last 150 thousand years are not much greater than during historic times.

To extend this evaluation to older Mauna Loa lavas, we constructed a composite section for the submarine southwest rift dive lavas. The composite section sample depths (Table 7) were plotted against the Ba/Y in the glass, which displays a large but coherent variation relative to other trace element ratios. The section shows large variations over short intervals, especially for the ridge lavas, which are assumed to have been erupted during a limited time interval (Figure 14). Similar but smaller variations were observed for ratios of other incompatible elements (e.g., La/Yb). The Ba/Y variation in the section appears cyclic, which is similar to patterns observed in stratigraphic sections from other Hawaiian volcanoes (Koolau- *Frey et al., 1994*; Kahoolawe- *Leeman et al., 1994*). Significant variations in trace element and isotope ratios can occur over relatively short periods (<100 years for Mauna Loa; *Rhodes and Hart, this volume*; ~200 years for Kilauea; *Pietruszka and Garcia, 1993*). These variations are apparently related to source heterogeneity, since Pb and Sr isotopes display similar variations [*Pietruszka and Garcia, 1993*; *Leeman et al., 1994*; *Rhodes and Hart, this volume*]. The cause of these cyclic variations is problematic. Perhaps the source components have different solidus temperatures. Thus, during progressive melting, the proportions of the components in the melt vary as the low temperature component is consumed.

Five of the southwest rift lavas are geochemically distinct from all other analyzed Mauna Loa lavas. They have incompatible trace element ratios similar to the neighboring Kilauea volcano (Figure 10). These lavas also have lower $^{87}Sr/^{86}Sr$ ratios than subaerial Mauna Loa lavas (0.70359-0.70365 vs. 0.70368-0.70397; *Gurriet, 1988*; *Kurz and*

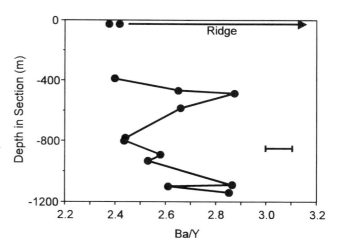

Fig. 14. Depth in section (table 7) vs. Ba/Y ratio for Mauna Loa submarine rift zone glasses. Note the large variations in the Ba/Y ratio over short depth intervals; this indicates rapid compositional changes in Mauna Loa magmas. The analytical error bar for the Ba/Y ratio is given in the lower right corner of the plot.

Kammer, 1991; *Kurz et al., this volume*), which are within the range of typical Kilauea lavas (0.70347-0.70367; *Hofmann et al., 1984*; *Pietruszka and Garcia, 1993*). As mentioned above, however, the major element abundances of these samples have a Mauna Loa signature, especially for Ti (Figure 10), which is a good element for distinguishing Mauna Loa from Kilauea lavas [*Rhodes et al., 1989*]. Thus, separate processes control the major element contents and the trace element and isotope ratios. The major element contents are probably controlled by partial melting processes [e.g., *Watson and McKenzie, 1991*]. The trace element and isotope ratio variations require source heterogeneity. The absence of the Kilauea-like source component in subaerial Mauna Loa lavas is an indication that it has been exhausted, perhaps because it has a lower melting temperature than the other source components.

SUMMARY

A giant submarine landslide on the southwest flank of Mauna Loa provided the best available window into the magmatic history of this volcano. It created the deepest exposure into the volcano: a 1.3 km thick section into the interior of its southwest rift zone. A new bathymetric map, submersible observations and a new magnetic survey [*Smith et al., in press*] all indicate that the axis of the submarine portion of the southwest rift zone is two to three kilometers west of the topographic ridge, which was presumed to be the axis of the rift. The glassy pillow lavas collected from the submarine rift lavas are partially degassed, which may be explained by at least 400 m of subsidence. If the current

TABLE 7. ICP-MS Analyses of Glasses from Mauna Loa Southwest Rift Zone

Sample	Depth	La	Ce	Pr	Nd	Sm	Eu	Gd	Tb	Dy	Ho	Er	Tm	Yb	Lu	Ba	Th	Nb	Y	Hf	Ta	U	Rb
182-6	1140	8.25	20.5	3.16	15.4	4.22	1.50	4.64	0.74	4.35	0.83	2.22	0.33	1.87	0.26	66	0.57	8.4	23.1	2.96	0.53	0.18	5.0
182-7	1100	8.22	20.9	3.14	14.6	4.12	1.49	4.70	0.76	4.21	0.81	2.23	0.32	1.86	0.26	64	0.69	8.8	24.5	2.88	0.53	0.23	5.1
183-3	935	7.17	18.1	2.85	13.5	3.93	1.39	4.44	0.69	4.16	0.80	2.10	0.31	1.97	0.25	56	0.47	7.0	22.1	2.76	0.44	0.16	4.2
183-7	795	7.67	19.1	3.01	14.6	4.18	1.53	4.92	0.76	4.49	0.86	2.33	0.33	1.92	0.27	59	0.50	7.6	24.2	2.90	0.47	0.16	4.5
183-12	585	7.72	19.6	3.02	14.2	4.21	1.49	4.89	0.74	4.43	0.85	2.27	0.33	1.92	0.26	62	0.49	7.6	23.3	2.92	0.47	0.17	4.5
183-14	485	9.08	23.1	3.53	16.8	4.88	1.72	5.61	0.86	5.08	0.95	2.53	0.36	2.15	0.29	75	0.63	9.2	26.1	3.42	0.58	0.24	5.3
183-15	465	8.66	22.3	3.50	16.4	4.57	1.65	5.47	0.82	4.94	0.94	2.48	0.35	2.07	0.29	70	0.52	9.2	26.4	3.26	0.53	0.17	4.9
184-2	1090	8.11	20.4	3.11	14.8	3.98	1.43	4.53	0.69	4.22	0.79	2.08	0.30	1.84	0.25	65	0.56	7.8	22.7	2.93	0.50	0.18	5.0
184-5	895	7.81	19.9	3.05	14.7	4.29	1.15	4.94	0.75	4.52	0.88	2.33	0.35	1.92	0.27	62	0.52	7.7	24.0	3.07	0.48	0.17	4.9
184-6	785	8.13	20.7	3.16	14.8	4.50	1.56	5.01	0.76	4.70	0.92	2.45	0.35	1.98	0.28	64	0.55	8.0	26.2	3.15	0.51	0.18	4.5
184-8	390	8.82	22.5	3.52	17.1	4.92	1.17	5.57	0.86	5.14	0.99	2.67	0.39	2.26	0.31	68	0.59	-	28.3	3.62	-	0.20	5.4
185-1	10-25	14.47	34.4	4.90	21.9	5.58	1.92	6.27	0.93	5.57	1.04	2.71	0.39	2.30	0.32	104	1.02	14.3	30.0	4.09	0.89	0.31	8.0
185-3	10-25	14.20	33.5	4.82	21.9	5.65	1.92	6.08	0.92	5.55	1.05	2.80	0.40	2.37	0.32	103	1.01	14.3	28.9	4.20	0.87	0.30	7.9
185-7	10-25	8.49	22.3	3.53	16.6	4.82	1.71	5.38	0.83	5.05	0.96	2.55	0.37	2.17	0.31	67	0.51	8.3	27.7	3.34	0.52	0.17	4.8
185-11	10-25	8.74	22.4	3.44	16.6	4.82	1.71	5.41	0.83	5.21	1.00	2.54	0.37	2.13	0.32	68	0.58	8.6	28.6	3.55	0.54	0.19	5.1

*Depth is in meters for reconstructed composite submarine southwest rift zone section.

subsidence rate of the volcano (2.6 mm/yr; *Moore et al., 1990*) has not changed significantly in the last few hundred thousand years, then the lavas in this section were erupted between 150 to ~300 ka.

Olivine-rich basalts are extremely abundant among the submarine lavas from the rift zone (~65% contain >10 vol.% phenocrysts). Their abundance decreases dramatically in the upper part of the two southwest rift sections examined in this study and are much less common on the subaerial portion of the volcano (<20%). The decrease in abundance of olivine-rich basalts in the southwest rift sections may be related to a decrease in the volcano's magma supply rate during the last hundred thousand years. A magma density filter may operate within the volcano to control where olivine-rich basalts are erupted. This filter may force the dense, olivine-rich magmas to bypass the shallow summit reservoir of the volcano and to intrude into the deeper portions of the rift zones. Although many of these olivine-rich lavas are cumulates, they do contain high forsterite content olivine phenocrysts. These phenocrysts grew in high MgO magmas (up to 17.5 wt%). Thus, Hawaiian tholeiitic magmas are some of the most mafic and hottest magmas erupted during the Cenozoic.

Our results for these submarine Mauna Loa lavas indicate that there has been no apparent temporal change in major element characteristics over perhaps the last several hundred thousand years. There are, however, some submarine lavas with distinct trace element ratios that probably reflect a different source that is similar geochemically to that of neighboring Kilauea. Thus, although the conditions of melt extraction for the volcano may not have changed, the source has varied. The Kilauea-like source represents a minor component for exposed Mauna Loa lavas. If it was more common in earlier Mauna Loa lavas, it may be a lower melting component that has been nearly or completely exhausted during the formation of this giant volcano.

Acknowledgments. We would like to thank the Pisces V diving team from the Hawaii Undersea Research Lab led by Terry Kerby, the captain and crew of the R/V Kila, and our diving companions Pete Lipman and Mark Kurz for their efforts in our successful dives on Mauna Loa's southwest rift zone. The assistance of Jennifer Parker in sample preparation, Bill Chadwick for merging our bathymetic data with those from previous NOAA cruises, Terri Duennebier with preparation of the color map, John Smith for showing us his unpublished submarine magnetic data for Mauna Loa, Aaron Pietruszka with measuring the Kahuku Pali section, and the Estate of S.M. Damon and personnel at the Kahuku Ranch for access to the Kahuku Pali section is gratefully acknowledged. Reviews by Di Henderson, Dan Fornari, and Tom Wright substantially improved this paper. This research was supported by NSF-Ocean Sciences. This is SOEST contrib. no 3942.

REFERENCES

Bargar, K. E., and E. D. Jackson, Calculated volumes of individual shield volcanoes along the Hawaiian-Emperor Chain, *U. S. Geol. Surv. J. Res.*, 2, 545-550, 1974.

Basaltic Volcanism Study Project, Pergamon Press, New York, 1286 p., 1981.

Budahn, J. R., and R. A. Schmitt, Petrogenetic modeling of Hawaiian tholeiitic basalts. A geochemical approach, *Geochim. Cosmochim. Acta*, 49, 67-87, 1985.

Byers, C., M. Garcia, and D. Muenow, Volatiles in pillow rim glasses from Loihi and Kilauea volcanoes, Hawaii, *Geochim. Cosmochim. Acta*, 49, 1887-1896, 1985.

Chadwick, W. W., J. G. Moore, M. Garcia, and C. G. Fox, Bathymetry-topography of south Mauna Loa volcano, Hawaii, *U. S. Geol. Survey Misc. Field Invest. Map* MF-2233, 1993.

Clague, D., and R. P. Denlinger, Role of olivine cumulates in destabilizing the flanks of Hawaiian volcanoes, *Bull. Volcanol.*, 56, 425-434, 1994.

Clague, D., W. S. Weber, and J. E. Dixon, Picritic glasses from Hawaii, *Nature*, 353, 553-556, 1991.

Clague, D., J. G. Moore, J. E. Dixon, and W. B. Friesen, Petrology of submarine lavas from Kilauea's Puna Ridge, Hawaii, *J. Petrol.*, in press.

Decker, R. W., R. Y. Koyanagi, J. J. Dvorak, J. P. Lockwood, A. T. Okamura, K. M. Yamashita, and W. R. Tanigawa, Seismicity and surface deformation of Mauna Loa volcano, Hawaii, *Eos Trans. AGU*, 64, 545-547, 1983.

Delaney, P. T., R. S. Fiske, A. Miklius, A. T. Okamura, and M. K., Sako, Deep magma body beneath the summit and rift zones of Kilauea volcano, Hawaii, *Science*, 247, 1311-1316, 1990.

Dixon, J. E., D. Clague, and E. M. Stolper, Degassing history of water, sulfur, and carbon in submarine lavas from Kilauea volcano, Hawaii, *J. Geology*, 99, 371-394, 1991.

Eggins, S. M., Petrogenesis of Hawaiian tholeiites: 1. Phase equilibria constraints, *Contrib. Mineral. Petrol.*, 110, 387-397, 1992.

Fornari, D. J., D. W. Peterson, J. P. Lockwood, A. Malahoff, and B. C. Heezen, Submarine extension of the southwest rift zone of Mauna Loa volcano, Hawaii: Visual observations from U.S. Navy Deep Submergence Vehicle DSV Sea Cliff, *Geol. Soc. Amer. Bull.*, 90, 435-443, 1979.

Fornari, D. J., J. P. Lockwood, P. W. Lipman, M. Rawson, and A. Malahoff, Submarine volcanic features west of Kealakekua Bay, Hawaii, *J. Volcanol. Geotherm. Res.*, 7, 323-337, 1980.

Francis, D., The Baffin Bay lavas and the value of picrites as analogues of primary magmas, *Contrib. Mineral. Petrol.*, 89, 144-154, 1985.

Frey, F., M. Garcia, and M. Roden, Geochemical characteristics of Koolau volcano: Implication of intershield geochemical differences among Hawaiian volcanoes, *Geochim. Cosmochim. Acta*, 58, 1441-1462, 1994.

Garcia, M., D. Muenow, K. Aggrey, and J. O'Neil, Major element, volatile and stable isotope geochemistry of Hawaiian submarine tholeiitic glasses, *J. Geophys. Res.*, 94, 10525-10538, 1989.

Garcia, M., B. Jorgenson, J. Mahoney, E. Ito, and T. Irving, Temporal geochemical evolution of the Loihi summit lavas: Results from ALVIN submersible dives, *J. Geophys. Res.*, 98, 537-550, 1993.

Grove, T. L. and W. B. Bryan, Fractionation of pyroxene-phyric MORB at low pressure: An experimental study, *Contrib. Mineral. Petrol.* 84, 293-309, 1983.

Gurriet, P. C., Geochemistry of Hawaiian dredged lavas, unpubl. M.S. thesis, 171 pp., M.I.T., Cambridge, Mass., 1988.

Hill, D. D., and J. J. Zucca, Geophysical constraints on the structure of Kilauea and Mauna Loa volcanoes, *U. S. Geol. Surv. Prof. Paper*, 1350, 903-917, 1987.

Hofmann, A. W., M. D. Feigenson, and I. Raczek, Case studies on the origin of basalt, III. Petrogenesis of the Mauna Ulu eruption, Kilauea 1969-1971, *Contrib. Mineral. Petrol.*, 88, 24-35, 1984.

Jurewicz, A., and E. B. Watson, Cations in olivine, Part 1, Calcium, Contrib. Mineral. Petrol., 99, 176-185, 1988.

Killingley, J. S., and D. W. Muenow, Volatiles in Hawaiian submarine basalts determined by dynamic high temperature mass spectrometry, *Geochim. Cosmochim. Acta*, 39, 1467-1473, 1975.

Kurz, M. D., and D. P. Kammer, Isotopic evolution of Mauna Loa volcano, *Earth Planet. Sci. Lett.*, 103, 257-269, 1991.

Leeman, W., D. Gerlach, M. Garcia, and H. West, Geochemical variation in lavas from Kahoolawe volcano, Hawaii, *Contrib. Min. Petrol.* 116, 62-77, 1994.

Lipman, P. W., The southwest rift zone of Mauna Loa: Implications for structural evolution of Hawaiian volcanoes, *Am. J. Sci.*, 280-A, 752-76, 1980.

Lipman, P.W., Growth of Mauna Loa during the last hndred thousand years, rates of lava accumulation verus gravitional subsidence, this volume.

Lipman P. W., and A. Swenson, Generalized geologic map of the southwest rift zone of Mauna Loa volcano, Hawaii, *U. S. Geological Survey Misc. Inves. Map* I-1312, scale 1:100,000, 1984.

Lipman, P. W., W. R. Normark, J. G. Moore, J. B. Wilson, and C. E. Gutmacher, The giant submarine Alika debris slide, Mauna Loa, Hawaii, *J. Geophys. Res.*, 93, 4279-4299, 1988.

Lipman, P. W., J. M. Rhodes, and M. A. Lanphere, The Ninole Basalt - implications for the structural evolution of Mauna Loa volcano, Hawaii, *Bull. Volcanol.*, 53, 1-19, 1990.

Lockwood, J. P., and P. W. Lipman, Holocene eruptive history of Mauna Loa volcano, *U.S. Geol. Survey Prof. Paper* 1350, pp. 509-535, 1987.

Lockwood, J. P., P. Lipman, L. Petersen, and F. Warshauer, Generalized ages of surface lava flows of Mauna Loa volcano, Hawaii, *U. S. Geol. Survey Misc. Invest. Map* I-1908, 1988.

Lonsdale, P., A geomorphological reconnaissance of the submarine part of the East Rift Zone of Kilauea volcano, Hawaii, *Bull. Volcanol.*, 51, 123-144, 1989.

Maaloe, S., Compositional range of primary tholeiitic magmas evaluated from major element trends, *Lithos*, 12, 59-72, 1979.

Maaloe, S., R. B. Pedersen, and D. James, Delayed fractionation of basaltic lavas, *Contrib. Mineral. Petrol.*, 98, 401-407, 1988.

Macdonald, G. A., Hawaiian petrographic province, *Geol. Soc. Am. Bull.*, 60, 1541-1596, 1949.

Monierth, C., A. D. Johnston, and K. V. Cashman, An empirical glass-composition-based geothermometer for Mauna Loa lavas, this volume.

Moore, J. G., Rate of palagonitization of submarine basalt adjacent to Hawaii, *U.S. Geol. Survey Prof. Paper* 550-D, D163-D171, 1966.

Moore, J. G., Subsidence of the Hawaiian Ridge, *U. S. Geol. Surv. Prof. Paper* 1350, 85-100, 1987.

Moore, J. G., and J. F. Campbell, Age of tilted reefs, Hawaii, *J. Geophys. Res.*, 92, 2641-2646, 1987.

Moore, J. G., and D. Clague, Coastal lava flows from Mauna Loa and Hulalai volcanoes, Kona, Hawaii, *Bull. Volcanol.*, 49, 752-764, 1987.

Moore, J. G., and D. Clague, Volcanic growth and evolution of the island of Hawaii, *Bull. Geol. Soc. Am.*, 104, 1471-1484, 1992.

Moore, J. G., and B. P. Fabbi, An estimate of the juvenile sulfur content of basalt, *Contrib. Min. Petrol.*, 33, 118-127, 1971.

Moore, J. G and W. W. Chadwick, Jr., Offshore geology of Mauna Loa and adjacent areas, Hawaii, this volume.

Moore, J. G., D. Clague, R. Holcomb, P. W. Lipman, W. Normark, and M. Torresan, Prodigious submarine landslides on the Hawaiian ridge, *J. Geophys. Res.*, 9, 17,465-17,484, 1989.

Moore, J. G., W. R. Normark, and B. J. Szabo, Reef growth and volcanism on the submarine southwest rift zone of Mauna Loa, Hawaii, *Bull. Volcanol.*, 52, 375-380, 1990.

Moore, J. G, W. B. Bryan, M. H. Beeson, and W. R. Normark, Giant blocks in the south Kona landslide, Hawaii, *Geology*, 23, 125-128, 1995.

Pearce, T. H., The analysis of zoning in magmatic crystals with emphasis on olivine, Contrib. Mineral. Petrol., 86, 149-54, 1984.

Pietruszka, A. J., and M. Garcia, Geochemical variations of historical summit lavas from Kilauea volcano, Hawaii: Evidence from Pb, Sr and Nd isotopes and trace elements, *Eos Trans. AGU*, 74, 642, 1993.

Raleigh, C. B., Mechanisms of plastic deformation of olivine, *J. Geophys. Res.*, 73, 5391-5406, 1968.

Rhodes, J. M., Homogeneity of lava flows: Chemical data for historic Mauna Loan eruptions, *J. Geophys. Res.*, 88, A869-79, 1983.

Rhodes, J. M., Geochemistry of the 1984 Mauna Loa eruption: Implications for magma storage and supply, *J. Geophys. Res.*, 93, 4453-4466, 1988.

Rhodes, J. M., The 1852 and 1868 Mauna Loa Picrite eruptions: Clues to parental magma compositions and the magmatic plumbing system, this volume.

Rhodes, J. M., and S. R. Hart, Episodic trace element and isotopic variations in historic Mauna Loa lavas: Implications for magma and plume dynamics, this volume.

Rhodes, J. M., K. Wenz, C. Neal, J. Sparks, and J. P. Lockwood, Geochemical evidence for invasion of Kilauea's plumbing system by Mauna Loa magma, *Nature*, 337, 257-60, 1989.

Roeder, P. L., and R. F. Emslie, Olivine-liquid equilibrium, *Contrib. Mineral. Petrol.*, 29, 275-289, 1970.

Rubin, M., L. K. Gargulinski, and J. P. McGeehin, Hawaii radiocarbon dates, *U. S. Geol. Survey Prof. Paper* 1350, pp. 213-242, 1987.

Ryan, M. P., Neutral buoyancy and the mechanical evolution of magmatic systems, *Geochem. Soc. Spec. Pub.* 1, 259-287, 1987.

Sleep, N. H., Hotspots and mantle plumes: Some phenomenology, *J. Geophys. Res.*, 95, 6715- 6736, 1990.

Smith, J. R, A. N. Shor, A. Malahoff, and M. E. Torresan, HAWAII MR1 sidescan reflectivity, Seabeam bathymetry, and magnetic anomalies on the submarine southeast flank of Hawaii Island, Hawaii Seafloor Atlas, Sheet 4, Hawaii Institute of Geophysics and Planetology, Honolulu, in press.

Stormer, J., Calcium zoning in olivine and its relationship to silica activity and pressure, *Geochim. Cosmochim. Acta*, 27, 1815-21, 1973.

Tilling, R. I., T. L. Wright, H. P. Millard, Jr., Trace-element chemistry of Kilauea and Mauna Loa lava in space and time: A reconnaissance, *U. S. Geol. Surv. Prof. Paper* 1350, 641-689, 1987.

Ulmer, P., The dependence of the Fe2+-Mg cation-partitioning between olivine and basaltic liquid on pressure, temperature and composition, *Contrib. Mineral. Petrol.*, 101, 261-73, 1989.

Van Heerden, L. A., and A.P. Le Roex, 1988, Petrogenesis of picrite and associated basalts from the southern mid-Atlantic ridge, *Contrib. Mineral. Petrol.*, 100, 47-60.

Watson, S., and D. McKenzie, Melt generation by plumes: A study of Hawaiian volcanism, *J. Petrol.*, 32, 501-537, 1991.

Wilkinson, J. F. G., Classification and average chemical composition of common basalts and andesites, *J. Petrol.*, 27, 31-62, 1986.

Wilkinson, J. F. G., and H. D. Hensel, The petrology of some picrites from Mauna Loa and Kilauea volcanoes, Hawaii, *Contrib. Mineral. Petrol.*, 98, 326-345, 1988.

Wright, T. L., Chemistry of Kilauea and Mauna Loa lava in space and time, *U. S. Geol. Survey Prof. Paper* 735, 39 p., 1971.

Wright, T. L., Origin of Hawaiian Tholeiite: A Metasomatic Model, *J. Geophys. Res.*, 89, 3233- 3252, 1984.

Wright, T. L., and R. S. Fiske, Origin of the differentiated and hybrid lavas of Kilauea volcano, *J. Petrol.* 12, 1-65, 1971.

Yang, H.-J., F. A. Frey, M. Garcia, and D. Clague, Submarine lavas from Mauna Kea volcano, Hawaii: implications for Hawaiian shield stage processes, *J. Geophys. Res*, 99, 15,577-15,594, 1994

Michael O. Garcia and Thomas P. Hulsebosch, Hawaii Center for Volcanology, Geology and Geophysics Department, University of Hawaii, Honolulu, HI 96822

J. Michael Rhodes, Department of Geology and Geography, University of Massachusetts, Amherst, MA 01003

The 1852 and 1868 Mauna Loa Picrite Eruptions: Clues to Parental Magma Compositions and the Magmatic Plumbing System.

J. M. Rhodes

Department of Geology/Geography, University of Massachusetts, Amherst

Picritic lavas were erupted on Mauna Loa in 1852 and again in 1868. They comprise only about 7% by volume of lavas erupted in historical (1843-1984) times. In both instances the picrites were preceded by the eruption of less MgO-rich lavas. Since most Mauna Loa lavas are thought to originate from a shallow, long-lived, continuously replenished, mixed and homogenized magma reservoir, the eruption of picritic lavas is an unusual and potentially informative event. Either they simply reflect the accumulation of olivine into these "reservoir lavas" or they are indicative of more MgO-rich parental, or possibly primary, magmas. Geochemical data, combined with mineral analyses, indicate that the picrites do not reflect melt compositions. They result from the accumulation of 10-27% olivine (Fo_{87-89}) into a parental magma with a lower MgO content. This magma was not typical of the low MgO (7-8%) lavas commonly erupted on Mauna Loa. It was more MgO-rich, with an MgO content of about 13%, and was less oxidized than most common subaerial lavas. The associated low-MgO lavas are not the complementary differentiates to the picrites, but probably reflect mixing between the picritic magmas and magma from the shallow magma reservoir. Because of the close geochemical coherence between the picrites and the systematic temporal variations in the composition of the more normal "reservoir lavas", it seems likely that they are derived from a deeper portion of a compositionally zoned magma column. Their eruption may be related to high magma supply rates, combined with vigorous activity in the magma column.

INTRODUCTION

Picritic magmas have achieved an importance in petrology far in excess of their volumetric significance within the volcanic record. This is because they are widely interpreted as being primary, mantle-derived melts from which most other basaltic magmas are derived through a variety of fractionation processes [e.g. *O'Hara*, 1968; *Clarke*, 1970; *Cox and Jamieson*, 1974; *Clarke and O'Hara*, 1979; *Elthon*, 1979; *Stolper*, 1980]. An alternative interpretation, advocated by *Bowen* [1928], is that picrites do not correspond to actual melt compositions, but reflect the accumulation of olivine phenocrysts in melts with lower MgO contents [e.g. *Green and Ringwood*, 1967; *Hart and Davis*, 1978; *Presnall et al.*, 1979; *Francis*, 1985]. Hawaiian lavas have figured prominently in this controversy. Those favoring an accumulative origin for Hawaiian picrites

include *Macdonald* [1949], *Powers* [1955], *Murata and Richter* [1966], *Hart and Davis* [1978], and *Wilkinson and Hensel* [1988]. Others have proposed that primary Hawaiian magmas are picritic in composition [e.g. *Macdonald*, 1968; *Wright*, 1971, 1984; *Maaløe*, 1979].

Picrites, however, are also relatively uncommon in the Hawaiian volcanic record. *Powers* [1955] estimated that they comprise about 15 percent of the volume of the Kilauea and Mauna Loa volcanic shields. On Mauna Loa, picrites have erupted only twice during the relatively short historical record, once in 1852 and again in 1868. Using the volume estimates of *Lockwood and Lipman* [1987] for historical (1843-1984) Mauna Loa eruptions, these picrites account for only about 7 percent of the volume of lava erupted in the past 152 years.

Previous studies have shown that most historical (1843-1984) and prehistoric (<30,000 years) Mauna Loa lavas have compositions that are controlled by the crystallization and accumulation of olivine, resulting in almost constant major element and compatible and moderately incompatible trace element abundances for a given MgO content [*Powers*, 1955; *Wright*, 1971; *Rhodes*, 1983, 1988; *Rhodes and Hart*, this

Mauna Loa Revealed: Structure, Composition, History, and Hazards
Geophysical Monograph 92

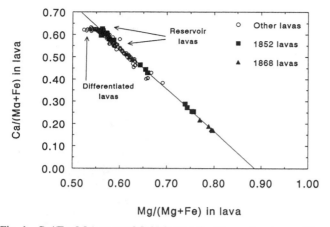

Fig. 1. Ca/(Fe+Mg) versus Mg/(Mg+Fe) for Mauna Loa lavas. The 1852 lavas are shown as solid squares (■), the 1868 lavas as solid triangles (▲), and all other lavas as open circles (☺). The regression line is fitted to the 1852 and 1868 data and intersects the Mg/(Mg+Fe) axis at about 0.88.

volume]. Most historical Mauna Loa lavas have MgO contents that range between 6.7 and 10 percent, with corresponding Mg-values from 0.55 to 0.65 (Figure 1). The vast majority of these, however, cluster at the low MgO end of an olivine-control trend between about 6.7 and 8.0 percent MgO, having Mg-values from 0.55 to 0.60. Although many of these lavas are at, or close to, multiple saturation, crystallizing olivine, clinopyroxene, plagioclase and occasionally low-Ca pyroxenes, surprisingly few have fractionated beyond olivine-control. Consequently, in contrast with neighboring Kilauea volcano, "differentiated" lavas (in the sense of *Wright and Fiske*, 1971) are rare (Figure 1). Because of this failure to differentiate, and because of the remarkable homogeneity of individual eruptions with compositions "perched" at the low-MgO end of an olivine-control trend, Rhodes and coworkers [*Rhodes*, 1983; *Rhodes and Sparks*, 1984; *Rhodes*, 1988; *Rhodes and Hart*, this volume] have proposed that most Mauna Loa lavas of this type are erupted from a shallow, long-lived, steady-state magma reservoir located at a depth of about 3 km beneath the summit caldera [*Decker et al.*, 1983]. They are prevented from differentiating, and their composition is "buffered", because of continuing replenishment by a more primitive, possibly picritic, parental melt [*O'Hara*, 1977; *Rhodes*, 1988]. Throughout the rest of this paper, lavas clustering at the low MgO end of an olivine-control trend with MgO between 6.7 and 8.0 percent will be referred to as *"reservoir lavas"*, to distinguish them from the picrites and their associated tholeiites that are the topic of this study. Picrites are traditionally identified on the basis of a high modal olivine content, the minimum volume varying from 15 percent [*Powers*, 1955; *Garcia et al.*, this volume] to as high as 25 percent [*Wilkinson and Hensel*, 1988]. Mauna Loa and Kilauea tholeiites with over 15 percent modal olivine typically contain over 14 weight percent MgO [e.g. *Wilkinson and Hensel*, 1988, Fig. 6]. Consequently, in this paper lavas with

more than 14 weight percent MgO are classified as picrites.

Almost all historical lavas with more MgO than the *reservoir lavas* belong either to the 1852 and 1868 picrite eruptions or are early erupted variants of the large-volume 1859 and 1950 eruptions. Both large-volume eruptions produced a wide range of lava compositions [*Rhodes*, 1983], possibly because of mixing of *reservoir lava* with a picrite. Thus, the olivine-control trend shown in Figure 1 and subsequent figures is largely defined by the 1852 and 1868 picrite data. An advantage of this kind of diagram is that the intersection of the olivine-control trend with the abscissa provides an estimate of the olivine composition dominating the trend [*Irvine*, 1979]. The regression line for the picrite data intersects the tight cluster of *reservoir lava* compositions and an olivine composition along the abscissa of about Fo_{87-88}. Trend lines for other major oxides (SiO_2, TiO_2, Al_2O_3, FeO, CaO) and incompatible trace elements (Sr, Zr, Y) against MgO intersect olivine compositions with a similar forsterite content. These estimates, together with microprobe data, emphasize the importance of olivine (Fo_{87-88}) to the origin of these rocks.

In this paper, I examine the geochemistry and mineralogy of the 1852 and 1868 Mauna Loa picrites and associated lavas, combined with what can reasonably be deduced about their eruptive histories, in order to evaluate their origin, possible primary characteristics, and the role of picrites in the Mauna Loa magmatic plumbing system.

GEOLOGICAL SETTING AND ERUPTIVE RECORD

Mauna Loa is the world's largest volcano, rising 4170 m above sea level. Most eruptive activity occurs in a large summit caldera, Mokuaweoweo, and along two narrow rift zones that trend to the northeast and southwest. The southwest rift zone is the longer of the two: over 70 km in length and continuing below sea level. The northeast rift zone extends for about 40 km, where it becomes diffuse and is buried by younger flows from neighboring Kilauea volcano. Typical eruptions on Mauna Loa initiate within the summit caldera and may be followed within a short period of time by much more voluminous, longer-lived eruptions along one of the two rift zones. The interval between summit and rift zone activity can range from a few hours or days to several months [*Macdonald and Abbot*, 1970; *Lockwood et al.*, 1976; *Lockwood and Lipman*, 1987]. The rift zone lavas are typically identical in composition with the earlier summit lavas [*Rhodes and Lockwood*, 1980; *Rhodes*, 1983; 1988], and appear to be tapping the same magma reservoir. This is in contrast with neighboring Kilauea volcano, where rift zone lavas tend to be more differentiated than those erupted at the summit [e.g. *Wright*, 1971; *Wright and Fiske*, 1971; *Moore*, 1983; *Garcia et al.*, 1989].

The 1852 Picrite Eruption

In all probability, the 1852 series of eruptive events began on August 8, 1851 with a small summit eruption that lasted for four days and produced lavas that flowed down the upper part of Mauna Loa's west flank (Figure 2). The lava was a typical *reservoir lava* with an MgO content of about 7.0 percent (Table 1). This was followed on February 17, 1852, by another, very short-lived summit eruption that lasted for a single day. Although witnessed from Hilo, and described by the Rev. Titus Coan [*Brigham*, 1909], the exact location of this phase of the eruption has remained in doubt, having been overshadowed by the subsequent events. *Lockwood* [1978], on the basis of

mapping and careful investigation of Coan's description of the location, has proposed that the aa flow upon which the NOAA Mauna Loa Weather Observatory is built was produced by the 1852 summit event (Figure 2). The chemical data presented here support this interpretation. Samples from the flow and its eruptive vent at 3810 - 3930 m. are very similar in composition to the earlier 1851 summit lava, and also to spatter from the uppermost fissure of the subsequent, much larger and better-documented, main phase of the 1852 eruption. This occurred

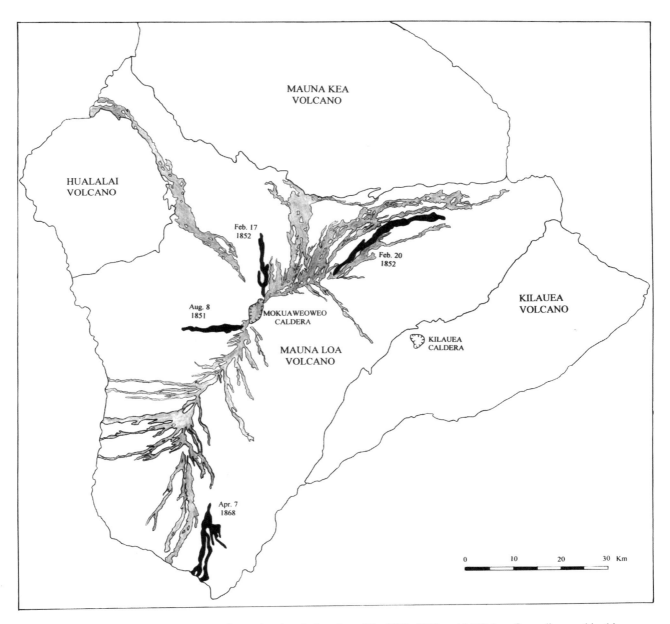

Fig. 2. Sketch map of Mauna Loa volcano showing the location of the 1851, 1852 and 1868 lava flows discussed in this paper. Other historical flows are shown by lighter shading. Adapted from *Lockwood et al.* [1988].

Table 1a. Chemical composition of 1852 Mauna Loa lavas.

Sample	1851 Lavas		Upper 1852 Observatory Flow					Lower Main 1852 Lavas		
	ML-80	ML-74	ML-70	ML-15	ML-288	ML-223	ML-52	ML-109	ML-284	ML-285
Type	Spatter	Aa	Aa	Spatter	Spatter	Aa	Spatter	Spatter	Spatter	Spatter
Location	L. Hohonu	L. Hou	SWR	Upper Vent	Upper Vent	Jeep trail	Lower	Lower	Lower	Lower
Elevation (m.)	3,960	3,900	2,800	3,920	3,810	3,350	2,630	2,590	2,590	2,560
SiO_2	51.61	51.58	51.36	51.38	51.37	51.43	51.34	51.95	50.71	50.82
TiO_2	2.14	2.13	2.14	2.15	2.09	2.13	2.11	2.04	1.96	1.96
Al_2O_3	13.75	13.70	13.73	13.80	13.63	13.75	13.72	13.80	12.82	12.77
Fe_2O_3*	11.83	11.83	11.79	11.66	11.84	11.74	11.82	11.74	11.88	11.94
MnO	0.16	0.16	0.16	0.17	0.17	0.17	0.16	0.17	0.17	0.16
MgO	6.85	7.03	6.94	6.82	7.65	7.12	7.33	6.83	10.23	10.47
CaO	10.15	10.47	10.55	10.55	10.43	10.52	10.24	10.60	9.66	9.49
Na_2O	2.15	2.22	2.21	2.32	2.41	2.34	2.12	2.29	2.16	2.14
K_2O	0.45	0.46	0.45	0.44	0.43	0.44	0.44	0.42	0.42	0.42
P_2O_5	0.26	0.26	0.26	0.26	0.25	0.25	0.26	0.25	0.25	0.25
Total	99.35	99.84	99.59	99.54	100.27	99.89	99.54	100.09	100.26	100.42
Mg-Value	0.560	0.567	0.564	0.563	0.587	0.572	0.577	0.561	0.655	0.659
Rb	6.8	7.4	7.0	6.7	6.4	6.6	7.1	6.5	6.7	6.5
Sr	342	351	359	341	339	341	345	337	329	323
Nb	10.7	11.4	10.7	10.4	10.1	10.8	10.2	10.4	9.5	9.3
Zr	140	142	140	140	137	143	139	130	129	130
Y	23	23	24	24	24	25	23	23	22	22
Zn	115	115	113	115	113	110	115	109	112	112
Ga	20	19	20	17	20	19	20	19	19	19
Ni	99	107	103	82	103	106	127	97	280	297
Cr	274	277	273	257	323	293	310	250	504	443
V	278	274	260	277	268	273	272	261	258	259

Fe_2O_3 is total Fe expressed as Fe_2O_3. Mg-Value is Mg/(Mg+Fe) after adjusting Fe^{3+}/total Fe to 0.1.

Table 1a. (continued)

Sample.	Lower Main 1852 Lavas					
	ML-286	ML-287	ML-14	ML-31	ML-19	ML-31g
Type	Spatter	Spatter	Spatter	Aa	Aa	Glass
Location	Lower	Lower	Lower	Jeep Trail	Distal end	Jeep Trail
Elevation (m.)	2,545	2,530	2,540	2,030	1,160	2,030
SiO_2	50.82	48.56	48.63	48.68	49.01	51.22
TiO_2	2.00	1.62	1.60	1.66	1.70	2.13
Al_2O_3	12.98	10.50	10.30	10.64	10.78	13.42
Fe_2O_3*	11.87	12.11	12.03	12.25	12.06	11.84
MnO	0.16	0.16	0.16	0.15	0.17	0.17
MgO	9.72	16.73	16.93	16.06	15.45	8.06
CaO	9.77	7.89	7.91	8.17	8.37	10.26
Na_2O	2.13	1.80	1.75	1.70	1.74	1.95
K_2O	0.43	0.35	0.34	0.34	0.35	0.43
P_2O_5	0.25	0.21	0.20	0.20	0.21	0.28
Total	100.12	99.92	99.85	99.85	99.84	99.76
Mg-Value	0.643	0.753	0.756	0.743	0.738	0.600
Rb	6.5	5.4	5.5	5.9	5.5	-
Sr	330	264	260	275	285	-
Nb	9.1	7.7	-	8.3	8.2	-
Zr	131	108	103	109	115	-
Y	22.3	18.7	17.6	18.2	18.8	-
Zn	113	112	115	117	109	-
Ga	19	16	15	16	16	-
Ni	263	725	774	699	645	-
Cr	460	968	1050	958	907	-
V	262	221	203	206	218	-

three days later, on February 20, 1852, on the northeast rift zone at an elevation of between 2620 and 2530 m. This phase of the eruption was much more voluminous (182 x 10^6 m³), [*Lockwood and Lipman*, 1987], lasted 20 days, and produced mostly picrites. There is, however, a gradation in spatter composition along the length of this eruptive fissure. Spatter with an MgO content of about 7.3 percent, closely resembling the earlier lavas from near the summit, erupted from the uppermost part of the fissure at 2620 m. The MgO content increased downrift to between 15 and 17 percent at the main picritic spatter cone at 2530 m. (Table 1). Presumably, the low

MgO magma was erupted initially, and was followed by increasingly picritic melts as the activity propagated downrift, coalescing to form a picritic cinder cone that was the focus of all subsequent activity and the source of the 25 km long 1852 picrite flow (Figure 2).

The 1868 Picrite Eruption

This series of events began with a short, one day, summit eruption on March 27, 1868. Regrettably, there are no samples available for study as all trace of the activity has been

Table 1b. Chemical composition of 1868 Mauna Loa lavas.

	Moderate MgO Flows				Later Picrite Flows				Groundmass Separates		
Sample	ML-86	ML-83	ML-88	ML-120	HAW-5	ML-87	ML-84	ML-85	ML-85g	ML-86g	ML-88g
Type	Pahoehoe	Aa	Spatter	Spatter	Aa	Pahoehoe	Aa	Spatter	Glass	Glass	Glass
Location	Hwy. 11	Puu Hou	U. Vent	L. Vent	Hwy. 11	Hwy. 11	Puu Hou	Puu Hou	Puu Hou	Hwy. 11	U. Vent
Height (m)	2,050	60	3,000	2,100	2,050	2,050	60	60	60	2,050	3,000
SiO_2	50.74	50.97	48.32	46.95	47.68	47.38	46.90	46.88	50.89	51.04	51.25
TiO_2	2.04	2.04	1.58	1.30	1.40	1.49	1.30	1.32	1.97	2.04	2.06
Al_2O_3	12.82	12.71	10.10	8.40	8.95	9.40	8.29	8.55	12.73	13.12	13.14
Fe_2O_3	12.13	12.08	12.28	12.43	12.45	12.55	12.57	12.53	11.82	11.74	11.86
MnO	0.17	0.16	0.16	0.17	0.16	0.15	0.16	0.16	0.16	0.17	0.17
MgO	8.97	9.18	17.01	22.20	21.02	19.02	22.21	21.91	9.91	8.60	8.43
CaO	10.05	9.96	7.91	6.58	6.99	7.46	6.58	6.66	9.94	10.21	10.24
Na_2O	2.07	2.09	1.53	1.48	1.48	1.49	1.16	1.21	2.35	1.89	1.86
K_2O	0.40	0.40	0.33	0.25	0.27	0.28	0.25	0.26	0.39	0.40	0.41
P_2O_5	0.23	0.24	0.19	0.15	0.17	0.18	0.15	0.15	0.24	0.24	0.26
Total	99.63	99.83	99.41	99.92	100.57	99.41	99.57	99.63	100.40	99.45	99.67
Mg value	0.619	0.626	0.753	0.797	0.788	0.769	0.795	0.794	0.649	0.617	0.602
Rb	6.3	6.7	5.3	4.1	4.4	4.9	4.2	4.1	6.3	-	7.2
Sr	315	316	248	206	213	232	205	205	308	-	317
Nb	10.5	10.2	7.8	6.9	7.2	7.2	6.4	6.3	9.1	-	-
Zr	132	133	104	87	90	95	84	85	126	-	-
Y	22.8	22.9	17.9	14.3	15.3	16.1	14.3	14.5	21.9	-	22.9
Zn	111	112	112	112	108	108	109	111	116	-	-
Ga	19	18	15	14	14	16	12	13	18	-	19
Ni	199	220	689	1003	1003	846	1115	1097	260	-	-
Cr	464	489	957	1327	1236	1072	1375	1395	972	-	-
V	252	251	214	185	174	189	167	171	260	-	-

obliterated by lava flows. A major, destructive earthquake centered on the southern flank of the volcano followed on April 2, 1868. On April 7, the main phase of the eruption began, low on the southwest rift zone from vents at an elevation between 600 to 900 m (Figure 1). The initial lavas were characterized by moderate MgO contents (about 9%), followed, and overlain, by more voluminous picritic lavas with about 19-22 weight percent MgO (Table 1). The eruption continued for four days and flowed more than 18 km into the sea, producing about 123 x 10^6 m^3 of lava [*Lockwood and Lipman*, 1987].

MINERAL COMPOSITIONS

Mineral compositions (Tables 2a-d) were obtained on phenocrysts set in a glassy to microcrystalline groundmass. Microprobe analyses were done in T. L. Grove's laboratory at M.I.T. using the 4-spectrometer JEOL 733 Superprobe, following analytical procedures outlined in *Grove and Juster* [1989]. The digital mineral compositional maps were obtained at UMass. on a Cameca SX50 microprobe using a 2 micron grid.

Olivine

Olivine is a ubiquitous phase in the picrites, occurring as large, homogeneous, euhedral and subhedral phenocrysts from 1 to 4 mm in diameter. Many contain glass inclusions, and inclusions of Cr-spinel. Core compositions range from Fo_{83} to Fo_{89}, the vast majority being within the narrow range Fo_{87-89}. Zoning is poorly developed; rims are very narrow, and typically about Fo_{83} in composition. There are no obvious differences in the compositions of olivines from the 1852 and 1868 picrites despite the differences in the whole-rock MgO contents, implying that the two parental magmas had very similar Fe-Mg characteristics. Olivine phenocrysts in the low-MgO spatter erupted at the lower 1852 vent are uniformly Fo_{82} in composition, within the range of most *reservoir lavas* (Fo_{77-82}) [*BVSP*, 1981; *Duggan*, 1987; *Rhodes*, unpublished data]. In contrast, the early, moderate-MgO, 1868 lavas exhibit a wide range of phenocryst compositions, similar to those of the picrites, from Fo_{82} to Fo_{89}. Phenocrysts of Fo_{87-89} are especially abundant. Nickel in the olivines ranges between 0.21 and 0.34 weight percent NiO (1700-2700 ppm Ni), and broadly correlates with forsterite content.

Figure 3 explores the familiar Fe-Mg partitioning relationships between olivine and melt [*Roeder and Emslie*, 1970] in order to place constraints on the melt compositions from which the olivines crystallized. In particular, it is important to assess whether the olivine could be in equilibrium with melts corresponding to the whole-rock picrite compositions, or whether they crystallized from, and accumulated, in a melt more closely comparable to *reservoir lavas*. In this diagram, olivines

Table 2a. Representative olivine phenocryst analyses from the 1852 and 1868 lavas.

Sample	ML-52 (1852)		ML-19 (1852)				ML-31 (1852)				ML-83 (1868)			ML-85 (1868)		
	3.1	4.1	1.1	4.2	3.2	1.2	1.3	1.4	2.1	1.4	1.1	3.1	1.1	2.1	4.1
	Core	Core	Core	Core	Core	Core	Rim	Rim	Core	Core	Rim	Core	Core	Core	Core
SiO_2	39.20	39.08	39.58	39.91	40.01	39.93	40.20	39.01	39.08	40.43	39.97	39.59	39.64	40.42	40.26
TiO_2	0.04	0.04	0.00	0.00	0.00	0.00	0.00	0.03	0.00	0.02	0.00	0.02	0.02	0.04	0.00
Al_2O_3	0.02	0.02	0.06	0.06	0.06	0.05	0.07	0.06	0.03	0.04	0.05	0.03	0.09	0.09	0.06
Cr_2O_3	0.03	0.02	0.14	0.14	0.09	0.12	0.08	0.08	0.02	0.10	0.17	0.07	0.02	0.03	0.09
FeO	16.84	16.66	13.01	11.78	11.19	11.17	11.27	16.44	13.69	10.57	12.54	14.30	15.86	11.32	11.34
MnO	0.26	0.23	0.23	0.16	0.16	0.11	0.12	0.17	0.14	0.08	0.09	0.21	0.22	0.16	0.15
NiO	0.23	0.21	0.27	0.21	0.27	0.27	0.27	0.19	0.26	0.34	0.32	0.22	0.31	0.31	0.30
MgO	43.37	43.52	45.50	47.04	46.77	48.00	48.08	43.84	46.32	47.67	46.40	44.80	44.28	48.24	47.47
CaO	0.27	0.27	0.23	0.20	0.20	0.21	0.22	0.25	0.23	0.21	0.21	0.25	0.25	0.22	0.21
Total	100.26	100.05	99.02	99.50	98.75	99.86	100.31	100.07	99.77	99.46	99.75	99.49	100.69	100.83	99.88
Fo%	81.9	82.1	86.0	87.5	88.0	88.3	88.3	82.5	85.7	88.9	86.7	84.6	83.1	88.2	88.1

Table 2b. Representative pyroxene analyses from the 1852 and 1868 lavas.

	ML-52 (1852)			ML-83 (1868)							
	Cpx.	Cpx.	Opx.	Zoned Opx. Crystal			Complex Zoned Crystal			Cpx.	Cpx.
	1.1	1.2	1.1	2.1	2.2	2.3	3.1	3.2	3.3	4.1	6.1
	Phen.	Phen.	Micro-	Core	Outer	Rim	Core	Inner	Outer	Phen.	Grnd.m
SiO_2	51.52	52.22	53.92	54.12	54.75	54.34	53.86	51.17	51.58	53.11	50.60
TiO_2	0.75	0.74	0.39	0.42	0.30	0.42	0.45	0.79	0.63	0.75	1.04
Al_2O_3	3.16	2.46	2.79	1.33	1.96	1.28	1.28	2.44	2.76	2.04	3.80
Cr_2O_3	0.51	0.20	0.74	0.10	0.55	0.15	0.15	0.31	0.63	0.10	0.85
FeO	8.62	9.32	11.02	14.17	10.37	13.86	15.31	8.99	7.22	7.06	8.29
MnO	0.20	0.22	0.21	0.32	0.23	0.31	0.24	0.17	0.16	0.19	0.21
MgO	18.04	19.42	28.95	27.29	29.83	27.72	26.87	16.72	18.53	20.84	18.08
CaO	17.18	15.17	2.47	2.16	2.44	2.44	2.01	19.03	17.66	15.46	16.30
Na_2O	0.26	0.21	0.04	0.03	0.03	0.00	0.00	0.29	0.26	0.09	0.19
Total	100.24	99.98	100.53	99.94	100.46	100.52	100.17	99.91	99.43	99.64	99.36
Wo%	35.0	30.7	4.8	4.2	4.7	4.7	3.9	38.6	36.0	30.9	34.0
En%	51.3	54.6	78.4	74.2	79.7	74.4	72.8	47.2	52.5	58.0	52.5
Fs%	13.7	14.7	16.7	21.6	15.6	20.9	23.3	14.2	11.5	11.0	13.5

Table 2c. Representative Cr-spinel analyses.

Sample	ML-83	ML-19	ML-31	ML-85
	1.1	1.3	2.4	2.3
Host	Grndmass	Fo_{86}	Melt Inc.	Fo_{88}
SiO_2	0.14	0.08	0.11	0.11
TiO_2	1.28	1.38	1.59	1.41
Al_2O_3	13.93	13.50	15.44	14.12
Cr_2O_3	48.60	48.63	43.26	49.19
FeO	24.39	23.82	27.78	21.51
MnO	0.25	0.29	0.19	0.17
NiO	0.16	0.15	0.14	0.19
MgO	10.84	12.77	11.57	13.30
CaO	0.00	0.00	0.09	0.00
Total	99.59	100.62	100.17	100.00
Mg/(Mg+Fe)	0.51	0.59	0.53	0.61
Cr/(Cr+Al)	0.70	0.71	0.65	0.70

Table 2d. Representative plagioclase analyses.

Sample	ML-52 (1852)		ML-83 (1868)	
	7.1	8.1	1.1	3.1
	Ground mass	Ground mass	Micro-phen	Ground mass
SiO_2	52.36	52.74	52.63	50.89
Al_2O_3	29.24	28.88	29.46	30.68
FeO	0.93	0.96	0.75	0.64
MgO	0.22	0.22	0.29	0.22
CaO	13.56	13.27	13.17	14.38
Na_2O	3.73	3.87	3.77	3.34
K_2O	0.09	0.12	0.10	0.09
Total	100.13	100.06	100.17	38.50
An%	66.4	65.1	65.5	70.1

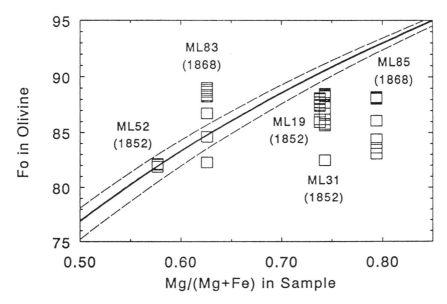

Fig. 3. Plot of forsterite in olivine versus Mg/(Mg+Fe) in the host rock. The curved lines are for Fe-Mg partitioning between olivine and melt assuming a K_D of 0.3 +/- 0.03 [*Roeder and Emslie*, 1970].

in equilibrium with melts corresponding to whole-rock compositions will plot along the curved band delineated by K_D values between 0.27 and 0.33 (the normal range for basaltic magmas). Only olivine in the low-MgO 1852 lava (ML-52) meets this criterion, and is in equilibrium with the host glassy spatter. In contrast, the moderate-MgO lava (ML-83) that preceded the eruption of the 1868 picrites exhibits a range of olivine compositions, from some that are too low (Fo_{82}), through those that appear to be in equilibrium with the whole-rock composition (Fo_{84}), to the greater number that are much too high (Fo_{88-99}). Since the high forsterite olivines correspond in composition with those in the accompanying picrites, an argument for a mixing relationship between the picrite magma and one with an MgO content similar to *reservoir lavas* appears plausible. Other disequilibrium characteristics, to be discussed later, support this interpretation. Picrite olivines are all too low in forsterite to be in equilibrium with melts corresponding to the whole-rock picrite compositions. The implication is that the picrites do not reflect melt compositions, and that they have accumulated varying amounts of olivine, mostly within the range Fo_{87-89}, into magma with a lower MgO content. Such an origin for Hawaiian picrites has been proposed by earlier workers [e.g. *Macdonald*, 1949; *Powers*, 1955] and, more recently, by *Wilkinson and Hensel* [1988], who used arguments similar to those presented here.

If the picrites do not reflect melt compositions, what can we learn about the parental melt from which the olivines crystallized? First, the olivines are too forsteritic to have crystallized from, and accumulated in, a *reservoir lava*, which typically would crystallize liquidus olivine between Fo_{77-82}. The

maximum forsterite content observed in the picrite olivines is about Fo_{88-89}. Assuming a low-pressure K_D of 0.3 for Mg/Fe partitioning between olivine and melt [*Roeder and Emslie*, 1970], Figure 3 shows that the olivine probably crystallized from a melt with an Mg-value close to 0.7. This is less than the Mg-value of the 1852 picrites which have the lowest MgO contents, but higher than any of the associated low-MgO lavas (Figures 1, 3). Obviously, variation in the actual K_D value from that assumed will influence the Mg-value. Nevertheless, the conclusions will remain essentially the same.

Cr-Spinel

Cr-spinel invariably occurs as small inclusions within olivine phenocrysts. A few occur as isolated phenocrysts in the groundmass, or as microphenocrysts within glass inclusions enclosed by olivine. All have rather constant Cr/(Cr+Al) and $Fe^{3+}/(Cr+Al+Fe^{3+})$ ratios of about 0.70 and 0.09 respectively. The Mg/(Mg+Fe) ratio is highly variable and, as observed by *Wilkinson and Hensel* [1988], reflects the composition of the host material. Cr-spinel inclusions in Fo_{88} olivine have the highest values of 0.60-0.61, whereas discrete Cr-spinel phenocrysts, in either the groundmass or glass inclusions, have values between 0.51 and 0.53.

Pyroxenes

I did not identify pyroxene phenocrysts within the picrites, only in the associated low to moderate MgO lavas. *Wilkinson and Hensel* [1988, Table 3, #6], however, report bronzite phenocrysts in the 1868 picrite. Augite is present in the 1852 low-MgO lavas, with an Mg/(Mg+Fe) ratio of 0.78, similar to that of most *reservoir lavas*, which typically range between 0.77 and 0.80 [*BVSP*, 1981; *Duggan*, 1987; *Rhodes*, unpublished data]. Pyroxene relationships in the moderate-MgO 1868 lavas are much more complex and are clearly indicative of disequilibrium conditions. Phenocrysts of both augite and orthopyroxene are present and both are reversely zoned. Some phenocrysts have a core of orthopyroxene mantled by reversely zoned augite (Table 2b, #3). Figure 4 shows digital maps of Ca and Mg variations to illustrate these relationships. The Mg/(Mg+Fe) ratios vary between 0.75-0.83 for the orthopyroxene and 0.76-0.83 for the augite. These values compare closely with those found in *reservoir lavas* (0.79-0.83) for the cores of low-Ca pyroxene phenocrysts, but extend to slightly higher values for augite. Using a K_D of 0.23 for Fe-Mg relationships between liquidus augite and melt [*Grove and Bryan*, 1983] implies that the Mg-value of the melt from which these pyroxene phenocrysts crystallized ranged between about 0.46-0.54. These are appropriate values for *reservoir lavas* and their differentiates, but not for liquids corresponding to the bulk rock composition, or to the picrites (Figure 1). Furthermore, we know from experimental studies [e.g. *Helz and Thornber*, 1987; *Montierth et al.*, this volume] and chemographic relationships (e.g. *Wright*, 1971; *Wright and Fiske*, 1971; *BVSP*, 1981; *Rhodes*, 1988) that augite does not begin to crystallize in Hawaiian tholeiites at low pressures until the MgO content of the melt is below 7-8 percent, corresponding to a temperature of about 1160-1170 °C. Thus, neither augite nor orthopyroxene should be stable liquidus phases in these lavas with MgO

contents of about 9 percent. This point is confirmed by the absence of phenocrysts or microphenocrysts of either augite or orthopyroxene in the glassy or microcrystalline groundmass of the picrites. As will be discussed later, the groundmass is very close in composition to the 1868 low-MgO lavas (Table 1). Thus, either the 1868 low-MgO lavas have accumulated pyroxenes, and possibly olivine, in a melt with an even lower MgO content (possibly a *reservoir lava*) or they represent mixed magmas. In view of the wide range in olivine compositions and the disequilibrium pyroxenes, the latter alternative appears most plausible, possibly involving a *reservoir lava* and the picritic magma.

Plagioclase

Plagioclase is absent from the picrites, but occurs as microphenocrysts in both the 1852 and 1868 low to moderate MgO lavas. It is very uniform in composition in the 1852 lava (An_{65-66}), but more variable (An_{65-72}) in the 1868 lava. These are similar to values for *reservoir lavas*, which typically range between An_{65-70}, most commonly An_{67} [*BVSP*, 1981; *Duggan*, 1987; *Rhodes*, unpublished data].

LAVA COMPOSITIONS

Whole rock major and trace element analyses by X-ray fluorescence (XRF) for the 1851, 1852 and 1868 lavas and separated groundmass are presented in Table 1. The methods are modifications of those of *Norrish and Chappell* [1967] and *Norrish and Hutton* [1969]. Details and estimates of accuracy and precision are given by *Rhodes* [1988]. A small subset of

Table 3. Abundances of rare-earth and other elements

Sample	ML-80	ML-52	ML-14	ML-86	ML-83	ML-85
Eruption	1851	1852	1852	1868	1868	1868
La	10.9	10.3	8.1	10	9.1	5.7
Ce	26.2	27.7	20.9	24.7	25.0	15.0
Nd	17.5	18.4	13.8	16.6	16.5	11.1
Sm	4.82	5.02	3.88	4.82	4.87	2.86
Eu	1.77	1.73	1.34	1.68	1.66	1.05
Tb	0.60	n.d.	n.d.	0.63	0.79	0.60
Ho	0.00	0.94	0.81	0.91	0.82	0.57
Yb	2.08	1.92	1.53	2.04	1.89	1.22
Lu	0.27	0.27	0.23	0.26	0.28	0.18
Hf	3.32	3.32	2.69	3.36	3.12	1.99
Th	0.47	0.41	n.d.	0.27	0.41	n.d.
Cr	263	327	996	443	494	1315
Sc	29.9	29.7	23.9	28.7	28.9	20.1

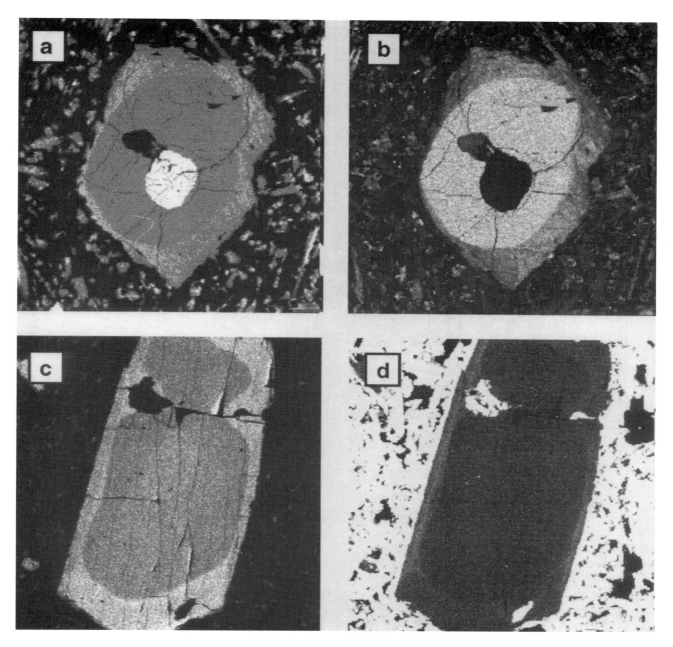

Fig. 4. Digital maps of Mg and Ca distribution in zoned pyroxene phenocrysts in ML-83, a moderate-MgO (9.2%) 1868 lava. (a) Mg distribution in a central rounded core of orthopyroxene (Wo:En:Fs 3.9:72.8:23.3) which is surrounded by reversely -zoned augite (38.6:47.2:14.2 to 36.0:52.5:11.5). (b) CaO distribution in the same phenocryst. (c) Mg distribution in an orthopyroxene which is reversely- zoned from Wo:En:Fs 4.2:74.2:21.6 to 4.7:79.7:15.6. (d) CaO distribution in the same phenocryst. See Table 2b for details.

these samples have been analyzed by Instrumental Neutron Activation Analysis (INAA) in F. A. Frey's laboratory at M.I.T. for the rare-earth and other trace elements (Table 3). The methods are those of *Ila and Frey* [1984].

In Table 1, the total iron is presented as Fe_2O_3, and adjusted so that (Fe^{3+}/total Fe) = 0.1 in calculating Mg-values (Mg/(Mg+Fe)) and for subsequent calculations and figures. This convention warrants some discussion. It was introduced by

Ringwood [1975] following the observation that this was about the lowest (Fe^{3+}/total Fe) ratio for the least oxidized glassy MORB and Hawaiian lavas, the assumption being that lavas with higher ratios may have been subsequently oxidized through a variety of processes. According to the experimentally determined relationship between ferric/ferrous ratios and oxygen fugacity developed by *Kilinc et al.* [1983], such (Fe^{3+}/total Fe) ratios are appropriate for conditions close to the magnetite-wustite (MW) buffer. It is now recognized that basaltic magmas may vary widely in oxygen fugacity [e.g. *Carmichael and Ghiorso*, 1986: *Christie et al.*, 1986; *Carmichael*, 1991]. *Carmichael and Ghiorso* [1986] show that most ferric/ferrous ratios reported in the literature for subaerial Kilauea glasses are appropriate for the fayalite-magnetite-quartz (FMQ) buffer. They propose that these lavas initially may have been even more oxidized and subsequently reduced upon eruption by sulfur degassing. Similarly, *Helz and Thornber* [1987] propose that the 1959 Kilauea Iki lava was close to the nickel-nickel oxide (NNO) buffer upon eruption. If these estimates are correct and apply equally to Mauna Loa, the (Fe^{3+}/total Fe) ratios used here will be too low (about 0.13 would be more appropriate), with the consequence that Mg-values and the calculated Fo content of olivines will also be slightly low. On the other hand, titrimetric measurement of FeO in hand-picked glassy matrix from ML-85, an 1868 picrite, indicates an Fe_2O_3/FeO of 0.11. This value is consistent with an oxygen fugacity very close to the magnetite-wustite (MW)

buffer [*Kilinc et al.*, 1983]. The results of *Christie et al.* [1986] on MORB glasses and pillow interiors show very clearly that the ferric/ferrous ratios of lavas, and therefore their inferred oxygen fugacity, is critically sensitive to cooling history. Quenched glassy pillow exteriors have average Fe^{3+}/total Fe ratios of about 0.07 compared with 0.15 for the pillow interiors, implying a difference in oxygen fugacity of about 2 log units. Given these uncertainties, and the measured Fe^{3+}/Fe^{2+}, coupled with an indication from Ni partitioning (to be discussed later) that these picritic lavas were initially reduced, an (Fe^{3+}/total Fe) ratio of 0.1 will be retained.

The overall relationships between the 1852 and 1868 picrites, their associated low-MgO tholeiites, separated matrix, and Mauna Loa *reservoir lavas* are clearly illustrated in Figure 5. This is an olivine-clinopyroxene-silica pseudoternary phase diagram modified from *Grove et al.* [1982] by incorporation of the experimental results of *Grove and Bryan* [1983]. The picrites plot well within the olivine field, producing a linear trend from olivine towards the field for *reservoir lavas*. The 1868 picrites are on the whole more olivine-rich than the 1852 picrites. The 1852 low-MgO lavas fall within two groups. Those inferred to have been erupted on February 17 at the summit and a single sample from the uppermost part of the February 20 vents at 2620 m. are similar to the earlier 1851 lavas and overlap the compositional field of *reservoir lavas*. This field, defined by over 220 analyses of lava erupted between 1843 and 1984, lies astride the olivine-plagioclase-

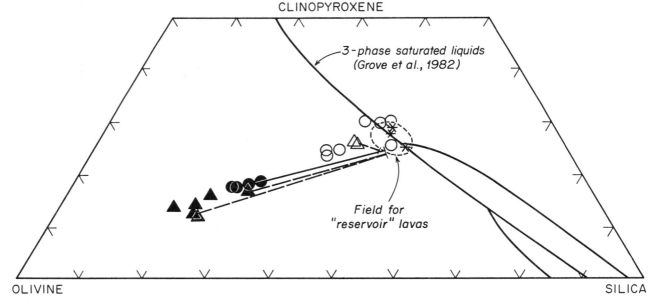

Fig. 5. Plot of 1851, 1852 and 1868 lavas in an olivine-clinopyroxene-silica pseudoternary projection (modified from *Grove et al.* [1982]). The 1851 lavas are represented by asterisks (✳); the 1852 picrites by solid circles (●) and the associated low-MgO lavas by open circles (O); the 1868 picrites by solid triangles (▲) and the associated moderate-MgO lavs as open triangles (△). Tie-lines connect whole-rock samples with separated glass and matrix. Most historical lavas plot within a tight cluster designated "reservoir lavas".

clinopyroxene saturation surface close to the reaction point olivine + liquid = pigeonite + clinopyroxene + plagioclase. The remainder of the low-MgO lavas, from the lower part of the February 20 vents, plot within the olivine field between the picrites and the *reservoir lavas*. The 1868 low-MgO lavas also plot within the olivine field on a trend between the picrites and the *reservoir lavas*. Groundmass samples, hand-picked from the 1852 and 1868 picrites and from an 1868 low-MgO lava, are all similar in composition and plot close to the field for *reservoir lavas* (Table 1). The tie-lines between the host picrite and its groundmass, however, do not intersect the low-MgO lavas. The groundmass samples are lower in the clinopyroxene component than are the low-MgO lavas, perhaps indicating clinopyroxene crystallization in the groundmass, or, more likely, that the low-MgO lavas are not simple differentiates of the picrites.

Rare-earth and incompatible element abundances for these lavas are presented in Tables 1 and 3, and plotted against a primitive mantle composition [*Sun and McDonough*, 1989] in Figure 6a. The patterns are typical of historical Mauna Loa lavas [*BVSP*, 1981; *Tilling et al.*, 1987; *Rhodes*, 1988; *Rhodes et al.*, 1989; *Lipman et al.*, 1990; *Rhodes and Hart*, this volume], the abundances reflecting differences in the MgO content of the lavas. All are depleted in Y and the heavy REE, a characteristic of Hawaiian lavas and attributed to the presence of residual garnet in the source [e.g. *BVSP*, 1981; *Hofmann et al.*, 1984]. In comparison with average Mauna Loa lavas (Figure 6b), adjusted to an MgO content of 13%, the 1852 and 1868 lavas are slightly enriched in incompatible elements. Although not readily apparent from Figure 6, the 1852 lavas are slightly more enriched in these elements than the 1868 lavas. The 1852 lavas have (La/Yb)$_{CH}$ of about 3.2, similar to the related 1851 lavas, whereas the 1868 ratios are lower, about 2.9. These differences reflect a steady decline in incompatible element abundances and in incompatible to moderately incompatible element ratios in Mauna Loa lavas between 1843

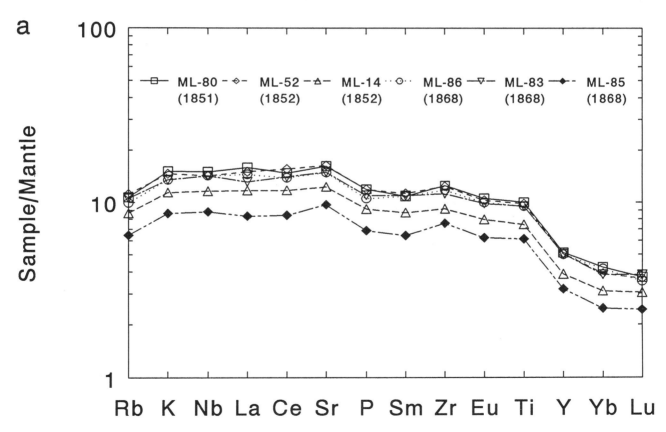

Fig. 6. Spidergrams of selected lavas from the 1851, 1852 and 1868 eruptions. In (a) the samples are referenced to primitive mantle [*Sun and McDonough*, 1989], whereas in (b) they are referenced to an average Mauna Loa lava [*Rhodes et al.*, 1989], adjusted through addition of olivine (Fo$_{88}$) to an MgO content of 13.0%. The reference values (at 13.0% MgO) are Rb - 5.3; K$_2$O - 0.34%; Nb - 8.6; La - 8.1; Ce - 22; Sr - 274; P$_2$O$_5$ - 0.21%; Sm - 4.2; Zr - 117; Eu - 1.49; TiO$_2$ - 1.82%; Y -20.7; Yb - 1.79; Lu - 0.25.

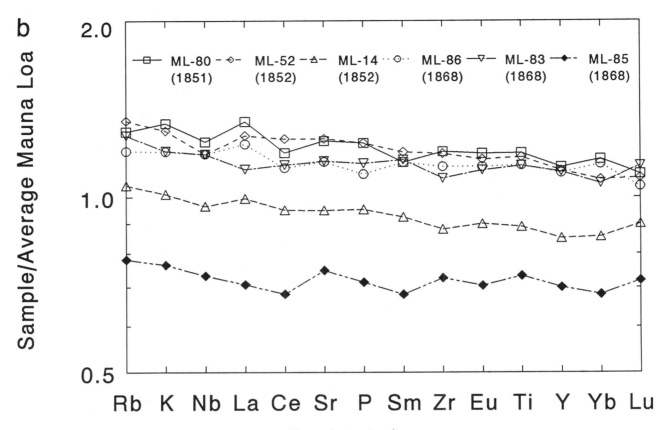

Figure 6. (continued)

and 1887 [*Rhodes et al.*, 1982; *Rhodes and Sparks*, 1984; *Tilling et al.*, 1987]. *Rhodes and Hart* [Figures 4-5, this volume] attribute these declines to the recharge of the magma reservoir by a relatively depleted parental magma during a period of unusually high magma supply rate.

An important characteristic of the 1852 and 1868 picrites and their associated low to moderate MgO tholeiites is the existence of strong linear relationships between most elements (Figures 7-8). From these relationships, if the concentration of one variable (such as MgO) is known, the abundance of most other variables may be calculated with reasonable certainty. Estimates for the highly incompatible elements (e.g. K, Rb, Sr) tend to be less precise because of the subtle, time-related differences in the 1852 and 1868 parental magma compositions outlined above. The equations involving strongly correlated elements that are excluded from olivine (e.g. Ti, Al, Ca, P, Nb, Zr, Y) confirm the importance of olivine-control. The intersection of the extension of these trends with the MgO axis occurs between 45.2 and 47.7 weight percent: values that are consistent with addition or removal of Fo_{86-88} olivine.

Figure 9 explores this further using the MgO-FeO relationship. The data define a linear trend from the low-MgO lavas to the picrites that is sub-parallel to olivine-control trends involving Fo_{87-89}, the dominant olivine compositions within the picrites. Similarly, the trends between three host picritic lavas and hand-picked glassy matrix closely follow the Fo_{87-89} control trends. Extrapolation of the regression line to olivine compositions results in a calculated olivine end-component of $Fo_{88.3}$. Olivine-control involving a lower forsteritic olivine (Fo_{82-84}) would produce too steep a trend, rather than the observed overall trend. Within the cluster of low-MgO lavas there is, however, some suggestion of a sub-trend involving Fo_{82}.

DISCUSSION

Origin of the Picrites

Of foremost importance is whether the 1852 and 1868 picrites reflect the composition of a high-MgO parental, possibly primary, liquid, or whether they are the products of olivine accumulation from a melt with a lower MgO content. Figure 3 shows that melts with compositions comparable to the picrites should crystallize a range of olivine compositions from about Fo_{90} to Fo_{93}, depending on the Mg-value of the picrite and

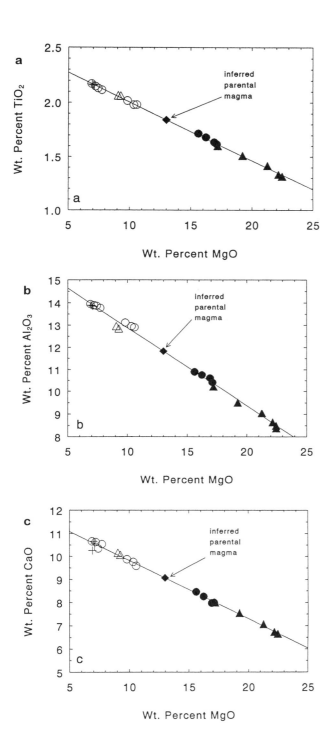

Fig. 7. TiO₂, Al₂O₃ and CaO abundances versus MgO for 1851, 1852 and 1868 Mauna Loa lavas. The 1851 lavas are shown as crosses (+); the 1852 picrites by solid circles (●) and the associated low-MgO lavas by open circles (O); the 1868 picrites by solid triangles (▲) and the associated moderate-MgO lavs as open triangles (△). An inferred parental magma is represented by a solid diamond (◆). The regression line is for the combined 1852 and 1868 data.

the appropriate selection of K_D for Fe-Mg partitioning between olivine and melt. Such olivines are not observed; the most forsteritic are about Fo_{87-89} and fall within the same compositional range irrespective of the Mg-value of the host picrite. Note that if a higher (Fe^{3+}/total Fe) ratio had been assumed in calculating the Mg-value of the picrites, the difference between the observed and calculated olivine compositions would be even larger. The implications are that the olivines in both the 1852 and 1868 picrites crystallized from a melt with a lower, and very similar, Mg-value. Extrapolation in Figure 3 from the most forsteritic olivines observed (Fo_{88-89}) to the equilibrium curves suggests that the Mg-value of the parental melt was close to 0.7. This, of course, rules out the possibility that the picrites simply reflect the accumulation of olivine crystallizing from the associated low-MgO tholeiites or from *reservoir lavas*, since the accumulating phenocrysts would have to be much less forsteritic (Fo_{82-84}). This point is also emphasized by the strongly linear compositional trends (Figures 7-9) which are consistent with the accumulation of Fo_{87-89} but not Fo_{82-84}.

The accumulative origin of the picrites is further emphasized in Figure 10. In this type of log-log plot involving a compatible element versus an incompatible element, crystal fractionation trends will plot as straight lines with a slope that is proportional (1-D) to the bulk distribution coefficient for the compatible element [*Allègre et al.*, 1977]. Mixing relationships involving accumulation of crystals in a melt will plot as a curve that trends away from the fractionation line [*Cocherie*, 1986]. Yttrium was chosen as the incompatible element in Figure 10 because, in contrast with other, more incompatible elements (e.g. K, Sr, La), its concentration was essentially the same in both the 1852 and 1868 parental magmas. Data from both eruptions can therefore be considered together. As will be shown subsequently, this is not the case for other incompatible elements. Furthermore, Y is to all intents and purposes an incompatible element during olivine crystallization. Inspection of Figure 10 reveals a distinct break in slope between 11 and 15 percent MgO. A similar dog-leg is seen in the Y-Ni relationship between 300 and 600 ppm Ni. These changes in slope could imply either a sudden change in the value of D's for MgO and Ni or the accumulation of olivine into a melt containing between 11 and 15 percent MgO. Given that olivine is the only phase crystallizing, the latter explanation appears the more plausible.

The Composition of the Parental Magmas

The preceding section has shown that the 1852 and 1868 picrites reflect the accumulation of olivine into a parental magma with a lower MgO content than the picrites themselves (15-22%). Furthermore, the *reservoir lavas* that are so commonly erupted on Mauna Loa cannot be this magma. Their MgO contents are too low, between 6.7-8.0% , and they should,

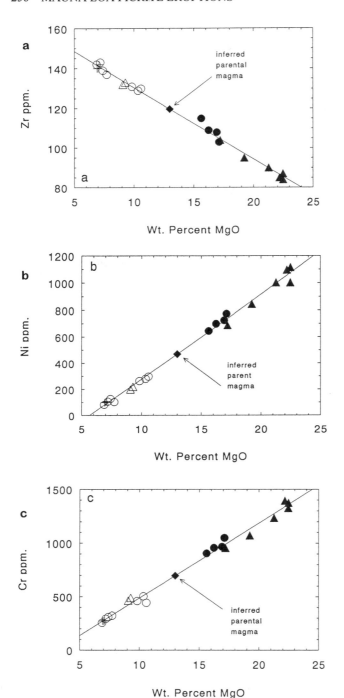

Fig. 8. Zr, Ni and Cr abundances versus MgO for 1851, 1852 and 1868 Mauna Loa lavas. The 1851 lavas are shown as crosses (+); the 1852 picrites by solid circles (●) and the associated low-MgO lavas by open circles (O); the 1868 picrites by solid triangles (▲) and the associated moderate-MgO lavs as open triangles (△). An inferred parental magma is represented by a solid diamond (◆). The regression line is for the combined 1852 and 1868 data.

therefore, crystallize and accumulate olivines with Fo_{77-82}. Accumulation of olivine of this composition is inconsistent with the FeO-MgO relationships (Figure 9) and with other whole-rock trends (Figures 1, 7-8). In the log-log plot of Figure 10, there is an obvious change in slope between 11 and 15 percent MgO. The simplest explanation for this dog-leg is that the MgO content of the parental magma lies within this range and that lavas with higher MgO contents reflect olivine accumulation into this magma, whereas those with lower values reflect either the products of crystal fractionation or mixing between the parental magma and one with an even lower MgO content, such as a *reservoir lava*.

The olivine data and the FeO-MgO relationships (Figure 9) permit further refinement of this estimate. The most forsteritic olivine phenocrysts in these lavas are Fo_{87-89}. They are also the most abundant. The chemographic data are consistent with the accumulation of olivine with these compositions, the best-fit line for FeO-MgO indicating an average forsterite content of $Fo_{88.3}$. Assuming a K_D of 0.3, the Mg-value of a melt in equilibrium with this olivine should be about 0.7, and the FeO/MgO (wt. %) should be about 0.77. From the regression line shown in Figure 9, these values correspond to an MgO content in the parental melt of about 13.0 weight percent.

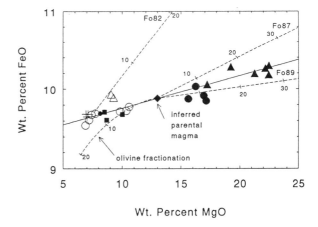

Fig. 9. FeO versus MgO for 1851, 1852 and 1868 Mauna Loa lavas. The FeO is calculated by assuming that $(Fe^{3+}/total Fe) = 0.1$. The 1851 lavas are shown as crosses (+); the 1852 picrites by solid circles (●) and the associated low-MgO lavas by open circles (O); the 1868 picrites by solid triangles (▲) and the associated moderate-MgO lavs as open triangles (△). Glass and matrix separates from the picrites are shown as solid squares (■). An inferred parental magma is represented by a solid diamond (◆). The regression line for the combined 1852 and 1868 data, shown as a single line, intersects the compositional trend for olivines at FeO-MgO values that are equivalent to $Fo_{88.3}$. The short dashed lines illustrate the effects of olivine addition to the inferred parental magma, and also the addition of olivine (Fo_{82}) to a typical "reservoir lava" composition. Also shown, by a dashed curve, is an olivine fractionation trend from the inferred parental magma.

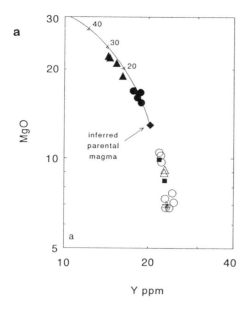

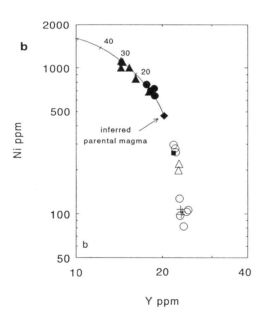

Fig. 10. Log-log plots of MgO and Ni versus Y for 1851, 1852 and 1868 Mauna Loa lavas. The 1851 lavas are shown as crosses (+); the 1852 picrites by solid circles (●) and the associated low-MgO lavas by open circles (O); the 1868 picrites by solid triangles (▲) and the associated moderate-MgO lavas as open triangles (△). An inferred parental magma is represented by a solid diamond (◆). The regression line is for the combined 1852 and 1868 data. Glass and matrix separates from the picrites are shown as solid squares (■). An inferred parental magma is represented by a solid diamond (◆). The curved lines illustrate the effects of varying amounts of olivine (Fo$_{88}$) accumulation in the inferred parental magma.

Because both the 1852 and 1868 picrites have accumulated olivine of similar composition, their parental magmas must also have possessed similar FeO/MgO characteristics, and therefore an MgO content of about 13 weight percent. Given this value, it is an easy matter to calculate other compositional characteristics of the parental magmas from the highly linear regression relationships or from appropriate element ratios (e.g. Na/Ca, K/Y). These are given in Table 4. Most elements are similar in abundance in the two parental magmas. The highly incompatible elements, however, are slightly higher in the 1852 lavas than in the 1868 lavas, reflecting the steady decline in incompatible element abundances between 1843 and 1887

Table 4. Estimated compositions of the parental magmas.

	1852	1868
SiO$_2$	50.29	50.29
TiO$_2$	1.85	1.86
Al$_2$O$_3$	11.83	11.83
Fe$_2$O$_3$	1.22	1.22
FeO	9.88	9.88
MnO	0.17	0.17
MgO	13.00	13.00
CaO	9.07	9.07
Na$_2$O	1.96	1.85
K$_2$O	0.39	0.36
P$_2$O$_5$	0.23	0.22
Mg-Value	0.70	0.70
Rb	6.2	5.9
Sr	301	286
Nb	9.1	9.1
Zr	121	119
La	9.1	8.1
Ce	24.3	22.1
Sm	4.5	4.2
Eu	1.5	1.5
Yb	1.7	1.7
Y	20.3	20.3
Ga	17.2	17.2
Ni	469	469
Cr	697	697
V	235	235

[*Rhodes et al.*, 1982; *Rhodes and Sparks*, 1984; *Tilling et al*, 1987; *Rhodes and Hart*, this volume]. Neither parental magma is strictly picritic in composition. They are olivine tholeiites [*Wilkinson and Hensel*, 1988], but much more MgO-rich than *reservoir lavas* or any other glassy or aphyric lava erupted within the historical record. According to the glass-based geothermometer for Mauna Loa lavas [*Montierth et al.*, this volume] a magma with an MgO content of 13% should have an eruption temperature of about 1311°C. The erupted picrites reflect the accumulation of olivine (mostly Fo_{87-89}) into these parental melts, up to 10% for the 1852 picrites, and up to 27% for those of 1868 (Figures 7-9).

Origin of the Associated Low to Moderate MgO Tholeiites

Both the 1852 and 1868 eruptions produced low to moderate MgO tholeiites prior to the eruption of the picrites. The range in composition for these lavas during the 1852 eruption is extensive, from lavas with about 7% MgO (similar to *reservoir lavas*) to up to about 10% MgO. The 1868 moderate-MgO tholeiites are more restricted in composition, with about 9% MgO. Given that the parental magmas, with about 13% MgO, have crystallized and accumulated olivine, it is reasonable to suppose that the associated low-MgO tholeiites may be the complementary derivatives, following crystal fractionation of the picrites. This assumption gains support from the observation that hand-picked glassy or fine-grained matrix from both the 1852 and 1868 picrites is similar in composition to the associated low-MgO tholeiites with about 9-10% MgO. In addition, all of the low-MgO tholeiites plot along linear trends from the parental magmas in log-log plots (Figure 10), a necessary condition if they are derived by crystal fractionation (*Allègre et al.*, 1977).

These observations, however, are also compatible with mixing involving the parental magmas and low-MgO *reservoir lavas* and the two alternatives cannot be resolved in Figure 10. For example, in Figure 10b the trend for the low-moderate MgO tholeiites can be reproduced either by mixing of the parental magma with a *reservoir lava* (similar to 1851), or by olivine fractionation in which the olivine/melt distribution coefficient (D) for Ni increases from about 6 to 12 (see below). Perusal of the FeO-MgO relationships (Figure 9), however, shows that the low-MgO lavas are too rich in FeO to result from the crystal fractionation of olivine from the parental melt composition. A mixing relationship appears to be more likely. Mixing is also indicated by the mineralogical data discussed earlier. The wide range in olivine compositions from Fo_{82} to Fo_{89} in the moderate-MgO tholeiites, together with the reversed- and complexly-zoned pyroxenes (Figure 4), is indicative of mixing between magmas similar in composition to the parental magmas and *reservoir lavas*.

Nickel Partitioning and Oxidation State

The relationship between Ni and MgO is shown in Figure 8, and the Ni content of the parental melt and an equilibrium olivine of $Fo_{88.3}$ can be estimated from the equation:-

$$Ni\ (ppm) = 63.97 * MgO\ (wt.\%) - 362.4\quad (R^2 = 0.994)$$

The parental melt with 13.0 percent MgO should therefore contain 469 ppm Ni (+/- 14 ppm at the 99% confidence level) and the equilibrium olivine 2718 ppm (+/- 86 ppm). Similar results are obtained based on the regression of Ni with oxides excluded from olivine (e.g. TiO_2, CaO, Al_2O_3) and *Wilkinson and Hensel* [1988] report between 2670 and 2950 ppm Ni in olivine separates from Mauna Loa picrites, including those erupted in 1852. Adopting 469 and 2718 ppm Ni for the parental magma and the olivine phenocrysts, respectively, results in a K_d of 5.8 +/- 0.4 for the partitioning of Ni between olivine and melt. This value is much lower than is considered typical for natural or synthetic basaltic melts. For example, the relationship developed by *Hart and Davis* [1978] for predicting the partitioning of Ni between olivine and melt gives a K_d of 8.6 for the parental magma and that of *Kinzler et al.* [1990] yields a K_d of 8.2.

Morse et al. [1991], using the data presented above together with data from the Kiglapait intrusion, proposed that the partitioning of Ni between olivine and melt was influenced by the redox state of the magma. They attributed low Ni K_d's to reducing conditions close to, or below, the magnetite-wustite (MW) buffer, rising very rapidly in a step-function with increasing oxygen fugacity to more "normal" conditions close to, or above, the fayalite-magnetite-quartz (FMQ) buffer. These predictions have been largely confirmed experimentally by *Ehlers et al.* [1992]. Furthermore, a measured Fe_2O_3/FeO ratio of 0.11 in the glassy matrix in one of the picrites is entirely consistent with an oxygen fugacity at, or close to, the MW buffer [*Kilinc et al.*, 1983]. The low K_d obtained from the olivine-melt relationships may well reflect the crystallization of the bulk of the olivine phenocrysts from the parental melt under reducing conditions deep within the volcano's magmatic plumbing system. In this respect, it is interesting to note that *Clague et al.*, [1991] estimated a k_d for Ni between olivine and melt in submarine Kilauea picritic glasses of 4.4.

Both *Morse et al.* [1991] and *Ehlers et al.* [1992] attributed the dependence of Ni partitioning on oxygen fugacity to a valency state for Ni lower than Ni^{2+} under reducing conditions, so that some of the Ni in the melt is not available to olivine. This view has recently been challenged [*Dingwell et al.*, 1994 and *Holzheid et al.*, 1994] on the basis of experiments in the synthetic system Di-An-CaO-SiO$_2$. These authors find no evidence for Ni^0 under reducing conditions in this system. This system, however, does not contain FeO, does not crystallize olivine, and is far removed in composition from a natural basaltic melt. Furthermore, the mechanism proposed by *Morse et al.* [1991] and *Ehlers et al.* [1992] is not dependent on Ni^0; all that is required is the presence of any species of Ni of valency less than Ni^{2+}.

Although the actual mechanism for the dependence of Ni partitioning between olivine and melt may be uncertain and controversial, the observations based on both experimental and natural systems appear robust, and may provide useful insights into the origin and oxidation state of magmas. For example, *Hart and Davis* [1978] presented damaging arguments against primary picritic magmas based on Ni partitioning between olivine and melt. They argued that basalts and picrites containing more Ni than their calculated melting curves (based on a model mantle composition and their experimentally determined Ni partition coefficients) could not be primary and must therefore have accumulated olivine. Most picrites and high-MgO tholeiites that had been proposed as potential primary magmas failed this test. Only basalts with between 11 and 13 percent MgO appeared appropriate. If, however, most primary magmas are more reduced than those erupted at the surface and have lower Ni K_d's than those found by Hart and Davis, then the criteria used by these authors may not be valid here. Using the K_d of 5.4 found for the 1852 and 1868 parental magmas and the same model mantle and melting calculations used by Hart and Davis, yields primary magmas containing between 530 and 600 ppm Ni at 13 percent MgO. These are well above the Hart and Davis melting curves, indicating that under reducing conditions primary picritic magmas are quite possible.

Implications for the Magmatic Plumbing System

I have proposed earlier and in previous publications [e.g. *Rhodes and Sparks*, 1984; *Rhodes*, 1988; *Rhodes and Hart*, this volume] that the vast majority of the lavas erupted on Mauna Loa, with MgO contents between 6.7 and 8.0 weight percent, are *reservoir lavas*, erupted from a long-lived, continuously replenished, well-mixed and homogenized, shallow magma reservoir. If this is the case, then the eruption of picrites or their high-MgO (≈ 13%) parental magmas is an unusual, and potentially informative, event. One might expect that these magmas would invade and become incorporated into the reservoir magma. That this did not happen in 1852 and 1868 implies one of three possibilities: (a) there was no long-lived magma reservoir at the time of their eruption; (b) somehow the picrites bypassed the magma reservoir; or (c) the picrites and their parental magma were erupted from a lower portion of a compositionally-stratified magma reservoir.

The argument for an ephemeral magma reservoir is difficult to sustain. This was a period of high eruption frequency and extremely high eruption rates [*Lockwood and Lipman*, 1987]. Furthermore, reservoir-like lavas were erupted in 1851 and early 1852, immediately before the 1852 picrites, again in 1855 and 1859 in the interval between the two picrite eruptions, and again in 1877 and 1880 following the 1868 picrites. It seems equally unlikely that the picrites simply bypassed the magma reservoir. If this were the case, one would not expect to find any close geochemical affinities between the picrites and the *reservoir lavas*, since the *reservoir lavas* reflect homogenization, and any systematic change in their composition must reflect the gradual influence of the influx of parental magmas with markedly different geochemical characteristics. *Rhodes and Hart* [this volume] attribute the systematic decrease in incompatible element abundances and $^{87}Sr/^{86}$ ratios from 1843 to 1887 to the steady influx of a relatively depleted parental magma during this period. If the 1852 and 1868 picrites, and their parental magmas, had bypassed the magma reservoir then it is likely that they would correspond in composition with this depleted parental magma. This is not the case, as noted earlier. They differ in incompatible element abundances and ratios, and these differences are completely consistent with the systematic temporal decrease in these abundances and ratios between 1843 and 1887 [*Rhodes and Hart*, this volume]. The most satisfactory explanation of these observations is that the picrites and their parental magmas are part of a continuous, compositionally stratified, magma column, and that the denser high-MgO magmas and picrites come from a deeper part of the column [e.g. *Stolper and Walker*, 1980; *Sparks and Huppert*, 1984]. In its simplest form, such a stratified magma column may result from olivine fractionation in a parental or primary picritic melt. Perhaps, more realistically, a spectrum of parental or primary magma compositions, ranging from picrites to high-MgO tholeiites are delivered from the mantle to the volcano's magmatic plumbing system, where they organize in the magma column according to density differences. The overall result will be similar, although the proportion of magma types will probably differ. Recent studies of melt migration in the mantle suggest that magma compositions may be influenced by the mode of transport, either by porous flow or in a network of channels [e.g. *McKenzie*, 1985; *Ribe*, 1988; *Eggins*, 1992; *Spiegelman and Kenyon*, 1992]. For example, *Kellemen et al.* [1992] propose that as magma percolates through the mantle, interaction between melt and wall-rock occurs, resulting in the dissolution of pyroxene and crystallization of olivine. If correct, this must inevitably result in a change in magma composition from picrite towards a more basaltic composition, with the consequence that basaltic magmas will be delivered to the magma column. On the other hand, if the flow is channelized, reaction will be inhibited and a picritic magma may be delivered to the volcano's magmatic plumbing system.

These assumptions imply that most of Mauna Loa's lavas come from the upper "head" of this stratified magma column, the shallow magma reservoir of *Decker et al.* [1983]. Why then are picrites and their parental magmas ever erupted? Perhaps, the *reservoir lavas* that occupy the upper part of the magma column or reservoir are depleted in volume through frequent eruption during periods of high magma supply and eruption rates, thereby allowing high-MgO magmas and picrites to ascend to higher levels than usual in the magma column. As noted above, 1843 to 1887 was a period of very high eruption rates (0.047 km^3/year), compared with subsequent years (1887-1984), during which the eruption rate averaged only about 0.02 km^3/year [*Lockwood and Lipman*, 1987]. In this respect, it is interesting to note that picrites are quite common in prehistoric lavas between about 750 and 1500 B.P., another period of high eruptive activity [*Lockwood and Lipman*, 1987].

It follows that if high-MgO magmas and picrites are to be found at deeper levels within a compositionally-stratified magma column, then even more primitive magma compositions may be present at deeper levels within the column. The most likely location in which to find these more primitive magmas and associated picrites, and to test the concept of a stratified magma column will be at low elevation on the two rift zones during periods of high magma supply and eruption rates, and along the submarine portion of the southwest rift zone. Recent discoveries of submarine picritic glass on Kilauea [*Clague et al.*, 1991] and extremely abundant picrites on the faulted submarine extension of Mauna Loa's southwest rift zone [*Garcia et al.*, this volume], are in accord with these predictions.

CONCLUSIONS

The 1852 and 1868 picritic lavas do not reflect picritic melt compositions. They result from the accumulation of between 10 to 27 percent olivine (Fo$_{87-89}$) into a parental magma with a lower MgO content. This magma was not typical of the abundant low MgO (e.g. 7-8% MgO) lavas that characterize most of Mauna Loa's voluminous and uniform eruptions. Mineralogical and whole-rock chemical relationships indicate that the parental magma was more MgO-rich than these *reservoir lavas*, with MgO contents of about 13%. Similar conclusions were reached by *Wilkinson and Hensel* [1988], largely on the basis of mineralogical arguments, in a study of Kilauea and Mauna Loa picrites including those investigated here.

The associated low to moderate MgO lavas were erupted prior to the eruption of the picrites during both the 1852 and 1868 eruptions. They range in composition from lavas that have all the characteristics of *reservoir lavas* to lavas with compositions that are intermediate between picrites and *reservoir lavas*. Although they may possibly be the complementary differentiates to the olivine-accumulated picrites, their bulk compositions are also consistent with mixing of *reservoir lavas* with the picritic magmas. This interpretation is to be preferred because of the distinct disequilibrium mineral assemblages in those with intermediate compositions.

Because of the well-defined linear chemical relationships, the composition of the parental magmas can be estimated fairly precisely. Using these estimates, together with estimates of the Ni content of the most forsteritic olivine, the K$_d$ for Ni between olivine and melt is shown to be about 5. *Morse et al.* [1991] suggest that such low K$_d$'s are a consequence of low oxygen fugacity, possibly close to the magnetite-wustite (MW) buffer. The Fe^{3+}/Fe^{2+} in the glassy matrix of an 1868 picrite is consistent with this assessment, indicating an oxygen fugacity (log fO$_2$) of about -9.2. If the oxidation state of most Mauna Loa lavas is similar to that of Kilauea lavas [*Helz and Thornber*, 1987; *Carmichael*, 1991], then the 1852 and 1868 lavas may reflect more reduced magmas, possibly from deeper levels of the magmatic plumbing system. Although similar in most major, compatible and moderately incompatible trace element abundances, the abundances of the incompatible elements are slightly higher in the 1852 lavas than in the 1868 lavas. This is consistent with the steady decline in incompatible element abundances in Mauna Loa lavas between 1843 and 1887 [*Rhodes and Hart*, this volume].

If most Mauna Loa lavas are erupted from a shallow, continuously replenished and homogenized magma reservoir [e.g. *Rhodes*, 1983; *Decker et al.*, 1983; *Rhodes*, 1988; *Rhodes and Hart*, this volume], then the eruption of MgO-rich lavas and picrites is an unusual and significant event. It implies either that these lavas represent random batches of more primitive magma that have managed to bypass the shallow reservoir, or that they were erupted from the lower levels of a compositionally-stratified magmatic plumbing system. The compositional coherence with the systematic temporal changes in incompatible abundances and ratios in Mauna Loa lavas is compelling evidence in support of the latter interpretation. It may well be that they are erupted at times when the magma supply rate is high and there is vigorous activity within the magma column. If this interpretation has merit, then we might expect to find high-MgO lavas and picrites during former periods of high eruptive activity, or on the deeper, submarine extension of the southwest rift zone. The recent discoveries by *Clague et al.* [1991] of submarine picritic glasses off Kilauea, and by *Garcia et al.* [this volume] of exceptionally abundant submarine picritic lavas along Mauna Loa's southwest rift zone provides support for this interpretation and affords exciting opportunities for further studies.

Acknowledgements. This work was funded by the National Science Foundation. Thoughtful and thorough reviews by M. Garcia, J. Lockwood, S. Mattox, C. Rhodes and J. Wilkinson are greatly appreciated. J. Sparks, P. Dawson and M. Chapman assisted with the XRF analyses. F. A. Frey and P. Ila are thanked for access to the INAA laboratory at MIT and assistance with analyses. T. Grove kindly provided access to the microprobe at MIT, and at UMass, D. Leonard, D. Snoeyenbos assisted with the digital mineral maps. Special thanks to Jack Lockwood and Pete Lipman for introducing me to Mauna Loa.

REFERENCES

Allègre, C. J., M. Treuil, J. F. Minster, J. B. Minster, and F. Albarède, Part I. Fractional crystallization processes in volcanic suites, *Contrib. Mineral. Petrol.*, *60*, 57-75, 1977.

Basaltic Volcanism Study Project, *Basaltic Volcanism on the Terrestrial Planets*, Pergamon Press, New York, 1286 p., 1981.

Bowen, N. L., *The Evolution of Igneous Rocks*, Dover, 334 pp., 1928.

Brigham, W. T., The volcanoes of Kilauea and Mauna Loa on the Island of Hawaii, *Bishop Museum Memoirs, V.2*, 222 p., Honolulu, 1909.

Carmichael, I. S. E., The redox state of basic and silicic magmas: a reflection of their source regions? *Contrib. Mineral. Petrol.*, *106*, 129-141, 1991.

Carmichael, I. S. E. and M. S. Giorso, Oxidation-reduction relations in basic magma: a case for homogeneous equilibria, *Earth Planet. Sci. Lett.*, *78*, 200-210, 1986.

Christie, D. M., I. S. E. Carmichael, and C. H. Langmuir, Oxidation state of mid-ocean ridge basalt glasses, *Earth Planet. Sci. Lett.*, *79*, 397-411, 1986.

Clague, D. A., W. S. Weber and J. E. Dixon, Picritic glasses from Hawaii, *Nature, 353*, 553-556, 1991.

Clarke, D. B., Tertiary basalts of Baffin Bay: possible primary magma from the mantle, *Contrib. Mineral. Petrol.*, *25*, 203-224, 1970.

Clarke, D. B. and M. J. O'Hara, Nickel and the existence of high-MgO liquids in nature, *Earth Planet. Sci. Lett.*, *44*, 153-158, 1979.

Cocherie, A., Systematic use of trace element distribution patterns in log-log diagrams for plutonic suites, *Geochim. Cosmochim. Acta*, *50*, 2517-2522, 1986.

Cox, K. G. and B. G. Jamieson, The olivine-rich lavas of Nuanetsi: a study of polybaric magmatic evolution, *J. Petrol.*, *15*, 269-301, 1974.

Decker, R. W., R. Y. Koyanagi, J. J. Dvorak, J. P. Lockwood, A. T. Okamura, K. M. Yamashita, and W. R. Tanigawa, Seismicity and surface deformation of Mauna Loa volcano, Hawaii, *Eos Trans. AGU*, *64*, 545-547, 1983.

Dingwell, D. B., H. St. C. O'Neill, W. Ertel and B. Spettel, The solubility and oxidation state of nickel in silicate melt at low oxygen fugacity: Results using a mechanically assisted equilibrium technique, *Geochim. Cosmochim. Acta*, *58*, 1967-1974, 1994.

Duggan, T. J., *Petrography and mineral chemistry of Mauna Loa lavas: 1843 to 1984*, M.S. Thesis, University of New Mexico, Albuquerque, New Mexico, 1987.

Eggins, S. M., Petrogenesis of Hawaiian tholeiites: 2, aspects of dynamic melt segregation, *Contrib. Mineral. Petrol.*, *110*, 398-410, 1992.

Ehlers, K., T. L. Grove, T. W. Sisson, S. I. Recca and D. H. Zervas, The effect of oxygen fugacity on the partitioning of nickel and cobalt between olivine, silicate melt and metal, *Geochim. Cosmochim. Acta*, *56*, 3733-3743, 1992.

Elthon, D., High magnesia liquids as the parental magma for ocean floor basalts, *Nature, 278*, 514-518, 1979.

Francis, D., The Baffin Bay lavas and the value of picrites as analogues of primary magmas, *Contrib. Mineral. Petro.*, *89*, 144-154, 1985.

Garcia, M. O., R. A. Ho, J. M. Rhodes, and E. W. Wolfe, Petrologic constraints on rift-zone processes: Results from episode 1 of the Puu Oo eruption of Kilauea Volcano, Hawaii, *Bull. Volcanol.*, *52*, 81-96, 1989.

Garcia, M. O., T. P. Hulsebosch and J. M. Rhodes, Glass and mineral chemistry of olivine-rich submarine basalts, southwest rift zone, Mauna Loa volcano: Implications for magmatic processes, this volume.

Green, D. H. and A. E. Ringwood, The genesis of basaltic magmas, *Contrib. Mineral. Petrol.*, *15*, 103-190, 1967.

Grove, T. L. and W. B. Bryan, Fractionation of pyroxene-phyric MORB at low pressure: An experimental study, *Contrib. Mineral. Petrol.*, *84*, 293-309, 1983.

Grove, T. L. and T. C. Juster, Experimental investigations of low-Ca pyroxene stability and olivine - pyroxene liquid equilibria at 1-atm in natural basaltic and andesitic liquids, *Contrib. Mineral. Petrol.*, *103*, 287-305, 1989.

Grove, T. L., D. C. Gerlach, and T. W. Sando, Origin of calc-alkaline series lavas at Medicine Lake volcano by fractionation, assimilation and mixing, *Contrib. Mineral. Petrol.*, *80*, 160-182, 1982.

Hart, S. R. and K. E. Davis, Nickel partitioning between olivine and silicate melt, *Earth Planet. Sci. Lett.*, *40*, 203-219, 1978.

Helz, R. T. and C. R. Thornber, Geothermometry of Kilauea Iki lava lake, Hawaii, *Bull. Volcanol.*, *49*, 651-658, 1987.

Hofmann, A. W., M. D. Feigenson, and I. Raczek, Case studies on the origin of basalt: III. Petrogenesis of the Mauna Ulu eruption, Kilauea, 1969-1971, *Contrib. Mineral. Petrol.*, *88*, 24-35, 1984.

Holzheid, A., A. Borisov and H. Palme, The effect of oxygen fugacity and temperature on solubilities of nickel, cobalt and molybdenum in silicate melts, *Geochim. Cosmochim. Acta*, *58*, 1975-1981, 1994.

Ila, P., and F. A. Frey, Utilizauon of neutron activation analysis in the study of geologic materials, *Atomkernenerg. Kerntech.*, *44*, 710-718, 1984.

Irvine, T. N., Rocks whose composition is determined by crystal accumulation and sorting, in *The Evolution of the Igneous Rocks: Fiftieth Anniversary Perspectives*, edited by H. S. Yoder, Jr., pp. 245-306, Princeton University Press, Princeton, New Jersey, 1979.

Kelemen, P., H. J. B. Dick and J. E. Quick, Formation of harzburgite by pervasive melt/rock reaction in the upper mantle, *Nature, 358*, 635-641, 1992.

Kilinc, A., I. S. E. Carmichael, M. L. Rivers and R. O. Sack, The ferrous-ferric ratio of natural silicate liquids equilibrated in air, *Contrib. Mineral. Petrol.*, *83*, 136-140, 1983.

Kinzler, R. J., T. L. Grove, and S. I. Recca, An experimental study on the effect of temperature and melt composition on the partitioning of nickel between olivine and silicate melt, *Geochim. Cosmochim. Acta, 54*, 1255-1265. 1990.

Lipman, P. W., J. M. Rhodes, and M. A. Lanphere, The Ninole Basalt - implications for the structural evolution of Mauna Loa volcano, Hawaii, *Bull. Volcanol.*, *53*, 1-19, 1990.

Lockwood, J. P., The volcanic environment of Mauna Loa Observatory, Hawaii - history and prospects, *in* Miller, J. ed.,

Mauna Loa Observatory - a 20th Anniversary Report, *National Oceanic and Atmospheric Administration (NOAA) Special Report*, 28-34, 1978.

Lockwood, J. P., and P. W. Lipman, Holocene eruptive history of Mauna Loa volcano, *U.S. Geol. Surv. Prof. Pap., 1350*, 509-535, 1987.

Lockwood, J. P., R. Y. Koyanagi, R.I. Tilling, R. T. Holcomb and D. W. Peterson, Mauna Loa threatening, *Geotimes, 21*, 12-15, 1976.

Lockwood, J. P., P. W. Lipman, L. D. Petersen and F. R Warshaur, Generalized ages of surface lava flows of Mauna loa Volcano, Hawaii, *U.S. Geol. Surv. Misc. Invest. map I-1908*, 1988.

Maaløe, S., Compositional range of primary tholeiitic magmas evaluated from major element trends, *Lithos, 12*, 59-72, 1979.

Macdonald, G. A., Hawaiian petrographic province, *Bull. Geol. Soc. Amer., 60*, 1541-1598, 1949.

Macdonald, G. A., Composition and origin of Hawaiian lavas, *Geol. Soc. Amer. Memoir, 116*, 477-522, 1968.

Macdonald, G. A. and A. T. Abbott, *Volcanoes in the Sea*, University of Hawaii Press, Honolulu, 441 p., 1970.

McKenzie, D. 1985, The extraction of magma from the crust and mantle, *Earth Planet. Sci. Lett., 74*, 149-157, 1985.

Montierth, C., A. D. Johnston and K. V. Cashman, An empirical glass-composition-based geothermometer for Mauna Loa lavas, this volume.

Moore, R. B., Distribution of differentiated tholeiitic basalts on the lower east rift zone of Kilauea volcano, Hawaii: A possible guide to geothermal exploration, *Geology, 11*, 136-140, 1983.

Morse, S. A., J. M. Rhodes and K. A. Nolan, Redox effect on the partitioning of nickel in olivine, *Geochim. Cosmochim. Acta, 55*, 2373-2378, 1991.

Murata, K. J., and D. H. Richter, Chemistry of the lavas of the 1959-60 eruption of Kilauea volcano, Hawaii, *U.S. Geol. Surv. Prof. Pap., 537-A*, 26 pp., 1966.

Norrish, K., and B. W. Chappell, X-ray fluorescent spectrography, in *Physical Methods in Determinative Mineralogy*, edited by J. Zussman, pp. 161-214, Academic, Orlando, Fla., 1967.

Norrish, K., and J. T. Hutton, An accurate X-ray spectrographic method for the analysis of a wide range of geological samples, *Geochim. Cosmochim. Acta, 33*, 431-454, 1969.

O'Hara, M. J., Are ocean floor basalts primary magmas?, *Nature, 220*, 683-686, 1968

O'Hara, M. J., Geochemical evolution during fractional crystallization of a periodically refilled magma chamber, *Nature, 266*, 503-507, 1977.

Powers, H. A., Composition and origin of basaltic magmas on the Hawaiian islands, *Geochim. Cosmochim. Acta, 7*, 77-107, 1955.

Presnall, D. C., J. R. Dixon, T. H. O'Donnell and S. A. Dixon, Generation of mid-ocean ridge tholeiite, *J. Petrol., 20*, 3-35, 1979.

Rhodes, J. M., Homogeneity of lava flows: Chemical data for historic Mauna Loa eruptions, *J. Geophys. Res., 88A*, 869-879, 1983.

Rhodes, J. M., Geochemistry of the 1984 Mauna Loa eruption: Implications for magma storage and supply, *J. Geophys. Res., 93*, 4453-4466, 1988.

Rhodes, J. M. and S. R. Hart, Episodic trace element and isotopic variations in historical Mauna Loa lavas: Implications for magma and plume dynamics, (this volume).

Rhodes, J. M., and J. P. Lockwood, Chemistry of paired eruptions on the northeast rift zone of Mauna Loa, Hawaii (abstract), *Annual Meeting Geol. Soc. Amer.*, p. 508, Atlanta, 1980.

Rhodes, J. M., and J. W. Sparks, Contrasting styles and levels of magma mixing, Mauna Loa volcano, Hawaii (abstract), *Proc. Inst. Earth. Man Conf: Open Magmatic Systems*, p.135, 1984.

Rhodes, J. M., K. P. Wenz, C. A. Neal, J. W. Sparks and J. P. Lockwood, Geochemical evidence for invasion of Kilauea's plumbing system by Mauna Loa magma, *Nature, 337*, 257-260, 1989.

Rhodes, J. M., J. P. Lockwood, and P. W. Lipman, Episodic variation in magma chemistry, Mauna Loa volcano, Hawaii (abstract), *Generation of Major Basalt Types*, Int. Assoc. of Volcanol. and Chem. of the Earth's Inter., Reykjavik, Iceland, 1982.

Ribe, N. M., Dynamical geochemistry of of the Hawaiian plume, *Eart Planet. Sci. Lett., 88*, 37-46, 1988.

Ringwood, A. E., *Composition and Petrology of the Earth's mantle*, McGraw Hill, 618 pp. 1975.

Roeder, P. L. and R. F. Emslie, Olivine-liquid equilibrium, *Contrib. Mineral. Petrol., 29*, 275-289, 1970.

Sparks, R. S. J. and H. E. Huppert, Density changes during fractional crystallisation of basaltic magmas: Fluid dynamic implications, *Contrib. Mineral. Petrol., 85*, 300-309, 1984.

Spiegelman, M. and P. Kenyon, The requirements for chemical disequilibrium during melt migration, *Earth Planet. Sci. Lett., 109*, 611-620, 1992.

Stolper, E., A phase diagram for mid-ocean ridge basalts: preliminary results and implications for petrogenesis, *Contrib. Mineral. Petrol., 74*, 13-28, 1980.

Stolper, E. and D. Walker, Melt density and the average composition of basalt, *Contrib. Mineral. Petrol., 74*, 7-12, 1980.

Sun, S.-S. and W. F. McDonough, Chemical and isotopic systematics of oceanic basalts: implications for mantle composition and processes. In *Magmatism in the Ocean Basins*, edited by A.D. Saunders and M. J. Norry, *Geol. Soc. Lond. Spec. Publ., 42*, 313-345, 1989.

Tilling, R. I., J. M. Rhodes, J. W. Sparks, J. P. Lockwood, and P. W. Lipman, Disruption of the Mauna Loa magma system by the 1868 Hawaiian earthquake: Geochemical evidence, *Science, 235*, 196-199, 1987.

Wilkinson, J. F. G., and H. D. Hensel, The petrology of some picrites from Mauna Loa and Kilauea volcanoes, Hawaii, *Contrib. Mineral. Petrol., 98*, 326-345, 1988.

Wright, T. L., Chemistry of Kilauea and Mauna Loa lavas in space and time, *U.S. Geol. Surv. Prof. Paper, 735*, 1-49, 1971.

Wright, T. L., Origin of Hawaiian tholeiite: A metasomatic model, *J. Geophys. Res., 89*, 3233-3252, 1984.

Wright, T. L., and R. S. Fiske, Origin of the differentiated and hybrid lavas of Kilauea volcano, Hawaii, *J. Petrol., 12*, 1-65, 1971.

J. M. Rhodes, Department of Geology and Geography, University of Massachusetts, Amherst, MA 01003

Episodic Trace Element and Isotopic Variations in Historical Mauna Loa Lavas: Implications for Magma and Plume Dynamics

J. M. Rhodes

Department of Geology/Geography, University of Massachusetts, Amherst, Massachusetts

S. R. Hart

Department of Geology and Geophysics, Woods Hole Oceanographic Institute,
Woods Hole, Massachusetts

Over the past 152 years, Mauna Loa volcano has erupted lavas with almost constant major element, and compatible and moderately incompatible trace element abundances at a given MgO content. This uniformity is attributed to continuing replenishment of a shallow magma reservoir. In contrast, incompatible element abundances and ratios, together with Sr, Nd and Pb isotopic ratios, vary systematically with time. The greatest rate of change occurred at a time (1843-1887) when Mauna Loa was vigorously active with high eruption rates, presumably a consequence of a high magma supply rate. Detailed analysis confirms what is evident from the isotopic data: that this open-system magmatism requires two or more parental magmas. One has the compositional attributes of lavas erupted in 1843, the other the characteristics of lavas erupted at the summit early in 1880. All other historical lavas can be considered as mixtures of these two end-members, modified by contemporaneous eruption and olivine crystallization. Both parental magmas have Sr, Pb and Nd isotopic ratios typical of magmas in the Hawaiian tholeiitic array, and intermediate between those of Kilauea and Koolau lavas, the end-members of the array. The 1843 parental magma has incompatible element ratios that are similar to, and overlap with the Koolau and Kilauea data. The inferred 1880 parental magma, however, is more depleted than the 1843 parental magma (and most other Hawaiian lavas), and is also isotopically closer to the Kilauea end-member of the tholeiitic array. The origin of these parental magmas is discussed in terms of melting within a radially heterogeneous plume in which the heterogeneity may develop at the source or through subsequent mantle entrainment. Two models are explored, both depend on the location of Mauna Loa at, or close to the plume margin. In the simplest case the parental magmas are produced by progressive melting of the heterogeneous outer plume. The second model is more dynamic, involving melt production and re-equilibration in a diverging, or inclined, plume.

INTRODUCTION

Mauna Loa Volcano, on the island of Hawaii, is the world's largest basaltic shield volcano. Still in the tholeiitic shield-building stage of its development [*Stearns, 1946: Peterson and Moore,* 1987], Mauna Loa is thought to have been active for between 500-800 ka [*Lockwood and Lipman,* 1987; *Lipman,*

this volume]. The age of the oldest sub-aerial lava flows, the Ninole and Kahuku Basalts, though poorly constrained, is about 100-200 ka [*Lipman,* 1980; *Lipman et al.,* 1990]. Most surface flows are considerably younger than 31 ka [*Lipman,* 1980; *Lockwood and Lipman,* 1987]. This paper focuses on the recorded historical activity, which extends for a mere 152 years from 1843 to 1995, the latest eruption having occurred in 1984.

It is essential to document both long- and short-term secular variation in lava compositions at individual volcanoes in order to achieve a better understanding of magma generation processes within the Hawaiian plume, and its subsequent transport, storage and eruption. In particular, this information is necessary to resolve the relative importance of inter- and

Mauna Loa Revealed: Structure,
Composition, History, and Hazards
Geophysical Monograph 92

intra-volcano variations [*Frey and Rhodes*, 1993]. Systematic isotopic and compositional variations have been documented for several individual Hawaiian volcanoes, from the waning tholeiitic shield-building stage through later stages of post-caldera and post-erosional alkali basalt eruptions [e.g. *Chen and Frey*, 1983; *Feigensen et al.*, 1983; *Feigensen*, 1984; *Frey and Roden*, 1987; *Lanphere and Frey*, 1987]. Systematic compositional changes, however, within the voluminous and long-lived tholeiitic stages of these volcanoes are less thoroughly understood, largely because of inadequate temporal control, limited sampling, and the smaller magnitude of the compositional changes. Yet it is from the tholeiitic stage that we are likely to learn most about the dynamics of mantle plumes and associated magma generation and volcanism [*Frey and Rhodes*, 1993].

Early petrologic studies of Mauna Loa and Kilauea volcanoes have emphasized the importance of olivine fractionation in controlling most of the variation of the major element composition of their lavas [*Powers*, 1955; *Murata and Richter*, 1966; *Wright*, 1971]. On Kilauea, Wright and coworkers [*Wright*, 1971; *Wright and Fiske*, 1971; *Wright et al.*, 1975; *Wright and Tilling*, 1980] recognized small but significant, short-term, major element variation at a fixed MgO content for olivine-controlled lavas from among successive historical eruptions. They attributed these changes to the supply of discrete magma batches from the mantle to Kilauea's shallow magmatic plumbing system. *Wright* [1971] also showed that, at a given MgO content, Kilauea lavas tend to increase in incompatible minor elements (e.g. K_2O, TiO_2, P_2O_5) from prehistoric lavas (pre-1750), through 18-19th century lavas, to those of the 20th century. More recent trace element studies tend, broadly, to support this trend [e.g. *Budahn and Schmitt*, 1985; *Tilling et al.*, 1987a,b], although the work of *Casadevall and Dzurisin* [1987] and *Rhodes et al.* [1989] show that relationships must be far more complex than a simple secular trend. Over the long term, however, Kilauea magmas have remained relatively similar in composition, perhaps for as long as 75-100 ka [*Chen et al.*, in press].

The temporal compositional relationships for Mauna Loa lavas, both long- and short-term, are even less clearly understood. *Wright* [1971] found no evidence for either long- or short-term temporal changes in major element compositions. Rhodes and coworkers [e.g. *Rhodes and Lipman*, 1979; *Rhodes et al.*, 1982; *Rhodes*, 1983; *Rhodes and Sparks*, 1984; *Rhodes*, 1988; *Sparks*, 1990] identified small, but significant temporal variations in abundances of highly incompatible elements in historical (1843-1984) lavas, stressing that, apart from episodic fluctuations similar in magnitude to the historical variations, there have been no major, long-term changes in magma chemistry over the last 31,000 years. Others, however, have proposed that, like Kilauea (but in an opposite sense), Mauna Loa lavas are subtly but systematically changing with time from

prehistoric to 20th century lavas, becoming progressively depleted in the highly incompatible elements [*Budahn and Schmitt*, 1985; *Tilling et al.*, 1987a,b]. Similarly, *Kurz and Kammer* [1990] show that lavas erupted in the past 10,000 years have lower $^3He/^4He$ and higher $^{87}Sr/^{86}Sr$ ratios than do older lavas. The implication of these studies is that the source for Mauna Loa magmas was more "plume-like" in the past, and that it is becoming progressively more depleted with time. *Lipman* [this volume], using various forms of evidence, persuasively argues that magma production rates are in decline as Mauna Loa moves away from the plume axis towards its periphery. Similarly, *Watson* [1993] has proposed that Mauna Loa is in transition from the tholeiitic shield-building stage to post-shield volcanism.

In order to evaluate critically both the long- and short-term variations in lava composition, and the implications they have for plume dynamics and the processes of magma generation, supply, storage, and eruption, it is necessary to understand clearly the scale, nature and cause of changes in lava composition that have occurred within the historical record (1843-1984). Within this relatively short period, we have reliable records of the location, time, duration, nature and volume of eruptions [*Brigham*, 1909; *Jaggar*, 1947; *Macdonald and Abbott*, 1970; *Lipman and Swenson*, 1984; *Lockwood and Lipman*, 1987; *Lockwood et al.*, 1987]. With detailed sampling and precise geochemical and isotopic data, it is possible to document accurately short-term temporal variations in lava composition, and to explore in detail the evolution of Mauna Loa's magmatic system during this time period. With this information we can then evaluate long-term fluctuations in the volcano's magmatic history, where temporal constraints are much less certain, and where a well-documented and continuous, eruptive record is not available.

SAMPLING AND ANALYSIS

There have been 33 recorded eruptions on Mauna Loa between 1843 and 1984 [*Macdonald and Abbot*, 1970; *Lockwood and Lipman*, 1987]. With the exception of a large eruption on the NW flank in 1859, all of these eruptions have occurred either within Mokuaweoweo, the summit caldera, or along one of the two prominent rift zones, aligned NE and SW respectively. In many instances, eruptions are initiated in the summit caldera, followed by more voluminous rift zone eruptions within the space of a few days, weeks, or months [*Lockwood and Lipman*, 1987]. A few small-volume eruptions, and those confined to the summit caldera, have been buried by subsequent flows and are no longer accessible. For some eruptions, separate flow fields have developed from vents many kilometers apart at different elevations along the two rift zones (e.g. 1843, 1852, 1880-81, 1926, 1935, 1942, 1984). In this study, all 19 of the eruptions that are currently accessible have been sampled. Typically, this would include spatter from the vents or chain of vents, aa and pahoehoe flow units, and distal, intermediate and proximal samples from each flow field. The number of samples from each eruption varies from as few as 3 for some of the smaller lava flows (1851, 1916), to 67 for the 1984 eruption. Only the 1984 lava was sampled during the course of eruption [*Rhodes*, 1988]. Table 1 presents a summary of these eruptive events and the source of the samples analyzed in this study. Individual eruptions are subdivided on the basis of spatially separate flow fields (e.g. 1942), or on the recognition of geochemically distinct lava types with different MgO contents (e.g. 1950).

About 30-200 g of material was coarse-crushed in a steel percussion mortar, followed by grinding in a Spex tungsten carbide shatterbox. Aliquants of this powder were taken for X-ray Fluorescence Analysis (XRF) and Instrumental Neutron Activation Analysis (INAA). The major elements and the trace elements Rb, Sr, Nb, Zr, Y, Pb, Zn, Ga, Ni, Cr, V were measured by XRF at the University of Massachusetts using modifications of the methods of *Norrish and Chapell* (1967) and *Norrish and Hutton* (1969). All samples were analyzed in duplicate. One or more samples were selected from most eruptive events, depending on the compositional diversity found by the XRF analyses, and analyzed for the rare earth elements (REE) and other trace elements (Sc, Hf, Th, Cr) by INAA. These were analyzed in F. A. Frey's laboratory at the Massachusetts Institute of Technology using the methods of *Ila and Frey* [1984]. Estimates of the analytical precision for both the XRF and INAA data, together with data for the U.S.G.S. basalt standard BHVO-1, are given in *Rhodes* [1988]. A total of 250 samples have been analyzed for major and trace elements by XRF, and 35 of these have been analyzed for the REE and other trace elements by INAA. Table 2 contains analyses of samples for which we have comprehensive XRF and INAA data.

Seven samples, exhibiting maximum compositional diversity and broadly spanning the historical record, were selected for Sr, Nd and Pb isotopic analyses at the Massachusetts Institute of Technology. This isotopic data is given in Table 3. For three of these samples new powders were prepared at M.I.T. under ultraclean conditions, and leached overnight in 6N HCl prior to analysis. The agreement between these separate powder preparations is excellent. Analytical conditions for Sr, Nd and Pb are as given in *Morris and Hart* [1983], with a Pb blank of 180-220 pg. The agreement of the Pb isotopic data with that of *Tatsumoto* [1978] for the 1907, 1926 and 1950 flows is very good, after allowing for differences in absolute values used for $^{208}Pb/^{204}Pb$ in the NBS standards.

COMPOSITION OF MAUNA LOA LAVAS

Previous studies established that the majority of historical (1843-1984) and prehistoric Mauna Loa lavas have olivine-controlled compositions, with almost constant major element abundances at a given MgO content [e.g. *Wright*, 1971; *Rhodes*, 1983; *Tilling et al.*, 1987a; *Rhodes*, 1988; *Rhodes et al.*, 1989]. This observation led *Wright* [1971] to propose that, in contrast to Kilauea, historical Mauna Loa lavas belonged to one, presumably much larger, magma batch. In addition, *Rhodes* [1983, 1988] showed that lavas from the majority of Mauna Loa eruptions are remarkably homogeneous in composition (e.g. 1843, 1851, 1880-81, 1887, 1899, 1907, 1916, 1935, 1940, 1942, 1949, 1975, and 1984), even when erupted from vents up to 20 km apart [*Rhodes and Lockwood*, 1980; *Rhodes*, 1988]. Exceptions are the picritic eruptions of 1852 and 1868, and the very large volume eruptions of 1855, 1859 and 1950. In addition, during the 1926 eruption, "differentiated" lavas were erupted from vents high on the southwest rift zone, and more normal, olivine-controlled lavas from vents at a lower elevation.

The effects of olivine-control are clearly shown in Figure 1, where most analyses plot along a well-defined trend extending towards an olivine composition of about Fo_{87-89}. The vast majority of these analyses are tightly clustered at the low MgO end of this trend between 6.6 and 10.0 weight percent, though most contain less than 8 weight percent MgO, with Mg-values less than 0.6. Rare picritic lavas erupted in 1852 and 1868 extend the olivine-control trend to the high MgO end. Both eruptions produced lavas that have accumulated abundant large olivine phenocrysts (Fo_{84-89}) into less MgO-rich tholeiite melts [*Wilkinson and Hensel*, 1988; *Rhodes*, this volume]. Displaced from the low MgO end of the trend are a few lavas that show the compositional effects of clinopyroxene and plagioclase crystallization as well as that of olivine. These multiply-saturated lavas, termed "differentiated" lavas by Wright and coworkers [e.g. *Wright*, 1971; *Wright and Fiske*, 1971], have MgO contents of less than 6.8 percent and Mg-values below 0.55 (Figure 1). They are rare in Mauna Loa's historical record, and are volummetrically insignificant. They include components of the 1926, 1933, 1950 and 1984 eruptions. All were erupted high on the southwest rift zone, near or within the summit caldera, and presumably reflect fractionation in small, isolated magma pockets. They will not be discussed further in this paper.

Because most Mauna Loa lavas plot on, or close to, an olivine-control line, adjusting the lava chemistry to a constant MgO content readily compensates for the effects of olivine crystallization and accumulation. Following *Wright* [1971] and *Tilling et al.* [1987a], we have adjusted the lava compositions to 7.0 weight percent MgO through the addition or subtraction of olivine of Fo_{87} composition. No immediate petrological significance is attached to these values, other than that they represent opposite ends of the olivine-control trend (Figure 1). *Tilling et al.* [1987a] tested this approach by comparing MgO-adjusted compositions of a picrite and a tholeiite from the same

Table 1: Summary information on eruptive units studied in this paper

Year	Day-Month	Location	Vent Elevation (ft)	Eruptive Volume ($m^3 \times 10^6$)	Basalt Type	MgO Content	Number Samples	Comments
1843	1-10	upper NER	11,500	202	low MgO	6.5	2	upper of two separate flow fields
		middle NER	9,700		low MgO	6.7	3	lower of two separate flow fields
1851	8-08	summit	13,000	35	low MgO	6.9	4	very similar to early 1852 lavas
1852	2-17	summit	12,850	182	low MgO	6.8 to 7.2	4	upper of two flow fields
	2-20	middle NER	8,300		picrites	7.3 to 16.9	8	highly variable, but mostly picritic
1855	8-11	middle NER	10,680	280	var. MgO	7.3 to 9.1	5	variable, high-volume eruption
1859	1-23	north flank	10,530	380	low MgO	6.9 to 7.6	13	variable, large volume eruption
					high MgO	8.1 to 9.8	8	
1868	3-27	lower SWR	2,000	130	high MgO	9.0	2	early lavas
					picrites	16.9 to 22.2	6	late lavas
1877	2-14	submarine	?	?	low MgO	7.8	1	eruption in Kealokekuo Bay
1880 to 1881	5-01	summit	13,120	1	high MgO	10.7	4	most depleted Mauna Loa lava
	11-05	middle NER	10,660	130	low MgO	7.8	3	Kau lobe
			9,800		low MgO	7.6	11	Hilo lobe
1887	1-16	lower SWR	5,950	128	var. MgO	7.5 to 8.5	11	
1899	7-01	middle NER	11,280	81	low MgO	6.9	3	
1907	1-10	lower SWR	6,000	121	low MgO	7.5	8	
1916	5-19	lower SWR	7,000	31	low MgO	7.8	3	
1919	9-26	middle SWR	8,000	183	low MgO	7.2	2	

Table 1: (continued)

Year	Day-Month	Location	Vent Elevation (ft)	Eruptive Volume (m³x10⁶)	Basalt Type	MgO Content	Number Samples	Comments
1926	4-10	upper SWR	12,960	121	low MgO	6.6	3	upper evolved lavas
		lower SWR	7,600		low MgO	7.7	4	lower lavas
1933	12-02	caldera	13,080	100	low MgO	7.1	1	beneath 1949 flow
1935	11-21	upper NER	12,130	87	low MgO	7.0	5	upper flow unit
		middle NER	8,820		low MgO	7.0	3	lower flow unit
1940	4-17	summit	13,300	110	low MgO	7.0	4	
1942	4-26	summit	12,960		low MgO	7.0	3	upper flow unit
		middle NER	9,220	176	low MgO	7.2	5	lower flow unit
1949	1-06	summit	13,400	116	low MgO	7.0	4	
1950	6-01	summit	13,000	376	low MgO	6.3	4	evolved lavas
		upper SWR	12,640		low MgO	6.8	15	large volume, variable composition flow
		middle SWR	10,550		int. MgO	7.5 to 8.4	8	
		middle SWR	9,260		high MgO	8.6 to 12.1	16	
1975	7-05	summit	13,080	30	low MgO	6.5	7	uniform flow
1984	3-26	upper SWR	12,760	220	low MgO	6.4	3	evolved lavas
		upper SWR	13,320		low MgO	6.7	8	very uniform eruption
		caldera	13,100		low MgO	6.7	11	
		upper NER	13,000		low MgO	6.7	12	
		middle NER	9,500		low MgO	6.8	33	

Information on eruption dates and volumes are from *Lockwood and Lipman* [1987]. Vent elevations reflect the highest sample elevation from a vent system. MgO data are averages for uniform eruptions, or ranges where there is significant variability.

Table 2. Selected analyses of historical Mauna Loa lavas.

Sample	ML-17	ML-20	ML-80	ML-288	ML-52	ML-284	ML-14	ML-46	ML-71
Sample type	spatter	aa	spatter	spatter	spatter	spatter	spatter	spatter	spatter
Eruption Year	1843	1843	1851	1852	1852	1852	1852	1855	1859
Date	Jan-10	Jan-10	Aug-08	Feb-17	Feb-20	Feb-20	Feb-20	Aug-11	Jan-23
Location	NER	NER	Summit	NER	NER	NER	NER	NER	N. Flank
SiO_2	51.70	51.60	51.61	51.37	51.34	50.71	48.63	50.98	50.10
TiO_2	2.17	2.15	2.14	2.09	2.11	1.96	1.60	2.07	1.98
Al_2O_3	13.73	13.90	13.75	13.63	13.72	12.82	10.30	13.30	12.57
Fe_2O_3*	11.99	11.91	11.83	11.84	11.82	11.88	12.03	12.10	12.36
MnO	0.16	0.17	0.16	0.17	0.16	0.17	0.16	0.17	0.16
MgO	6.58	6.70	6.85	7.65	7.33	10.23	16.93	8.27	9.64
CaO	10.39	10.50	10.15	10.43	10.24	9.66	7.91	10.21	9.81
Na_2O	2.20	2.04	2.15	2.41	2.12	2.19	1.75	2.28	1.95
K_2O	0.46	0.47	0.45	0.43	0.44	0.42	0.34	0.41	0.37
P_2O_5	0.28	0.26	0.26	0.25	0.26	0.25	0.20	0.25	0.22
Total	99.67	99.70	99.35	100.27	99.54	100.29	99.85	100.05	99.16
Mg-Value	0.547	0.553	0.560	0.587	0.577	0.655	0.756	0.601	0.632
Rb	7.3	7.0	6.8	6.4	7.1	6.7	5.5	6.7	6.1
Sr	354	365	342	339	345	329	260	331	299
Nb	11.5	10.1	10.7	10.1	10.2	9.5	7.7	10.7	9.8
Zr	142	147	140	137	139	129	103	135	128
Y	23.1	23.7	23.3	24.4	23	22.2	17.6	22.7	21.7
Zn	127	103	115	113	115	112	115	117	118
Ga	19	21	20	21	20	19	15	19	19
Ni	95	99	99	103	127	280	774	141	203
Cr	213	214	274	323	310	504	1050	414	560
V	274	268	278	268	272	258	203	271	257
La	12.1	10.7	10.9	10.0	10.3	9.6	8.1	9.8	8.8
Ce	27.2	28.3	26.2	27.4	27.7	24.9	20.9	25.4	24.1
Nd	17.4	18.2	17.5	17.2	18.4	16.3	13.8	17.2	15.3
Sm	4.93	4.77	4.82	4.71	5.02	4.42	3.88	4.89	4.37
Eu	1.76	1.77	1.77	1.69	1.73	1.62	1.34	1.70	1.61
Yb	2.02	1.99	2.08	1.92	1.92	1.81	1.53	1.95	1.86
Lu	0.27	0.27	0.27	0.27	0.27	0.25	0.23	0.28	0.27
Hf	3.33	3.49	3.32	3.35	3.32	3.11	2.69	3.30	3.03
Th	0.5	0.6	0.5	0.5	0.4	0.5	n.d.	0.6	0.3
Cr	205	213	263	316	327	516	996	405	566
Sc	30.1	29.7	29.9	29.7	29.7	27.9	23.9	29.8	29.0

Fe_2O_3* is total Fe expressed as Fe_2O_3. Mg-Value is Mg/(Mg+Fe) after adjusting Fe^3/total Fe to 0.1. Lower Cr by INAA.

Table 2. (continued)

Sample	ML-86	ML-83	ML-85	ML-41	ML-45	ML-95	ML-47	ML-91	ML-62A
Sample type	pahoehoe	aa	spatter	spatter	spatter	spatter	spatter	spatter	spatter
Eruption Year	1868	1868	1868	1880	1880	1887	1899	1907	1916
Date	Mar-27	Mar-27	Mar-27	May-01	Nov-05	Jan-16	Jul-01	Jan-10	May-19
Location	SWR	SWR	SWR	Summit	NER	SWR	NER	SWR	SWR
SiO_2	50.74	50.97	46.88	50.54	51.94	51.93	51.65	51.76	51.40
TiO_2	2.04	2.04	1.32	1.75	1.96	1.99	2.10	2.04	2.04
Al_2O_3	12.82	12.71	8.55	12.02	13.26	13.29	13.56	13.51	13.25
Fe_2O_3*	12.13	12.08	12.53	12.43	12.00	12.16	12.13	11.97	11.93
MnO	0.17	0.16	0.16	0.17	0.17	0.17	0.16	0.17	0.16
MgO	8.97	9.18	21.91	10.75	7.75	7.55	6.89	7.51	7.63
CaO	10.05	9.96	6.66	9.17	10.13	10.15	10.54	10.43	10.42
Na_2O	2.07	2.09	1.21	1.85	2.13	2.02	2.13	1.84	2.09
K_2O	0.40	0.40	0.26	0.28	0.34	0.33	0.36	0.36	0.36
P_2O_5	0.23	0.24	0.15	0.17	0.21	0.21	0.22	0.22	0.22
Total	99.63	99.83	99.63	99.14	99.89	99.80	99.73	99.80	99.50
Mg-Value	0.619	0.626	0.794	0.656	0.587	0.577	0.556	0.580	0.585
Rb	6.3	6.7	4.1	4.3	5.3	5.2	5.6	5.3	5.1
Sr	315	316	205	244	278	275	294	293	299
Nb	10.5	10.2	6.3	8.0	8.7	9.1	9.9	9.5	9.2
Zr	132	133	85	108	123	123	132	132	131
Y	22.8	22.9	14.5	20.9	22.7	23.3	23.8	22.9	23.0
Zn	111	112	111	119	115	116	117	116	115
Ga	19	18	13	18	20	19	19	20	19
Ni	199	220	1097	227	111	110	92	127	130
Cr	464	489	1395	741	384	330	262	365	339
V	252	251	171	236	253	258	267	265	260
La	10.0	9.1	5.7	6.7	8.2	8.0	8.8	8.5	8.7
Ce	24.7	25.0	15.0	19.5	20.7	21.5	22.0	24.4	23.6
Nd	16.6	16.5	11.1	13.5	14.9	15.0	15.3	16.1	16.4
Sm	4.82	4.87	2.86	4.09	4.34	4.52	4.76	4.73	4.61
Eu	1.68	1.66	1.05	1.45	1.61	1.60	1.63	1.70	1.64
Yb	2.04	1.89	1.22	1.85	2.00	2.02	2.12	2.02	2.01
Lu	0.26	0.28	0.18	0.28	0.27	0.28	0.28	0.31	0.28
Hf	3.36	3.12	1.99	2.88	3.08	3.19	3.34	3.13	3.30
Th	0.3	0.4	n.d.	0.2	0.4	0.3	0.3	0.2	0.3
Cr	443	494	1315	689	384	331	251	331	347
Sc	28.7	28.9	20.1	27.7	29.3	30.0	29.9	30.5	30.6

Table 2. (continued)

Sample	HAW-8	HAW-9	ML-34	ML-129	ML-37	ML-81	ML-72	ML-269	ML-65
Sample type	aa	aa	aa	spatter	glass	spatter	spatter	spatter	spatter
Eruption Year	1919	1926	1935	1940	1942	1949	1950	1950	1950
Date	Sep-26	Apr-10	Nov-21	Apr-17	Apr-26	Jan-06	Jun-01	Jun-01	Jun-01
Location	SWR	SWR	NER	Summit	NER	Summit	Summit	SWR	SWR
SiO_2	51.94	51.69	51.30	51.73	51.34	51.55	51.78	51.32	51.18
TiO_2	2.10	2.04	2.06	2.05	2.05	2.04	2.30	2.05	1.94
Al_2O_3	13.81	13.48	13.54	13.70	13.58	13.62	13.72	13.59	13.08
Fe_2O_3*	12.04	11.47	12.05	12.06	12.13	12.10	12.27	11.95	12.14
MnO	0.17	0.17	0.17	0.17	0.17	0.16	0.16	0.17	0.17
MgO	7.14	7.81	6.99	7.03	7.07	6.99	6.33	7.14	8.71
CaO	10.64	10.49	10.62	10.59	10.52	10.51	10.33	10.44	10.08
Na_2O	2.31	2.25	2.23	2.17	2.18	2.08	2.54	2.35	2.00
K_2O	0.38	0.35	0.36	0.37	0.36	0.37	0.46	0.37	0.36
P_2O_5	0.23	0.24	0.22	0.23	0.22	0.22	0.27	0.23	0.22
Total	100.76	99.99	99.54	100.10	99.62	99.64	100.16	99.61	99.87
Mg-Value	0.566	0.600	0.561	0.562	0.562	0.560	0.532	0.568	0.612
Rb	5.6	5.5	6.0	5.7	5.8	5.6	7.2	5.8	5.0
Sr	307	303	305	303	304	303	319	306	291
Nb	10.7	9.7	9.2	9.1	9.5	9.3	11.6	8.9	9.1
Zr	135	131	129	132	130	132	159	134	123
Y	23.5	22.9	23.4	23.6	23.1	23.6	28	24	22.3
Zn	108	131	109	125	116	117	121	120	116
Ga	22	20	19	18	19	19	21	20	19
Ni	96	127	87	86	91	89	87	111	191
Cr	289	342	278	284	290	296	200	295	420
V	257	258	258	274	270	269	287	276	258
La	8.9	8.5	8.9	8.8	9.1	8.9	11.0	8.9	8.4
Ce	24.2	25.5	23.7	24.6	23.9	23.9	31.4	24.3	24.4
Nd	16.6	1.7	16.7	17.4	15.3	16.2	20.6	16.4	15.6
Sm	4.87	4.66	4.76	4.71	4.59	4.84	5.33	4.52	4.46
Eu	1.72	1.70	1.66	1.67	1.64	1.69	1.84	1.64	1.59
Yb	2.01	1.99	1.99	1.97	2.15	2.04	2.34	2.02	1.88
Lu	0.31	0.27	0.29	0.26	0.29	0.29	0.31	0.27	0.27
Hf	3.25	3.27	3.13	3.29	3.24	3.34	3.84	3.06	3.19
Th	0.4	0.3	0.2	0.4	0.5	0.3	0.7	n.d.	0.3
Cr	307	356	272	270	280	278	194	274	420
Sc	30.9	30.5	30.6	30.4	29.7	31.0	29.8	29.8	29.6

Table 2. (continued)

Sample	ML-28	ML-66	ML-76	ML-164	ML-160	ML-177	ML-190	ML-210
Sample type	aa	aa	spatter	spatter	spatter	spatter	quench	quench
Eruption Year	1950	1950	1975	1984	1984	1984	1984	1984
Date (Mo-Day)	Jun-01	Jun-01	Jul-05	Mar-25	Mar-25	Mar-25	Mar-25	Apr-15
Location	SWR	SWR	Summit	SWR	Summit	NER	NER	NER
SiO_2	50.78	50.31	51.81	51.76	51.38	51.29	51.29	51.44
TiO_2	1.82	1.71	2.10	2.18	2.08	2.08	2.08	2.06
Al_2O_3	12.56	11.99	13.72	13.57	13.64	13.63	13.69	13.63
Fe_2O_3*	11.83	11.90	12.30	12.24	12.04	12.09	12.03	12.10
MnO	0.16	0.17	0.17	0.17	0.18	0.17	0.19	0.19
MgO	10.77	12.10	6.55	6.42	6.67	6.66	6.65	6.74
CaO	9.65	9.34	10.48	10.40	10.53	10.53	10.49	10.46
Na_2O	2.12	2.15	2.11	2.47	2.69	2.65	2.68	2.55
K_2O	0.34	0.31	0.38	0.43	0.38	0.38	0.38	0.39
P_2O_5	0.20	0.18	0.23	0.28	0.26	0.24	0.23	0.25
Total	100.24	100.16	99.85	99.92	99.85	99.73	99.70	99.80
Mg-Value	0.667	0.691	0.540	0.536	0.549	0.548	0.549	0.551
Rb	4.7	4.6	5.8	6.7	5.7	5.4	5.5	5.5
Sr	283	266	306	317	315	315	315	317
Nb	8.2	8.2	9.9	10.4	9.5	9.3	9.3	8.6
Zr	123	110	137	151	138	137	136	133
Y	21.3	19.5	24.0	25.2	23.8	24.4	24.1	24.0
Zn	111	109	120	120	119	117	120	126
Ga	19	17	20	20	20	19	20	20
Ni	341	425	83	80	90	91	89	86
Cr	689	788	194	197	231	225	222	234
V	235	222	277	275	275	270	277	270
La	8.0	7.4	9.3	10.2	9.0	9.0	9.0	9.1
Ce	22.7	19.5	24.9	29.5	25.6	25.6	25.7	25.1
Nd	14.4	13.3	17.1	20.1	17.8	17.6	17.2	17.2
Sm	4.01	3.71	4.79	5.48	4.70	4.93	4.85	4.68
Eu	1.46	1.34	1.72	1.82	1.69	1.70	1.69	1.70
Yb	1.90	1.66	2.09	2.25	2.10	2.09	2.12	2.07
Lu	0.24	0.22	0.29	0.31	0.28	0.27	0.27	0.28
Hf	2.83	2.65	3.30	3.81	3.43	3.29	3.22	3.35
Th	0.4	0.3	0.5	0.6	0.5	0.4	0.3	0.5
Cr	645	724	194	217	233	228	231	246
Sc	27.8	26.7	30.9	30.7	31.4	31.0	31.1	31.0

Table 3:   Isotopic data for historical Mauna Loa lavas.

Sample	Eruption	Location		$^{87}Sr/^{86}Sr$	$^{143}Nd/^{144}Nd$	$^{206}Pb/^{204}Pb$	$^{207}Pb/^{204}Pb$	$^{208}Pb/^{204}Pb$
ML-76	05-Jul, 1975	N.E.R.		0.703824 +/- 24(2)	0.512891 +/- 9(3)	18.142	15.459	37.856
ML-81	06-Jan, 1949	Summit		0.703857 +/- 20(2)	0.512900 +/- 12(2)	18.132	15.469	37.886
ML-95	16-Jan, 1887	S.W.R.		0.703808 +/- 26(1)	0.512891 +/- 11(3)	18.184	15.466	37.876
ML-45	11-Nov, 1880	N.E.R.		0.703783 +/- 18(2)	0.512906 +/- 14(2)	18.141	15.457	37.872
			(L)	0.703800 +/- 30(1)	0.512907 +/- 20(1)			
				0.703792 +/- 15	0.512906 +/- 12			
ML-41	01-May, 1880	Summit		0.703810 +/- 30(1)	0.512943 +/- 20(1)	18.156	15.443	37.843
			(L)	0.703830 +/- 30(1)	0.512901 +/- 20(1)			
				0.703820 +/- 21	0.512922 +/- 14			
ML-52	20-Feb, 1852	N.E.R.		0.703906 +/- 27(1)	0.512838 +/- 18(1)	18.074	15.462	37.868
			(L)	0.703940 +/- 29(1)	0.512876 +/- 14(1)			
				0.703923 +/- 20	0.512857 +/- 11			
ML-20	10-Jan, 1843	N.E.R.		0.703916 +/- 16(3)	0.512862 +/- 19(1)	18.076	15.450	37.830

Numbers in parentheses are the number of separate mass spectrometer runs averaged for each listed value. (L) are separate powders prepared at M.I. T. and leached overnight in 6N HCl. Sr and Nd ratios are relative to 0.70800 for the E & A standard, and 0.512640 for the BCR-1 syandard. Quoted errors are 2-sigma. Pb ratios have been corrected for mass fractionation by comparison to NBS-981, using the absolute values of Todt et al. (1989). Reproducibility of the Pb data is better than 0.04%/amu.

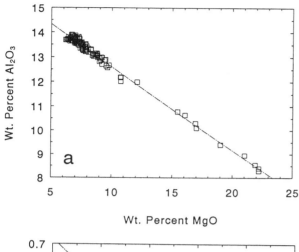

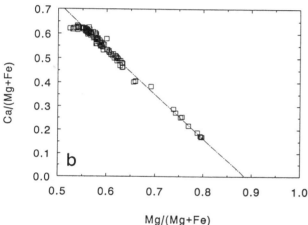

Fig. 1. (a) Al$_2$O$_3$ versus MgO for individual samples (188) of historical Mauna Loa lavas. Only 5 of the 67 samples analyzed from the 1984 eruption [*Rhodes*, 1988] are included in this plot to avoid domination by the very uniform compositions that characterize this eruption. The regression line fitted to the data (R^2 = 0.99) defines an olivine-control trend. (b) Ca/(Mg + Fe) versus Mg/(Mg + Fe) for individual samples of historical Mauna Loa lavas. Most samples plot along a well-defined olivine-control trend towards Fo$_{87-89}$. A few lavas with low Mg/(Mg + Fe) show the effects of clinopyroxene and plagioclase fractionation. These are the "differentiated" lavas of [*Wright* 1971].

1868 eruption, and found that the adjusted compositions agreed very closely.

When MgO-adjusted lavas are examined for temporal variations, we find little significant variation in the other major elements or the compatible and moderately incompatible trace elements. Figure 2, a plot of MgO-adjusted Y and Sc abundances versus the year of eruption, illustrates this point very clearly. Both Y and Sc remain essentially constant throughout the eruptive record.

The more incompatible elements, on the other hand, show greater diversity, even when adjusted to a constant MgO

content. Figure 3 is a spidergram in which the MgO-adjusted average data for a few selected eruptions are compared with an MgO-adjusted average Mauna Loa composition [*Rhodes et al.*, 1989]. Clearly, there is a greater spread in the data for the highly incompatible elements than for less incompatible elements. These differences are systematic. Lavas erupted prior to 1880 have higher incompatible element abundances, whereas those erupted after 1880 have abundances that range from below average to about average. *Tilling et al.* [1987a], using data from *Wright* [1971] and *Rhodes* [1983], showed similar differences in the MgO-adjusted data between lavas erupted prior to 1868 and those erupted afterwards. They speculated that this apparently abrupt change in magma composition was in some way related to disruption of the magmatic plumbing system by the great 1868 earthquake which preceded the eruption.

K, Sr and La are the most effective elements for evaluating temporal changes in incompatible element abundances. The low concentrations of Rb, Th and Nb, and therefore larger relative errors (+/- 0.4 ppm), make them less useful. It also follows from Figure 3 that ratios of highly incompatible to less incompatible elements (e.g. K/Y, K/Ti, Sr/Y, La/Yb) will provide information comparable to that of the MgO-adjusted data.

The temporal changes in incompatible element abundances and ratios are explored with the aid of Figures 4 and 5. In these

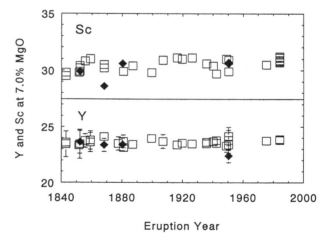

Fig. 2. MgO-normalized Sc and Y abundances versus time for historical Mauna Loa lavas. The data have been normalized to a constant MgO content of 7.0% through addition or subtraction of olivine (Fo$_{87}$) [*Wright*, 1971; *Tilling et al.*, 1987a]. The Sc data reflect analyses of individual samples (Table 2). Errors (2σ) for single analyses are less than the symbol size. The Y data are averages for each eruption, or sub-unit of an eruption, summarized in Table 1. The standard deviation (2σ) for each unit or sub-unit is shown as a vertical bar. These bars are absent when the standard deviation (2σ) is less than the symbol size. Olivine-controlled lavas are shown as open squares (□) and picritic and high MgO (>10%) lavas as solid diamonds (◆).

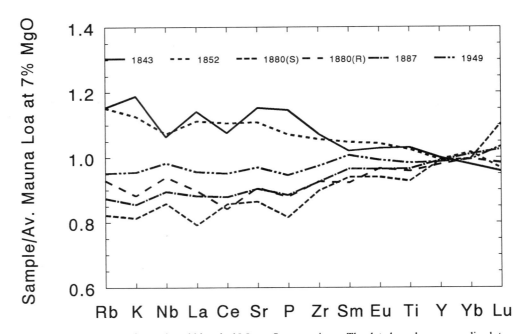

Fig. 3. Spidergram showing lavas from selected historical Mauna Loa eruptions. The data have been normalized to a constant MgO content of 7.0% through addition or subtraction of olivine (Fo$_{87}$), and then referenced to an average Mauna Loa composition [*Rhodes et al.*, 1989]. The reference values (at 7.0% MgO) are Rb - 6.0; K$_2$O - 0.39%; Nb - 9.8; La - 9.3; Ce - 25.1; Sr - 313; P$_2$O$_5$ - 0.24%; Zr - 134; Sm - 4.8; Eu - 1.7; TiO$_2$ - 2.08%; Y - 23.6; Yb - 2.05; Lu - 0.28.

diagrams, lavas with more than 10 weight percent MgO and picrites are distinguished from the other, more "normal", olivine-controlled lavas. This distinction is made because the mineralogy and bulk composition of the picrites and MgO-rich lavas are such that it appears that they may have bypassed the shallow magma reservoir from which most Mauna Loa lavas are thought to have been erupted [*Rhodes and Sparks*, 1984; *Wilkinson and Hensel*, 1988; *Rhodes*, this volume]. If this is the case, the picrites and their associated lavas could be more diverse in composition, and will not necessarily exhibit the same time-related variations as magmas that have been mixed and homogenized in the magma reservoir.

Systematic changes in composition with time are readily apparent for the olivine-controlled lavas from Figures 4 and 5. In both diagrams there is a progressive decrease with time in MgO-adjusted incompatible element abundances and incompatible element ratios from 1843 to the period 1880-1887. The most depleted lava is a summit lava that has been correlated with the May, 1880 eruption [*Macdonald*, 1971]. After 1887 there is an increase in these abundances and ratios until about 1919, after which they remain roughly constant up to the most recent eruption in 1984. Similar, but less dramatic, variations are found for K/Sr, La/Sm, Zr/Y ratios. It is important to note that there are also small variations in ratios involving elements of similar incompatibility (e.g. K/Nb, La/Nb,

Zr/Nb), whereas others (e.g. K/Rb, Sr/Nb and La/Ce) remain roughly constant.

The greatest change occurs in only 44 years, between 1843 and 1887. The relative decrease in the MgO-adjusted K$_2$O content is about 0.8%/year, and Sr decreases by about 0.6%/year. Similarly, changes in the incompatible element ratios range from 1.0%/year for chondrite-normalized La/Yb to 0.8%/year for K/Y. The rate of increase in these elements and their ratios from 1887 to 1919 is only about half the rate of decrease in the preceding 44 years. Consequently, the higher incompatible element abundances that characterize lavas erupted between 1843 and 1868 are never again realized within the historical record (Figures 4 and 5). It is important to stress that within this 44 year span the range in composition is as large as the differences upon which *Budhan and Schmitt* [1985] and *Tilling et al.* [1987a,b] based their proposed secular trends from prehistoric to historical lavas.

Tilling et al. [1987a], with a less comprehensive data set at their disposal, drew attention to some of these compositional differences, proposing a major disruption in MgO-adjusted incompatible element abundances following the 1868 earthquake and eruption. Our data do not confirm this suggestion. There is a steady decline in incompatible element abundances and ratios prior to, and after 1868 (Figures 4-5). Lavas erupted between 1868 and 1880 are critical in resolving

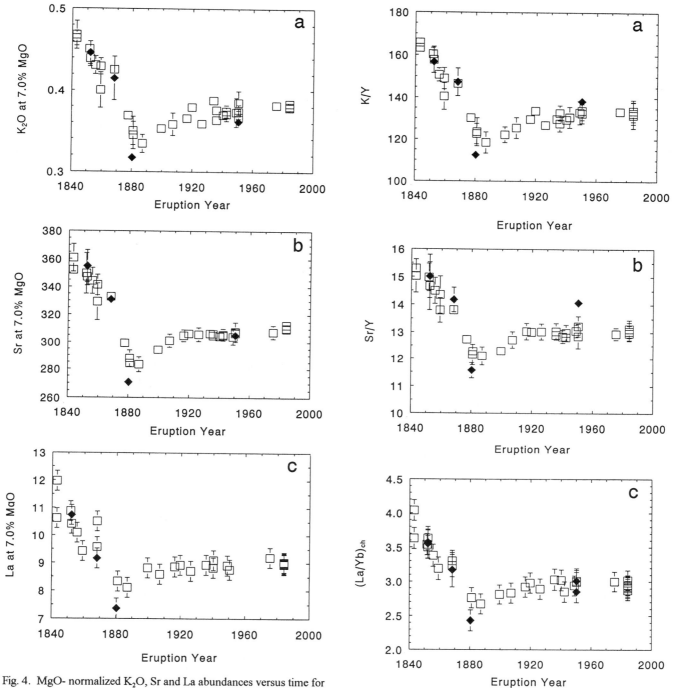

Fig. 4. MgO- normalized K₂O, Sr and La abundances versus time for historical Mauna Loa lavas. The data have been normalized to a constant MgO content of 7.0% through addition or subtraction of olivine (Fo₈₇). The K₂O and Sr data are averages of each eruption or sub-unit (Table 1) and the vertical bars reflect the standard deviation (2σ) for each unit or sub-unit. These bars are absent when the standard deviation (2σ) is less than the symbol size. The La data reflect individual analyses (Table 2), with (2σ) error bars. The symbols are the same as in Fig. 2.

Fig. 5. K/Y, Sr/Y and chondrite-normalized La/Yb versus time for historical Mauna Loa lavas. The K/Y and Sr/Y data are averages of each eruption or sub-unit (Table 1) and the vertical bars reflect the standard deviation (2σ) for each unit or sub-unit. These bars are absent when the standard deviation (2σ) is less than the symbol size. The La/Yb data are for individual analyses (Table 2), with (2σ) error bars. The chondrite abundances are from *Boynton* [1984]. The symbols are the same as in Fig. 2.

these two interpretations. Although there was continuous and vigorous lava lake activity in Mokuaweoweo caldera between 1872 and 1876, regrettably, none of this remains exposed because of later filling of the caldera [*Lockwood and Lipman*, 1987]. There was, however, a shallow submarine flank eruption in 1877 in Kealakekua Bay [*Moore et al.*, 1985]. The composition of a single sample from this eruption is intermediate between 1880 and earlier lavas, confirming that the change in composition between 1843 and 1880-1887 is gradational, and was in progress both before and after the 1868 earthquake.

Of more significance is the sudden change from steadily declining incompatible element abundances and ratios through 1843 and 1887, to an increase in these values between 1887 and about 1919. We note that this change coincides with a marked decrease in the eruption rate from about 0.047 km^3/year to about 0.021 km^3/year [*Lockwood and Lipman*, 1987].

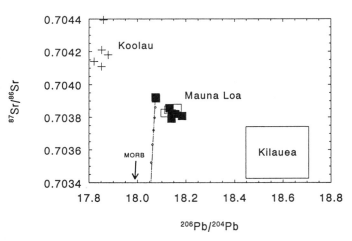

Fig. 7. ^{87}Sr/^{86}Sr versus ^{206}Pb/^{204}Pb for historical Mauna Loa lavas. Solid symbols are from this paper (Table 3). Open symbols are from *Kurz and Kammer* [1991]. Shown for comparison are data for Koolau volcano [*Roden et al.*, 1994] and a field for historical Kilauea lavas [*O'Nions et al.*, 1977; *Tatsumoto*, 1978; *White and Hofmann*, 1982; *Hofmann, et al.*, 1984; *Stille et al.*, 1986]. Note how the Mauna Loa trend follows the trend for Hawaiian tholeiitic shield-building lavas [*West et al.*, 1987] between Kilauea and Koolau volcanoes. A hypothetical mixing line between the most enriched 1843 Mauna Loa lavas and a depleted MORB source (DMM, *Zindler and Hart*, 1986) is shown for reference. Because of alteration problems associated with the Koolau lavas only samples with K$_2$O/P$_2$O$_5$ > 1.5 [*Frey et al.*, 1994] have been plotted in this and subsequent figures.

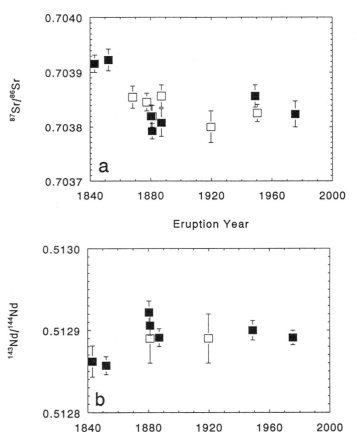

Fig. 6. ^{87}Sr/^{86}Sr and ^{143}Nd/^{144}Nd versus time for historical Mauna Loa lavas. Solid symbols are from this paper (Table 3). Error bars are from Table 3. Open symbols are from *O'Nions et al.* [1977] and *Kurz and Kammer* [1991].

We analyzed seven historical lavas for Sr, Nd, and Pb isotopic ratios (Table 3). The data compare closely with the very limited isotopic data previously available for historical Mauna Loa lavas [*O'Nions et al.*, 1977; *Tatsumoto*, 1978; *Kurz and Kammer*, 1990]. A ^{87}Sr/^{86}Sr ratio of 0.70331 reported for a sample of the 1881 flow by *O'Nions et al.* [1977] was clearly in error, and has been reanalyzed giving a ratio of 0.70382 [R. K. O'Nions pers. comm. to S. R. Hart, 1989]. In Figure 6, the ^{87}Sr/^{86}Sr and ^{143}Nd/^{144}Nd essentially mimic the changes in incompatible trace element data with time. ^{87}Sr/^{86}Sr drops sharply between 1843 and 1880, rising slightly in the more recent lavas, whereas ^{143}Nd/^{144}Nd, although not as pronounced, shows an inverse relationship. Note that ^{87}Sr/^{86}Sr data from the literature are consistent with these trends, and that several samples from the submarine 1877 eruption, and a single sample of the 1868 flow [*Kurz and Kammer*, 1990], have values intermediate between our data for the 1843/1852 and 1880/1887 eruptions. Temporal changes in the Pb isotopic data are not as readily resolved, but the oldest, most incompatible element-enriched lavas (1843 and 1852) have the lowest ^{206}Pb/^{204}Pb (Table 3). In a plot of ^{87}Sr/^{86}Sr versus ^{206}Pb/^{204}Pb (Figure 7), there is a distinct linear trend. This trend follows that of the trend established for Hawaiian shield-building lavas [e.g. *Staudigal et al.*, 1984; *West et al.*, 1987], from Koolau lavas at one extreme to Kilauea and Loihi lavas at the other.

Several points should be made concerning the isotopic data. First, $^{87}Sr/^{86}Sr$ is positively correlated with the increase in incompatible elements and their ratios (Figure 8), whereas $^{143}Nd/^{144}Nd$ and $^{206}Pb/^{204}Pb$ are both negatively correlated. Although this is what one might expect intuitively, these correlations are opposed to the overall trend for several mature Hawaiian volcanoes, where incompatible element-enriched post-shield alkali basalts have lower $^{87}Sr/^{86}Sr$ than earlier shield-building tholeiites [e.g. *Chen and Frey*, 1983; *Feigensen*, 1984; *Frey and Roden*, 1987; *Lanphere and Frey*, 1987]. Second, the temporal changes are not large relative to analytical error, and it is doubtful if any significance would have been attached to these differences without the corroborating evidence of the correlated trace element data. Third, the

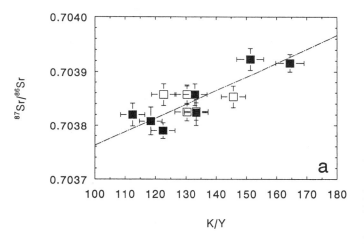

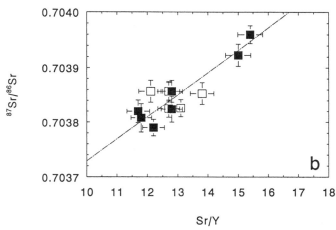

Fig. 8. Correlation of $^{87}Sr/^{86}Sr$ with K/Y and Sr/Y for historical Mauna Loa lavas. Solid symbols are from this paper (Table 3). Open symbols are from *Kurz and Kammer* [1991]. Error bars are at 2σ. The regression line is for data from this paper.

difference in $^{87}Sr/^{86}Sr$ in only 44 years (1843-1887) is almost as large (70%) as the entire range reported by *Kurz and Kammer* [1990] over a 12,000 year period. Only much older lavas (>28 ka) appear to have significantly lower (<0.70375) $^{87}Sr/^{86}Sr$ ratios.

DISCUSSION

From the preceding section it is clear that although Mauna Loa lavas have remained relatively uniform in composition throughout the past 152 years in terms of major element, and compatible and moderately incompatible trace element abundances, varying mostly in response to olivine fractionation and accumulation, there have been small, but systematic changes in the abundances of the incompatible elements and in the isotopic ratios of Sr, Nd, and possibly Pb. What, then, are the magmatic processes responsible for these short-term temporal variations? Surely it is not a coincidence that the most dramatic changes occurred at a time when Mauna Loa was enjoying its most vigorous magmatic activity within the historical record? Eruptions were frequent between 1843 and 1880, and towards the end of this period, between 1872 and 1876, there was sustained lava lake activity in the summit caldera [*Lockwood and Lipman*, 1987]. The eruption rate between 1843 and 1880 of 0.047 km³/year is about twice the eruption rate (0.021 km³/year) of subsequent years [*Lockwood and Lipman*, 1987, Figure 18.7], and may well reflect an unusually high magma supply rate.

In the following discussion we explore the magmatic processes that could give rise to these short-term changes in magma composition. In particular, we will emphasize the importance of open-system magmatism, the nature of the parental magma(s), and their implications for magma generation in the Hawaiian plume.

Crystal Fractionation of a Parental Magma

As noted above, most lavas are on an olivine-control trend, with the consequence that any change in incompatible element concentration should be accompanied by even larger correlated changes in compatible elements such as Mg, Ni, and Cr. These are not observed, and furthermore any effects of olivine fractionation have been removed or minimized by adjusting the data to a constant MgO value or by using element ratios. Only if the lavas had differentiated beyond olivine-control might it be possible to account for the changes in composition with time by simple crystal fractionation. In contrast with Kilauea Volcano, such differentiated lavas are rare on Mauna Loa and imply quite different magmatic plumbing systems for the two volcanoes.

Recently, *Langmuir* [1989] has proposed that *in situ* crystallization occurring along the walls and roofs of magma reservoirs is much more effective in changing magma compositions than is Rayleigh crystal fractionation, or even "open-system" processes [*O'Hara*, 1977]. The efficacy of this process is dependent on multi-phase crystallization, particularly of late-stage minerals, in a boundary layer between the magma and the walls of the reservoir, with the return of the residual liquid to the magma reservoir. If plagioclase, clinopyroxene, and possibly ilmenite, magnetite and apatite, were crystallizing in a boundary layer, changes in Sr, Ti and P abundances would be expected. These elements should not correlate with changes in highly incompatible elements, as they would be preferentially partitioned into the crystallizing phases. Similarly, we would expect to see the effects of clinopyroxene and plagioclase fractionation in the major elements accompanying any increase in incompatible elements or change in element ratios. These predictions are perhaps realized in the rare examples of "differentiated" lavas, where *in situ* crystallization may have played a role in their development. The vast majority of lavas, however, are olivine-controlled. They reveal no evidence of significant multi-phase crystallization in their major element compositions, and changes in abundances of Sr, Ti and P correlate positively with changes in other incompatible elements. Therefore, *in situ* crystallization cannot be responsible for the compositional changes observed at Mauna Loa. In summary, then, the continuing fractionation of a single parental magma reservoir may serve to account for the relative simplicity and uniformity of major element variations, but it cannot appropriately account for the short-term temporal changes in incompatible element abundances and ratios or in isotopic ratios.

Open-System Magmatism

Following *O'Hara* [1977], the concept of a periodically replenished, tapped and fractionated (RTF) magma chamber has gained wide acceptance. Rhodes and coworkers [*Rhodes et al.*, 1982; *Rhodes and Sparks*, 1984; *Rhodes*, 1988] have maintained that most lavas erupting on Mauna Loa over the past 152 years are supplied from just such a homogenized, well-mixed and continuously replenished, shallow magma reservoir located beneath the summit caldera. Possible exceptions include the rare picritic lavas, the large volume 1950 eruption, and differentiated lavas (MgO < 6.7%). The evidence supporting this assumption is detailed in *Rhodes* [1988], but will be briefly reviewed here.
(1) Most Mauna Loa eruptions produce lavas with remarkably homogeneous compositions [*Rhodes*, 1983], even when erupting from vents at different elevations and locations along the summit caldera and active rift zones, or from vent systems many km. in length. The best documented of these is the recent 1984 eruption [*Rhodes*, 1988]. Here, the variations in lava composition are not much larger than the analytical error, even

though about $220 \times 10^6 \, m^3$ of lava was erupted along a 20 km. vent system over a three week period.
(2) The majority of lavas have compositions that are "perched" at the low MgO end of an olivine control trend (Figure 1). This concentration of lava compositions at the intersection of the olivine-control trend and a trend for differentiated lavas [*Rhodes*, 1988, Figures 7-8] is all the more remarkable in view of the fact that, upon eruption, the lavas tend to be multiplely-saturated, crystallizing plagioclase, clinopyroxene and occasional low-Ca pyroxene, as well as olivine. The failure of Mauna Loa magma to fractionate beyond this point is attributed to the "buffering" effects of continuing magma replenishment [*O'Hara*, 1977; *O'Hara and Mathews*, 1981; *Defant and Nielsen*, 1990].
(3) Similarly, these low-MgO lavas plot in a tight cluster, around the pigeonite reaction point, in an olivine-clinopyroxene-silica pseudoternary phase diagram [*Grove et al.*, 1982; *Rhodes*, 1988, Figure 8]. This is exactly where one would expect a long-lived steady-state magma to plot in this projection since, following replenishment, subsequent fractionation of the magma must proceed inexorably towards the reaction point [*Rhodes*, 1988].
(4) Gases from Mauna Loa's summit have low CO_2/SO_2 ratios [*Greenland*, 1987], lower than the CO_2/SO_2 ratios typical of Kilauea's summit, but similar to gas ratios from Kilauea's east rift zone. Such low ratios are thought to result from degassing and attaining equilibrium over a long period of time at shallow levels within the volcano's plumbing systems.
(5) *Decker et al.* [1983] showed that long-term geodetic and seismic data were consistent with the continuing accumulation of magma in a shallow storage reservoir about 3-4 km beneath the summit caldera. Since it is the locus of inflation and deflation, this reservoir has all the characteristics of a continually replenished magma system. Immediately following the 1984 eruption, Mauna Loa began to inflate, presumably in response to continuing magma supply [*Lockwood et al.*, 1987], and has continued to do so to the present [*Okamura et al.*, 1992].

Most RTF models assume a single parental magma, attributing changes in composition of the erupted lavas to fluctuations in the recharge rate, and to variations in the proportions of magma that erupt and crystallize. Although powers bordering on the occult have been invoked to achieve changes in magma composition by open-system magmatism [*O'Hara*, 1977; *O'Hara and Mathews*, 1981], recent studies question the efficacy of the process [e.g. *Cox*, 1988; *Langmuir*, 1989; *Defant and Nielsen*, 1990]. Clearly, the changes in isotopic ratios preclude such a simple model for Mauna Loa's historical lavas. Nevertheless, given the importance of open-system magmatism to petrology, and considering the compelling evidence outlined above for open-system magmatism on Mauna Loa, we believe it would be instructive to evaluate the process further. We have good control on such parameters as eruption rates and volumes, rates of compositional change and crystallization paths, and can make reasonable assumptions about magma supply rates. A more detailed analysis will be presented elsewhere [*Rhodes*, in preparation]. The obvious question is what, if any, of the temporal compositional changes found in the historical lavas could be a consequence of open-system processes involving a single parental magma? Alternatively, as the isotopic data indicate, are two or more magmas required?

Starting with a single parental magma, three points need to be examined: (1) Can the incompatible element-enriched 1843-1852 lavas be produced from this parental magma; (2) will a higher recharge rate of this same parental magma lead to the progressive reduction in incompatible elements, and change in element ratios, observed between 1843 and 1887; and (3) will a subsequently reduced recharge rate from 1887 onwards lead to an increase in incompatible element abundances and ratios, and to the eventual establishment of a quasi steady-state magma between 1935 and 1984? The marked reduction in the eruption rate after 1887 (which presumably reflects a reduction in the magma supply rate) from about 0.047 km³/year to about 0.021 km³/year [*Lockwood and Lipman*, 1987 Figure 18.7] is consistent with these ideas. Furthermore, with the exception of the large-volume 1950 eruption and occasional "differentiated" lavas, lavas erupted from 1935 onwards are remarkably similar in composition, and have all the earmarks of a steady-state magma. The minor differences could be attributed to slight fluctuations in the balance between magma supply and crystallization.

The most suitable candidate for the parental magma is the 1880 summit lava, or perhaps a picritic precursor to this lava. It is fine-grained, high in MgO (10.7%), and does not appear to have accumulated olivine. It is the most depleted lava erupted on Mauna Loa (including prehistoric lavas, Rhodes, unpublished data) and, although broadly similar, is compositionally distinct from preceding and subsequent lavas, including the 1880-1881 rift zone lavas erupted later in the same year. For these reasons, we believe that it has bypassed the shallow magma reservoir in a manner similar to the 1959 eruption on Kilauea [*Wright*, 1973; *Helz*, 1987].

The approach we adopt is that of *O'Hara and Mathews* [1981] modified according to the conventions of *Defant and Nielsen* [1990], where the amount of recharged magma and the amount of magma lost from the system through eruption or dike intrusion, are referenced to the amount of magma that crystallizes in a given time. The effects of assimilation are assumed to be unimportant or unresolvable. Since we are assuming a steady-state system, the magma lost from the system through crystallization, eruption or intrusion is compensated for by recharge of the parental magma. For example, a [4,0,3] model implies a recharge rate that is four times the

crystallization rate, with no assimilation, and an eruption (or intrusion) rate three times that of crystallization. The ratio of magma crystallizing to that which is erupted, or intruded, (x/y) is 0.33.

The problem of generating the incompatible element-enriched magmas erupted in 1843 and shortly afterwards from an 1880 parental magma by open-system fractionation is formidable. The major difficulty is that, in order to achieve the necessary increase in incompatible element abundances (e.g. K, La), very low recharge rates and correspondingly high crystallization to eruption (or intrusion) rates (<[2,0,1] are required, with x/y ratios >1. If the simulation by *Defant and Nielsen* [1990] of open-system magmatism for Kilauea is appropriate for Mauna Loa, then, at these low recharge rates, phases other than olivine will dominate the crystallization process, and the lavas should depart from well-defined olivine-control trends. This is clearly not the case. At higher recharge rates, consistent with olivine-controlled lavas, steady-state compositions are produced before the high incompatible element abundances of the 1843-1852 lavas can be attained. Furthermore, as pointed out by *Cox* [1988], the high x/y ratios required to effectively change incompatible element abundances and ratios will also increase the less incompatible elements. This results in calculated abundances of Zr, Y and Yb that are too high, and K/Y, K/Zr and La/Yb ratios that never approach the values in the lavas.

Can the rapid decrease in incompatible element abundances and ratios between 1843 and 1887 be simulated by recharge of the 1843 magma reservoir by the 1880 parental magma? It is only when very high recharge rates (>[8,0,7]) and low x/y ratios (0.1-0.14) are used that appropriate concentrations and ratios are obtained. At these high recharge rates, olivine alone, or olivine plus clinopyroxene, is the likely crystallizing phase [*Nielsen*, 1990]. In contrast with the previous discussion, the targeted 1887 lava can be simulated very closely, as can the progressive reduction in incompatible element abundances and ratios between 1843 and 1887. The required high recharge rate does not seem unreasonable. This was a period of intense eruptive activity; there was an active lava lake in the summit caldera between 1872 and 1876, and the eruption rate of 0.047 km³/year was twice that from 1887 onwards [*Lockwood and Lipman*, 1987, Figure 18.7] and well above the value of ≈0.02 km³/year estimated by *Lipman* [this volume] for the last 4 ka of Mauna Loa's history.

If the recharge rate lessened after 1887, perhaps by half, was this sufficient to reverse the decline in incompatible element abundances and ratios, leading to a gradual increase in these values, and ultimately to the establishment of a quasi steady-state magma from 1935 onwards? In order to achieve a steady-state magma with incompatible element abundances similar to those of lavas erupted between 1935-1984, moderate recharge rates (≈[4,0,3]) are required. Lower rates result in steady-state magmas with much higher incompatible element abundances, whereas higher rates never achieve the concentrations of the 1935-1984 magmas. This estimate does not seem unreasonable in view of the reduced eruption rate during this period. There are, however, problems with this model also. Somewhat higher recharge rates are required by the *Defant and Nielsen* [1994] simulation of open-system magmatism for Kilauea in order to maintain steady-state magmas on, or close to, the olivine control trend. More importantly, in order to achieve appropriate abundances for the highly incompatible elements (K and La), the abundances of Zr, Y and Yb are again overestimated. Conversely, Sr is too low. This is because plagioclase should be an important crystallizing phase at these low to moderate recharge rates [*Nielsen*, 1990].

In summary, in consort with the isotopic evidence, it appears that open-system magmatism involving a single parental magma, alone is incapable of accounting for the systematic temporal changes in incompatible element abundances and ratios in Mauna Loa's historical lavas. On the other hand, there are compelling arguments (summarized earlier) that are supportive of the process. Also, we have shown that the rapid decrease from 1843 to 1887 in incompatible element abundances and their ratios can be explained as a consequence of continuing recharge of a magma reservoir containing magma corresponding in composition to the 1843 lavas by a parental magma similar in composition to the 1880 summit magma. The process, however, is not a simple mixing of the 1843 magma with the 1880 magma, since this would lead to correlated increases in MgO and Ni, accompanying the decrease in incompatible element abundances and ratios. As these are not observed, we infer that crystallization, dominated by olivine, has accompanied the recharge of 1880 parental magma. In other words, an RTF process. Since neither the 1843 magma nor the quasi-steady state magmas erupted after 1935 can be produced from the 1880 summit magma by RTF processes, and since these magmas are isotopically distinct, it seems inescapable that more than one parental magma is required. In the following section, we will test and explore the consequences of this assertion.

Multiple parental magmas

The most extreme compositions erupted during Mauna Loa's 152 year historical eruptive record in terms of incompatible element abundances and isotopic ratios are the 1843 lavas and the May 1880 summit lavas. The 1880 lavas are the most depleted in incompatible elements, whereas the 1843 lavas are the most enriched. We showed earlier that the lavas erupted between 1843 and 1887 can largely be considered to be mixtures of these two magmas, modified by contemporaneous eruption and crystallization. Figures 9-10 show that all of the historical lavas can be described within close limits as mixtures of these two end-component magmas. Element ratios are used

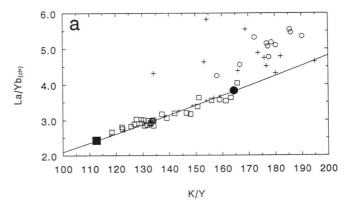

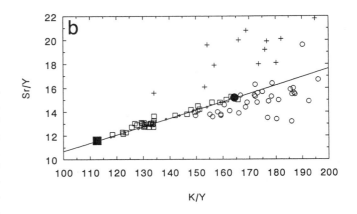

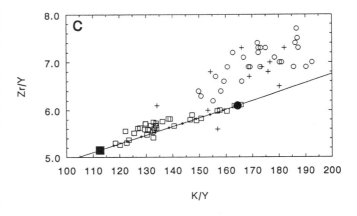

Fig. 9. Chondrite-normalized La/Yb, Sr/Y and Zr/Y versus K/Y for historical Mauna Loa lavas. The K/Y and Sr/Y data are averages of each eruption or sub-unit (Table 1), but the La/Yb data reflect individual analyses (Table 2). The chondrite abundances are from *Boynton* [1984]. The data closely follow a mixing trend calculated between inferred 1843 (●) and 1880 (■) parental magmas (Table 4). The small open diamonds along this trend reflect 10% intervals of mixing. Shown for comparison are data for Koolau volcano (+) [*Frey et al.*, 1994] and historical lavas from Kilauea volcano (O) [*Rhodes et al.*, 1989, and unpublished data].

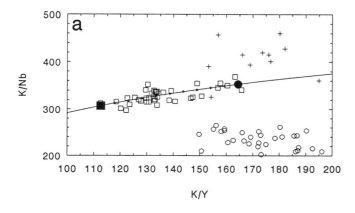

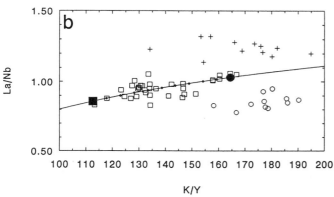

Fig. 10. K/Nb and La/Nb versus K/Y for historical Mauna Loa lavas. The K/Nb ratios are averages of each eruption or sub-unit (Table 1), whereas the La/Nb data reflect individual analyses (Table 2). The mixing curve and data for Koolau and Kilauea volcanoes are the same as in Fig. 9.

in preference to abundances to remove the complicating effects of recharge and crystallization. We do not know, of course, that the end-components of these mixing curves necessarily have exactly the same compositional characteristics as the 1843 and 1880 summit lavas. They could be more extreme in composition, but there is no indication of such magmas in the eruptive record. Indeed, the 1843 and 1880 summit lavas are comparable in incompatible element abundances and ratios to the most extreme compositions erupted during the past 31 ka of Mauna Loa's eruptive history [Rhodes, unpublished data].

These observations can be reconciled with the concept of open system magmatism, argued for above, by assuming that at least two parental magmas supply the volcano, and that the proportions in which these magmas are supplied must fluctuate. This fluctuation appears to be related to the eruption rate, and therefore possibly to the supply rate. When eruption rates are high, such as during the period 1843 to 1887, the depleted parental magma is dominant; at lower eruption rates there will be a more equal balance between the two. For example, the long period of quasi steady-state magmas erupted between 1935

and 1984 can be simulated by assuming a 40:60 supply of enriched to depleted magmas. The fact that highly enriched lavas were erupted between 1843 and 1852 implies that prior to this time the supply must have been dominated by the enriched parental magma. This in turn could imply a low eruption and supply rate before 1843, an inference that is consistent with eruption rates deduced from estimates of areal coverage during the very recent prehistoric record [*Lockwood and Lipman*, 1987; *Lockwood*, this volume].

Alternatively, the volcano could be supplied by a range of magmas, with compositions that fluctuate between the extremes of the 1843 and 1880 summit lavas. The consequences of these two alternatives are indistinguishable.

Melting, Source Heterogeneity, and Implications for the Hawaiian Plume

Since the compositional and temporal trends of the historical lavas can be thought of as mixtures of end-member parental 1843 and 1880 summit magmas in an open-magma system, we need to understand how these two magmas are produced and the implications for magmatic processes and source heterogeneity in the Hawaiian plume. Their compositional characteristics are summarized in Table 4, where they have been normalized to a nominal 13% MgO through olivine (Fo_{88}) addition. This is done to facilitate comparison, and in recognition of the fact that the most MgO-rich melts identified among the historical lavas are in equilibrium with Fo_{87-89} olivine and contain close to 13 weight percent MgO [*Wilkinson and Hensel*, 1988; *Rhodes*, this volume]. We do not infer that these are necessarily primary magma compositions. Indeed, *Rhodes* [this volume] speculates that the magma column may be stratified, becoming more picritic with depth.

Inspection of Table 4 shows that there is little difference in major element abundances between the two inferred parental magmas. The greatest differences are observed for the incompatible elements, these differences decreasing in the order La, K≈Rb, P, Sr, Nb≈Ce, Zr, Ti, Sm, Y, Yb. This ranking is very similar to the ranking of D's for garnet peridotite [e.g. *Hart and Dunn*, 1993; *Johnson*, 1993] and could be indicative of melting processes in the presence of garnet. Similarly, the almost constant Y and Yb abundances of Mauna Loa magmas (Figures 3-4) and similar abundances of these elements in the two parental magmas (Table 4) provide strong evidence for residual garnet [*Hofmann et al.*, 1984]. Residual garnet is also required for Hawaiian tholeiites from an analysis of Sm/Nd and Lu/Hf isotopic data in oceanic basalts [*Salters and Hart*, 1991].

The inferred 1843 and 1880 parental magmas are also isotopically distinct (Table 3), with implications for heterogeneity in the plume source of Mauna Loa magmas. From an isotopic perspective, mantle heterogeneity is frequently discussed as mixtures of four end-member components [e.g.

Table 4: Estimate of Parental Magma Compositions.

	1843	1880	Percent Difference
SiO_2	49.73	50.21	-0.9
TiO_2	1.83	1.65	+10.3
Al_2O_3	11.59	11.41	+1.5
Fe_2O_3	12.11	12.29	-1.5
MnO	0.17	0.17	0
MgO	13.00	13.00	
CaO	8.84	8.65	+2.1
Na_2O	1.87	1.89	-1.1
K_2O	0.40	0.27	+31.8
P_2O_5	0.24	0.17	+29.2
Rb	6.0	4.2	+30.0
Sr	303	230	+24.1
Nb	9.3	7.3	+21.5
Zr	122	103	+16.0
Y	20.0	19.9	+0.5
Zn	96	117	-21.9
Ga	17	17	0
La	9.59	6.27	+34.6
Ce	23.4	18.3	+21.6
Sm	4.18	3.84	+8.1
Yb	1.69	1.74	-2.9

Zindler and Hart, 1986]: depleted MORB mantle (DMM); two enriched mantle components (EM1 and EM2); and a high U/Pb component (HIMU). Hawaiian shield-building tholeiites form a well-defined linear array within this tetrahedral isotopic space [e.g. Zindler and Hart, 1986; West et al., 1987; Hart et al., 1992]. Tholeiitic lavas from Koolau Volcano define one end of this trend, close to estimates of the composition of "bulk-earth" and EM1, and lavas from Kilauea and Loihi volcanoes define the other end [e.g. *Staudigel et al.*, 1984; *West et al.*, 1987; *Frey and Rhodes*, 1993; *Roden et al.*, 1994]. This heterogeneity in the Hawaiian plume may have originated in several ways. It could be intrinsic, inherited at the plume source, or the plume may be radially zoned because of entrainment [e.g. *Griffiths and Cambell*, 1991; *Hart et al.*, 1992; *Hauri et al.*, 1994].

It is beyond the scope of this paper to attempt to integrate the historical Mauna Loa data with the entire Hawaiian data base. Consequently, in the subsequent discussion we use data

generated in our laboratories for historical Kilauea lavas [*Rhodes et al.*, 1989, and unpublished data] and from Koolau [*Frey et al.*, 1994; *Roden et al.*, 1994] as end-members to illustrate the compositional and isotopic range for the Hawaiian tholeiitic data array. Historical Mauna Loa lavas have isotopic ratios that are intermediate between the Koolau and Kilauea end-members and, although restricted in range, their trend closely follows the overall trend for shield-building tholeiites (Figure 7). Lavas that are most strongly influenced by the relatively enriched 1843 parental magma plot closer to the Koolau end of the trend, whereas the more depleted lavas, influenced by the 1880 parental magma, plot closer to the Kilauea end. Figure 7 also shows that the Mauna Loa trend, like the trend for Hawaiian tholeiites in general, and unlike the trend for post-erosional lavas, is not influenced by a depleted MORB mantle [*West et al.*, 1987].

There is, however, no such simple Hawaiian tholeiitic trend for the incompatible trace elements and their ratios. *Frey and Rhodes* [1993] have emphasized that, although there are significant and persistent long-term major and trace element abundance differences between Hawaiian tholeiitic shields, these differences do not, in general, correlate with the differences in isotopic ratios. An important exception is Nb. Abundances of Nb, adjusted to 7% MgO, decrease from an average of 17 ppm in historical Kilauea lavas [*Rhodes et al.*, 1989] to an average of 9 ppm in Koolau lavas [*Frey et al.*, 1994]. Consequently, element ratios involving Nb (especially Sr/Nb and Zr/Nb) correlate positively with $^{87}Sr/^{86}Sr$ and inversely with $^{206}Pb/^{204}Pb$, and provide a useful guide to source differences [*Frey and Rhodes*, 1993; *Roden et al.*, 1994; *Yang et al.*, 1994; *Chen et al.*, in press; *Rhodes*, in press]. There are very small, but nonetheless significant differences in ratios involving Nb (e.g. Zr/Nb, K/Nb and La/Nb, Figure 10) in Mauna Loa lavas as a consequence of the differences between the 1843 and 1880 parental magmas. Nevertheless, these ratios, like the Sr, Nd and Pb isotopic ratios, imply source components for both parental magmas that are intermediate in composition between those of Koolau and Kilauea. This is best illustrated in terms of Sr/Nb versus $^{87}Sr/^{86}Sr$ (Figure 11a).

Inspection of Figures 9 and 11b shows that other incompatible element ratios do not reflect the differences in the source isotopic ratios. The data fields for Koolau tholeiites and historical Kilauea lavas have similar ratios of La/Yb, K/Y and Zr/Y and are frequently overlapping. Furthermore, although lavas influenced by the 1843 parental magma plot close to the fields for Koolau and historical Kilauea lavas, the progressive influence of the 1880 parental magma results in trends that move away from the cluster of Koolau and Kilauea data towards a more depleted composition. Not only does this depletion apply to element ratios that are considered relatively easy to fractionate through melting processes (e.g. K/Y, La/Yb, Sr/Y, Zr/Y, Figure 9), but it is also evident in the ratios of elements

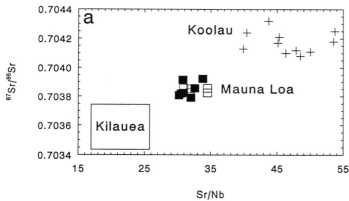

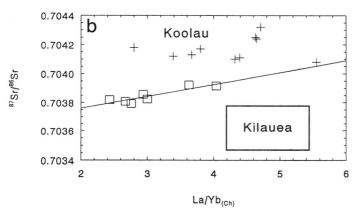

Fig. 11. Sr/Nb and chondrite-normalized La/Yb versus $^{87}Sr/^{86}Sr$ in Hawaiian lavas. Data for Mauna Loa are from this paper, (■) and *Kurz and Kammer* [1991], (□). Data for Koolau volcano (+) are from *Frey et al.*[1994] and *Roden et al.* [1994], and data for the fields for Kilauea volcano are from *O'Nions et al.* [1977]; *White and Hofmann* [1982]; *Hofmann, et al.* [1984]; *Stille et al.* [1986]; *Rhodes et al.* [1989, and unpublished data].

that have similar D's, making fractionation less likely (e.g. La/Nb, K/Nb and K/La, Figure 10). These element ratios, moreover, are thought to be relatively constant in the oceanic mantle (*Sun and McDonough*, 1989).

In summary, the following observations need to be accounted for in order to explain the origin of the proposed 1843 and 1880 parental magmas:-

1) They have essentially identical major element and compatible and moderately incompatible trace element abundances.

2) Isotopic ratios of Sr, Pb and Nd, and trace ratios involving Nb are intermediate between end-member Koolau and Kilauea ratios along the Hawaiian tholeiitic array. The 1880 parental magma is closer to the Kilauea end of the array.

3) The 1880 parental magma is more depleted than the 1843 magma, and is more depleted than is typical for Hawaiian tholeiites in general.

4) High eruption rates (and therefore magma supply rates?) appear to be related to the supply of the depleted 1880 parental magma to the magmatic plumbing system.

The crux of the problem is to produce an 1880 parental magma that is isotopically closer to Kilauea (stronger plume influence?) and yet is more depleted than typical Hawaiian tholeiites. Two possible explanations are explored. Both depend on the location of Mauna Loa at, or close to, the plume periphery, and both require heterogeneity within the plume. One involves the progressive melting of small scale heterogeneities near the plume margin. The other, a more dynamic approach, involves melt production and re-equilibration in a diverging or inclined plume.

Progressive melting of small scale heterogeneities. Uniform major element concentrations accompanied by changes in incompatible element abundances are to be expected when melts are produced by varying amounts of melting within the mantle. If the amount of melting is relatively small, one might anticipate the correlated changes in La/Yb. Sr/Y, K/Y etc. that are seen in Figure 9 between the inferred 1843 and 1880 parental magmas (the trends in Figure 9 for individual historical Mauna Loa eruptions are attributed to mixing of these parental magmas) On the other hand, isotopic ratios should remain constant throughout the range of melting. On the basis of such observations, *Hofmann et al.* [1984] interpreted a monotonic decrease in MgO-adjusted incompatible element abundances throughout the 1969-1971 Mauna Ulu eruption as reflecting a progressive 20 percent increase in equilibrium partial melting of a uniform mantle source. If the differences in the MgO-adjusted incompatible element abundances in the 1843 and 1880 Mauna Loa parental magmas were produced in a similar manner, it follows that melt production has fluctuated markedly in the last 152 years. *Watson and McKenzie* [1991] estimate that the average amount of aggregated tholeiitic melt produced in a Hawaiian melting column is about 6-7%. If the changes in the concentrations of the more incompatible elements are inversely proportional to changes in the melt fraction, it follows that the 1880 parental magma was produced by about 40 percent more melting than the 1843 parental magma. In other words, following Watson and McKenzie, there was a difference of about 6-9 percent in the amount of melt produced. This inference is in keeping with the observed high eruption rates during the period (1843-1887) during which the 1880 parental magma was dominating the magma system, these rates being roughly double those of ensuing years.

On Mauna Loa, however, unlike Mauna Ulu, the isotopic ratios have also changed, albeit slightly, along with the incompatible element abundances (Figures 6, 8). The ratios of highly incompatible elements such as K/Nb and La/Nb have also changed slightly (Figure 10). These ratios, however, all involve Nb, which, as noted earlier, correlates with isotopic ratios, and they are therefore likely to be source dependant. The implications are that the sources for the two parental magmas must be somewhat different in composition, the source of the 1880 parental magma being isotopically closer to that of Kilauea. This conclusion must also be reconciled with our earlier inference that the 1880 parental magma is related to high magma supply rates. It seems likely, therefore, that the source for Mauna Loa magmas is heterogeneous on a scale that is small relative to the scale of melting, and that progressive melting produces correlated changes in isotopic and incompatible element ratios. The result is larger volumes of the depleted 1880 parental magma. This implies small-scale heterogeneity within the Hawaiian plume, either as a consequence of heterogeneity at the plume source, or through entrainment of surrounding mantle [e.g. *Griffiths and Cambell*, 1991; *Hart et al.*, 1992].

In terms of plume-mantle entrainment, debate continues about which end of the Hawaiian tholeiitic array is representative of the plume component, and what is the source and nature of the entrained material. The high ^{3}He/^{4}He in Loihi and Kilauea lavas have been used to argue that these magmas represent the plume component, and that they are formed close to the plume axis [e.g. *Staudigal et al.*, 1984; *Kurz and Kammer*, 1991; *Kurz et al.*, this volume]. Conversely, Hart and coworkers [*Hart et al.*, 1992; *Hauri et al.*, 1994] show that many plume arrays converge in Sr, Nd, and Pb isotopic space towards a focus zone (their FOZO component), which they attribute to entrained lower mantle at the plume periphery. According to this scenario, Kilauea and Loihi magmas, with Sr, Nd and Pb isotopic ratios close to the proposed FOZO component, would contain a greater proportion of the entrained lower mantle, whereas Koolau magmas may be closest to a plume component. *Lipman* [this volume], however, argues that magma production on Mauna Loa has declined dramatically during the past 100 ka as a consequence of migration away from the plume axis towards its periphery. This is in accord with a marked temporal decrease in ^{3}He/^{4}He [*Kurz and Kammer*, 1991; *Kurz et al.*, this volume], and would seem to imply that Kilauea (and Loihi) magmas are formed closer to the plume axis than Mauna Loa magmas. If this is the case, then the characteristics of the source of the Mauna Loa magmas could be ascribed to entrainment into the plume periphery of an EM1 component dispersed throughout the lower mantle [*Hauri et al.*, 1994]. Initial melting of this heterogeneous mixture might lead to the production of the 1843 parental magma as the entrained component was partially consumed, eventually leading to production of the 1880 parental magma as more of the plume component is progressively incorporated into the melt.

Melting and melt-matrix reaction in a diverging or inclined plume. Alternatively, the isotopic and geochemical characteristics of the source for Mauna Loa magmas, and in particular the depleted 1880 parental magma, might be a consequence of both the location of this source near the plume periphery and the dynamics of plume-lithosphere interaction. As the plume approaches the base of the lithosphere it will be deflected away from the plume axis. This will result in matrix flow-lines that become less steep, and a reduction in the rate of melt production [e.g. *Ribe,* 1988; *Eggins,* 1992]. For a volcano such as Mauna Loa thought to be currently close to the plume margin [*Lipman,* this volume] this could mean that the upper parts of its melting column (assumed to be vertical [*Ribe,* 1988]) will intersect and interact with material from the plume interior that has already undergone substantial prior melting, and melt extraction, closer to the plume axis. According to *Ribe* [1988], there is sufficient thermal energy in an ascending plume magma to actually melt a significant fraction of the overlying lithosphere. If correct, a similar mechanism might continue melting of this already depleted plume material, contributing a depleted melt component, but with a Kilauea-like isotopic signature, to the more fertile melts produced at deeper levels in the Mauna Loa melting column. Additionally, *Navon and Stolper* [1987] point out that as ascending melts traverse regions of the mantle with which they are not in equilibrium they will probably interact chemically with that mantle, given sufficient time and contact. The precise nature of this interaction is difficult to evaluate, and could include chromatographic exchange, melting of the matrix and crystallization of the melt, depending whether or not the melt and matrix are in chemical, thermal or phase equilibrium [e.g. *Eggins,* 1992; *Kelemen et al.,* 1994]. Thus, if melt extraction is highly efficient near the plume axis, on reaching this overlying barren zone percolating magma in Mauna Loa's ascending magma column might interact with the depleted matrix, possibly resulting in a more depleted magma with isotopic characteristics approaching those of the plume interior. According to these scenarios, the 1880 parental magma will contain a greater proportion of melt from the upper part of the melting column than the 1843 parental magma. If the plume is inclined as a result of lithospheric drag [*Frey and Rhodes,* 1993, Figure 6], the scheme outlined above might be even more effective for volcanoes like Mauna Loa now thought to be located above the trailing edge of the plume [*Lipman,* this volume].

Figure 12 attempts to illustrate schematically the concepts outlined above of plume heterogeneity and melting or melt-matrix reaction in a deflected or inclined plume as mechanisms for providing Mauna Loa with parental magmas with different trace element and isotopic signatures. Whichever scheme is appropriate, it must be capable of accounting for an increase in the magma supply rate when depleted parental magma dominates the supply of magma to Mauna Loa's magmatic plumbing system. This, of course, is inherent in the progressive melting model. In the second model, the supply of magma will depend on how much of the depleted 1880 magma is produced in the upper part of the melting column. This in turn may well

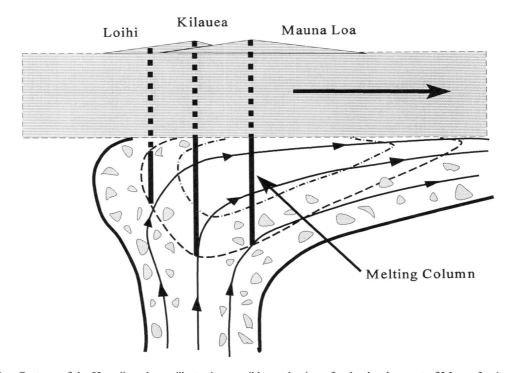

Loihi   Kilauea   Mauna Loa

Melting Column

Fig. 12.   Cartoon of the Hawaiian plume, illustrating possible mechanisms for the development of Mauna Loa's parental magmas. The model is loosely based on that of *Watson and McKenzie* [1991], with a 100 km diameter plume deflected by 70 km thick lithosphere, and melting initiated at a depth of about 130 km. The upper part of the plume is inclined as a consequence of lithospheric drag, and to provide an explanation for the origins of observed inter-shield geochemical characteristics [*Frey and Rhodes*, 1993, Fig. 6]. Flow-lines of the plume matrix are shown by thin arrows, and the near-vertical melt segregation - reaction paths (*Ribe*, 1988; *Eggins*, 1992) as thick lines. Heterogeneities within the plume, as a consequence of mantle entrainment or initiated at the plume source, are shown as irregular stippled blobs.

depend on melt segregation, either by porous flow or flow through a network of channels. Channelized flow in the upper part of the plume would inhibit the production of the depleted 1880 parental magma, resulting in low magma supply rates dominated by the 1843 parental magma.

CONCLUSIONS

Historical Mauna Loa lavas, erupted over the past 152 years, have compositions that are dominated by olivine-control, with almost constant major element, and compatible and moderately incompatible trace element abundances at a given MgO content. Most lavas have compositions that are "perched" at the low-MgO end of the olivine-control trend as a consequence of continuing replenishment of a shallow magma reservoir. Incompatible element abundances and ratios, however, together with Sr, Nd and Pb isotopic ratios, change systematically over time. The greatest change, involving a marked decrease in incompatible element abundances and ratios and $Sr^{87}Sr^{86}$ occurred between 1843 and 1887. These values increased from 1887 to about 1919, leveling off and subsequently becoming

almost constant. The greatest rate of change occurred at a time when Mauna Loa was vigorously active, with an eruption rate almost double that of subsequent years, presumably a consequence of a high magma-supply rate.   The range in incompatible element abundances and ratios, and in Sr and Pb isotopic ratios, in the 44 years between 1843 and 1887 are almost as large as the differences in the entire subaerial record (>28 ka and possibly as much as 100 ka). Consequently, any interpretations of long-term temporal variations [e.g. *Budahn and Schmitt*, 1985; *Tilling et al.*, 1987a,b] are tenuous at best, and will require much greater sampling densities for reliable evaluation.

Although the major element data point to open-system magmatism, the changes in isotopic and incompatible element abundances and ratios cannot be attributed to open-system magmatism involving a single parental magma. At least two parental magmas are required. One has the compositional attributes of the incompatible element-enriched lavas erupted in 1843, the other the characteristics of incompatible element-depleted lavas erupted at the summit early in 1880. These lavas are among the most extreme compositions erupted on Mauna

Loa during the past 31 ka. All the historical lavas can be considered as mixtures of these two end-component magmas, modified by contemporaneous eruption and olivine crystallization.

Sr, Nd and Pb isotopic ratios are intermediate between Koolau and Kilauea lavas, the isotopic end-members of the Hawaiian tholeiitic array. Ratios involving Nb (e.g. Sr/Nb, Zr/Nb, K/Nb) correlate with the isotopic data in Hawaiian tholeiites and are also intermediate, and therefore provide a useful guide to variation in source compositions. There is no such systematic relationship for other incompatible element abundances and ratios. In the 1843 parental magma, these ratios either overlap with the Koolau and Kilauea data (e.g. K/Y, Zr/Y, Sr/Y) or are slightly lower (e.g. La/Yb). The 1880 parental magma is more depleted than the 1843 parental magma, and is characterized by lower Sr isotopic ratios and K/Y, La/Yb, Zr/Y, Sr/Y, La/Nb, K/Nb and higher Nd and Pb isotopic ratios and Zr/Nb. Although some of these differences between the two parental magmas might be ascribed to progressive melting of a common source, the isotopic differences require source heterogeneity.

Two possible explanations are offered. Both depend on the location of Mauna Loa at, or close to, the plume periphery and both require heterogeneity within the plume. In the first, the plume develops radial heterogeneity through entrainment of lower mantle containing a dispersed EM1 component. This component is concentrated at the plume margin, and is intimately mixed with the plume material on a scale that is smaller than the scale of melting. Variable amounts of melting during adiabatic decompression result in magmas that range in composition between the inferred 1843 and 1880 parental magmas. In the second explanation, the plume is again assumed to be radially heterogeneous, possibly developed at its source or else through subsequent lower mantle entrainment. This model relies on the deflection of the plume from its axis by the lithosphere, at the most extreme, resulting, as a consequence of lithospheric drag, in an inclined plume (Figure 12). Melting close to the margin of the plume results in magma with the chemical and isotopic characteristics of the 1843 parental magma. If the melt segregation column is essentially vertical [Ribe, 1988], it will eventually intersect the deflected depleted material, originating from the center of the plume, that was the source of Kilauea or Loihi magmas. Continued melting of this depleted material or exchange with the percolating magma could lead to the more depleted 1880 parental magma, with isotopic characteristics that are closer to those of Kilauea magmas.

Acknowledgements. This research was supported by the National Science Foundation. We thank J. Sparks, P. Dawson and M. Chapman for their assistance with the XRF analyses, and S. Li and J. Blustajn for their help with the isotopic analyses. F. A. Frey and P. Ila are thanked for access to the INAA laboratory at MIT and assistance with analyses.

Thoughtful reviews by Fred Frey, Claudia Rhodes, Howard West and Tom Wright are appreciated and have improved the paper. JMR especially wishes to thank Jack Lockwood and Pete Lipman for introducing him to Mauna Loa many years ago, and for the pleasure and satisfaction that has ensued from the association with this most magnificent of volcanoes.

REFERENCES

Brigham, W. T., The volcanoes of Kilauea and Mauna Loa on the Island of Hawaii, *Bishop Museum Memoirs*, v.2, no.4, 222p., 1909.

Boynton, W. V., Cosmochemistry of the rare earth elements in meteorite studies. In *Rare Earth Element Geochemistry*, edited by P. W. Henderson, pp 63-114, Elsevier, 1984.

Budahn, J. R. and R. A. Schmitt, Petrogenetic modeling of Hawaiian tholeiitic basalts: A geochemical approach, *Geochim. Cosmochim. Acta, 49,* 67-87, 1985.

Casadevall, T. J. and D. Dzurisin, Stratigraphy and petrology of the Uwekahuna Bluff section, Kilauea caldera, *U. S. Geol. Surv. Prof. Paper, 1350,* 351-375, 1987.

Chen, C-Y. and F. A. Frey, Origin of Hawaiian tholeiites and alkali basalts, *Nature, 302,* 785-789, 1983.

Chen, C-Y., F. A. Frey, J. M. Rhodes and R. M. Easton, Temporal geochemical evolution of Kilauea Volcano: Comparison of Hilina and Puna Basalt, *Geophys. Monogr. Ser.*, edited by A. Basu and S. R. Hart, AGU, Washington, D.C. (in press).

Cox, K. G., Numerical modeling of a randomized RTF magma chamber: A comparison with continental flood basalt sequences, *J. Petrol, 29,* 681-697, 1988.

Decker, R. W., R. Y. Koyanagi, J. J. Dvorak, J. P. Lockwood, A. T. Okamura, K. M. Yamashita, and W. R. Tanigawa, Seismicity and surface deformation of Mauna Loa volcano, Hawaii, *Eos Trans. AGU, 64,* 545-547, 1983.

Defant, M. J. and R. L. Nielsen, Interpretation of open system petrogenetic processes: Phase equilibria constraints on magma evolution, *Geochim. Cosmochim. Acta, 54,* 87-102, 1990.

Eggins, S. M., Petrogenesis of Hawaiian tholeiites: 2, aspects of dynamic melt segregation, *Contrib. Mineral. Petrol., 110,* 398-410, 1992.

Feigensen, M. D., Geochemistry of Kauai volcanics and a mixing model for the origin of Hawaiian alkali basalts, *Contrib. Mineral. Petrol., 87,* 109-19, 1984.

Feigenson, M. D., A. W. Hofmann and F. J. Spera, Case studies on the origin of basalt II. The transition from tholeiitic to alkalic volcanism on Kohala Volcano, Hawaii, *Contrib. Mineral. Petrol., 84,* 390-405, 1983.

Frey, F. A. and J. M. Rhodes, Intershield geochemical differences among Hawaiian volcanoes: implications for source compositions, melting process and magma ascent paths, *Phil. Trans. Roy. Soc. Lond A., 342,* 121-136, 1993

Frey, F. A., M. O. Garcia and M. F. Roden, Geochemical characteristics of Koolau Volcano: Implications of intershield geochemical differences among Hawaiian volcanoes, *Geochim. Cosmochim. Acta, 58,* 1441-1462, 1994.

Greenland, L. P., Composition of gases from the 1984 eruption of Mauna Loa volcano, *U. S. Geol. Surv. Prof. Paper, 1350,* 781-790, 1987.

Griffiths, R. W. and I. H. Cambell, On the dynamics of long-lived plume conduits in the convecting mantle, *Earth. Planet. Sci. Lett., 103*, 214-227, 1991.

Grove, T. L., D. C. Gerlach, and T.W. Sando, Origin of calc-alkaline series lavas at Medicine Lake volcano by fractionation, assimilation and mixing, *Contrib. Mineral. Petrol. , 80*, 160-182, 1982.

Hart, S. R. and T. Dunn, Experimental cpx/melt partitioning of 24 trace elements, *Contrib. Mineral. Petrol., 113*, 1-8, 1993.

Hart, S. R., E. H. Hauri, L. A. Oschmann and J. A. Whitehead, Mantle plumes and entrainment: Isotopic evidence, *Science, 256*, 517-520, 1992.

Hauri, E. H., J.A. Whitehead and S. R. Hart, Fluid dynamic and geochemical aspects of entrainment in mantle plumes, *J. Geophys. Res., 99*, 24,275-24,300, 1994.

Helz, R. T., Differentiation behavior of Kilauea Iki lava lale, Kilauea Volcano, Hawaii: An overview of past and current work, in *Magmatic Processes:Physiochemical Principles*, editor B. O. Mysen, Geochem. Soc. Specl. Pub., 1, 241-258, 1987.

Hofmann, A. W., M. D. Feigenson, and I. Raczek, Case studies on the originof basalt: III. Petrogenesis of the Mauna Ulu eruption, Kilauea, 1969-1971, *Contrib. Mineral. Petrol., 88*, 24-35, 1984.

Ila, P. and F. A. Frey, Utilization of neutron activationanalysis in the study of geologic materials, *Atomkernegie Kerntechnik, 44*, 710-718, 1984.

Jaggar, T. A., Origin and development of craters, *Geol. Soc. Amer. Memoir 21*, 508p., 1947.

Johnson, K. T. M., Experimental cpx/ and garnet/melt partitioning of REE and other trace elements (abstract), *Eos, 74*, p.658, 1993.

Kelemen, P. B., H. J. B. Dick and J. E. Quick, Production of harzburgite by pervasive melt-rock reaction in the upper mantle, *Nature, 358*, 635-641, 1992.

Kurz, M. D. and D. P. Kammer, Isotopic evolution of Mauna Loa volcano, *Earth Planet. Sci. Lett., 103*, 257-269, 1991.

Kurz, M. D., T. C. Kenna, D. P. Kammer, J. M. Rhodes and M. O. Garcia. Isotopic evolution of Mauna Loa Volcano: A view from the submarine southwest rift zone, (this volume).

Langmuir, C. H., Geochemical consequences of *in situ* crystallization, *Nature, 340*, 199-205, 1989.

Lanphere, M., and F. A. Frey, Geochemical evolution of Kohala Volcano, Hawaii, *Contrib. Mineral. Petrol., 95*, 100-112, 1987.

Lipman, P. W., The southwest rift zone of Mauna Loa: Implications for structural evolution of Hawaiian volcanoes, *Amer. J. Sci., 280-A*, 752-776, 1980.

Lipman, P. W., Declining growth of Mauna Loa Volcano during the last 100,000 years: Rates of lava accumulation Vs. gravitational subsidence, (this volume).

Lipman, P. W. and A. Swenson,Generalized geologic mao of the southwest rift zone of Mauna Loa Volcano, Hawaii, *U.S. Geological Survey Miscellaneous Investigations Map, I-1323*, scale 1:100,000, 1984.

Lipman, P. W., J. M. Rhodes, and M. A. Lanphere, The Ninole Basalt - implications for the structural evolution of Mauna Loa volcano, Hawaii, *Bull. Volcanol., 53*, 1-19, 1990.

Lockwood, J. P., Mauna Loa eruptive history - the radiocarbon record, (this volume).

Lockwood, J. P., and P. W. Lipman, Holocene eruptive history of Mauna Loa volcano, *U. S. Geol. Surv. Prof. Paper, 1350*, 509-535, 1987.

Lockwood, J. P., J. J. Dvorak, T. T. English, R. Y. Koyanagi, A. T. Okamura, M. L. Summers, and W. R. Tanigawa, Mauna Loa 1974-1984: A decade of intrusive and extrusive activity, *U. S. Geol. Surv. Prof. Paper., 1350*, 537-570, 1987.

Macdonald, G. A., Geologic map of the Mauna Loa Quadrangle, Hawaii, *U.S. Geological Survey Geologic Quadrangle Map, GQ-897*, scale 1:24,000, 1971.

Macdonald, G. A. and A. T. Abbott, *Volcanoes in the Sea*, University of Hawaii Press, Honolulu, 441 p., 1970.

Moore, J. G., D. J. Fornari, and D. A. Clague, Basalts from the 1877 submarine eruption of Mauna Loa, Hawaii: new data on the variation of palagonitization rate with temperature, *U. S. Geol. Surv. Bull., 1663*, 11 p., 1985.

Morris, J. D. and S. R. Hart, Isotopic and incompatible element constraints on the genesis of island arc volcanics: Cold Bay and Amak Islands, Aleutians, *Geochim. Cosmochim. Acta, 47*, 380-392, 1983.

Murata, K. J. and D. H. Richter, Chemistry of the lavas of the 1959-60 eruption of Kilauea volcano, Hawaii, *U. S. Geol. Surv. Prof. Paper, 537-A*, 26 p., 1966.

Navon O. and E. Stolper, Geochemical consequences of melt percolation: The upper mantle as a chromatographic column, *J. Geology, 95*, 285-307, 1987.

Nielsen, R. L. Simulation of igneous differentiation processes, in *Modern Methods of Igneous Petrology: Understanding Magmatic Processes*, edited by J. Nicholls and J. K. Russell, *Reviews in Mineralogy, 24*, 65-105, Mineralogical Society of America, Washington, D.C., 1990.

Norrish, K. and B. W. Chappell, X-ray fluorescent spectrography, in *Physical Methods in Determinative Mineralogy*, edited by J. Zussman, pp. 161-214, Academic Press, New York, 1967.

Norrish, K. and J. T. Hutton, An accurate X-ray Spectrographic method for the analysis of a wide range of geological samples, *Geochim. Cosmochim. Acta, 33*, 431-454, 1969.

O'Hara, M. J., Geochemical evolution during fractional crystallization of a periodically refilled magma chamber, *Nature, 266*, 503-507, 1977.

O'Hara, M. J. and R. E. Mathews, Geochemical evolution in an advancing, periodically replenished, periodically tapped, continuously fractionating magma chamber, *J. Geol. Soc. Lond., 138*, 237-277, 1981.

Okamura, A. T., A. Miklius, P. Okuba and M. K. Sato, Forecasting eruptive activity of Mauna Loa volcano, Hawaii (abstract), *Eos, 73*, p.343, 1992.

O'Nions, R. K., P. J. Hamilton, and N. M. Evensen, Variations in $^{143}Nd/^{144}Nd$ and $^{87}Sr/^{86}Sr$ ratios in oceanic basalts, *Earth Planet. Sci. Lett., 34*, 13-22, 1977.

Peterson, D. W. and R. B. Moore, Geologic history and evolution of geologic concepts, Island of Hawaii, *U. S. Geol. Surv. Prof. Paper, 1350*, 149-189, 1987.

Powers, H. A., Composition and origin of basaltic magmas on the Hawaiian islands, *Geochim. Cosmochim. Acta, 7*, 77-107, 1955.

Rhodes, J. M., Homogeneity of lava flows: Chemical data for historic Mauna Loa eruptions, *J. Geophys. Res., 88A*, 869-879, 1983.

Rhodes, J. M., Geochemistry of the 1984 Mauna Loa eruption: Implications for magma storage and supply, *J. Geophys. Res., 93*, 4453-4466, 1988.

Rhodes, J. M., The 1852 and 1868 Mauna Loa Picrite eruptions: Clues to parental magma compositions and the magmatic plumbing system, (this volume).

Rhodes, J. M., The geochemical stratigraphy of lava flows sampled by the Hawaiian Scientific Drilling Project, *J. Geophys. Res.,* (submitted).

Rhodes, J. M. and P. W. Lipman, Chemistry of prehistoric lavas erupted along the southwest rift zone of Mauna Loa (abstract), *Hawaii Symposium on Intraplate and Submarine Volcanism,* p. 97, Hilo, 1979.

Rhodes, J. M. and J. P. Lockwood, Chemistry of paired eruptions on the northeast rift zone of Mauna Loa, Hawaii (abstract), *Annual Meeting Geol. Soc. Amer.,* p. 508, Atlanta, 1980.

Rhodes, J. M., and J. W. Sparks, Contrasting styles and levels of magma mixing, Mauna Loa volcano, Hawaii (abstract), *Proc. Inst. Earth. Man Conf: Open Magmatic Systems,* p.135, 1984.

Rhodes, J. M., J. P. Lockwood, and P. W. Lipman, Episodic variation in magma chemistry, Mauna Loa volcano, Hawaii (abstract) *Generation of Major Basalt Types,* International Association of Volcanology and Chemistry of the Earth's Interior, Reykjavik, Iceland, 1982.

Rhodes, J. M., K. P. Wenz, C. A. Neal, J. W. Sparks, and J. P. Lockwood, Geochemical evidence for invasion of Kilauea's plumbing system by Mauna Loa magma, *Nature, 337,* 257-260, 1989.

Ribe, N. M., Dynamical geochemistry of the Hawaiian plume, *Earth. Planet. Sci. Lett., 88,* 37-46, 1988.

Roden, M. F., T. Trull, S. R. Hart and F. A. Frey, New He, Nd, Pd and Sr isotopic constraints on the constitution of the Hawaiian plume: Results from Koolau Volcano, Oahu, Hawaii, USA, *Geochim. Cosmochim. Acta, 58,* 1431-1440, 1994.

Salters, V. J. M. and S. R. Hart, The Hf-paradox, and the role of garnet in the MORB source, *Nature, 342,* 420-422, 1991.

Sparks, J. W., Long-term compositional and eruptive behavior of Mauna Loa volcano: Evidence from prehistoric caldera basalts, Ph.D. thesis, University of Massachusetts, Amherst, 1990.

Staudigel, H., A. Zindler, S. R. Hart, T. Leslie, C.-Y. Chen and D. Clague, The isotopic systematics of a juvenile intraplate volcano: Pb, Nd and Sr isotope ratios of basalts from Loihi seamount, Hawaii, *Earth Planet. Sci. Lett., 69,* 13-29, 1984.

Stearns, H. T., Geology of the Hawaiian Islands, *Bull. Hawaii Division of Hydrography, 8,* 112 p., 1946.

Stille, P., D. M. Unruh and M. Tatsumoto, Pb, Sr, Nd and Hf isotopic constraints on the origin of Hawaiian basalts and evidence for a unique mantle source, *Geochim. Cosmochim. Acta, 50,* 2303-2319, 1986.

Sun, S.-S. and W. F. McDonough, Chemical and isotopic systematics of oceanic basalts: implications for mantle composition and processes. In *Magmatism in the Ocean Basins,* edited by A.D. Saunders and M. J. Norry, *Geol. Soc. Lond. Spec. Publ., 42,* 313-345, 1989.

Tatsumoto, M., Isotopic composition of lead in oceanic basalt and its implication to mantle evolution, *Earth Planet. Sci. Lett., 38,* 13-29, 1978.

Tilling, R. I., J. M. Rhodes, J. W. Sparks, J. P. Lockwood, and P. W. Lipman, Disruption of the Mauna Loa magma system by the 1868 Hawaiian earthquake: Geochemical evidence, *Science, 235,* 196-199, 1987a.

Tilling, R. I., T. L. Wright, and H. P. Millard, Jr., Trace-element chemistry of Kilauea and Mauna Loa lava in space and time: A reconnaissance, *U. S. Geol. Surv. Prof. Paper, 1350,* 641-689, 1987b.

Todt, W., R.A. Cliff, A. Hanser and A.W. Hofmann, $^{202}Pb+^{205}Pb$ double-spike for lead isotopic analyses, *Terra Cognita, 4,* p.209, 1984.

Watson, S., Rare earth element inversion and percolation models for Hawaii, *J. Petrol., 34,* 763-783, 1993.

Watson, S. and D. McKenzie, Melt generation by plumes: A study of Hawaiian volcanism, *J. Petrol., 32,* 501-537, 1991.

West, H. B., D. C. Gerlach, W. P. Leeman and M. O. Garcia, Isotopic constraints on the origin of Hawaiian lavas from the Maui Volcanic Complex, Hawaii, *Nature, 330,* 216-220, 1987.

White, W. and A. W. Hofmann, Sr and Nd isotope geochemistry of oceanic basalts and mantle evolution, *Nature, 294,* 821-825, 1982.

Wilkinson, J. F. G., and H. D. Hensel, The petrology of some picrites from Mauna Loa and Kilauea volcanoes, Hawaii, *Contrib. Mineral. Petrol., 98,* 326-345, 1988.

Wright, T. L., Chemistry of Kilauea and Mauna Loa lavas in space and time, *U. S. Geol. Surv. Prof. Paper, 735,* 1-49, 1971.

Wright, T. L., Magma mixing as illustrated by the 1959 eruption, Kilauea volcano, Hawaii, *Geol. Soc. Amer. Bull., 84,* 849-858, 1973.

Wright, T. L. and R. S. Fiske, Origin of the differentiated and hybrid lavas of Kilauea volcano, Hawaii, *J. Petrol., 12,* 1-65, 1971.

Wright, T. L. and R. I. Tilling, Chemical variations in Kilauea eruptions 1971-1974, *Am. J.Sci., 280-A,* 777-793, 1980.

Wright, T. L., D. A. Swanson, and W. A. Duffield, Chemical compositions of Kilauea east-rift lava, 1968-1971, *J. Petrol., 16,* 110-133, 1975.

Yang, H.-J., F. A. Frey, M. O. Garcia and D. A. Clague, Submarine lavas from Mauna Kea volcano, Hawaii: Implications for shield-stage processes, *J. Geophys. Res., 99,* 15,577-15,594, 1994.

Zindler, A. and S. R. Hart, Chemical geodynamics, *Rev. Earth Planet. Sci., 14,* 493-571, 1986.

J. M. Rhodes, Department of Geology/Geography, University of Massachusetts, Amherst, MA 01003.

S. R. Hart, Department of Geology and Geophysics, WoodsHole Oceanographic Institution, Woods Hole, MA 02543.

Isotopic Evolution of Mauna Loa Volcano: A View From the Submarine Southwest Rift Zone

Mark D. Kurz, T. C. Kenna, D. P. Kammer

Department of Marine Chemistry and Geochemistry, Woods Hole Oceanographic Institution, Woods Hole, MA

J. Michael Rhodes

University of Massachusetts, Amherst, MA

Michael O. Garcia

University of Hawaii, Honolulu, HI

New isotopic and trace element measurements on lavas from the submarine southwest rift zone (SWR) of Mauna Loa continue the temporal trends of subaerial Mauna Loa flows, extending the known compositional range for this volcano, and suggesting that many of the SWR lavas are older than any exposed on land. He and Nd isotopic compositions are similar to those in the oldest subaerial Mauna Loa lavas (Kahuku and Ninole Basalts), while $^{87}Sr/^{86}Sr$ ratios are slightly lower (as low as .7036) and Pb isotopes are higher ($^{206}Pb/^{204}Pb$ up to 18.30). The coherence of all the isotopes suggests that helium behaves as an incompatible element, and that helium isotopic variations in the Hawaiian lavas are produced by melting and mantle processes, rather than magma chamber or metasomatic processes unique to the gaseous elements. The variations of He, Sr, and Nd are most pronounced in lavas of approximately 10 ka age range [*Kurz and Kammer*, 1991], but the largest Pb isotopic variation occurs earlier. These variations are interpreted as resulting from the diminishing contribution from the upwelling mantle plume material as the shield building ends at Mauna Loa. The order of reduction in the plume isotopic signature is inferred to be Pb (at >100 ka), He (at ~14 ka), Sr (at ~9 ka), and Nd (at ~8 ka); the different timing may relate to silicate/melt partition coefficients, with most incompatible elements removed first, and also to concentration variations within the plume. Zr/Nb, Sr/Nb, and fractionation-corrected Nb concentrations, correlate with the isotopes and are significantly higher in some of the submarine SWR lavas, suggesting temporal variability on time scales similar to the Pb isotopes (i.e. ~ 100 ka). Historical lavas define trace element and isotopic trends that are distinct from the longer term (10 to 100 ka) variations, suggesting that different processes cause the short term variability. The temporal evolution of Mauna Loa, and particularly the new data from the submarine SWR, suggest that the isotopic composition of the upwelling plume mantle is best represented by data from Loihi seamount tholeiites. The temporal evolution suggests that the mantle source of the latest stage of Mauna Loa, which is characterized by radiogenic $^{87}Sr/^{86}Sr$ (up to .70395), unradiogenic $^{206}Pb/^{204}Pb$ (~18.0), $^3He/^4He$ ratios similar to MORB, and low Nb concentrations, is a small-volume contribution related to non-plume components (such as normal asthenosphere, or entrained mantle).

1. INTRODUCTION

Previous studies of Mauna Loa volcano have shown that isotopic compositions of lava flows have significantly

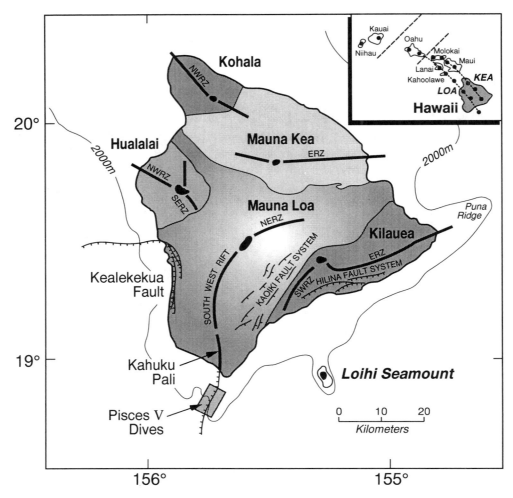

Figure 1. Map of Hawaii showing the principal rift zones of the Hawaiian volcanoes, and the study area on the submarine southwest rift zone (SWR) sampled during Pisces V dives 182-185 [see also *Garcia et al.*, this volume]. Dark areas are the volcano summits.

changed on eruptive time scales between 0.1 and 10 ka [*Kurz et al.*, 1987; *Kurz and Kammer*, 1991; *Kurz*, 1993]. A marked decrease in $^3He/^4He$ at approximately 10 ka, suggests a diminished hotspot influence beneath Mauna Loa, in accord with a decline in eruption rates since about 100 Ka inferred from geologic data [e.g. *Lipman et al.*, 1990; *Moore et al.*, 1990; *Lipman*, this volume]. If the hotspot supply to Mauna Loa diminished since 10 ka, as implied by the helium data, then this volcano offers a special opportunity to study a critical transition of Hawaiian volcano evolution.

Mauna Loa is an ideal place to study temporal evolution of volcanism because of its frequent historical eruptions and the excellent stratigraphic framework provided by geologic mapping combined with radiocarbon dating [e.g., *Lipman and Swenson*, 1984; *Lockwood and Lipman*, 1987; *Rubin et al.*, 1987; *Lockwood*, 1995]. The oldest

stratigraphically exposed Mauna Loa lavas are along the Kahuku and Kealekekua faults and in the Ninole Hills (see Figure 1). This paper presents new isotopic and trace element data from the submarine southwest rift zone of Mauna Loa (SWR), on the Kahuku fault, which extend the age range of analyzed lavas and provide insights into the causes of the geochemical variations. We also present additional trace element and Nd isotopic data for the samples discussed by *Kurz and Kammer* [1991], which demonstrate coherent variations among the trace elements and the Sr, Nd, Pb, and He isotopic systems.

The isotopic variations among Mauna Loa lavas are important to global geochemical models because such detailed stratigraphic sequences are available for few oceanic islands. Coherent variations in the isotopes of He, Sr, and Pb rule out many of the models that have been invoked to "decouple" helium from the radiogenic isotopes, such as metasomatism, magma chamber degassing, and radiogenic ingrowth of ^{4}He [e.g. *Condomines et al.*, 1983; *Zindler and Hart*, 1986; *Vance et al.*, 1989; *Hilton et al.*, 1995]. However, the complexity of the Mauna Loa isotopic variations may require the presence of three distinct mantle sources, combined with plate movement and waning of the hotspot [*Kurz and Kammer*, 1991]. The oldest Mauna Loa lavas have the highest ^{3}He/^{4}He ratios, while the historical lavas have low ^{3}He/^{4}He ratios similar to normal mid-ocean ridge basalts (MORB). The high ^{3}He/^{4}He ratios identify the source of the oldest Mauna Loa shield lavas as the upwelling mantle plume, and the other two sources may relate either to lithosphere, normal asthenosphere, or entrained asthenosphere [*Kurz and Kammer*, 1991; *Kurz*, 1993]. An alternative explanation is that the complex isotopic variations are produced by only two mantle sources (plume and normal asthenosphere), coupled with the effects of melt percolation through a porous medium and chromatographic separation of different elements [*McKenzie and O'Nions*, 1991; *Watson*, 1993]. Determining the geometry of mantle heterogeneities and melting processes remains one of the difficulties of reconciling geochemical and geophysical observations. The explanation for the short-time-scale Mauna Loa isotopic variations is important not just for understanding the present-day configuration of the largest volcano on earth, but to the processes of melting in the mantle.

2. SAMPLES AND ANALYTICAL METHODS

2.1. *Samples*

The ^{14}C age and Sr, Nd, Pb and He isotopic data for subaerial Mauna Loa lavas are presented in table 1 (radiocarbon dated and historical) and table 2 (Kahuku and Ninole); sample localities have been discussed elsewhere [*Kurz et al*, 1983; *Kurz et al.*, 1987; *Kurz and Kammer*, 1991]. The samples in tables 1 and 2 are assigned a stratigraphic order, as reflected in "unit number" which increases with age. For the Kahuku samples, stratigraphic order is simply related to depth in section on the Kahuku Pali; the samples range from the youngest (top of the section) to the oldest. The top of the Kahuku Pali is covered with Pahala ash, which is generally assumed to be older than 31 ka [*Lipman and Swenson*, 1984; *Lockwood and Lipman*, 1987]. Unit numbers 17 and 18 directly underlie the Pahala ash and occur at the top of the Kahuku fault, and are assumed to be >31 ka in age. Unit number 19 is from the base of the Kealekekua fault [*Kurz and Kammer*, 1991], placing it below the Pahala ash (and hence older than units 17 and 18); the placement of this unit with respect to the other Kahuku exposures is speculative, but it is

assumed here to be younger. Ninole Basalt, exposed on the southern flank of Mauna Loa, has been dated at 100 to 200 ka in age [*Lipman et al.*, 1990], and is assumed to be older than the subaerial Kahuku Basalts. Data for most of the Kahuku and Ninole lavas and their relative locations are given in *Kurz and Kammer* [1991]. The only new subaerial Mauna Loa samples in the tables are KS87-26 and KS87-27 (table 2), which were collected from 41 and 38 meters (above sea level) on the Kahuku fault, at the same location described in *Kurz and Kammer* [1991].

Pisces V dives 182-185 were carried out on the submarine portion of the Kahuku fault (see Figure 1) between 550 and 2000 meters below sea level [*Garcia et al.*, this volume]. A primary objective was to extend the stratigraphy back in time, which is possible because there has been little recent activity on the submarine portion of the Kahuku fault, and the exposed sections are not covered by recent lavas [*Fornari et al.*, 1979; *Moore et al.*, 1990; *Moore and Clague*, 1992; *Garcia et al.*, this volume]. *Lipman* [this volume] discusses attempts to K-Ar date these lavas, which have been unsuccessful due to the young ages, low K contents, and excess (mantle) ^{40}Ar. The subsidence rate for Hawaii of 2.5 mm/yr [*Moore*, 1987] and the depth would imply ages between 200 and 800 ka. However, because all of the flows have glassy, pillow morphology, it is certain that they were erupted under water, and the subsidence rates provide only upper age limits. Based on subsidence rates for a submerged shoreline on the SWR, *Lipman* [this volume] suggests a minimum age of 130-150 ka for the top of the submarine section. In summary, age estimates for the submarine Southwest rift lavas are 100-300 ka, but must be viewed as tentative [*Garcia et al.*, this volume; *Lipman*, this volume].

The dive samples from the SWR are considered to be a continuation of the subaerial Kahuku section, which represents a fault or landslide scar [*Garcia et al.*, this volume]. Although the age constraints do not exclude the possibility that there is age overlap between the subaerial and submarine samples, the dive samples represent an age sequence on a single stratigraphic section. Samples collected during dives 182 and 183 are a continuous depth (and hence stratigraphic) traverse of this sequence at one location. The two samples from dive 185 were collected from the top of the fault, and are assumed to be the youngest samples, even though they were collected from 1400 to 1700 meters water depth. The samples from dive 184 are placed relative to the others based on the reconstructed section presented by *Garcia et al.* [this volume]. As discussed below, the isotopic data, presented in table 3, suggest that at least some of the submarine SWR lavas are older than the subaerial Kahuku and Ninole Basalts, and therefore are consistent with ages greater than 100-200 ka. Further dis-

TABLE 1. Isotopic and Trace Element Data for Historical and Radiocarbon Dated Lava Flows

sample Unit #	ML55(1950) 1	ML82(1887) 2	185 3	187 3	202 3	ML84(1868) 4	k7-03# 5	k7-05 5	K7-15 5	k7-31# 6	ha82-01# 7
age(yrs)*	0	63	73	73	73	82	640	640		890	910
^{3}He/^{4}He	8.6	8.6	8.2	8.2	8.3	8.0	8.5	8.6		8.8	8.1
±	0.4	0.2	0.2	0.2	0.2	0.8	0.1	0.1		0.3	0.1
^{87}Sr/^{86}Sr	0.703826	0.703857	0.703857	0.703825	0.703857	0.703853	0.703930	0.703936	0.703942	0.703957	0.703950
^{206}Pb/^{204}Pb	18.11	18.16	18.13	18.11	18.14	18.14		18.08		18.10	18.09
^{207}Pb/^{204}Pb	15.46	15.44	15.44	15.43	15.45	15.45		15.45		15.44	15.46
^{208}Pb/^{204}Pb	37.91	37.88	37.85	37.82	37.88	37.88		37.85		37.85	37.87
^{143}Nd/^{144}Nd	0.5128945	0.512894		0.5128843	0.512903	0.5128775	0.5128327	0.512838	0.5128373	0.512827	0.5128373
MgO	8.65	8.53				22.21	15.54			8.28	15.42
Sr/Y	13.5	12.3				14.3	13.7			13.3	12.1
K/Y	133.4	122.7				145.7	160.9			154.2	156.8
Sr/Nb	31.8	30.9				32.0	26.9			29.3	26.9
K/Sr	9.9	10.0				10.2	11.7			11.6	12.9
Zr/Nb	13.2	13.5				13.1	12.3			13.2	13.4
Zr/Y	5.6	5.4				5.9	6.3			6.0	6.0
K/P							2.6			2.8	2.8
Rb	5.2	5.0				4.2	5.9			7.4	6.5
Sr	296	275				205	266			345	272
Nb	9.3	8.9				6.4	9.9			11.8	10.1
Zr	123	120				84	122			156	135
Y	21.9	22.4				14.3	19.4			26.0	22.4

cussion of the field relations, geological context, and petrology of these samples is presented by *Garcia et al.* [this volume] and *Lipman* [this volume] in this volume.

2.2. *Analytical Methods*

Analytical methods for the measurement of Sr, Pb, and He have been presented elsewhere [*Kurz et al.*, 1987; *Kurz and Kammer*, 1991]. The Nd isotopic measurements were carried out on aliquots of the same whole rock sample previously used for the Sr and Pb measurements. The Nd chemistry was carried out on either the unused fraction from the Pb chemistry, or separate dissolutions, using the methylactic acid technique. The Nd was loaded onto the Ta side filament of Ta-Re double filament, and the isotopes measured on a VG 354 thermal ionization mass spectrometer using dynamic multicollection. The ^{143}Nd/^{144}Nd ratios are fractionation-corrected using a ^{146}Nd/^{144}Nd ratio of .7219, and normalized to a La Jolla value of .51184. During the course of this study, the average measured value for the La Jolla standard was .511851 (± .000008, n=25). Typical statistics on an individual sample run are 10 ppm on ^{143}Nd/^{144}Nd (± .000005). Sr and Pb isotopes were measured using methods described previously [*Kurz and Kammer*, 1991]. During the course of the SWR measurements the average measured value of the NBS 987 standard was .710264 (± .000019, n=18), and the measured values for NBS 981 were ^{206}Pb/^{204}Pb = 16.904 ± .016,

^{207}Pb/^{204}Pb = 15.450 ± .021, and ^{208}Pb/^{204}Pb = 36.568 ± .060 (n=16). Pb isotopic ratios are normalized to the values of *Catanzaro et al.* [1968] for NBS 981. The reproducibility for the standards given above (and used for normalization purposes) are conservative estimates for the two standard-deviation uncertainty of the Sr, Nd, and Pb isotopic measurements given in tables 1-3. The Sr, Nd, and Pb blanks were typically less than 40, 70, and 150 picograms, respectively, and are a negligible contribution to all the measured ratios. All of the isotopic measurements were carried out in the Isotope Geochemistry Facility at Woods Hole Oceanographic Institution.

Helium isotopes were measured on olivine phenocrysts that were hand picked from crushed and sieved 1 to 2 mm size fractions of the lavas. Blanks were roughly 3.5 x 10^{-11} ccSTP ^{4}He with atmospheric ^{3}He/^{4}He. Because subaerial lava flows themselves are extensively degassed, the olivines contain the only remnant of magmatic gases. The bulk of the helium within the olivines resides within the melt inclusions [*Kurz*, 1993] and is released by crushing *in vacuo*; all measurements in tables 1-3 were obtained by crushing *in vacuo*. Helium measurements were carried out in olivines, while the Sr, Nd, and Pb were measured on co-existing groundmass. *Garcia et al.* [this volume] have shown that many olivines in the SWR lavas are out of equilibrium with the coexisting basaltic compositions, and that some of them may be xenocrysts. The coherence between the isotopes of Sr, Nd, Pb and He suggests that this

TABLE 1 (continued). Isotopic and Trace Element Data for Historical and Radiocarbon Dated Lava Flows

sample Unit #	k7-14 8	T87-8 8	k5-05 8	k7-04 9	k7-13 10	k5-28 11	k7-01 12	k7-07 13	k7-08 13	ha82-05 14	ks87-02 14	k5-32 15
age(yrs)	2180	2180	2180	2300	2940	5660	7230	7750	7750	9020	9020	11780
^{3}He/^{4}He	8.6	9.2	9.1	8.4	8.3	8.7	9.2	10.1	8.8	13.1	13.1	11.0
±	0.1	0.1	0.2	0.2	0.1	0.2	0.1	0.1	0.1	0.1	0.1	0.2
^{87}Sr/^{86}Sr	0.703950	0.703913	0.703936	0.703917		0.703905	0.703830	0.703779	0.703967	0.703821	0.703838	0.703775
^{206}Pb/^{204}Pb	18.15	18.10	18.11	18.10	18.19	18.06	18.14	18.10	18.12		18.07	18.09
^{207}Pb/^{204}Pb	15.46	15.47	15.47	15.45	15.43	15.46	15.45	15.45	15.46		15.43	15.45
^{208}Pb/^{204}Pb	37.93	37.93	37.90	37.85	37.86	37.84	37.84	37.84	37.88		37.82	37.84
^{143}Nd/^{144}Nd	0.512814	0.512852	0.51283	0.512853	0.512915	0.512835	0.512878	0.512883	0.51289	0.512864	0.512868	0.512863
MgO			10.04	6.75	16.04	9.45		12.44	7.59	15.59	15.46	7.57
Sr/Y			13.6	13.5	11.5	15.7		13.2	15.6	12.5	12.6	12.7
K/Y			143.5	145.2	119.5	102.5		124.5	138.4	116.0	115.0	131.0
Sr/Nb			31.7	32.4	27.5	35.0		33.8	37.6	34.1	29.9	36.8
K/Sr			10.5	10.8	10.4	6.5		9.4	8.9	9.3	9.1	10.3
Zr/Nb			13.3	13.7	13.2	13.3		14.9	13.2	14.6	13.4	16.7
Zr/Y			5.7	5.7	5.5	6.0		5.8	5.5	5.4	5.7	5.8
K/P			3.2	3.1	2.5	2.8		2.3	1.7	2.9	3.2	3.2
Rb			5.5	6.2	4.1	5.0		3.9	5.6	4.0	4.1	5.2
Sr			301	321	217	308		257	361	242	236	302
Nb			9.5	9.9	7.9	8.8		7.6	9.6	7.1	7.9	8.2
Zr			126	136	104	117		113	127	104	106	137
Y			22.1	23.8	18.9	19.6		19.4	23.1	19.4	18.7	23.7

# Denotes samples for which the trace element data were obtained from a different subsample of the same lava flow.
Most of the Sr, Pb and He measurements from Kurz and Kammer (1991).
*Both radiocarbon and ^{14}C ages are reported relative to calendar year 1950.

TABLE 2: Isotopic and Trace Element Data From Kahuku and Ninole Basalts.

sample Unit #	k5-30 16	k5-11# 17	k5-23# 18	k7-34 19	k7-24 20	k7-26 21	k7-27 22	K7-29 23	k7-28 24	K7-19 25	k7-20 26
age(yrs)	28150	>31000	>31000	Kealekeku	Kahuku	Kahuku	Kahuku	Kahuku	Kahuku	Ninole	Ninole
^{3}He/^{4}He	18.0	14.8	20.0	15.7	16.0	17.4	16.8		15.4		18.5
±	0.4	0.1	0.2	1.0	0.4	0.1	0.2		0.2		0.1
^{87}Sr/^{86}Sr	0.703691	0.703689	0.703736	0.703724	0.703678	.703669	.703675	0.70367	0.703692	0.703740	0.703756
^{206}Pb/^{204}Pb	18.17	18.17	18.19	18.20	18.13			18.18	18.20	18.19	18.18
^{207}Pb/^{204}Pb	15.46	15.45	15.45	15.43	15.43			15.47	15.45	15.44	15.43
^{208}Pb/^{204}Pb	37.93	37.91	37.95	37.92	37.84			37.98	37.96	37.94	37.93
^{143}Nd/^{144}Nd	0.512831	0.512915	0.512920	0.512925	0.512938	.512927	.512921	0.512934	0.512917		0.512889
MgO(wgt%)		8.70	9.51	19.77	17.96			6.70	19.65	10.40	11.33
Sr/y		12.2		11.7	12.2			10.8	11.1	11.4	12.3
K/Y		117.8		50.7	113.4			109.6		27.8	39.4
Sr/Nb		30.2		31.1	36.1			27.7	28.0	30.0	28.1
K/Sr		9.6		4.3	9.3			10.1		2.4	3.2
Zr/Nb		13.9		13.6	15.4			13.3	13.9	12.9	11.5
Zr/Y		5.6		5.1	5.2			5.2	5.5	4.9	5.0
K/P		2.9		1.4	3.0			2.5		0.9	1.1
Rb (ppm)	5.1	4.3		0.7	3.8			5.4	3.4	0.8	1.4
Sr (ppm)	281	261		190	213			291	199	270	267
Nb (ppm)	9.3			6.1	5.9			10.5	7.1	9.0	9.5
Zr (ppm)	129			83	91			140	99	116	109
Y (ppm)	23.0	22.1		16.2	17.5			26.9	18.0	23.6	21.7

# Denotes samples for which the trace element data were obtained from a different subsample of the same lava flow.
Most of the Sr, Pb and He measurements from Kurz and Kammer (1991).

TABLE 3. Isotopic and Trace Element Data for Submarine South West Rift Lavas

Sample unit #	p5182-1 40	p5-182-4 39	p5-182-6A 38	p5-182-7 37	p5-182-8 36	p5-183-3 35	p5-183-6 34	p5-183-11 32	p5-183-14 31	p5-183-15 30	p5-184-6 33	p5-185-5 29	p5-185-11 28
water depth	1420	1375	1310	1265	1235	1100	1015	790	650	630	1425	1670	1505
depth in section	1250	1205	1140	1095	1065	935	850	625	485	465	785	0-50	0-50
^{3}He/^{4}He (R/Ra)	17.2	18.9	20.0	19.1	16.6	18.7	19.9	13.3	15.9	17.2	18.3	16.9	16.4
±	0.1	0.2	0.3	0.2	0.3	0.3	0.2	0.1	0.1	0.1	0.3	0.2	0.1
^{4}He(ccSTP/g)	6.68E-08	8.21E-09	5.07E-09	9.08E-09	4.01E-09	2.30E-09	1.13E-08	4.16E-08	3.21E-08	1.91E-08	7.34E-09	6.70E-09	1.45E-08
^{87}Sr/^{86}Sr	0.703688	0.703676	0.703660	0.703652	0.703682	0.703693	0.703684	0.703623	0.703645	0.703609	0.703682	0.703743	0.703651
^{206}Pb/^{204}Pb	18.18	18.25	18.24	18.33	18.25	18.15	18.19	18.23	18.29	18.28	18.19	18.31	18.18
^{207}Pb/^{204}Pb	15.46	15.45	15.43	15.52	15.46	15.43	15.43	15.46	15.47	15.45	15.46	15.51	15.47
^{208}Pb/^{204}Pb	37.96	37.97	37.94	38.16	38.02	37.89	37.89	38.00	38.03	37.98	37.98	38.21	37.99
^{143}Nd/^{144}Nd	0.512932	0.512911	0.512913	0.512928	0.512915	0.512910	0.512911	0.512917	0.512937	0.512928	0.512912	0.512879	0.512880
MgO(wgt %)	21.84	32.28	27.93	32.81	23.28	21.83	11.97	15.13	13.34	17.41	15.78	9.85	9.84
Sr/Y	10.6	11.5	11.8	12.4	11.0	11.4	12.0	12.1	12.1	11.6	11.9	12.2	11.8
K/Y	124.8	125.0	137.2	130.0	122.9	117.9	118.6	118.3	122.1	116.3	134.8	126.1	124.2
Sr/Nb	25.5	22.7	23.9	21.5	23.2	26.7	27.6	28.8	27.8	24.8	28.7	22.8	29.0
K/Sr	11.8	10.9	11.6	10.5	11.1	10.3	9.9	9.8	10.1	10.0	11.3	10.4	10.5
Zr/Nb	13.0	11.1	11.2	9.8	11.2	12.5	12.3	12.7	12.6	11.6	13.1	10.7	13.1
Zr/Y	5.4	5.6	5.6	5.7	5.3	5.3	5.3	5.3	5.5	5.4	5.4	5.7	5.3
Rb (ppm)	4.1	2.3	2.3	2.0	3.5	2.9	4.2	3.9	4.7	4.2	4.5	6.5	4.3
Sr (ppm)	181	102	136	103	172	171	243	216	245	211	221	314	264
Nb (ppm)	7.1	4.5	5.7	4.8	7.4	6.4	8.8	7.5	8.8	8.5	7.7	13.8	9.1
Zr (ppm)	92	50	64	47	83	80	108	95	111	99	101	148	119
Y (ppm)	17.1	8.9	11.5	8.3	15.6	15.0	20.3	17.9	20.2	18.2	18.6	25.8	22.4

is a minor problem, and that even if the olivines are xenocrysts with respect to major elements, the helium within them is closely related to the lavas which carry them.

Trace element data for the Mauna Loa samples, presented in the tables, were measured using X-ray fluorescence (XRF) spectrometry at University of Massachusetts [*Rhodes*, 1983; 1988].   Only key trace element data are presented here; a more complete discussion of major and trace element abundances will be presented elsewhere.

3. RESULTS AND DISCUSSION

3.1. *Temporal Isotopic Evolution of Mauna Loa*

The Sr, Nd, Pb and He isotopic data from tables 1-3 are summarized in Figure 2. Because the exact ages of the Kahuku and Ninole Basalts and the submarine samples are not known, the x axis in this diagram is the stratigraphic order, as represented by the unit number. We emphasize that there are considerable uncertainties in the stratigraphic order, and that use of unit number for plotting purposes does not imply that the age sequence is continuous. As discussed above, without age determinations, it is impossible to evaluate the assumption that the Ninole basalts are older than the Kahuku Basalts, or that there may be overlap between subaerial Kahuku and the submarine SWR samples. The youngest flows are historical or radiocarbon dated, and the ages are given in table 1.

There are coherent temporal variations with respect to all the isotope systems, particularly in the age range near 10 ka (indicated by the dotted line in Figure 2). The ^{3}He/^{4}He ratios are highest in lavas older than 10 ka, which are assumed here to relate to the decrease of the hotspot signal at that time. The new Nd isotopic data show that low ^{143}Nd/^{144}Nd ratios correlate with high ^{87}Sr/^{86}Sr, that the oldest samples have the highest Nd isotope ratios, and that Nd also displays an important shift at ~10 ka. However, all the isotopic systems *except helium* display important shifts between the historical and radiocarbon dated flows, which demonstrates that the variations do not simply reflect a simple "shut off" of the hotspot, as would be inferred from the helium isotopes alone.

The isotopic data strongly support the inferred stratigraphy, because the data from the SWR samples extend the isotopic trends found for the ^{14}C dated and Kahuku Basalt. This is illustrated in Figure 3, where the arrows indicate the temporal isotopic evolution [see *Kurz and Kammer*, 1991; *Kurz*, 1993]. The isotopic data for the submarine SWR lavas are consistent with the hypothesis that many of them are older than the Kahuku and Ninole basalts. There is considerable overlap in isotopic compositions, and due to

the age uncertainties, it is quite possible that the youngest SWR lavas overlap in age with the oldest Kahuku and Ninole Basalts. However, the Mauna Loa isotopic trends are extremely well defined, suggesting that isotopic measurements are useful as an age constraint. Figure 3 also illustrates that the isotopic compositions of the youngest lavas deviate significantly from the trend found for the SWR and other Kahuku lavas. The "bend" in the Sr-He diagram defined by samples younger than 0.6 ka was previously interpreted to reflect lithospheric influence [*Kurz*, 1993].

Some of the submarine SWR samples differ significantly from the subaerial samples in Pb isotopes, and to a lesser degree in Sr isotopes (Figures 2 and 3). The He and Nd isotopic values are similar to the Kahuku and Ninole Ba-

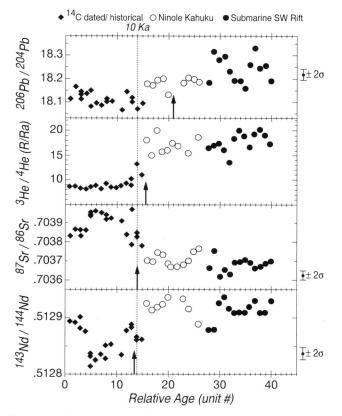

Figure 2.   Isotopic composition of Mauna Loa lavas placed in a stratigraphic framework. Numerical ages are known only for those that are less than 30 ka of age (solid diamond symbols, see tables 1 and 2). The x axis is the unit number from tables 1-3, with higher unit numbers having greater age (dotted line indicates 10 ka). The arrows indicate the plume decay half life, which is empirically defined as the age at which the plume isotopic signature has decreased to half its maximum value (or increased in the case of Sr). Note that the oldest lavas have the most radiogenic Pb. The transition at ~10 ka is attributed to diminished influence of the plume.

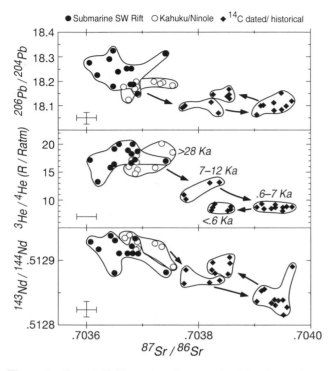

Figure 3. Sr vs. Nd, He, and Pb isotopes for Mauna Loa lavas, with different fields denoting different age ranges (see Figure 2). The complex temporal evolution is indicated by the arrows which indicate time evolution toward the present day. The position of the SWR samples in all of these diagrams is consistent with their age assignment as older than the Kahuku and Ninole samples. Error bars are ± 2σ (error on ^{3}He/^{4}He is typically smaller than symbols).

salts, but the Pb isotopic ratios found in the SWR samples are the most radiogenic ever reported for Mauna Loa. Assuming that the decrease in ^{3}He/^{4}He and increase in ^{87}Sr/^{86}Sr near 10 ka is due to the shut off of the hotspot, and that the submarine SWR lavas are most representative of the plume isotopic composition, then the Pb isotopic signature of the hotspot disappeared *earlier* than the other isotopic signals. This is illustrated by the arrows in Figure 2, which indicate the time at which the "plume signal" decreased to 1/2 of its maximum (i.e. the unit number which has the value closest to 1/2 the maximum minus the minimum isotopic ratio). The placement of the arrows in Figure 2 highly sensitive to the choice of maximum and minimum values, and there is significant overlap between the SWR and Kahuku lavas. However, assuming that the highest Pb isotopic compositions represent the plume, the Pb isotopes were the first to decrease, followed by He, Sr, and Nd isotopes. For example, at 10 ka (dashed line in Figure 2), the Pb isotopic values are similar to those in

younger lava flows, while Sr, Nd, and He differ significantly. The timing derived from this approach is also speculative, due to the lack of geochronology for the older flows. The ages of the Kahuku and Ninole lavas are assumed to be 100 -200 ka [*Lipman et al.*, 1990], and so we can only state that the major shift for Pb isotopes occurred at >100 ka. The major shift in the isotopes therefore occurred at very different times: greater than 100 ka for Pb, at ~14 ka for He, at 9 ka for Sr, and 7.7 ka for Nd.

There are a number of uncertainties in establishing timing differences between the isotopes. First, the hypothesis that the hotspot "shut off' at different times, is based on the assumption that high ^{3}He/^{4}He ratios are diagnostic of the plume, and that the geochemical transition near 10 ka in the isotopes represents disappearance of the plume component. Figures 2 and 3 show that Sr, Nd and Pb isotopes also varied significantly *after* the "transition", and the assumption that the plume component disappeared

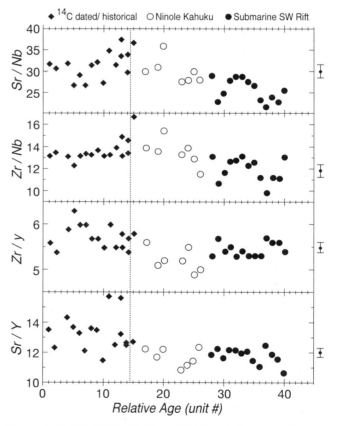

Figure 4. Sr/Nb, Zr/Nb, Zr/Y, and Sr/Y variations as a function of stratigraphic age. Note that both Sr/Nb and Zr/Nb are significantly lower in some of the submarine southwest rift lavas, which are assumed here to be the oldest samples from Mauna Loa. Error bars are ± 2σ.

after 10 ka is based solely on the helium data. With the exception of helium, the isotope and trace element ratios (see below) do not show a monotonic decrease with time. In addition, the Pb isotopes are highly variable within the submarine SWR lavas and do not show a simple stratigraphic decrease. Finally, the timing is based on a relatively sparse data set between approximately 7 and 28 ka.

The apparent timing differences can also be sensitive to the contrast between the plume and the other mantle sources, which relates not just to the differences in isotopic composition, but also to elemental concentrations in the different sources. The total Pb isotopic variation is only 8.5 standard deviations of the measurement uncertainty, and may be more easily affected by concentration differences between the different mantle sources.

3.2. Temporal Trace Element Variations

Incompatible trace elements also vary systematically on similar time scales as the isotopes. Figure 4 shows that Zr/Nb and Sr/Nb are lower in some of the SWR lavas than in any of the radiocarbon dated or historical samples. The Kahuku Basalts are transitional between the oldest radiocarbon dated flows and the SWR lavas, and the oldest radiocarbon dated flows have the highest Zr/Nb and Sr/Nb. As with the isotopes, there are also important variations in the younger lava flows, on time scales as short as 100 years [see also *Rhodes and Hart,* this volume]. Ratios involving the more compatible elements such as Zr/Y and Sr/Y (Zr/Y should correlate with Sm/Yb) show less variability in the oldest lavas. Most of the variability within these ratios is found within the youngest lavas, and the radiocarbon dated flows have significantly *higher* Sr/Y and Zr/Y.

The variations with respect to Zr/Nb and Sr/Nb in the older flows are primarily related to Nb concentrations, because there appears to be little temporal variation in the fractionation corrected concentrations of the other elements in the SWR lavas. Figure 5 shows the fractionation normalized (to 7% MgO) concentrations for Sr, Zr, Y, and Nb, based on the assumption that olivine removal and addition are the only controls on the primary trace element concentration [e.g. *Wright,* 1971]. The SWR lavas have higher (MgO normalized) Nb concentrations, while the other elements show little variability in this age range. The youngest lavas, however, have higher fractionation corrected Sr and Zr concentrations.

One possible reason for the differences between the trace elements is that they have differing degrees of incompatibility with respect to silicate/melt partitioning, and are extracted at different rates from the upwelling plume. The

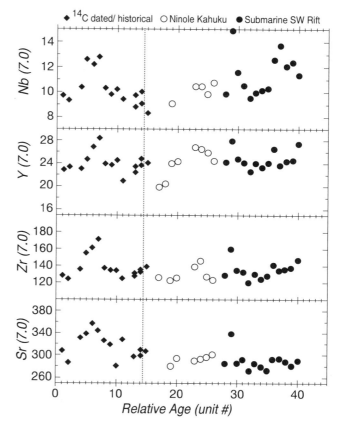

Figure 5. Variations in the fractionation corrected concentrations of Nb, Y, Zr, and Sr as a function of stratigraphic age. The trace elements are normalized to MgO content of 7% for consistency with *Rhodes and Hart* [this volume], although it is possible that the parental magmas are significantly more magnesian than this [*Garcia et al.,* this volume]. The calculations assume that the olivine crystallized have Fo(87), and that they contain no trace elements. Measurement uncertainties are similar in size to the symbols for the elemental abundances.

order of the partition coefficients (Kd) for the elements discussed here should be Nb< Pb < Sr < Nd < Zr < Y, assuming that melting occurs in the garnet stability field [see *Hart and Dunn,* 1993; *Hauri et al.,* 1994]. This order is consistent with Nb (and the isotopic variations), but not with the Zr, Y and Sr data. For example, the Sr/Y and Zr/Y ratios are higher in the youngest lavas, which is the opposite of what would be expected from melting depletion of the plume. The isotopic data suggest different mantle sources for these lavas, and the trace element variations may also be caused by different source concentrations, mixing processes, or variations in melting parameters. Consideration of the Rb/Sr ratios and Sr isotopes suggests mixing of sources or melts plays a role in the variations

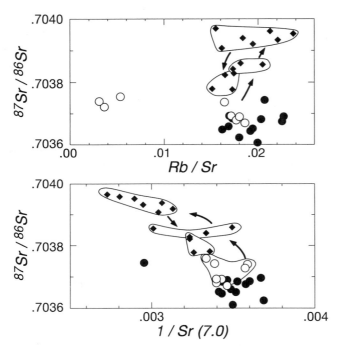

Figure 6. $^{87}Sr/^{86}Sr$ as a function of Rb/Sr and 1/Sr. The lack of a relationship between $^{87}Sr/^{86}Sr$ and Rb/Sr demonstrates that the Rb/Sr ratios in the lavas do not reflect the time integrated source values. The correlation between $^{87}Sr/^{86}Sr$ and 1/Sr suggests that mixing of mantle sources plays a role in the isotopic and trace element variations. Note that higher Sr concentrations are found in the younger lavas (see also Figure 5). Several of the Kahuku lavas have very low Rb/Sr ratios due to Rb loss during subaerial weathering. Arrows denote temporal evolution for the different age groups, as in Figure 3.

(Figure 6). The absence of a relationship between $^{87}Sr/^{86}Sr$ and Rb/Sr demonstrates that the Rb/Sr ratios in the lavas do not reflect the time integrated source values, and must have been altered by mixing and melting processes. The correlation between $^{87}Sr/^{86}Sr$ and 1/Sr supports the hypothesis that mixing of mantle sources or melts plays a role, and shows that higher Sr concentrations characterize the mantle source of the younger lavas. Thus, the high Sr/Y and Zr/Y found in the youngest lavas may reflect higher Sr and Zr concentrations in the source region, which is also consistent with the fractionation corrected Sr and Zr (see Figure 5).

Although the change near 10 ka is referred to here as a "transition", data presented by *Rhodes and Hart* [this volume], and the data from the youngest lavas in Figures 2 and 4 (including some historical flows), provide clear evidence for shorter time scale variations. *Rhodes and Hart* [this volume] found correlations within the historical flows that are distinct from those illustrated by Figure 7, i.e. be-

tween $^{87}Sr/^{86}Sr$ and trace element ratios involving the alkalis, such as Sr/Y and K/Y, but *not* with trace element ratios involving Nb. This suggests that the trace element variations in the youngest Mauna Loa flows may be produced by different mechanisms than those governing the variations found in older lavas. Figure 7 shows that there is an overall correlation between Zr/Nb and the isotopes, but that the trend *within* individual age groups is orthogonal to the overall trend. The samples with highest Zr/Nb (the youngest ones) have the highest $^{87}Sr/^{86}Sr$, but the reverse is true within each of the age groups (see for example the Kahuku samples). The distinction between the historical and long term trends is further illustrated by the correlation between $^{87}Sr/^{86}Sr$ and Sr/Y observed for the historical flows [*Rhodes and Hart*, this volume] superimposed on data for the longer time scale presented here (see Figure 8). The data for the older samples, including those from the SWR (see also *Garcia et al.*, [this volume]), dis-

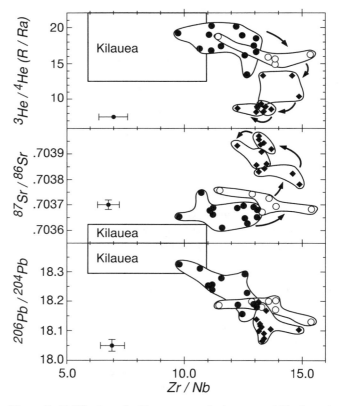

Figure 7. Zr/Nb plotted with respect to the isotopes of Pb, Sr and He for Mauna Loa and Kilauea. The arrows in the top two diagrams denote the temporal evolution, with arrows pointing toward younger ages (as in Figure 3); arrows and distinct fields are omitted from the lower diagram for clarity. Error bars in lower left are ± 2σ.

play considerably more scatter, which also probably relates to the different sampling intervals. *Rhodes and Hart* [this volume] attribute the linear trend for the historical data (Figure 8) to mixing between two primary melts. It is possible that the linear historical trend reflects variations in melting parameters (depth and degree of partial melting) primarily reflected in the trace element ratios, while the larger isotopic variability over the longer time scale primarily reflects changes in the mantle source.

In summary, the "transition" in Mauna Loa isotope geochemistry is also found with respect to some trace elements (see Figures 4 and 5). There is considerable scatter in the data, and there are clearly shorter time scale variations, but the Nb concentrations are highest in the submarine SWR lavas. The Nb variations occur with a timing similar to the Pb isotopes, and the transition to low Nb concentrations *precedes* those found for He, Sr, and Nd isotopes. [Despite the scatter within the data from submarine SWR lavas, a two-way t-test suggests that, as a population, they are distinct from the subaerial Kahuku lavas (at the 99% confidence interval) with respect to Nb, Zr/Nb, Sr/Nb, and Pb isotopes.] Therefore, the change in Mauna Loa geochemistry occurs first for Pb isotopes and Nb concentrations, followed by the isotopes of He, Sr, and Nd (in that order). The transition for He, Sr, and Nd occurs at approximately 10 ka; the transition in Pb and Nb occurred at least 100 ka earlier, assuming that the ages of the Kahuku Basalts are between 100 and 200 ka [*Lipman et al.*, 1990]. In contrast to Nb, the concentrations of Sr, Zr, and Y display the largest variability within the youngest lava flows, suggesting that the temporal trace element and isotopic variability is not simply related to degree of incompatibility.

3.3. *Mechanisms for Temporal Variability*

The most reasonable explanation for the large temporal isotopic variations at Mauna Loa, at least those found near 10 ka in age, is that they relate to movement of Mauna Loa over the hotspot. The Pacific plate is moving at an average speed of ~ 10 cm/year, and the typical formation time for Hawaiian volcanoes is between 500 ka and 1 Ma [*Lockwood and Lipman*, 1987; *Moore and Clague*, 1992; *Lipman*, this volume], so it is not unreasonable that Mauna Loa has recently been removed from plume influence. There are many geological arguments suggesting a recent decrease in eruption rates [*Lipman*, this volume] and there is a large change in eruption rates at 10 ka that coincides with the timing of the geochemical variations [*Lockwood*, this volume]. Based on $^3He/^4He$, the plume influence diminished significantly at ~10 ka. The other data (e.g. Figures 2 and 5) shows that the largest variations in Nb

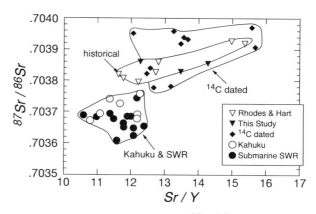

Figure 8. A comparison of Sr/Y and $^{87}Sr/^{86}Sr$ for Mauna Loa lavas. The field for historical encloses both data from table 1 and those of *Rhodes and Hart* [this volume], and shows the distinction between the historical and longer term trends.

concentrations and Pb isotopes are earlier than those for He, Sr and Nd isotopes. There is considerable scatter with respect to Nb concentration and Pb isotopes, but the highest values are found in samples from the submarine SWR. We suggest that the largest variations relate to the decrease in the plume signal indicated by the helium isotopes. The different timing could be partially related to degree of incompatibility, because Nb and Pb are highly incompatible elements. This would suggest a relationship between efficiency of extraction from the upwelling melting column, with the most incompatible elements removed first. The Sr, Y, and Zr concentration data are not consistent with this explanation, because these elements show more variation in the younger lavas (and in the opposite direction expected from melting effects), and no correlation to the isotopic variations in the older lavas. However, the fractionation corrected concentration data are affected by parameters that have lesser affects on the isotope ratios, such as initial concentrations in the different sources, and degree and depth of partial melting. It is possible, for example, to mix two mantle sources having different $^{87}Sr/^{86}Sr$ but very similar Sr concentrations, which could result in an apparent decoupling of the isotope ratio and concentration data.

Os isotopic data provide an important argument that the timing of the isotopic (and Nb) variations are not entirely caused by differing degrees of incompatibility and selective removal by melting. Os is a compatible element with respect to silicate melting, and if incompatibility (and resultant extraction efficiency) were the primary control on the isotopic variations, then Os would be the last isotopic signal to be removed from the plume. However, the Os isotopic data for Mauna Loa show a correlation with Sr, and

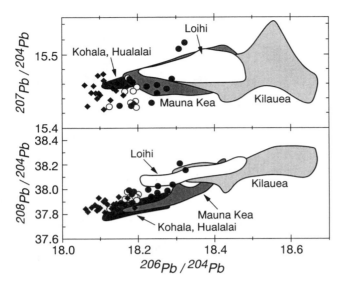

Figure 9. Pb isotopic data for the island of Hawaii. Only data for shield tholeiites are plotted here, in order to allow comparison between the different shields. Data sources: Mauna Loa (this study) Kilauea and Hualalai (unpublished data from this laboratory), Mauna Kea [*Yang et al.*, 1994], Kohala [*Tatsumoto*, 1978], Loihi seamount [*Staudigel et al.*, 1984].

similar temporal variations to those shown in Figure 2 for Sr [*Hauri and Kurz*, 1994; *Hauri et al.*, 1996].

The distinct isotopic and trace element trends observed for the historical flows (Figure 8), as well as the scatter within the submarine SWR lavas strongly suggest that there are several time scales of variability. The 10 to 100 ka variations are attributed to movement of Mauna Loa away from the hotspot. The shorter time scale variations may relate to the geometry of the heterogeneous plume or to the processes of melting and melt migration. The historical lavas have trace element variations that are almost as large as the entire Mauna Loa range, but the isotopic variations are much smaller (see Figure 8), which may imply a greater influence of the melting processes. The variations in Sr/Y may relate to both differing amounts of residual garnet within the source or to variations in Sr/Y within the different sources. Separating the effects of source heterogeneity versus melting parameters will require additional major and trace element data, and is beyond the scope of this paper. *Garcia et al.* [this volume] have shown that glasses from the SWR have similar major element compositions to subaerial Mauna Loa, which would suggest only minor changes in melting parameters over the time period discussed here.

It has been suggested that melt percolation and mixing between two distinct sources could explain the isotopic and trace element variations in Hawaii [*McKenzie and*

O'Nions, 1991; *Watson*, 1993]. As mentioned above, although some of the elements show timing variations in order of compatibility (Nb, Pb, He, Sr, Nd), others do not. The Os isotopes present severe problems for percolation models, as discussed above, because Os is a compatible element but correlates with Sr [*Hauri et al.*, 1996]. The two-source melt percolation model also cannot explain the Pb isotopes. This is shown in Figure 9, which illustrates that Loihi seamount lavas are intermediate between Mauna Loa and Kilauea, and therefore cannot represent one of the two end-members, contradicting helium isotopic evidence that Loihi is an end-member. Additionally, any two component mixing in a Pb-Pb diagram will produce a straight line, and percolation cannot produce "loops" (of the kind shown in Figure 3 for the other isotopes) with respect to Pb isotopes. Figure 9 shows that the Mauna Loa data alone do not define a straight line (see the $^{208}Pb/^{204}Pb$ variations). Consideration of the Pb isotopic data for other Hawaiian volcanoes (not shown in Figure 9, such as Haleakala and Koolau) also show significant deviations from a simple mixing line, which excludes a two source mixing model. Additional trace element constraints on melt percolation model have been discussed by *Frey and Rhodes* [1993], and will not be reviewed here. It is sufficient to conclude that a two component melt percolation model of the type discussed by *McKenzie and O'Nions* [1991] cannot explain the temporal evolution of Mauna Loa.

These arguments do not exclude melt migration and solid-melt interaction as important processes within the Hawaiian mantle. Some of the temporal variability on Mauna Loa may be related to degree of incompatibility and hence to melt percolation, the efficacy with which melting removes elements from the plume, and also the extent to which the melts interact with the overlying mantle column. The isotopic data suggest that the temporal variability may also relate to concentration differences between the various sources. It seems most likely that plate motion carries the volcano away from the hotspot source, and that the temporal geochemical variations are produced by a combination of these effects. The largest Pb isotopic variations occur ~100 ka prior to the Sr, Nd, and He decay, and the Pacific plate would have moved only 10 km in this time period, which requires a relatively small scale of mantle heterogeneity.

3.4. *Comparison with Adjacent Volcanoes*

Because the new SWR data extend the known isotopic and trace element range for Mauna Loa, it is important to compare these data with adjacent Hawaiian volcanoes. The trace element and isotope geochemistry of Hawaiian shield tholeiites has been extensively studied [e.g. *Wright*, 1971; *Tatsumoto*, 1978; *Hofman et al.*, 1984; *Budahn and Schmitt*, 1986; *Tilling et al.*, 1987; *Garcia et al.*, 1989; *Frey and Rhodes*, 1993; *Frey et al.*, 1994; *Leeman et al.*, 1994; *Yang et al.*, 1994], and each shield has apparently distinct geochemical characteristics which may relate to differences in both the mantle source and the physical processes involved. The subaerial lavas of Kilauea and Mauna Loa, the only two presently active volcanoes on the island, are geochemically distinct, even though they are only 30-40 kilometers apart [*Rhodes et al.*, 1989]. However, based on the trace element data presented here, the submarine SWR lavas, which are presumed to be the oldest, are the closest to the values found for Kilauea (see Figure 7). The oldest Mauna Loa samples also have isotopic signatures closest to those of Kilauea, and the age trend suggests a convergence in the geochemistry for older samples. The overlap between the Mauna Loa SWR and Kilauea compositions is supported by the trace element data presented by *Garcia et al.* [this volume], but does not necessarily extend to the major elements. The data presented here illustrate the importance of temporal evolution to comparison between the different shield volcanoes.

Based on the high ^{3}He/^{4}He found at Loihi seamount, we have proposed that Loihi isotopic data represent the closest approximation to an undegassed plume "end-member" for Hawaii [*Kurz et al.*, 1982; *Kurz et al.*, 1983; *Kurz and Kammer*, 1991; *Kurz*, 1993]. Loihi is also the youngest volcano in the archipelago, suggesting that the high ^{3}He/^{4}He ratios are related to volcano evolution [*Kurz et al.*, 1983; *Kurz et al.*, 1987]. The new data from SWR of Mauna Loa support this hypothesis because the oldest lavas have the highest ^{3}He/^{4}He (see Figure 10), and define a trend which is closer to the isotopic composition of Loihi seamount than any of the younger lavas. In addition, all Hawaiian volcanoes have higher ^{3}He/^{4}He in the oldest tholeiitic lavas than in the associated alkali basalts. At present these generalizations are based on stratigraphic studies at Haleakala, Hualalai, Mauna Kea, Kilauea and Mauna Loa [*Kurz et al.*, 1987; *Kurz and Kammer*, 1991; *Kurz*, 1993; unpublished data this laboratory]. All of these results strongly support the mantle plume model as originally formulated by *Wilson* [1963] and *Morgan* [1971], particularly if high ^{3}He/^{4}He ratios reflect undegassed mantle sources (i.e. those having high He/(Th+U)) deriving from the lower mantle [e.g. *Kurz et al.*, 1982].

The new data from the SWR of Mauna Loa have a bearing on the relationship between the two parallel chains that form the Hawaiian archipelago. In the islands south east of the Molokai fracture zone, the volcanoes have alternated in time between the two chains, with Loihi,

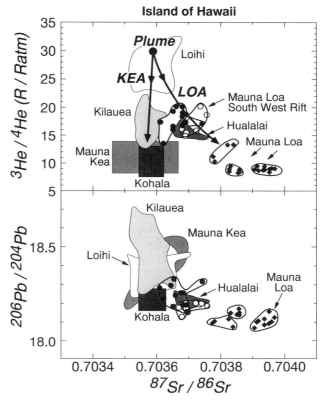

Figure 10. A comparison of the Sr, Pb, and He isotopic compositions for tholeiites on the island of Hawaii. Note that this figure excludes the alkali basalt data for Mauna Kea, Hualalai, and Kohala. Data sources for Loihi seamount: *Kurz et al.*, 1983; *Staudigel et al.*, 1984; Kilauea and Hualalai: unpublished data this laboratory; Mauna Kea: *Kurz et al.*, 1994; Kohala: *Stille et al.*, 1986; *Graham et al.*, 1990]. Although there have been more recent isotopic studies of Loihi seamount [*Garcia et al.*, 1993], the data plotted in this figure (and in Figure 9) are those samples for which helium and other isotopic data coexist for the same samples.

Mauna Loa, Hualalai, Mahukona, Kahoolawe, Lanai and West Molokai representing the southern "Loa" trend, and Kilauea, Mauna Kea, Kohala, Haleakala, West Maui, and East Molokai representing the northern "Kea" trend (e.g., *Jackson et al.*, [1972]; see inset Figure 1). *Tatsumoto* [1978] noticed that the Kea trend volcanoes on the island of Hawaii had more radiogenic Pb isotope ratios, and was the first to relate isotope geochemistry to the existence of the two chains. He suggested that the Loa volcanoes represented the primary expression of the upwelling plume, and that the volcanoes in the Kea trend were more affected by lithospheric contamination (which by inference had high U/Pb) due to lower upwelling and eruption rates and greater interaction with the lithosphere. At present, this hypothesis remains speculative; no comprehensive explanation has ever been offered for the existence of the two chains, and it is unclear if they are produced by lithospheric loading or mantle processes.

The isotopic data from Mauna Loa support Tatsumoto's suggestion that the Loa and Kea trends are distinguished with respect to isotopic composition on the island of Hawaii (with Kea trend volcanoes having more radiogenic Pb). The new data from the SWR are particularly important because they illustrate that the isotopic and trace element compositions tend to converge for the older shield building tholeiites, and that the geochemical difference between the Loa and Kea trends is primarily a late stage, time dependent effect. The oldest lavas from Kilauea and Mauna Loa are most similar to Loihi, but the last 30 ka of Mauna Loa eruptions have diverged significantly, as reflected in higher $^{87}Sr/^{86}Sr$ and lower $^{3}He/^{4}He$ ratios (see Figure 10). Figure 10 suggests that the Sr and He isotopes can be used to distinguish the Loa and Kea trend volcanoes, with high $^{3}He/^{4}He$ indicative of plume influence. It is unclear how robust this observation is, because there is little data from the other volcanoes in the archipelago that might be used to evaluate it. The Mauna Loa data demonstrate how important the temporal evolution of a single volcano is to such comparisons; it is problematic to compare volcanoes at different stages of their evolution, even though they may be tholeiitic shields.

4. IMPLICATIONS

4.1. Geochemistry of Hawaiian Basalts

The temporal variations found at Mauna Loa, particularly with respect to Pb isotopes and Nb concentrations, demonstrate how misleading it can be to compare geochemical data from Hawaiian volcanoes at different stages in their evolution; even within shield building tholeiites, the stratigraphic framework is critical. For example, it is common to consider Koolau as representative of the plume end-member for Hawaiian volcanism, due to the high $^{87}Sr/^{86}Sr$ ratios found in the tholeiites [e.g. *Stille et al.*, 1986; *Leeman et al.*, 1994]. However, the isotopic compositions found at Koolau are similar to the late stage of Mauna Loa , because the high $^{87}Sr/^{86}Sr$ are coupled with unradiogenic Pb isotopes, intermediate $^{3}He/^{4}He$, and low Nb concentrations [*Stille et al.*, 1986; *Roden et al.*, 1994]. The Mauna Loa isotopic and temporal variations, combined with other data from the island of Hawaii, suggest that this set of geochemical characteristics does *not* relate to the plume, but may be a relatively small volume contribution, limited to the last stages of shield building. As dis-

cussed below, the "component" represented by isotopic compositions similar to Koolau and late stage Mauna Loa may be explained by entrained asthenosphere or lithosphere. Although this component is clearly associated with the plume and/or its interaction with the lithosphere, the data presented here suggest it is not the most volumetrically important contributor to Hawaiian shields, and probably does not represent the plume itself.

One important implication of the data presented here is that Nb and Pb may be the first elements to be removed from the upwelling plume. This conclusion must be viewed as highly speculative, in light of the isotopic evidence for mixing between different mantle sources, and the possibility that the different timing for changes in Pb and Nb (relative to He, Sr and Nd) is simply caused by concentration differences between the sources. Whether the order of depletion is caused by mixing or melt silicate partitioning, the coherence between helium and the other elements strongly suggests that the physical control on helium isotopes within the melting region is similar to other elements. Therefore, there is little reason to invoke unusual processes, such as vapor phase transfer or metasomatism [e.g. *Vance et al.*, 1990], to "decouple" helium. It seems extremely likely that helium behaves as an incompatible element with respect to melt/silicate partitioning, which is also suggested by phenocryst/glass partitioning [*Kurz*, 1993]. Therefore, the Mauna Loa geochemical data offer little support for models which discard the helium isotopic evidence [e.g. *Hofmann*, 1986].

4.2 Models for the Hawaiian Plume

Based on the arguments presented above, it is necessary to reconcile the requirement for three mantle sources, and the temporal geochemical variations, with a reasonable geometry for the upwelling mantle plume beneath Hawaii. Figure 11 shows one highly schematic example of such a model; many other geometries are possible, but this cartoon demonstrates the possibilities with respect to the different geochemical contributions to a Hawaiian volcano [see also *Frey and Rhodes*, 1993]. At least four distinct mantle reservoirs could serve as sources for Hawaiian volcanism: the upwelling plume, entrained material at the edge of the plume, normal asthenospheric mantle, and Pacific plate lithosphere. All of these are possible contributors; although the geochemistry alone cannot conclusively identify them, the isotopic data provide some important constraints.

The highest $^{3}He/^{4}He$ ratios (up to 20 times atmospheric), found in the oldest Mauna Loa flows, identify the mantle plume itself as a major contributor to the main shield stage. As mentioned above, all known temporal studies of Hawaiian volcanoes show that higher $^3He/^4He$ ratios are found in the earliest shield stages. Normal asthenosphere, and presumably normal lithosphere, have $^3He/^4He$ ratios close to 8 times atmospheric, based primarily on studies of MORB. The lower $^3He/^4He$ ratios in the younger Mauna Loa lavas were originally interpreted as mixing between the plume and entrained mantle at the plume edge [*Kurz and Kammer*, 1991; *Kurz*, 1993]. The geochemical reason for invoking entrained material, as opposed to lithosphere [e.g. *Liu and Chase*, 1991], is that high $^{87}Sr/^{86}Sr$ ratios should not be observed in normal lithosphere, because normal MORB typically have $^{87}Sr/^{86}Sr$ less than .7030 [e.g. *King et al.*, 1993]. However, studies of Hawaiian lherzolite xenoliths, which may represent fragments of the oceanic lithosphere, have $^{87}Sr/^{86}Sr$ of .7035 to .7042 [*Vance et al.*, 1989], making this a relatively weak argument.

Another argument that lithosphere is not extensively involved as a source of melting comes from seismic and modeling studies, which suggest minimal lithospheric thinning beneath Hawaii [*Davies*, 1992; *Ribe and Christensen*, 1994; *Woods et al.*, 1991]. Lithospheric thinning, and the resultant thermal buoyancy, was thought to have produced Hawaiian Swell [*Detrick and Crough*, 1978] which is now in doubt. The new data presented here are important because they demonstrate that the geochemistry of Mauna Loa has changed most dramatically within the last 30 to 100 ka, representing a small fraction of the erupted, and therefore mantle source volume. It is only during the recent eruptions (< 30 ka) that isotopic data indicate a non-hotspot source. The historical Mauna Loa lavas define a "bend" in all of the isotope systems (toward less radiogenic isotopic compositions, see Figure 3), which was previously attributed to the lithospheric component [*Kurz*, 1993].

The model shown in Figure 11 assumes that the geochemical variations are related to heterogeneity within the plume rather than lithospheric involvement, due to the seismic observations, the relatively radiogenic Sr isotopic compositions in recent Mauna Loa, as well as the Os isotopic evidence [*Hauri et al.*, 1996].

The order of element removal suggested by Figure 2 is consistent with the cartoon in Figure 11, if we assume that the end of shield building stage represents a fundamental change in the mantle source. The model presumes that all melting occurs within the garnet stability field (depths greater than 90 Km), and attributes the geochemical change between 10 and 100 ka to zonation within the plume. At the end of shield building, the melting rate decreases and the geochemical signature of the plume diminishes.

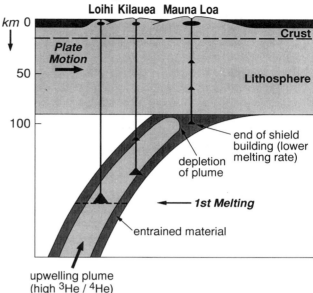

Figure 11. One highly schematic model for the Hawaiian plume, which could explain the temporal geochemical variations. The model illustrates that there are several possible mantle sources, including a heterogeneous plume, the surrounding asthenosphere, and the lithosphere. Melting begins in the garnet stability field. By the time shield building is complete, the plume isotopic signature may have been partly removed due to extensive melting of the plume, and the plume is assumed to be concentrically zoned. The last stage of shield building, as represented by recent Mauna Loa lavas, has contributions from entrained mantle material. It is inferred that the melting rate decreases at the end of shield building, which could imply slower melt migration rates through the lithosphere. Lithospheric involvement is assumed to be minimal, and limited to post-shield lavas. The absence of a depth scale below 100 km is intentional and is intended to emphasize the uncertainty in the depth for initiation of melting.

ishes. It is at this stage that the geochemical signature of the entrained material, which accumulates at the base of the lithosphere, may become dominant. It is possible that this component has been melting all along, but has been overwhelmed by melts from the closely associated, and volumetrically more important, plume material. The different timing implied by Figure 2 could then be explained by the concentration differences within the heterogeneous plume, combined with differing degrees of incompatibility (i.e. Pb and Nb having greatest concentration most effectively partitioned into the melt).

There is considerable evidence that melting beneath Hawaii occurs within the garnet stability field [*Budahn and Schmitt*, 1985; *Frey and Rhodes*, 1993; *Wright*, 1971;

Hofman et al., 1984; *Salters and Hart*, 1991]. In addition, the lack of Th-U disequilibrium within the historical tholeiites of Mauna Loa and Kilauea suggest relatively slow removal of melts from the garnet stability region [*Cohen and O'Nions*, 1993; *Hemond et al.*, 1994], at least within the context of existing models [e.g. *Spiegelman and Elliot*, 1993; *Iwamori*, 1993]. *Hemond et al.* [1994] suggest that this can be explained by the slow melt aggregation (via the fractal tree mechanism of *Hart*, [1993]), followed by rapid transit through the lithosphere. Identification of the mechanics, and site of, the mantle/melt interaction are critical to evaluating the use of mantle mixing schemes [e.g., *Hart et al.*, 1992], to explain global isotopic variations.

One serious problem with the model shown in Figure 11 is that it is not entirely consistent with major element data. Variations in depth of melting should result in systematic major element variations as a function of age. Although each volcano on the island of Hawaii has distinct major element chemistry [e.g. *Frey and Rhodes*, 1993], Mauna Loa does not show temporal major element variations on the time scale discussed here [*Garcia et al.*, this volume]. Further evaluation must await consideration of a more comprehensive major element data set.

Finally, the new data from Mauna Loa, and the general convergence toward the isotopic characteristics of Loihi seamount for the oldest Hawaiian shield tholeiites (see Figure 10), support the hypothesis that Loihi seamount represents the present day expression of the plume center. The hypothesis that Mauna Loa has been removed from the hotspot within the last 10 ka would require the radius of the plume to be less than 40 kilometers (the distance between Mauna Loa and Loihi). This is consistent with the simple calculation of plume radius from the shield building time of 500 ka, combined with plate motion of 10 cm/year, which yields 50 Km. Although the reason for the Loa and Kea trends remains enigmatic, the Mauna Loa data suggest that isotopes can be used to distinguish them (particularly Sr and He) and that these geochemical differences relate to the end of shield building.

5. CONCLUSIONS

1. The new He, Sr, Nd, and Pb isotopic data for the submarine SWR reported here extend the known range of isotopic values for Mauna Loa, particularly for Pb isotopes. The fact that these data continue the temporal trends found for subaerial (stratigraphically placed and ^{14}C dated) lava flows, strongly suggests that the SWR lava flows are older than the Kahuku and Ninole basalts. The coherence of the isotopic data also demonstrates that helium is removed

from the mantle by melting processes rather than metasomatic processes or vapor transfer, and is not affected by magma chamber contamination processes.

2. He, Sr, and Nd isotopic compositions in the submarine SWR lavas are similar to the Kahuku and Ninole Basalts, but Pb is significantly more radiogenic, suggesting that the Pb isotopic signature of the mantle plume decreased significantly earlier. The new data from the SWR lavas, combined with existing isotopic data from subaerial Mauna Loa, suggest that the order of removal of the plume isotopic signature is Pb (>100 ka), He (~14 ka), Sr (9 ka), and Nd (8 ka), which may relate to the order of silicate/melt partitioning and to concentration differences in the different mantle sources. Zr/Nb and Sr/Nb, and fractionation corrected Nb concentrations, correlate with the isotopes, and also show significant variations on the 100 ka time scale. The assignment of different timing for the different isotopes, particularly Sr, Nd, and He, must be viewed as speculative due to the limitations of the data set, but is a testable hypothesis, given the large number of Mauna Loa lava flows of known age [*Lockwood*, this volume].

3. Superimposed on the 10 to 100 ka temporal variations, there are also important shorter term variations, as demonstrated by the scatter within the submarine SWR samples, and data from historical samples [*Rhodes and Hart*, this volume]. The isotopic and trace element trends are distinct for the historical lavas, suggesting that different processes are responsible for the variability on the two different time scales. The historical variations (e.g. Figure 8) may be related to melting effects, while the longer term trend can be attributed to removal of the plume source from Mauna Loa.

4. The temporal isotopic evolution of Mauna Loa suggests that the Loa and Kea volcanoes can be distinguished based on isotopes, particularly He, Sr, and Pb, but that the differences are primarily found in the last stages of shield building. The model suggested here (Figure 11) relates the geochemical differences to a heterogeneous, zoned plume, and the changing dynamics of melting within the plume at the end of shield building (as it impinges on the lithosphere). The temporal isotopic evolution of Mauna Loa also supports the hypothesis that Loihi seamount isotopic compositions best represent the plume end member for Hawaiian volcanoes.

Acknowledgments. This work was supported by NSF OCE92-14006 and EAR93-18889 to MK and OCE90-12030 to MG. We thank the Pisces V diving team, led by Terry Kerby, for their assistance in sample collection, and Peter Lipman and Tom Hulsebosch for assistance with the dive program and numerous discussions. Constructive reviews by Fred Frey, Peter Lipman, and Gene Yogodzinski greatly improved the manuscript. MK gratefully acknowledges the encouragement of J. Lockwood and F. Truesdell at HVO, and the field mapping effort at HVO which has made detailed study of Mauna Loa possible. This is WHOI contribution number 9038, and SOEST contribution number 3940.

REFERENCES

Budahn, J.R. and R.A. Schmitt, Petrogenetic modeling of Hawaiian tholeiitic basalts: a geochemical approach. *Geochim. Cosmochim. Acta 49*, 67-87, 1986.

Catanzaro, E.J., T.J. Murphy, W.R. Shields, and E. Garner, Absolute isotopic abundance ratios of common, equal atom and radiogenic lead isotope standards. *J. Res. NBS. 72A*, 261, 1968.

Cohen, A.S., and R.K. O'Nions, Melting rates beneath Hawaii: evidence from uranium series isotopes in recent lavas. *Earth Planet. Sci. Lett. 120*, 169-175, 1993.

Condomines, M., K. Gronvold, P.J. Hooker, K. Muehlenbachs, R.K. O'Nions, J. Oskarsson, and E.R. Oxburgh, Helium, oxygen, strontium and neodymium relationships in Icelandic volcanics. *Earth Planet. Sci. Lett. 66*, 125-136, 1983.

Davies, G.F., Temporal variation of the Hawaiian plume flux. *Earth and Planetary Science Letters 113*, 277-286, 1992.

Detrick, R.S., and S. T. Crough, Island subsidence, hot spots and lithospheric thinning. *J. Geophys. Res. 83*, 1236-1244, 1978.

Fornari, Peterson, Lockwood, Malahoff, and Heezen, Submarine extension of the SWRZ of Mauna Loa volcano, Hawaii. *GSA Bull. 90*, 435-443, 1979.

Frey, F.A. and J.M. Rhodes, Intershield geochemical differences among Hawaiian volcanoes: implications for source compositions, melting process and magma ascent paths. *Phil. Trans. Roy. Soc. London A342*, 121-136, 1993.

Frey, F.A., M.O. Garcia, and M.F. Roden, Geochemical characteristics of Koolau volcano: implications of intershield geochemical differences among Hawaiian volcanoes. *Geochim. Cosmochim. Acta 58*, 1441-1462, 1994

Garcia, M.O, D. Muenow, K. Aggrey, and J. O'Neil, Major element, volatile and stable isotope geochemistry of Hawaiian submarine tholeiitic glasses. *J. Geophys. Res. 94*, 10525-10538, 1989.

Garcia, M.O., T.P. Hulsebosch, and J. M. Rhodes, Glass and Mineral Chemistry of Olivine-Rich submarine basalts, South West Rift Zone, Mauna Loa Volcano: Implications for magmatic processes, this volume.

Garcia, M.O., J.J. Mahoney, and E. Ito, An evaluation of temporal geochemical evolution Loihi summit lavas: results from Alvin submersible dives. *J. Geophys. Res. 98*, 537-550, 1993.

Graham, D., J. Lupton, M. Garcia, He isotopes in olivine phenocrysts from submarine basalts of Mauna Kea and Kohala, Island of Hawaii. *EOS (Trans. AGU) 71*, 657, 1990.

Hart, S.R., Equilibration during mantle melting: A fractal tree model. *Proc. National Acad. Sci. U.S.A. 90*, 11914-11918, 1993.

Hart, S.R., E.H. Hauri, L.A. Oschmann and J.A. Whitehead, Mantle plumes and entrainment: Isotopic evidence. *Science 256*, 517-520, 1992.

Hart, S.R., and T. Dunn, Experimental cpx/melt partitioning of 24 trace elements. *Contrib. Mineral. Petrol. 113*, 1-8, 1993.

Hauri, E.H., and M.D. Kurz, Osmium isotope systematics of the Hawaiian hotspot: contrasting time-series results from Kilauea and Mauna Loa. *EOS (trans. Am. Geophys. Union) 75*, 708, 1994.

Hauri, E.H., J.C. Lassiter, D.J. DePaolo, and J.M. Rhodes, Osmium isotope systematics of drilled lavas from Mauna Loa, Hawaii. *J. Geophys. Res. (submitted)*, 1996.

Hauri, E.H., T.P. Wagner, and T.L. Grove, Experimental and natural partitioning of Th, U, Pb and other trace elements between garnet, clinopyroxene and basaltic melts. *Chem. Geol.* 117-166, 1994.

Hemond, C., A.W. Hofmann, G. Heusser, M. Condomines, I. Raczek, J.M. Rhodes, U-Th-Ra systematics in Kilauea and Mauna Loa basalts, Hawaii. *Chem. Geol. 116*, 163-180, 1994.

Hilton, D. R., J. Barling, and G.E. Wheller, Effect of shallow level contamination on the helium isotope systematics of ocean island lavas. *Nature 373*, 330-333, 1995.

Hofmann, A.W., Nb in Hawaiian magmas: constraints on source composition and evolution. *Chem. Geol. 57*, 17-30, 1986.

Hofmann, A.W., M.D. Feigenson, and I. Raczek, Case studies on the origin of basalt: Petrogenesis of the Mauna Ulu eruption, Kilauea, 1969-1971, *Contrib. Mineral. Petrol. 88*, 24-35, 1984.

Hofmann, A.W., K.P. Jochum, M. Seufert, and W.M. White, Nb and Pb in oceanic basalts: New constraints on mantle evolution. *Earth Planet. Sci. Lett. 79*, 33-45, 1986.

Iwamori, H., Dynamic disequilibrium melting model with porous flow and diffusion controlled chemical equilibration. *Earth Planet. Sci. Lett 114*, 301-313, 1993.

Jackson, E.D., E.A. Silver, and G.B. Dalrymple, Hawaiian-Emperor Chain and its relation to Cenozoic circumpacific tectonics. *Bull. Geol. Soc. Am. 83*, 601-618. 1972.

King, A. J., D G. Waggoner, and M. O. Garcia, "Geochemistry and petrology of basalts from leg 136, central Pacific Ocean", in: R. H. Wilkens, J. Firth, J. Bender, et al, (1993) *Proceedings of the Ocean Drilling Program, Scientific Results*, Vol. 136: College Station, TX (Ocean Drilling Program), 1993.

Kurz, M. D., Mantle heterogeneity beneath oceanic islands: some inferences from isotopes. *Proceedings of the Royal Society of London A342*, 91–103, 1993.

Kurz, M.D., W.J. Jenkins, and S.R. Hart, Helium isotopic systematics of oceanic islands: implications for mantle heterogeneity. *Nature 297*, 43-47, 1982.

Kurz, M.D., W.J. Jenkins, S. Hart, and D. Clague, Helium isotopic variations in Loihi Seamount and the island of Hawaii. *Earth Planet. Sci. Lett. 66*, 388-406, 1983.

Kurz, M.D., M.O. Garcia, F.A. Frey and P.A. O'Brien, Temporal helium isotopic variations within Hawaiian volcanoes: basalts from Mauna Loa and Haleakala. *Geochim. Cosmochim. Acta 51*, 2905-2914, 1987.

Kurz, M.D., and D.P. Kammer, Isotopic evolution of Mauna Loa Volcano. *Earth Planet. Sci. Lett. 103*, 257-269, 1991.

Kurz, M.D., J.K. Lassiter, B.M. Kennedy, D.J. DePaolo, J.M. Rhodes, and F. A. Frey, Helium isotopic evolution of Mauna Kea volcano:first results from the 1 km drill core. *EOS (Trans. Am. Geophys. Union) 75*, 711, 1994.

Leeman, W.P., D.C. Gerlach, M.O. Garcia, and H.B. West, Gecohemical variations in lavas from Kahoolawe volcano, Hawaii: evidence for open system evolution of plume-derived magmas. *Contrib. Mineral. Petrol. 116*, 62-77, 1994.

Lipman, P.W., Growth of Mauna Loa volcano during the last hundred thousand years: rates of lava accumulation versus gravitational subsidence, this volume.

Lipman, P.W., and A. Swenson, Generalized geological map of the southwest rift of Mauna Loa, U.S. Geol. Surv. Map I-1323, 1984.

Lipman, P.W., J.M. Rhodes, and G.B. Dalrymple, The Ninole Basalt-implications for the structural evolution of Mauna Loa volcano, Hawaii. *Bull. Volcanol. 53*, 1-19, 1990.

Liu M. and C.G. Chase, Evolution of Hawaiian basalts: a hotspot melting model. *Earth Planet. Sci. Lett. 104*, 151-165, 1991.

Lockwood, J.P., Mauna Loa Eruptive History-the radiocarbon record, this volume.

Lockwood, J.P., and P.W. Lipman, Holocene eruptive history of Mauna Loa volcano. *U.S. Geological Survey Professional Paper 1350*, 509-536, 1987.

McKenzie, D. and R.K. O'Nions, Partial melt distributions from inversion of rare earth element concentrations. *J. Petrol. 32*, 1021-1091, 1991.

Moore, J.G., Subsidence of the Hawaiian Ridge. *U.S. Geological Survey Professional Paper 1350*, 85-100, 1987.

Moore, J.G., W.R. Normark, and B.J. Szabo, Reef growth and volcanism on the submarine southwest rift zone of Mauna Loa. *Bull. Volcanol. 52*, 375-380, 1990.

Moore, J.G., and D.A. Clague, Volcano growth and evolution of the island of Hawaii. *Bulletin of the Geological Society of America 104*, 1471-1484, 1992.

Morgan,W.J., Convection plumes in the lower mantle. *Nature 230*, 42-43, 1971.

Rhodes, J.M., Homogeneity of lava flows: chemical data for historical Mauna Loa eruptions. *J. Geophys. Res. Supp. 88*, A869-A879, 1983.

Rhodes, J.M., Geochemistry of the 1984 Mauna Loa eruption: implications for magma storage and supply. *J. Geophys. Res. 93*, 4453-4456, 1988.

Rhodes, J.M., K.P. Wenz, C.A. Neal, J.W. Sparks, and J.P. Lockwood, Geochemical evidence for invasion of Kilauea's plumbing system by Mauna Loa magma. *Nature 3376*, 257, 1989.

Rhodes, J.M., and S.R. Hart, Episodic trace element and isotopic variations in historic Mauna Loa lavas: implications for magma and plume dynamics, this volume.

Ribe, N.M, and U.R. Christensen, Three dimensional modeling of plume-ltihosphere interaction. *Journal Geophysical Re-search 99*, 669-682, 1994.

Roden, M.F., T. Trull, S.R. Hart, F.A. Frey, New He, Nd, Pb and Sr isotopic constraints on the constitution of the Hawaiian plume: Results from Koolau Volcano, Oahu, Hawaii, USA. *Geochim. Cosmochim. Acta. 58*, 1431-1440, 1994.

Rubin, M., L.K. Gargulinski, and J.P. McGeehin, Hawaiian radiocarbon dates. *U.S. Geological Survey Prof. Paper 1350*, 213-242, 1987.

Salters, V.J.M, and S.R. Hart, The mantle sources of ocean ridges, islands and arcs: the Hf-isotope connection. *Earth Planet. Sci. Lett. 104*, 364-380, 1991.

Spiegelman, M., and T. Elliot, Consequences of melt transport for uranium series disequilibrium in young lavas. *Earth Planet. Sci. Lett. 118,* 1-20, 1993.

Staudigel H., A. Zindler, S.R. Hart, T. Leslie, C.Y. Chen, and D. Clague, The isotope systematics of a juvenile volcano: Pb, Nd, and Sr isotope ratios of basalts from Loihi Seamount. *Earth Planet. Sci. Lett. 69*, 13-29, 1984.

Stille, P., D.M. Unruh, and M. Tatsumoto, Pb, Sr, Nd, and Hf isotopic constraints on the originof Hawaiian basalts and evidence for a unique mantle source. *Geochim. Cosmochim. Acta 50*, 2303-2319, 1986.

Tatsumoto, M., Isotopic composition of lead in oceanic basalt and its implication to mantle evolution. *Earth Planet. Sci. Lett. 38*, 63-87, 1978.

Tilling, R.I., T.L. Wright, and H.P Millard, Trace element chemistry of Kilauea and Mauna Loa is space and time: a reconnaissance. *U.S.G.S. Prof. Paper 1350*, 641-689, 1987.

Vance D., J.O.H. Stone, and R.K. O'Nions, He, Sr, and Nd isotopes in xenoliths from Hawaii and other oceanic islands. *Earth Planet. Sci. Lett. 96*, 147-160, 1989.

Watson, S. , Rare earth element inversions and percolation models for Hawaii. *Journal of Petrol. 34*, 763-783, 1993.

Wilson, J.T., A possible origin of the Hawaiian Islands. *Can. J. Phys. 41*, 863-870, 1963.

Woods, M.T., J.J. Leveque, and E.A. Okal, Two station measurements of Rayleigh wave group velocity along the Hawaiian swell. *Geophys. Res. Lett. 18*, 105-108, 1991.

Wright, T.L., Chemistry of Kilauea and Mauna Loa in space and time. *U.S. Geological Survey Professional Paper 735*, 1-49, 1971.

Yang, H, F.A. Frey, M.O. Garcia, D.A. Clague, Submarine lavs from Mauna Kea volcano, Hawaii: implications for Hawaiian shield stage processes. *J. Geophysical Research 99*, 15,577-15594, 1994.

Zindler, A. and S.R. Hart, Helium: Problematic primordial signals. *Earth Planet. Sci. Lett. 79*, 1-8, 1986.

M. D Kurz, T. C Kenna, D. P. Kammer, Department of Marine Chemistry and Geochemistry, MS 25, Woods Hole Oceanographic Institution, Woods Hole, MA 02543-1541.

J. Michael Rhodes, Department of Geology and Geography, University of Massachusetts, Amherst, MA 01003.

Michael O. Garcia, Hawaii Center for Volcanology, Geology and Geophysics Department, University of Hawaii, Honolulu, HI 96822.

Contrasting Th/U in Historical Mauna Loa and Kilauea Lavas

Klaus Peter Jochum, and Albrecht W. Hofmann

Max-Planck-Institut für Chemie, Mainz, Germany

High-precision measurements of Th and U concentrations in recent Kilauea and Mauna Loa lavas have been obtained using a new multi-ion counting technique in an otherwise conventional spark source mass spectrometer. This method combines advantages of solid-state sample preparation with the improved precision provided by ion counting (instead of photoplate detection). The results are significantly more precise than published data obtained by thermal-ionization isotope dilution requiring sample dissolution and spike equilibration. Our results resolve a previously undetected 3 % difference in Th/U between Mauna Loa (Th/U = 2.92± 0.02) and Kilauea (Th/U = 3.01± 0.01) lavas erupted during this century. In contrast, Pb isotope data indicate that the Mauna Loa source should have a slightly higher Th/U ratio than the Kilauea source. If we assume that Th/U of the Mauna Loa source is equal to or greater than that of the Kilauea source, the fractionation of the ratio during melt extraction must be equal to or greater than 3 %, while the extracted melt fraction changes by about a factor of two. Model calculations using experimentally determined partitioning between garnet lherzolite and melt show that partition coefficients near the upper limit of the range of published, experimentally determined coefficients are required to yield viable melt compositions.

1. INTRODUCTION

The use of trace elements for the purpose of deciphering chemical characteristics of the Earth's mantle depends largely on the "incompatible" character of some of these elements. When partial melting occurs in the mantle, highly incompatible elements are nearly quantitatively transferred from the solid source rock to the melt. Therefore, the concentration ratio of two such elements in the melt will be similar to that of the source, and the ratio can serve a tracer of (mantle) source composition. Thorium and uranium are of particular interest in this respect, because the relationship between $(Th/U)_{melt}$ and $(Th/U)_{source}$ can be explored using $^{230}Th/^{232}Th$ - $^{238}U/^{232}Th$ relationships [*Condomines et al.*, 1981; *Williams and Gill*, 1989; *Spiegelman and Elliott*, 1993]. Surprisingly, Th-U fractionations of up to about

30%, relative to the source, have been found in mid-ocean ridge basalts.

Previous studies have shown that, in contrast with MORB, the tholeiitic basalts from Mauna Loa and Kilauea volcanoes showed little or no deviation from radioactive equilibrium [*Cohen and O'Nions*, 1993; *Hémond et al.*, 1994]. Thus it appeared that the production of Hawaiian tholeiites does not significantly fractionate Th from U. Moreover, Th/U ratios in basalts from these two volcanoes appeared to be essentially indistinguishable at 2.95 ± 0.05 (Mauna Loa) and 3.01 ± 0.08 (Kilauea). This was surprising because isotopic compositions, including the Th/U-dependent $^{208}Pb/^{206}Pb$ ratios, and other incompatible element ratios, such as Ba/Th, of the two volcanoes differ significantly [*Hémond et al.*, 1994; *Frey and Rhodes*, 1993].

The present paper is an outgrowth of measurements originally intended as a test of a recently developed high precision multi-ion counting (MIC) technique for a spark source mass spectrometer [*Jochum et al.*, 1994]. For this purpose, we selected a Kilauea basalt standard (BHVO-1)

Mauna Loa Revealed: Structure,
Composition, History, and Hazards
Geophysical Monograph 92

and several other Kilauea and Mauna Loa tholeiites, because we expected all of them to have identical Th/U ratios, as previously documented by *Cohen and O'Nions* [1993] and *Hémond et al.* [1994]. An important advantage of this new MIC-SSMS technique over other analytical techniques (such as thermal ionization mass spectrometry, TIMS; or inductively coupled mass spectrometry, ICP-MS) is that it does not involve dissolution and chemical separation of samples. Therefore, uncertainties concerning the stability of solutions, equilibration with spike and contamination during chemical dissolution processes are eliminated.

2. TECHNICAL ASPECTS

2.1. *Sample selection and preparation*

We selected 12 very fresh tholeiites, erupted from Kilauea and Mauna Loa volcanoes between 1907 and 1983. The samples include the USGS reference material BHVO-1, which is a tholeiite from a 1919 flow in the Kilauea caldera [*Govindaraju*, 1994]. Locations, sample descriptions, major element, trace element and isotopic data are given by *Tatsumoto* [1978], *Hofmann et al.* [1986], and *Newsom et al.* [1986].

About 60 mg of sample powder was mixed with 30 mg of spiked graphite in an agate mixing mill for one hour. The spiked graphite which is usually used in our laboratory for multi-element isotope dilution measurements [*Jochum et al.*, 1988] contains 12 spike isotopes including a 99.82% enriched ^{235}U spike. The sample-graphite mixture was compressed to rod-shaped electrodes in a polyethylene slug.

2.2. *Mass spectrometry*

A spark source mass spectrometer, type AEI-MS 702R, was used for this work. General descriptions of analyses with this instrument have been given by *Taylor* [1965]. A vacuum discharge was generated between the sample electrodes by a pulsed high-voltage AC potential using the spark conditions: spark voltage = 25 kV, pulse repetition rate = 300 Hz, pulse length = 100 μs. The total ion charge (which is a measure for the number of ions entering the magnetic field before mass separation) was measured with a monitor.

To detect many ions simultaneously along the straight focusing plane given by the Mattauch-Herzog geometry of this instrument, we recently equipped the commercial mass spectrometer with a multi-ion counting (MIC) system designed ultimately to consist of 20 separate small (1.85 mm wide) channeltrons [*Laue et al.*, 1994]. In the initial experiments reported here, we tested the system with 5 channeltrons [*Jochum et al.*, 1994]. Each channeltron is connected to an individual high voltage power supply (2100

V), a separate preamplifier (dead time = 25 ns, input threshold = 20 mV), and a counter. A conversion plate (potential = - 1400 V) is incorporated in front of each channeltron. The different channeltrons were calibrated against each other by peak switching between the various channeltrons. To measure simultaneously ^{232}Th, ^{235}U and ^{238}U isotopes with the MIC system at a mass resolution of about 600, we used an exit slit system with 0.4 mm wide slits at the locations of the masses of interest. The mass lines of ^{232}Th, ^{235}U and ^{238}U are interference-free in all cases. An unspiked sample yielded a $^{238}U/^{235}U$ ratio of 139±2, which is within error limits identical to the natural ratio.

Uranium was analyzed directly by isotope dilution by measuring the difference in $^{238}U/^{235}U$ from natural uranium due to the spike present in the electrode materials. The thorium concentration was calculated from the $^{232}Th/^{238}U$ ratio using U as internal standard; a relative sensitivity factor was used for calibration. A single run represents the mean of 25 measurements, each corresponding to a total ion charge of 30 nC and a measuring time of about 2 min. We obtained a precision of about 1% for U abundances and Th/U ratios, and about 2% for Th concentrations (see also Table 2). Comparison of our data for the well-analyzed international reference materials BCR-1 and W-1 with compiled values [*Govindaraju*, 1994] indicate that accuracy is better than 2%, i.e. similar to the attainable precision.

3. RESULTS AND DISCUSSION

3.1. *Comparison with literature data*

Tables 1 and 2 list the MIC-SSMS results. In Table 1, we also list published TIMS analyses of aliquots of our samples. Absolute Th and U concentrations of both methods agree within 15% with no obvious systematic difference. The reason for some of the discrepancies may be sample heterogeneity caused by different amounts of olivine in the samples used for analysis. Because Th and U concentrations are influenced in the same way, Th/U ratios agree much better than the absolute concentrations. With the exception of one sample (Mauna Loa 1907), the agreement of Th/U measurements is better than 4%.

Tab. 2 shows the results for four independent analyses of the BHVO-1 basalt standard. The Th and U concentration data vary by less than 2%, whereas the Th/U ratio has a precision (1σ r.s.d.) of about 0.4%. Our results are compared with compilation data [*Govindaraju*, 1994] and some recently published values using modern analytical techniques. Our own and the published concentration data differ by up to 40%. Especially unsatisfactory is the very large range of Th/U ratios (from 2.56 to 3.67). It is clear that these differences cannot be ascribed to sample heterogeneity alone but are more likely attributable to analytical problems. Our MIC-SSMS values agree very

TABLE 1. Th and U concentrations (ppm) and Th/U ratios in historical Kilauea and Mauna Loa samples. The MIC-SSMS data are compared with literature values.

Sample	Th	U	Th/U	Anal. technique
KILAUEA				
1921	1.14	0.377	3.02	MIC-SSMS
	1.34	0.45	2.98	TIMS [a]
1955	1.88	0.626	3.00	MIC-SSMS
	1.69	0.57	2.96	TIMS [a]
1960	1.61	0.537	3.00	MIC-SSMS
	1.49	0.51	2.92	TIMS [a]
1963	1.29	0.428	3.01	MIC-SSMS
	1.23	0.41	3.00	TIMS [a]
DAS69-1-3 (1969)	1.10	0.365	3.02	MIC-SSMS
		0.334		SSMS [b]
KL-2 (1983)	1.03	0.342	3.02	MIC-SSMS
		0.332		SSMS [b]
		0.330		TIMS [b]
	1.04	0.333	3.12	TIMS [c]
MAUNA LOA				
1907	0.584	0.199	2.93	MIC-SSMS
	0.56	0.21	2.67	TIMS [a]
1926	0.530	0.183	2.90	MIC-SSMS
	0.56	0.20	2.80	TIMS [a]
1950	0.536	0.183	2.93	MIC-SSMS
	0.55	0.19	2.89	TIMS [a]
ML-3A (1975)	0.510	0.174	2.93	MIC-SSMS
ML-3B (1975)	0.551	0.190	2.90	MIC-SSMS
		0.197		SSMS [b]
		0.177		TIMS [b]

a: *Tatsumoto*, 1978; b: *Hofmann et al.*, 1986; c: *Hémond et al.*, 1994

well (better than 3%) with the high precision ID measurements using TIMS and ICP-MS by *White* [1993].

3.2. *Contrasting Th/U ratios*

Our results confirm previous results that Th and U concentrations in Kilauea lavas are about two times higher than in Mauna Loa lavas [e.g. *Cohen and O'Nions*, 1993; *Hémond et al.*, 1994]. The important new result is the small but significant difference in Th/U of about 3% between the lavas of both volcanoes. This is clearly demonstrated in Figure 1, where the Th/U ratios are plotted

against the year of eruption. Samples of both volcanoes show internally uniform ratios of 3.01±0.01 (Kilauea) and 2.92±0.02 (Mauna Loa), respectively, with no relationship to the year of eruption. Previous authors were not able to observe a significant difference in the Th/U ratios of historical Mauna Loa and Kilauea lavas, because the scatter in their data caused the results to overlap. In Figure 2, we compare our results with published TIMS data of historical Mauna Loa and Kilauea basalts [*Tatsumoto*, 1978; *Hémond et al.*, 1994; *Cohen and O'Nions*, 1993]. In contrast to our results, published TIMS data for the two volcanoes show a greater spread and overlap in the Th/U range between 2.86

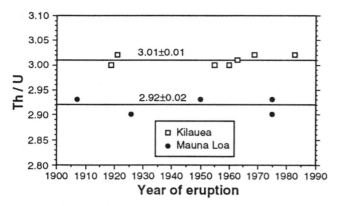

Fig. 1. Th-U ratios in recent Mauna Loa and Kilauea lavas determined by SSMS using the new multi-ion counting technique. Samples of both volcanoes show uniform ratios with no relationship to the year of eruption.

and 3.02. It is likely (though not certain) that the greater scatter of the published data is analytical in nature. For example, conventional thorium analyses by ID-TIMS are sometimes plagued by incomplete equilibration of spike and sample, unless special care is taken to avoid this problem. However, medians calculated from the two frequency distributions are consistent with higher Th/U ratios for Kilauea (2.99) than for Mauna Loa (2.95).

3.3. Discussion

Hémond et al. [1994] noted that Th/U ratios of Mauna Loa and Kilauea basalts with an overall mean value of 2.98 ± 0.07 (1σ s.d.) were identical within the scatter of their own as well as previously published data. For example, the mean value of 6 Kilauea and 1 Mauna Loa analyses of Cohen and O'Nions [1993] is 2.95 ± 0.03 and is thus indistinguishable from the results of Hémond et al. [1994]. This was surprising in view of the fact that other concentration ratios of similarly highly incompatible elements, particularly Ba/Th, do differ substantially (by about 25%) between the two volcanoes. In addition, the isotopic compositions of Sr, Nd and Pb are also different, so that it is clear that there are significant differences in source composition between Mauna Loa and Kilauea.

Our new data resolve a small but consistent difference in Th/U, which we use to evaluate whether this difference is caused by a source difference or a difference in melt fraction. One approach to answer this is to correlate the trace element ratios with isotopic compositions. If a correlation exists then one might conclude that the trace element ratios should also reflect source differences. However, in the present case this line of reasoning is not likely to be valid. The main reason for this inference is that Mauna Loa has consistently higher $^{87}Sr/^{86}Sr$ but lower Rb/Sr than historical Kilauea [Stille et al., 1986; Pietruszka and Garcia, 1994; Hofmann et al, 1995]. Thus, a strong

correlation exists, but its sense is opposite that expected.

If the Th-U ratios of the lavas represented long-lived Th/U differences of the two sources, one would expect to see this reflected by lead isotopic compositions. Particularly suitable for this comparison is the radiogenic $^{208}Pb*/^{206}Pb*$ ratio ($\equiv (^{208}Pb/^{204}Pb - 29.475)/ (^{206}Pb/^{204}Pb - 9.307)$), where the two constants are the respective primordial isotope ratios). This ratio is expected to correlate positively with Th/U (because ^{208}Pb is the daughter product of thorium, and ^{206}Pb is a daughter product of uranium). Figure 3 shows the available data for our samples. Clearly, the same "paradox" exists in the Th-U-Pb systematics as for the Rb-Sr system: Mauna Loa lavas have consistently higher $^{208}Pb*/^{206}Pb*$ ratios but lower Th/U ratios than historical Kilauea lavas. The reason for these "paradoxical" results is probably rather simple: Mauna Loa historical basalts consistently represent greater melt fractions than Kilauea historical basalts. Therefore, the concentration ratio of any element pair, in which the more highly incompatible element forms the numerator, should be greater in Kilauea than in Mauna Loa provided, of course, that the source ratios are identical. In reality, the isotopic differences indicate strongly, though not conclusively, that the source-Rb/Sr and Th/U ratios are actually greater in Mauna Loa than in Kilauea [Hémond et al., 1994]. (The argument is not completely conclusive because the negative correlation shown in Figure 3 might also represent mixtures of completely unrelated source end members. Nevertheless, if one of these end members is a relatively enriched plume

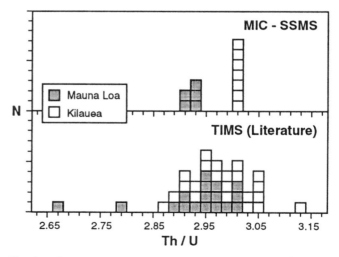

Fig. 2. Comparison of Th-U ratios in recent Mauna Loa and Kilauea lavas determined by spark source mass spectrometry using a multi-ion counting technique (MIC-SSMS) and thermal ionization mass spectrometry (TIMS) [Tatsumoto, 1978; Cohen and O'Nions, 1993; Hémond et al., 1994]. In contrast to published TIMS data, the MIC-SSMS results resolve a small but significant difference in Th/U of about 3% between the lavas of both volcanoes.

TABLE 2. Th and U concentrations (ppm) and Th/U ratios obtained from 4 independent analyses of the standard reference material BHVO-1. They are compared with recently published data.

Sample	Th	U	Th/U	Anal. technique
BHVO-1 (1)	1.199	0.399	3.005	MIC-SSMS
BHVO-1 (2)	1.198	0.400	2.995	MIC-SSMS
BHVO-1 (3)	1.195	0.400	2.988	MIC-SSMS
BHVO-1 (4)	1.245	0.413	3.014	MIC-SSMS
mean BHVO-1	1.209±0.024	0.403±0.007	3.001±0.011	
BHVO-1	1.08	0.42	2.57	Compilation [a]
BHVO-1	1.09	0.42	2.60	INAA [b]
BHVO-1	1.20	0.43	2.79	ICP-MS [c]
BHVO-1	1.10	0.43	2.56	ICP-MS [d]
BHVO-1	1.10	0.30	3.67	ICP-MS [e]
BHVO-1	1.234	0.409	3.02	ID-ICP-MS [f]
BHVO-1	1.247	0.408	3.06	ID-TIMS [f]
BHVO-1	1.17	0.418	2.80	HPLC [g]

a: *Govindaraju*, 1994; b: *Bedard and Barnes*, 1990; c: *Jenner et al.*, 1990; d: *Garbe-Schönberg*, 1993; e: *Garcia et al.*, 1993; f: *White*, 1993; g: *Rehkämper*, 1995

source and the other is a depleted mantle reservoir such as the MORB source or the so-called FOZO component of *Hart et al.* [1994], such mixtures should still be positively, not negatively correlated on Figure 3).

On the basis of the above arguments, we assume that the higher Th/U ratios of Kilauea lavas are generated by fractionation during partial melting. To evaluate the partial melting effects quantitatively, we assume the limiting case where the two sources have identical Th and U concentrations. We then consider qualitatively the more realistic case where the Kilauea source has lower Th, U, and Th/U values than the Mauna Loa source.

We compute the change in Th/U in the melt as a function of Th concentration (expressed as the Th enrichment factor of the melt relative to the source, which is roughly equivalent to the inverse of the extracted melt fraction), using experimental partitioning data for Th and U in peridotite-melt systems. These data require that garnet is present as a residual phase in order to generate melts with higher Th/U than their sources [*LaTourette et al.*, 1993; *Beattie*, 1993; *Hauri et al.*, 1994]. In Figure 4, we show (source-normalized) Th/U in the melt as a function of (source-normalized) Th concentration for three different melt extraction models, (1) equilibrium "batch" melting, (2) aggregated melts extracted from a porous source by either "continuous" or "dynamic" melting (using the nomenclature and equations of *Williams and Gill* [1989]), and (3) aggregated purely fractional melts (with zero porosity). For simplicity, we choose a standard garnet lherzolite source, even though there are good reasons to think that the Hawaiian plume is not just a simple, slightly depleted

mantle peridotite, but contains a significant amount of differentiated material, most likely recycled oceanic crust [*Hofmann and White*, 1980]. This is specifically indicated by the anomalously low $\delta^{18}O$ values [*Kyser et al.*, 1982; *Garcia et al.*, 1989] and several other geochemical "anomalies" (e.g. positive Sr anomalies; [*Hofmann et al.*, 1995]).

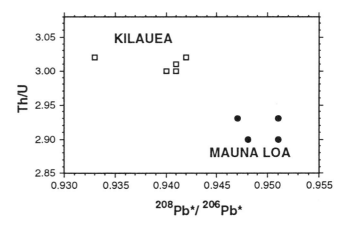

Fig. 3. Th/U versus radiogenic $^{208}Pb*/^{206}Pb*$ ($\equiv [^{208}Pb/^{204}Pb - (^{208}Pb/^{204}Pb)_i]/[^{206}Pb/^{204}Pb - (^{206}Pb/^{204}Pb)_i]$, where the subscript i signifies the primordial isotope ratio. In a single-stage evolution models, $^{208}Pb*/^{206}Pb*$ is approximately proportional to Th/U, with $^{208}Pb*/^{206}Pb* = 0.95$ and 0.94 corresponding to Th/U = 3.86 and 3.82, respectively. The data thus do not conform to single stage models, but $^{208}Pb*/^{206}Pb*$ is still likely to be positively correlated with *source* Th/U.

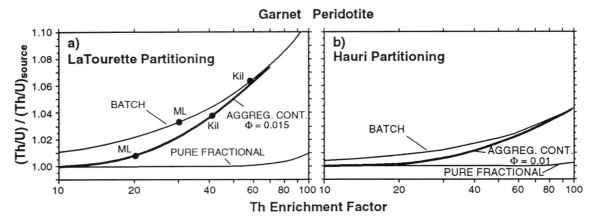

Fig. 4.   Th/U and Th concentrations in melts relative to the source values for three idealized melt extraction models. "Batch" refers to melts equilibrated with the residue; "pure fractional" refers to aggregate fractional melts extracted at zero porosity ; "aggreg. cont." refers to accumulated continuously extracted melts at constant porosity F. The latter curve also applies to "dynamic melting" (for details see *Williams and Gill* [1989]). The bulk partition coefficients used in Fig. 4a are from *LaTourette and Burnett* [1992] and *LaTourette et al.* [1993], in Fig. 4b from *Hauri et al.* [1994]; they pertain to a garnet peridotite source with 8 % clinopyroxene and 12 % garnet.   The filled dots are fits using the relative Th/U and Th concentration data for average historical Mauna Loa (ML) and Kilauea (KL) lavas.   No fit is possible for any of the melting models in Fig. 4b (except beyond the range of the diagram at unrealistic enrichment factors greater than 100).

Figure 4a shows the case for the partition coefficients of *LaTourette and Burnett* [1992] and *LaTourette et al.* [1993]. Purely fractional melting (with zero porosity) does not produce melts with significant Th/U fractionation except at very low aggregate melt fractions corresponding to very high Th concentrations in the melt. The reason for this lies in the extreme efficiency of fractional melting for extracting incompatible elements from the residue. This causes the aggregate melt to contain both Th and U nearly quantitatively, even at relatively low melt fractions. Batch melting fractionates Th/U more efficiently. Average Th concentrations differ approximately by a factor of two (after correction for olivine fractionation; see *Hémond et al.* [1994]), and Th/U ratios differ by 3 % between the Kilauea and Mauna Loa lavas. The dots marked ML (Mauna Loa) and Kil (Kilauea) represent fits of these relative Th/U and Th values to the melting models. For the case of batch melting, this requires Th enrichment factors of about 30 and 60 for Mauna Loa and Kilauea, respectively, implying melt factions as low as 3 and 1.5 %, respectively.

The most favorable, and possibly realistic, result is obtained for "dynamic melting" or "continuous melting" (which are mathematically equivalent, see *Williams and Gill* [1989]) with constant melting rates and melt porosity of 1.5%. Accumulated melts produced by this mechanism show the greatest Th/U fractionation for a given change in Th concentration. Thus, we obtain a 3% change in Th/U when the normalized Th concentration changes from about 20 (Mauna Loa) to 40 (Kilauea). The corresponding extracted melt fractions are approximately 5 and 2 %, and the absolute source concentrations of Th would be about

25 ppb (roughly a third the primitive mantle concentration).

If the Kilauea source actually has a slightly lower Th/U ratio than the Mauna Loa source, as discussed above, then the difference in Th-U fractionation generated by partial melting is larger than the measured difference of 3 % used in Figure 4a. If the source concentrations of the two volcanoes remain nearly the same (so that the relative enrichment factors remain unchanged), then the absolute enrichment factors for the melts of both volcanoes must increase, and both melt fractions must decrease. However, if in addition the Kilauea source also has significantly lower Th and U concentrations than the Mauna Loa source, then the difference between the melt enrichment factors will be greater than the factor of two used in Fig. 4a, and the effect of the increased Th/U will be partly compensated. In other words the spacing between the two dots representing the magmas of the two volcanoes, which are fitted to the partial melting curves, will increase both vertically and horizontally. In general, the results will be more sensitive to changes in source Th/U. For example, a source difference of 2% in Th/U would require approximately 25 to 30% lower absolute Th and U source concentrations for Kilauea, thus shifting the Kilauea Th enrichment factor from 60 to about 78 for batch melting, and from 40 to 55 for continuous or dynamic melting. In summary, given the above assumptions and partition coefficients, the calculated melt fractions are most likely maximum values for the various melting models.

Figure 4b shows the corresponding results for the experimental partition coefficients of *Hauri et al.* [1994]. Here, Th enrichments greater than 50 for Mauna Loa, and greater than 100 for Kilauea would be required even in the most favorable extraction model, in order to change the Th/U ratio by 3% or more. Such high enrichment factors would correspond to very low Th source concentrations of about 10 ppb, thus making the Hawaiian plume source similarly depleted in Th and U as the MORB source [*Jochum et al.*, 1983]. In view of the isotopic data of Hawaiian tholeiites, we consider this to be an unreasonably low limiting value. The results for the partitioning data of *Beattie* [1993] are intermediate to the cases shown in Figures 4a and b.

The above calculations illustrate a long-standing general problem in understanding fractionation of highly incompatible trace elements. Purely fractional melting, a mechanism favored by many recent authors [e.g., *McKenzie and O'Nions*, 1995], appears to be incapable of significantly fractionating highly incompatible elements at realistic enrichment factors. Even the more efficiently fractionating dynamic melting requires partition coefficients near the high end of the range of experimentally determined values. As noted above, these calculations were made for a standard garnet lherzolite source, which may be unrealistic but serves as a useful reference point. If the actual source is a mixture of normal peridotite and (recycled) eclogitic former ocean-crust basalt or gabbro, then both the bulk partition coefficients and the effective melt fractions should increase roughly proportionally, and the final effect on Th-U fractionation between source and melt may be rather similar. Quantitative modeling of such sources and the corresponding melt extraction models has been attempted for a somewhat similar case of MORB melting in the presence of ubiquitous garnet pyroxenites [*Hirschmann and Stolper*, 1995], but is beyond the scope of the present paper.

In summary, the Th/U data shown in this paper demonstrate a new and significant, though quantitatively small chemical difference between Mauna Loa and Kilauea lavas. This difference is consistent with the inference, obtained from the absolute concentrations of Th and U (and all other highly incompatible elements) that Kilauea lavas are formed by smaller degrees of partial melting than Mauna Loa lavas. The fractionation of Th/U is at least 3%, if one assumes on the basis of isotope data that Th/U of the Kilauea source is equal to or lower than that of the Mauna Loa source. Even this small degree of Th-U fractionation cannot be quantitatively modeled with purely fractional melting models but can be reconciled with models of continuous or dynamic melting.

Acknowledgments. We thank M. Tatsumoto for supplying some of the samples, C. Dienemann and H.M. Seufert for technical assistance with the SSMS analyses, and Steve Goldstein, Steve Galer, Dave Clague, Kevin Johnson, and Mike Rhodes for reviews and suggestions.

REFERENCES

Beattie, P., Uranium-thorium disequilibria and partitioning on melting of garnet peridotite, *Nature, 363,* 63-65, 1993.

Bedard, L.P., and S.-J. Barnes, Instrumental neutron activation analysis by collecting only one spectrum: Results for international geochemical reference samples, *Geostandards Newsletter 14,* 479 - 484, 1990.

Cohen, A.S., and R.K. O'Nions, Melting rates beneath Hawaii: evidence from uranium series isotopes in recent lavas, *Earth Planet. Sci. Lett., 120,* 169-175, 1993.

Condomines, M., P. Morand, and C.J. Allègre, ^{230}Th-^{238}U radioactive disequilibria in tholeiites from the FAMOUS zone (Mid-Atlantic ridge, 36° 50' N): Th and Sr isotopic geochemistry, *Earth Planet. Sci. Lett., 66,* 247-256, 1981.

Frey, F.A., and J.M. Rhodes, Intershield geochemical differences among Hawaiian volcanoes: implications for source compositions, melting process and magma ascent paths, *Phil. Trans. Roy. Soc. Lond. A., 342,* 121-136, 1993.

Garbe-Schönberg, C.-D., Simultaneous determination of thirty-seven trace elements in twenty-eight international rock standards by ICP-MS, *Geostandards Newsletter 17,* 81 - 97, 1993.

Garcia, M.O., D.W. Muenow, K.E. Aggrey, and J.R. O'Neil, Major element, volatile, and stable isotope geochemistry of Hawaiian submarine tholeiitic glasses. *J. Geophys. Res., 94,* 10,525-10,538, 1989.

Garcia, M.O., B.A. Jorgensen, and J.J. Mahoney, Temporal geochemical evolution of Loihi summit lava. Results from Alvin submersible dives, *J. Geophys. Res. 98,* 537 - 550, 1993.

Govindaraju, K., 1994 compilation of working values and sample description for 383 geostandards, *Geostandards Newsletter, 18,* 1-158, 1994.

Hart, S.R., E.H. Hauri, L.A. Oschmann, and J.A. Whitehead, Mantle plumes and entrainment: Isotopic evidence, *Science, 256,* 517-520, 1992.

Hauri, E.H., T.P. Wagner, and T.L. Grove, Experimental and natural partitioning of Th, U, Pb and other trace elements between garnet, clinopyroxene and basaltic melts, *Chem. Geol., 117,* 149-166, 1994.

Hémond, C., A.W. Hofmann, G. Heusser, M. Condomines, I. Raczek, and J.M. Rhodes, U-Th-Ra systematics in Kilauea and Mauna Loa basalts, Hawaii, *Chem. Geol., 116,* 163-180, 1994.

Hirschmann, M.M., and E.M. Stolper, A possible role for garnet pyroxenite in the origin of the "garnet signature" in MORB, *Contrib. Mineral. Petrol.,* ms. submitted, 1995.

Hofmann, A.W., C. Hémond, J.M. Rhodes, I. Raczek, K. Lehnert, and M.O. Garcia, Sources and partial melting characteristics of Mauna Loa and Kilauea, Hawaii, in prep., 1995.

Hofmann, A.W., K.P. Jochum, M. Seufert, and W.M. White, Nb and Pb in oceanic basalts: new constraints on mantle evolution, *Earth Planet. Sci. Lett., 79,* 33-45, 1986.

Hofmann, A.W., and W.M. White, The role of subducted oceanic crust in mantle evolution, *Carnegie Inst. Wash. Year Book, 79,* 477-483, 1980.

Jenner, G.A., H.P. Longerich, S.E. Jackson, and B.J. Fryer, ICP-MS - A powerful tool for high-precision trace-element analysis in Earth sciences: Evidence from analysis of selected U.S.G.S. reference samples, *Chem. Geol., 83,* 133-148, 1990.

Jochum, K.P., A.W. Hofmann, E. Ito, H.M. Seufert, and W.M. White, K, U, and Th in mid-ocean ridge basalt glasses and heat production, K/U, and K/Rb in the mantle, *Nature, 306,* 431-436, 1983.

Jochum, K.P., H.M. Seufert, S. Midinet-Best, E. Rettmann, K. Schönberger, and M. Zimmer, Multi-element analysis by isotope dilution-spark source mass spectrometry (ID-SSMS), *Fresenius Z .Anal .Chem., 331,* 104-110, 1988.

Jochum, K.P., H.-J. Laue, H.M. Seufert, and A.W. Hofmann, First analytical results using a multi-ion counting system of a spark source mass spectrometer, *Fresenius J. Anal. Chem., 350,* 642-644, 1994.

Kyser, T.K., J.R. O'Neil, and I.S.E. Carmichael, Genetic relationships among basic lavas and ultramafic nodules: Evidence from oxygen isotope compositions, *Contrib. Mineral. Petrol., 81,* 88-102, 1982.

LaTourette, T.Z., and D.S. Burnett, Experimental determination of U and Th partitioning between clinopyroxene and natural and synthetic basaltic liquid, *Earth Planet. Sci. Lett., 110,* 227-244, 1992.

LaTourette, T.Z., A.K. Kennedy, and G.J. Wasserburg, Thorium-uranium fractionation by garnet: evidence for a deep source and rapid rise of oceanic basalts, *Science, 261,* 739-741, 1993.

Laue, H.-J., H.M. Seufert, and K.P. Jochum, Konzept und erste Ergebnisse eines neuartigen Multiionenzähldetektors für ein Funkenmassenspektrometer mit Mattauch-Herzog-Geometrie (abstract), *Verh. DPG 4, 523,* 1994.

McKenzie, D., and R.K. O'Nions, The source regions of ocean island basalts, *J. Petrol., 36,* 133-159, 1995.

Newsom, H.E., W.M. White, K.P. Jochum, and A.W. Hofmann, Siderophile and chalcophile element abundances in oceanic basalts, Pb isotope evolution and growth of the Earth's core, *Earth Planet. Sci. Lett., 80,* 299-313, 1986.

Pietruszka, A.J., and M. Garcia, The historical summit lavas of Kilauea Volcano (1790-1982) (abstract), *EOS Trans. AGU, 75,* 712, 1994.

Rehkämper, M., A highly sensitive HPLC method for the determination of Th and U concentrations in geological samples, *Chem. Geol. , 119,* 1-12, 1995.

Spiegelman, M., and T. Elliott, Consequences of melt transport for uranium series disequilibrium in young lavas, *Earth Planet. Sci. Lett., 118,* 1-20, 1993.

Stille, P., D.M. Unruh, and M. Tatsumoto, Pb, Sr, Nd, and Hf isotopic constraints on the origin of Hawaiian basalts and evidence for a unique mantle source. *Geochim. Cosmochim. Acta, 50,* 2303-2319, 1986.

Tatsumoto, M., Isotopic composition of lead in oceanic basalts and its implication to mantle evolution, *Earth Planet. Sci. Lett., 38,* 119-138, 1978.

Taylor, S.R., Geochemical analysis by spark source mass spectrograpy, *Geochim. Cosmochim. Acta, 29,* 1243-1262, 1965.

White, W.M., $^{238}U/^{204}Pb$ in MORB and open system evolution of the depleted mantle, *Earth Planet. Sci. Lett., 115,* 211-226, 1993.

Williams, R.W., and J.B. Gill, Effects of partial melting on the uranium decay series, *Geochim. Cosmochim. Acta, 53,* 1607-1619, 1989.

K.P. Jochum, and A.W. Hofmann, Max-Planck-Institut für Chemie, Postfach 3060, 55020 Mainz, Germany

Applications of GIS to the Estimation of Lava Flow Hazards on Mauna Loa Volcano, Hawai`i

Jim Kauahikaua, Sandy Margriter, Jack Lockwood, and Frank Trusdell

U.S. Geological Survey, Hawai`i National Park, Hawai`i

Vector-based GIS computer software is used with preliminary digital geologic maps of the active volcano Mauna Loa, Hawai`i to assess the probability that lava flows from this volcano might adversely affect certain areas more than others. A few possible GIS strategies are explored with the maps completed thus far. Traditional coverage rates and probabilities of lava flow occurrence are calculated for target regions of different sizes and shapes. Probability of lava flow occurrence is the probability that one lava flow will enter the target region within the target time period. Probability of coverage is the probability that a particular point will be covered by one or more lava flows within the target time period. The results show that the two approaches yield similar results, with the occurrence of lava flows in a predefined region being more probable than coverage by lava flows. The probabilities of lava flow occurrence reflect all effects of lava flows, whereas the coverage rate estimates reflect only the most serious effect, that of coverage by lava flows.

1. INTRODUCTION

Assessment of volcanic hazards is necessary for any public or private venture on or around active volcanoes in order to minimize potential losses caused by volcanic activity. We may instinctively know not to build anything meant to last on an actively erupting volcanic vent, but we may not be so certain about how far away is "safe enough." To a large degree, "safe enough" depends on how long the venture is intended to last. Because active Hawaiian volcanoes take as long as a few thousand years to completely resurface themselves, there must be areas on these volcanoes where it might be "safe enough" to build structures meant to last 50 to 100 years, for example. The U.S. Geological Survey's Volcanic Hazards Program seeks methods to assess and mitigate these hazards on a variety of U.S. and foreign volcanoes [*Wright and Pierson*, 1992].

This paper explores methods by which hazards from lava flows on Mauna Loa volcano can be quantitatively assessed with the aid of Geographic Information System (GIS) software. The

Mauna Loa Revealed: Structure, Composition, History, and Hazards
Geophysical Monograph 92

basic portions of any GIS can be used for digitizing maps, subsequent editing of those maps, and printing or plotting at any scale. Much like computer spreadsheets that manipulate data tables, a typical GIS can also manipulate data tables where each row can be associated with a spatial element on a digital map. For example, within a GIS, all the flows that are within a certain distance of any point (such as a well), line (such as a fault or road or a political boundary), or a region (such as a town or development) can be selected, and the percentage of those flows that are of specific age periods, morphologies, or mineralogies can be tabulated. The more sophisticated GIS also allow computations using digital elevation models (DEM). For example, regions visible from a point or all areas that would drain water to a point can be plotted. Use of GIS and digital geologic maps provides a valuable set of tools that geologists, planners, and emergency managers can use to derive answers to specific questions about the mapped area.

The data used with the GIS in this application are digital geologic maps, now completed for a little more than one-half of Mauna Loa (Figure 1) at a sufficient level of detail to permit delineation of individual lava flows. Each lava flow may be represented by several polygons depending on whether it is overlain and/or crossed by younger flows, or whether portions of the flow are `a`a and others pahoehoe. Several layers of information can be used simultaneously, such as geology, slope of terrain, and position of major roads.

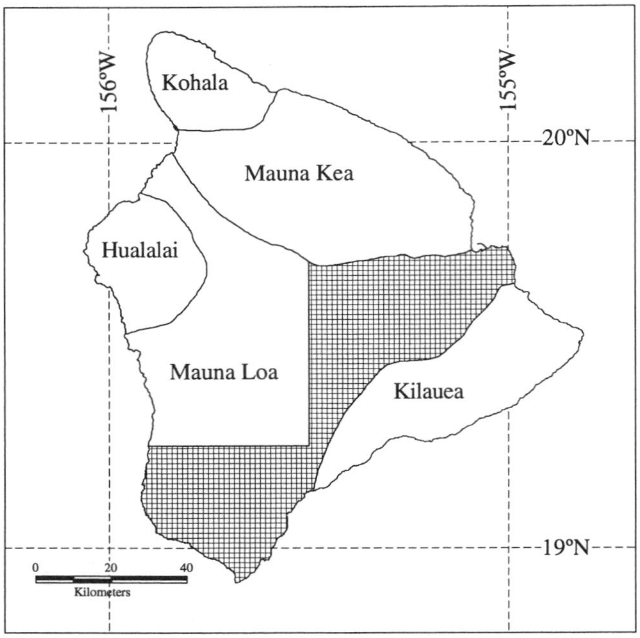

Fig. 1. Map of the island of Hawai'i showing boundaries of the five volcanoes which make up the island. Individual flow geologic mapping is complete in the stippled portion of Mauna Loa.

2. PREVIOUS ESTIMATES OF LAVA FLOW HAZARD IN HAWAI'I

Previous volcanic hazard analyses for Hawai'i used qualitative criteria to rate relative severity of hazard. Lava flow hazard zones in Hawai'i have been defined by *Mullineaux et al.* [1987], *Heliker* [1990], and *Wright et al.* [1992] on the basis of volcanic structure, terrain, and degree of coverage by lava flows. Figure 2 shows the Hazard Zone map of Hawai'i from *Wright et al.* [1992]. Hazard Zone 1 includes the summit calderas and rift zones of Kilauea and Mauna Loa volcanoes. More than 25% of Hazard Zone 1 has been covered by lava in the past 200 years [*Heliker*, 1990]. Between 15% and 25% of Hazard Zone 2 but only 5% of Hazard Zone 3 has been covered by lava in the past 200 years. Hazard Zones 4 through 9 reflect decreasing frequency of eruptions or areas shielded from lava flows by topography.

Volcanic hazard assessments have been part of Environmental Impact Statements (EIS) for various projects in Hawai'i. The EIS for the Kahauale'a Geothermal Project [*True/Mid-Pacific Geothermal Venture*, 1982] on Kilauea's east rift zone divided up the Kahauale'a tract into four zones of relative volcanic hazards potential, ranked qualitatively from "highest" to "low". The zone of highest volcanic hazards potential was about 2 km wide and the northernmost boundary coincided with the northernmost eruptive fissures that were active in the early- to mid-1960s. This assessment is particularly interesting because the ongoing eruption of the Pu'u 'O'o and Kupaianaha vents began in 1983 within this zone of highest volcanic hazards potential. As of 1995, 100% of the zone of "highest" volcanic hazards potential, 75% of the zone of "high" volcanic hazards potential, and 5% of the zone of "moderate" volcanic hazards potential, has been covered. The hazard assessment for this area was unusually prophetic [*Kauahikaua et al.*, 1994].

The existence of a hazard assessment does not unfortunately guarantee its implementation. *Belt Collins & Associates* [1987] Environmental Impact Statement for a planned resort development on the southern coast of Mauna Loa less than 5 km west of the southwest rift zone is also qualitative. Without additional information, their discussion starts with the statement that the entire resort is within Hazard Zone 2 as defined by the USGS, and continues with several arguments why this may not be appropriate to this development. The assessment ends with a justification for development in Zone 2.

A quantitative hazard assessment has recently been completed for a draft EIS covering the Conceptual Plan for the Commercial Satellite Launching Facility at Palima Point on the southwest rift zone (SWRZ) of Kilauea Volcano [*Legg*, 1993]. The authors

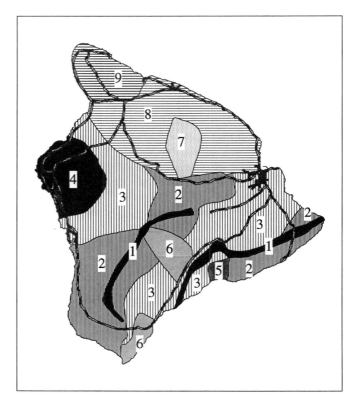

Fig. 2. Map of the island of Hawai'i showing the Hazard Zones of Wright et al. (1992) ranging from the highest hazard zone 1 to the lowest hazard zone 9.

concluded that the area downslope of the SWRZ has the highest potential for burial by lava flows, with a probability of burial of about 50% for a 50-year time span, while the area to the north and west of the SWRZ has a probability of burial of less than 5% for the same time span.

3. METHODS

3.1 *Previous Uses of GIS Methods to Estimate Lava Flow Hazard*

To date, only a few studies have used GIS methods to help estimate lava flow hazard on active volcanoes. Methods similar to those used in this paper were used to estimate volcanic hazard in the area around the town of Hilo, Hawai`i [*Kauahikaua et al.*, 1993], and in the areas involved in the generation and transmission of geothermal energy from Kilauea Volcano's lower east rift zone [*Kauahikaua et al.*, 1994]. *Jones* [1995 and in press] used GIS techniques to estimate the combined hazards to areas around Mount Etna from several different forms of volcanic activity and assessed risk to population centers. *Wadge et al.* [1994] used a computer program to simulate lava flows and their effects over a given period of time to generate a lava flow hazard map for Mt. Etna.

3.2 *Application of GIS Techniques to Estimating Lava Flow Hazard for Mauna Loa*

Hazard is defined by *Wright and Pierson* [1992] as " an event or process that is potentially destructive" whereas risk is defined as "the magnitude of a potential loss - of life, property, or productive capacity - within the area subject to hazard(s)." These definitions require those who estimate hazards to focus on the geologic processes that produce the hazard rather than the effects of those geologic process on any aspect of civilization. Hazard and risk would only be equal if the region for which hazards are to be estimated is uniformly populated and developed. A discussion of the risk from Mauna Loa lava flows can be found in *Trusdell* [this volume].

The primary basis for estimating potential hazards posed by volcanic activity in any area is the record of past volcanic activity in that area. Detailed information on the spatial and chronological distribution of lava flows is necessary for quantitative hazard assessment. Two of the authors (J.L. and F.T.) are currently mapping and dating individual lava flows on Mauna Loa at a 1:24,000 scale [e.g., *Buchanan-Banks*, 1993; *Lockwood et al.*, 1992] which updates a previous map [*Lockwood et al.*, 1988]. More than half of the mapping, including the northeast rift zone, west flank, and south flank of Mauna Loa, has been completed and converted to vector digital form (Figure 1). ARC/INFO, a vector-based GIS software package running on a Unix workstation, was used to digitize and edit hand drawn geologic maps. Each lava flow was assigned a "flow ID", "morphology code" (pahoehoe or 'a'a, for example), and an "age category" [*Lockwood et al.*, 1992]. After attributes were assigned on the UNIX ARC/INFO system, the maps were converted to the vector-based MapInfo format for use on a 486-based personal computer.

The application of GIS techniques to data in this form requires an algorithm with which to estimate hazard within a specific region, called a target region in this paper, which is also represented by a polygon. Examples of target regions might be a commercial or government development for which the hazard is being assessed [e.g., *Kauahikaua et al.*, 1994], or they might be hazard zonations produced by non-GIS techniques [e.g., *Wright et al.*, 1992] for which more quantitative estimates are desired. If the set of target regions are of equal area and are located on a regular grid, a contour map of probabilities can be generated. We present examples of two different estimates of lava flow hazard used on three different sets of target regions.

3.3 *Rate of Coverage as a Means of Estimating Lava Flow Hazard*

Cumulative rates of coverage by lava flows are easily computed from geological mapping of a volcano if the various geologic units are well dated. Coverage rates are commonly computed backwards from the present [e.g., *Lipman*, 1980; *Lockwood and Lipman*, 1987] by identifying all flows within predefined age limits, calculating the area covered within each geologic time period, and then plotting the cumulative coverage as a function of time. Implicit is the assumption that the future cumulative coverages will behave similarly allowing the reversal of the time scale to estimate future behavior. An equation for the theoretical fractional cumulative coverage, as used empirically in the above references, can be derived in a fashion similar to that for simple radioactive decay. Let us assume a constant rate of coverage which covers newly covered ground as well as previously uncovered ground. Both ground covered and uncovered since the time period began would be covered at the same rate. The rate of decrease in the amount of uncovered ground, -dx/dt, would then be proportional to the remaining amount of uncovered ground, x,

$$- \frac{dx}{dt} = ax \qquad (1)$$

The solution to this ordinary differential equation (ODE) is found in any ODE text and can be obtained by separation of variables followed by integration and substitution of the initial condition, $x = 1.0$ (normalized area of target region) when $t = 0.0$, and is

$$x = e^{-at} \qquad (2)$$

Table 1.Coverage by lava within the last 200 (Kilauea) and 150 (Mauna Loa) years.

Hazard Zone	Kilauea (%)	Kilauea (km²)	Mauna Loa (%)	Mauna Loa (km²)
1	40	110	57	20.2
2	34	144	33	186
3	3	20.4	2	15.5

Evaluation of the constant, a, can be done easily from the knowledge that after a time, T, a fractional amount of ground, fc, has been covered. Therefore, the amount of ground that is covered within time, t, is given by

$$\text{Cumulative Fractional Coverage} = 1.0 - x = 1.0 - e^{-at} \quad (3)$$

where   t is time in years from (before or after) present,
        $a = -\ln[1-fc]/T$, and
        fc is fractional coverage over time period, T.

Mapping data can be fit to this equation to determine the fractional coverage rate over any time period. For example, if a is determined to be 0.0051, then the coverage rate is 0.4 (40%) per century, which is the same as 0.005 (0.5%) per year.

Although coverage rates provide a general measure of a volcano's potential for hazardous activity, there are two problems with this method. The first problem is that coverage percentages are misleading if the areas, within the various regions for which the hazards are to be estimated, are significantly different. For example, if area, rather than the percentage area covered by lava, is measured (Table 1), we see that more area in Hazard Zone 2 was covered by lava than in Hazard Zone 1 during this period on both the Kilauea east rift zone and on the Mauna Loa northeast rift zone. Nearly ten times as many square kilometers of Mauna Loa northeast rift zone were covered within the past 150 years in Hazard Zone 2 than in Zone 1 because Hazard Zone 2 is bigger. Hazard Zone 1 has a higher hazard rating based on proximity to vents as well as on percentage coverage [Heliker, 1990; Wright et al., 1992]. Lockwood and Lipman [1987] provide another example in their conclusion that Mauna Loa coverage rates have been 40% over the last 1,000 years as compared with 20% for Hualalai and 90% for Kilauea. This gives the impression that Kilauea is more than twice as hazardous as Mauna Loa in terms of lava flow hazard and that Hualalai is half as hazardous as Mauna Loa. When the average coverage rates are computed in terms of areas, however, Mauna Loa is found to have covered 2056 km² in the last 1,000 years. Kilauea is second with 1330 km² and Hualalai a distant third with 160 km² covered in the same period. We conclude that areal coverage rates should be used for regions of unequal area, and percentage coverage rates should be used for regions of equal area.

The second problem using coverage rates based on surface geologic mapping is that they are biased towards low rates for older flows. Equation (3) models areal coverage for a given region over time, if new coverage is always placed randomly. Lava flows are deposited onto the surface of a volcano from linear, localized vent areas so their placement is spatially correlated. Coverage rates should be highest in areas near vents and lowest at large distances from vents. Older flows are more frequently exposed at greater distances from vent areas because more recent flows have preferentially covered areas closer to vents. Therefore, coverage rates for older flows will be biased towards low rates. To corroborate this point, percentage coverage for lava flows from 1843 and younger were computed for zones extending various distances from the mapped vent areas. The coverage ranged from 23.1% (93 km²) within 2 km of the vents, 17.1% (64 km²) between 2 and 4 km, 11.9% (81 km²) between 4 and 8 km, and 11.6% (116 km²) between 8 and 16 km from the vents. Although not a proof, this exercise does suggest that the percentage coverage rates calculated by Lipman [1980], Lockwood and Lipman [1987], and Trusdell [this volume] for age intervals several thousand years old may be too low. Because their coverage rate estimates generally decreased with increasing time from the present, applying these results would mean that the overall coverage rates may be more uniform through time.

3.4 *Probability of Lava Flow Occurrence as a Means of Estimating Lava Flow Hazard*

Hazards posed by lava flows can also be analyzed in terms of probability of the occurrence of a lava flow within a target region. Attention can be confined to that region's catchment, defined by topography as the region in which an eruption could conceivably affect the target region. For example, only the northeast rift zone of Mauna Loa is important when assessing the lava flow hazard in the vicinity of Hilo because other vent areas of Mauna Loa are not within the catchment of Hilo. Within the catchment for the target region, the theoretical probability of lava flow occurrence can be calculated using the equation [Kilburn, 1983]:

$$Prob = \sum P_v(t) \times P_v(b) \times P(c) \times T_c \times D_v \quad (4)$$

where $P_v(t)$ is the probability of an eruption for each vent within the catchment area;

t is the time interval for which the probability is being evaluated;

$P_v(b)$ is the probability of burial of the catchment by the flows from each vent;

$P(c)$ is the proportion of catchment covered by vents with density, D_v;

T_c is the total size of the catchment area; and

D_v is the vent density (vents/unit area) within the portion of the catchment covered by vents.

For any selected target region within a catchment, equation (4) simply states that the probability of lava flow occurrence is proportional to the probability of an eruption within the catchment within the time period of interest, $P_v(t)$, reduced by the probability that any lava flow produced by such an eruption will reach that area, $P_v(b)$. Although this is a complete formulation for estimating potential lava flow hazard, it is more direct to estimate probability of lava flow occurrence from the frequency of lava flows which have flowed into the target region in the past.

A probability of lava flow occurrence for any specified region can be evaluated using an estimate of the recurrence interval of lava flows in that region [*Kauahikaua et al.*, 1993]. This method is more specific than the theoretical approach of equation (4), which predicts the probability of a number of various factors influencing whether a lava flow will reach the selected area. Assuming that time intervals between lava flows are randomly distributed (Poisson distribution is the most common for interval data), the probability of the occurrence of n lava flows within a given time interval, t_i, can be estimated from the mean recurrence interval, T, of past lava flow occurrence in an area by the equation:

$$Prob_n = (\frac{t_i}{T})^n \frac{e^{-\frac{t_i}{T}}}{n!} \quad (5)$$

[*Scheidegger*, 1975, p. 71]. The probability of occurrence of a lava flow within the time interval, t, is the same as the probability of the recurrence interval being shorter than t, and one minus the probability of there being no eruptions (one minus equation 5 with $n=0$). This is expressed as:

$$Prob(t) = 1 - e^{-\frac{t}{T}} \quad (6)$$

[*Klein*, 1982]. In any given region, the probability of lava flow occurrence will increase with decreasing recurrence interval, T, and increasing evaluation interval, t.

Estimation of recurrence interval, T, is therefore the most

crucial step. A number of papers have been written on the estimation of recurrence intervals and the degree of randomness for volcanic eruption intervals [*Wickman*, 1966; *Klein*, 1982]. These techniques should apply equally well to the estimation of recurrence intervals for lava flows covering a limited area on a volcano. Recurrence intervals for limited areas should be longer than the recurrence interval for all eruptions of a given volcano because not all eruptions of a given volcano will affect the limited area.

The most statistically-sound method of estimating the probability of lava flow occurrence within 50 years, for example, would be to determine the number of lava flow occurrences in 50-year intervals as far back into the past as possible, calculate an average frequency per 50-year interval (equivalent to $1/T$), and then use equation (5) to estimate the probability of a single eruption within the next 50-year interval [*de la Cruz-Reyna*, 1991]. With this information, we can examine the distribution of flow frequencies (i.e., check whether they are really random and thereby justify the use of equations (5) and (6)). Ideally, the calculations would be based on a complete list of all lava flows that have flowed into the selected area for the last several thousand years and their ages. Such a list is not obtainable, but more limited information is available from other sources, including drill holes with detailed lithologic logs and geologic maps of individual surface flows. Unfortunately, even though all discernible flow units in the study area may have been mapped individually or have been logged in a drill hole, rarely are all units dated. All lava flows after 1843 (earliest known recorded observations by westerners) have reliable age determinations, but few earlier events can be dated with less than 50-year error. Therefore, it is not possible to obtain lava flow frequencies in suitably short, uniform-length time intervals for much of the geologic record. The shortest intervals in the recent past for which flow ages can be grouped confidently are generally 200 to 1,000 years.

Given the type of data available, the most direct way of estimating a recurrence interval is to divide the age of the oldest flow by the number of flows within the chosen area. Mathematically, this is exactly equivalent to taking the reciprocal of the mean of the number of flows within fixed-length time intervals [*Ho*, 1990; *Ho et al.*, 1991]. If any bias is present in this estimation, it would be caused by some flow units being completely covered and not represented at the surface. The number of flows within the area would be underestimated and the recurrence interval would be too large.

The best recurrence interval estimates are obtained from target regions in which the most flows have been tabulated. If there are few flows, or in the worst case, only one flow within the target region, the recurrence interval found using the method described above may be inaccurate. In these cases, the target region should therefore be enlarged to include more flows and the recurrence interval recalculated.

Our inability to confirm that the frequency of lava flows within each target region is random with time is a problem. Mauna Loa has erupted at highly variable intervals during its geologic history [Lockwood and Lipman, 1987; Lockwood, this volume]. We have some evidence to support this assertion because the eruption intervals of Mauna Loa are distinctly random over portions of the past 150 years [Wickman, 1966; Klein, 1982]. We must assume that recurrence intervals for Mauna Loa are random (Poisson distribution), and estimate a separate recurrence intervals for the entire geologic record of Mauna Loa and for its last 150 years.

The application of this idea is relatively simple. All flows that outcrop within each target region were first selected using the appropriate GIS operation and the number of different flow IDs was counted and tallied within age groups. Thereby, the total number of different flows and their age distribution was determined using either ARC/INFO or MapInfo software. The average recurrence interval was finally calculated by dividing the age of the oldest flow by the total number of flows.

We use equation (4) to calculate the probabilities for any time interval given the recurrence interval. For example, an average recurrence interval of 25 years means that there is a probability of 86% that at least one event will occur in 50 years, 70% in 30 years, and 4% in one year. Use of equation (4) is also essential to extrapolate to other time intervals because, as can be seen from this example, probabilities based on a Poisson distribution cannot be linearly interpolated or extrapolated. For example, linearly extrapolating a 4% probability of at least one event in one year into a 100% probability of at least one event in 25 years is clearly erroneous; equation (5) gives the correct probability of 63% for at least one event in 25 years. *Kauahikaua et al.* [1994] gives an example of a probabilistic lava flow hazard assessment using this method which illustrates the relationship between the hazard zones of *Wright et al.* [1992] and calculated probabilities of lava flow occurrence.

3.4.1 *Uncertainties.* Uncertainties in probability estimates are due to uncertainties in the estimation of the recurrence intervals and/or nonrandomness of the intervals. The latter factor has already been discussed. The recurrence interval, T, is the age of the oldest flow, or flow age category, divided by the total number of exposed flows; therefore, errors in T are due to errors in either of these quantities. The ages of flows less than 150 years old are known to within a month from historical records.

Radiocarbon ages used in the Mauna Loa mapping have standard errors of between 50 and 1,600 years [Rubin et al., 1987], with a median value of about 60 years. With such errors, it seems prudent to view our recurrence interval estimates as having a typical estimated error, σ_T, of about 1.5 to 2 years for 40-50 year recurrence intervals and 11 years for a 200 year recurrence interval. The uncertainty of the probability estimates can be obtained by differentiating equation (5) with respect to T,

$$\sigma_{Prob} = \frac{\partial Prob(t)}{\partial T} \sigma_T = \frac{te^{-t/T}}{T^2} \sigma_T . \qquad (7)$$

For t=50 years and T=40 years, the probability estimate would be 71% with an uncertainty of 1.7%. For t=50 years and T=187.5 years, the probability estimate would be 23% with an uncertainty of 1.1 %.

4. RESULTS

Probabilistic estimates of lava flow hazards on Mauna Loa were calculated for three different sets of target regions. The first and most general set of regions was a grid of 3 km squares, the second was a set of topographically-defined catchments around the flanks of Mauna Loa, and the third was Hazard Zones 1 and 2 from *Wright et al.* [1992]. We also calculated coverage rates for the last 150 years for the last two sets of target regions for comparison.

4.1 *Grid of 3 km squares*

Grid squares of a given size are the least biased target areas in which to estimate hazards. The choice of grid size was influenced by two factors: 1) the average width of a single flow, and 2) the smallest square that would minimize variability between adjacent squares. Test calculations were done for 1 and 2 km squares, and the results were found to be highly variable. The results for the 3 km squares began to exhibit spatial patterns that were more prominent than the variability between squares; the size of the squares is probably still too small to be optimal, however and further work is warranted.

The recurrence interval results are shown superimposed on the lava flow hazards map of *Wright et al.* [1992] (Figure 3). The average recurrence interval shown in this figure, based on the entire geologic record from Mauna Loa, confirms that Hazard Zones 1 and 2 are the highest hazard areas, but does not distinguish between the two zones. However, when the recurrence intervals are based only on flows erupted between 1843 and the present, Hazard Zones 1 and 2 can be distinguished (Figure 4). The difference between the two sets of recurrence intervals is the time span over which they are computed. *Kauahikaua et al.* [1994] argued that, when different, the most relevant recurrence interval estimate for the immediate future are those derived from the most immediate past. Therefore, the best estimate of the probability that a lava flow will impact these locations should be based on the recurrence intervals shown in Figure 4 and converted to probabilities using Table 2.

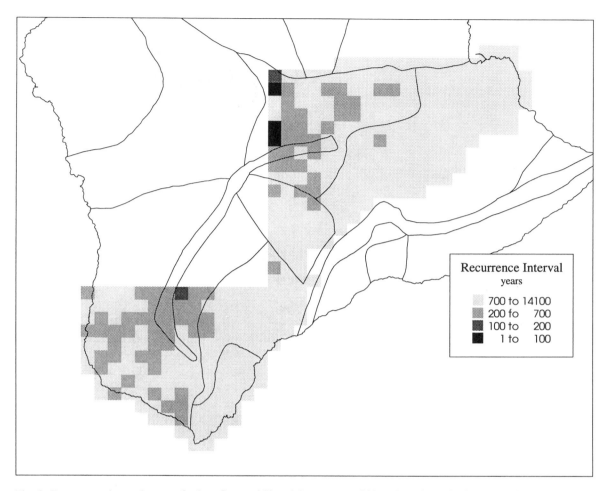

Fig. 3. Recurrence interval ranges for lava flows within a 3-km-square grid based on the entire known geological record of Mauna Loa, plotted over the hazard zone boundaries from Figure 2.

4.2 Catchments

The second set of target regions used in this study, called catchments, is based directly on topography. The effect of topographic slope on the path of lava from the vent to the sea is only indirectly reflected in probability of lava flow occurrence estimates. Topographic situations which would tend to concentrate lava flow paths would be indicated by a high number of flows in a given region and, probably, a low recurrence interval. A catchment, also known as a watershed, is the region in which water on the ground would drain downwards into a specified point or polygon. The ARC/INFO GRID module requires only the USGS 1:24,000 Digital Elevation Model (DEM) and a drain point or polygon to compute a catchment.

In order to select an unbiased set of drainpoints at the coast, downward paths were begun from each of the 30 m wide DEM pixels that fell on the boundary of Mauna Loa Hazard Zone 1 [Wright et al., 1992]. This choice of starting paths was convenient because this hazard zone boundary completely surrounded the more recent vents along both rift zones and summit caldera. Any flow from any of the included vents would have to cross this boundary on its way downslope. The downward paths from these several hundred starting points converged to form paths eventually reaching the coast at twenty locations within the region already geologically mapped. Note that there are several areas on the flanks of Mauna Loa that are not within any of the catchments defined in the above manner.

The number of lava flows for the south and east side of Mauna Loa between 1843 and the present within each catchment is

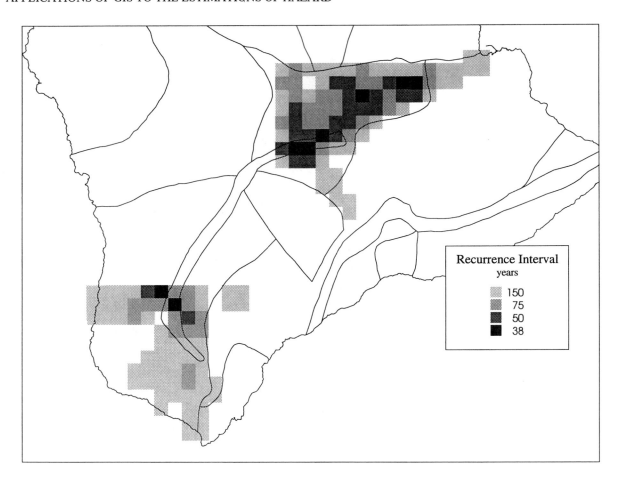

Fig. 4. Recurrence interval ranges for lava flows within a 3-km-square grid based on the geologic record of Mauna Loa from the last 150 years, plotted over the hazard zone boundaries from figure 2.

Table 2. Probability Estimates of lava flow occurrence in the next 100 years based on the last 150 years.

Number of flows since 1843	Average recurrence interval since 1843	Probability of lava flow occurrence in next 100 years
1	150	49%
2	75	73%
3	50	86%
4	37.5	93%
5	30	96%
6	25	98%

shown in Figure 5. The equivalent probability estimates (Table 2) reconfirm that the northeast rift zone and southwest rift zones are the most hazardous regions. Coverage rates for the last 150 years were also calculated for these catchments. In general, higher coverage rates corresponded to greater number of flows within the coverage. Figure 6a is a plot comparing number of flows since 1843 in a catchment with fractional coverage of that catchment (areal coverage divided by total area). Figure 6b is a plot comparing number of flows with areal coverage for each catchment. From the latter figure, we can calculate an average coverage per flow since 1843 per catchment of approximately 15 km^2.

It may be possible to also use these catchments to determine the general path of a lava flow. For example, Figure 7 shows a closer look at the catchments on the northeast rift zone plotted with the 1880-81 and the 1984 lava flows from Mauna Loa.

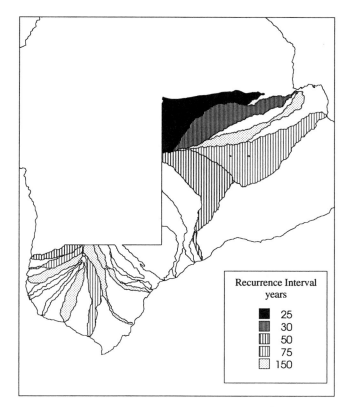

Fig. 5. Recurrence intervals for lava flows erupted between 1843 and the present within topographically-defined catchments.

Both flows generally stay within the same catchment suggesting that, if the 1984 eruption had continued for a longer period, the flows may have reached north Hilo nearly along the path of the 1881 flow. If a complete map of catchments is made, perhaps the paths of future flows may be predicted in this way.

4.3 Hazard Zones 1 and 2

In the mapped portion of hazard zones 1 and 2, there have been flows from 15 different eruptions from 1843 to the present for a recurrence interval estimate of about 10 years. The probability of another eruption affecting either zone within the next 100 years is, from equation (6), 99.99%. On the other hand, coverage rates are 48.9% (42 km^2) for 150 years or 36.1% (32 km^2) in 100 years for Hazard Zone 1, and 25.5% (307 km^2) for 150 years or 17.8% (214 km^2) in 100 years for Hazard Zone 2.

5. DISCUSSION

Coverage-rate estimates may be thought of as similar to probabilities because their formulas are similar in form (compare equation 3 and equation 6). By estimating that 25% of

an region will be covered by lava in the next century, we are thus estimating that there is a 25% chance that any point within that area will be covered with one or more lava flows in the same period.

Comparing the examples where both probability of lava flow occurrence and coverage rates were estimated, and assuming that a coverage rate is roughly equivalent to a probability of coverage, we see that the probability of lava flow occurrence estimates are significantly higher than probability of coverage estimates. This is a result of the way these parameters are defined. Probability of lava flow occurrence is the probability that one lava flow will enter the target region within the target time period. Probability of coverage is the probability that a particular point will be covered by one or more lava flows within the target region within the target time period.

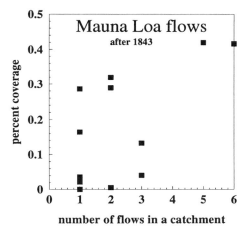

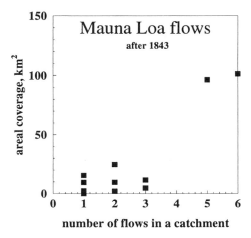

Fig. 6. Plots comparing a) fractional coverage with number of flows in each catchment since 1843, and b) areal coverage with number of flows in each catchment since 1843.

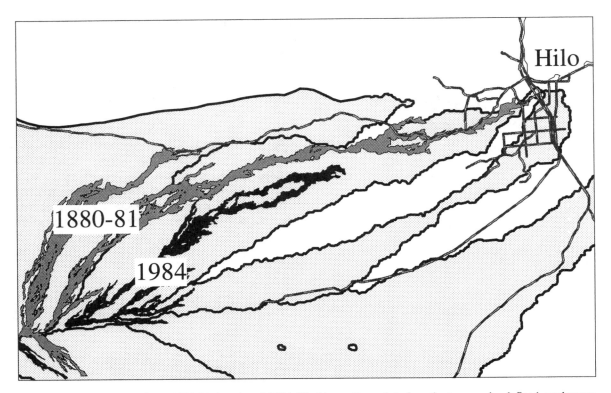

Fig. 7. Mauna Loa flows from 1880-81 (gray) and 1984 (black) eruptions plotted on the topography-defined catchments (stippled within heavy black lines). Also shown are major roads on the island and the location of the town of Hilo. If the 1984 eruption had continued to feed that flow, it would have gone towards the tip of the 1881 flow as it entered the northern edge of Hilo.

It is more likely that a flow will enter a target region than actually cover a significant portion of that region. Hazards from lava flows include fire, isolation from utilities, and access isolation as well as coverage. Probability of lava flow occurrence is a better indication of the probability that any point within the target region will be covered by a lava flow or affected by it's secondary effects. On the other hand, probability of coverage is the probability that any point will only be covered by lava. Both are valuable estimators of different degrees of hazard posed by lava flows.

In the case of lava flows, when risk is considered, both coverage and occurrence are estimators of risk to property more than of risk to life. They are based on frequency and size of lava flows, and not on predicted velocity of those flows or on estimated potential of advanced warning (proportional to distance from vents). A complete lava flow hazard assessment must also take these more dynamic factors into account. In addition to detailed flow-by-flow geologic mapping, the fuller assessment will require studies of the processes by which lava flows are emplaced in order to estimate dynamics, such as velocities, of past flows.

6. CONCLUSION

The results of this study suggest two methods to estimate degrees of lava flow hazard for Mauna Loa when the geologic mapping and digital conversion are completed. We have developed the necessary algorithms and determined some of their properties. Probability of lava flow occurrence seems to be a better estimate of the comprehensive effects of lava flows; the probability of coverage is restricted to only one effect. Choice of target regions is dependent on the purpose of the hazard assessment. Specific regions, such as towns, municipal utilities, or developments can be used for specific hazard assessments. Equal-area regions on a regular grid can be used for a contourable hazard assessment of a more general nature.

Acknowledgements. The authors thank Christina Heliker, Tari Mattox, Dave Clague, Mike Rhodes, and Steven Schilling for constructive reviews of the manuscript.

REFERENCES

Belt Collins & Associates, Hawaiian Riviera resort, Kahuku, Ka`u, Hawaii -- *Final Environmental Impact Statement*, pp. IV-9 - IV-11, 1987.

Buchanan-Banks, J.M., Geologic Map of Hilo 7 1/2' Quadrangle, Island of Hawaii. *U.S. Geological Survey Miscellaneous Investigations Series Map I-2274*, 1:24,000 scale, 1993.

de la Cruz-Reyna, S., Poisson-distributed patterns of explosive eruptive activity. *Bull. Volcanol., 54*, 57-67, 1991.

Heliker, C., Volcanic and seismic hazards on the Island of Hawaii, 48 pp, *U.S. Geological Survey General Interest Publication*, 1990.

Ho, C.-H., Bayesian analysis of volcanic eruptions, *J. Volcanol. Geotherm. Res., 43,* 91-98, 1990.

Ho, C.-H., Smith, E.I., Feuerbach, D.L., and Maumann, T.R., Eruptive probability calculation for the Yucca Mountain site, USA: statistical estimation of recurrence rates, *Bull. Volcanol., 54*, 50-56, 1991.

Jones, A., Improvement of volcanic hazard assessment techniques using GIS: a case study of Mt. Etna, Sicily, in *Innovations in GIS II*, edited by P. Fisher, Taylor and Francis: London, 1995.

Jones, A., Mitigating volcanic hazards: a role for Geographical Information Systems, *Natural Hazards,* in press.

Kauahikaua, J., Margriter, S., Lockwood, J., and Heliker, C., GIS applications for volcanic hazard assessment in Hawai'i, *invited talk at International Workshop on Geographical Information Systems in Assessing Natural Hazards, Perugia, Italy, Sept. 20-22, 1993*, pp. 108-111, 1993.

Kauahikaua, J., Moore, R.B., and Delaney, P., Volcanic activity and ground deformation hazard analysis for the Hawaii Geothermal Project Environmental Impact Statement, *U.S.G.S. Open-File Report 94-553*, 44 p, 1994.

Kilburn, C.R.J., Studies of lava flow development, in *Forecasting Volcanic Events*, edited by H. Tazieff and J.-C. Sabroux, pp. 83-98, Elsevier, Amsterdam, 1983.

Klein, F.W., Patterns of historical eruptions at Hawaiian Volcanoes, *J. Volcanol. Geotherm. Res., 12*, 1-35, 1982.

Legg, M.R., Probabilistic analysis of inundation by lava flow from volcanic activity, in *Conceptual Plan, Commercial Satellite Launching Facility, Palima Point, Ka'u, Hawaii, Draft Environmental Impact Statement*, v. 3, appendix to Appendix A, 19 pp, 1993.

Lipman, P.W., Rates of volcanic activity along the Southwest Rift Zone of Mauna Loa Volcano, Hawaii, *Bull. Volcanol., 43-4*, 703-725, 1980.

Lockwood, J.P., and Lipman, P.W., Holocene eruptive history of Mauna Loa volcano, in Volcanism in Hawaii, edited by R.W. Decker, T.L. Wright, and P.J. Stauffer, *U.S. Geological Survey Professional Paper 1350*, 509-535, 1987.

Lockwood, J.P., Lipman, P.W., Petersen, L.D., and Warshauer, F.R., Generalized ages of surface lava flows of Mauna Loa Volcano, Hawaii, *U.S. Geological Survey Miscellaneous Investigations Series Map I-1908*, Scale 1:250,000, 1988.

Lockwood, J.P., Margriter, S.C., and Trusdell, F.A., Digital geologic mapping of Mauna Loa Volcano, Hawaii [abs.], *Eos, Transactions of the American Geophysical Union suppl., 73*, 613, 1992.

Lockwood, J.P., Mauna Loa eruptive history - the radiocarbon record, *this volume.*

Mullineaux, D.R., Peterson, D.W., and Crandell, D.R., 1987. Volcanic hazards in the Hawaiian Islands, in Volcanism in Hawaii, edited by R.W. Decker, T.L. Wright, and P.J. Stauffer, *U.S. Geological Survey Professional Paper 1350*, 599-621, 1987.

Rubin, M., Lockwood, J.P., and Friedman, I., Hawaiian radiocarbon dates, in Volcanism in Hawaii, edited by R.W. Decker, T.L. Wright, and P.J. Stauffer, *U.S. Geological Survey Professional Paper 1350*, 213-242, 1987.

Scheidegger, A.E., *Physical aspects of natural catastrophes*, 289 pp., Elsevier Scientific Publishing Company, New York, 1975.

True/Mid-Pacific Geothermal Venture, *Environmental impact statement for the Kahauale'a geothermal project*, report prepared by R.M. Towill Corporation, Honolulu, Hawaii, 1982.

Trusdell, F.A., Lava flow hazards and risk assessment on Mauna Loa, Hawaii, *this volume.*

Wadge, G., Young, P.A.V., and McKendrick, I.J., 1994, Mapping lava flow hazards using computer simulation, *J. Geophys. Res., 99*, 489-504, 1994.

Wickman, F.E., Repose period patterns of volcanoes IV, Eruption histories of some selected volcanoes, *Arkiv for Mineralogi och Geologi, 4*, 337-350, 1966.

Wright, T.L., and Pierson, T.C., Living with Volcanoes: The U.S. Geological Survey's Volcano Hazards Program, 57 pp., *U.S. Geological Survey Circular 1073*, 1992.

Wright, T.L., Chun, J.Y.F., Esposo, J., Heliker, C., Hodge, J., Lockwood, J.P., and Vogt, S.M., Map showing lava-flow Hazard Zones, Island of Hawaii, *U.S. Geological Survey Miscellaneous Field Studies Map MF-2193*, scale 1:250,000, 1992.

J. Kauahikaua, S. Margriter, J.P. Lockwood, and F.A. Trusdell, U.S. Geological Survey, Hawaiian Volcano Observatory, P.O. Box 51, Hawaii National Park, HI 96718.
(email: jimk@tako.wr.usgs.gov)

Lava Flow Hazards and Risk Assessment
on Mauna Loa Volcano, Hawaii

Frank A. Trusdell

U.S. Geological Survey, Hawaiian Volcano Observatory, Hawaii National Park, Hawaii

"It is profoundly significant that the Hawaiians of Ka'u did not fear or cringe before, or hate, the power and destructive violence of Mauna Loa. They took unto them this huge mountain as their mother, and measured their personal dignity and powers in terms of its majesty and drama." (Pukui and Handy, 1952)
The Island of Hawai'i is the fastest-growing region in the State of Hawai'i with over 100,000 residents. Because the population continues to grow at a rate of 3% per annum, more and more construction will occur on the flanks of active volcanoes. Since the last eruption of Mauna Loa in 1984, $2.3 billion have been invested in new construction on the volcano's flanks, posing an inevitable hazard to the people living there. Part of the mission of The U.S. Geological Survey's Hawaiian Volcano Observatory is to make the public aware of these hazards. Recent mapping has shown that lava flows on Mauna Loa have covered its surface area at a rate of 30-40% every 1000 years. Average effusion rates of up to 12 million cubic meters per day during eruptions, combined with slopes >10 degrees, increase the risk for the population of South Kona. Studies of Mauna Loa's long-term eruptive history will lead to more accurate volcanic hazards assessments and enable us to refine the boundaries between the hazards zones. Our work thus serves as a guide for land-use planners and developers to make more informed decisions for the future. Land-use planning is a powerful way to minimize risk in hazardous areas.

INTRODUCTION

With an estimated volume of at least 80,000 km³ [*Lipman*, this volume, and *Garcia et al.*, this volume], Mauna Loa is the world's largest volcano. It comprises 5,100 km² of the island of Hawai'i-slightly more than half of the surface area of the island. Mauna Loa has erupted 33 times in the last 150 years, producing lava flows that have covered extensive areas on the flanks of the volcano and reached the ocean five times along the west coast of the island (Plate 1). In 1880, flows covered land that is now within the city limits of Hilo, the largest city and main port on the island. The population at risk on the slopes of

Mauna Loa Revealed: Structure,
Composition, History, and Hazards
Geophysical Monograph 92
Published in 1995 by the American Geophysical Union

the volcano is growing rapidly. Development on its flanks include several multi-million dollar resorts, as well as the city of Hilo (Figure 1).

Approximately 75% of Mauna Loa has been mapped in detail, and 60% of the information is now compiled in a Geographical Information System (GIS) format (Plate 1). A discussion of lava flow hazards based on the entire eruptive history of Mauna Loa is most appropriate, but would be premature until all the prehistoric lava flows have been mapped. The following discussion of lava flow hazards will thus be mainly limited to the well-documented historical period.

In conjunction with the International Decade for Natural Disaster Reduction (IDNDR), the Association of Volcanology and Chemistry of the Earth's Interior (IAVCEI) has designated Mauna Loa a Decade Volcano. The aim of the IDNDR program is to reduce losses from volcanic eruptions through intensive research and mitigation efforts on Decade Vol-

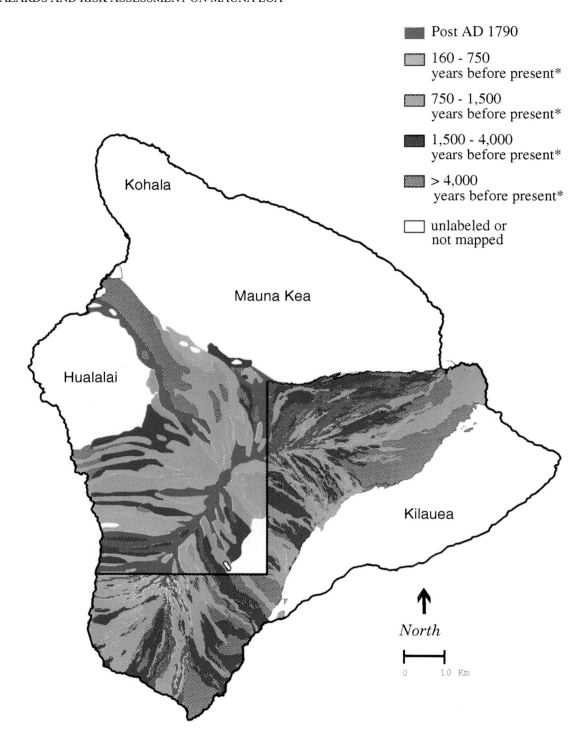

Post AD 1790

160 - 750
years before present*

750 - 1,500
years before present*

1,500 - 4,000
years before present*

> 4,000
years before present*

unlabeled or
not mapped

Kohala

Mauna Kea

Hualalai

Kilauea

North

0   10 Km

Plate 1. Map of Mauna Loa Volcano showing GIS-compiled mapping, combined with the generalized geologic map of Lockwood et al. (1988; the heavy line separates GIS map compilation on the south and east
from compilation by Lockwood et al.(1990) to the west.)

canoes. Volcanic hazards on Mauna Loa can be anticipated and risk substantially mitigated by monitoring activity of the volcano, by documenting the past activity in order to refine our knowledge of the hazards, and by educating the public and local government officials as to our findings and the implications for hazards assessments and risk.

GEOGRAPHY AND STRUCTURE OF MAUNA LOA

Mauna Loa has a summit caldera and two rift zones, one extending to the northeast and the other to the southwest (Figure 1). *Lockwood and Lipman* (1987) divided Mauna Loa into five geographic subdivisions: summit (SUM), Northeast Rift (NER), Southwest Rift (SWR), Moku'aweoweo North (MKN), which includes the north flank, and Moku'aweoweo South (MKS) (Figure 1). All of the geographic subdivisions (except for the summit) are large regions that include the source vents within the rift zones and the adjoining downslope areas. The "summit" is defined as all the area above the 3,500-m elevation. The rift zones and summit have been the source of the vast majority of Mauna Loa's eruptions.

Radial vents (RV) are a class of eruptive fissures that are oriented radially to Mauna Loa's summit and are located outside the defined rift and summit regions on the west, northwest and north flanks. Within the NER, SWR, and MKN geographic subdivisions, radial vents have erupted from the 3,355-m elevation to below sea level (e.g., 1877 Kealakekua Bay submarine eruption; *Moore et al.*, 1985, Figure 1). Thirty-three radial vents have been identified, 58% of which are within geographic subdivision MKN, 24% within the NER, and 18% within the SWR.

ERUPTIVE HISTORY OF MAUNA LOA

The eruptive history of the past 10,000 years has been arbitrarily broken down into five time intervals. Lava flows from the historical period (Group V; 1843 until the present) have covered 14% of the surface of Mauna Loa. Group IV lavas (0.15 to 0.75 ka) have covered an additional 14%; group III lavas (0.75 to 1.5 ka), 29%; group II lavas (1.5 to 4.0 ka), 28%; and group I lavas, (older than 4.0 ka), 15% [*Lockwood and Lipman*, 1987]. Although Mauna Loa is much older than 10,000 years, most of its surface has been covered during this interval.

During the historical period (since 1843), 33 erup-

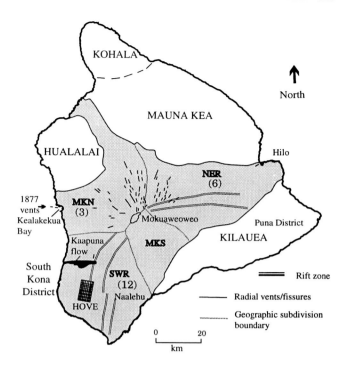

Figure 1. Map of Island of Hawai'i showing geographic subdivisions and rift zones on Mauna Loa. (NER, Northeast Rift; SWR, Southwest Rift; MKN, Mokuaweoweo North; and MKS, Mokuaweoweo South). The place names referred to in the text are shown and average effusion rates of historical eruptions are in parenthesis. Units are in millions of cubic meters per day. The combination of steep slopes and high productivity provide significant risk for the South Kona district.

tions have occurred on Mauna Loa [*Lockwood and Lipman*, 1987; Table 18.1]; historical lava flows cover 806 km^2 of its surface. Typical eruptive activity commences with a curtain of fire (1- to 2-km-long line of lava fountains) in the summit caldera, Moku'aweoweo. The eruption may consist solely of this summit activity within Moku'aweoweo or it may progress to the flank when the dike propagates down a rift zone and opens additional vents. In the meantime, the summit activity wanes and eventually dies off. In the historical period, the Northeast Rift Zone has been the eruptive locus for 31% of the eruptions (covering 202 km^2 of land, or 17% of the area in the NER geographic subdivision); the MKN (including radial vents) has been the source of 6%, covering 130 km^2, or 12%, of its area; and the SWRZ the source of 25% (covering 233 km^2, or 14%, of SWR geographic subdivision). The summit region has been the site of

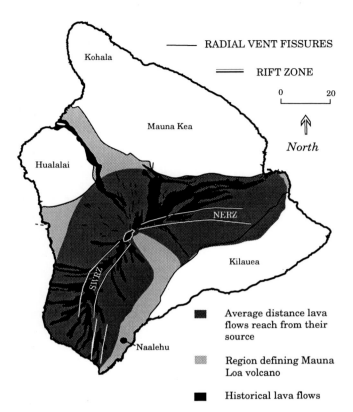

Figure 2. Map depicting the average flow length of historical lava flows. The region in dark gray represents the average distance that historical lava flows have traveled down its flanks from the vent regions of Mauna Loa. Historical lavas are shown in black.

the remaining eruptions, or 38%. The MKS geographic subdivision has no exposed eruptive vents, but lava spilling out of Mokuʻaweoweo has inundated 6.4 km², or 1%, of the total area.

HAZARDS POSED BY MAUNA LOA

The definitions of "volcanic hazards" and "volcanic risk" used in this paper are modified from *Fournier d'Albe*, 1979. "Volcanic hazard" refers to the probability of an area being affected by a potentially damaging volcanic event or process occurring within a given period of time. "Volcanic risk" refers to the expected consequence of volcanic activity in terms of deaths or injuries and the destruction or economic loss of property. Knowledge of volcanic hazards is used to assess volcanic risk. Volcanic risk has been defined thusly: risk = value x vulnerability x hazard [*Fournier d'Albe*, 1979]. Value is further defined as number of lives, amount of property, productive ca-

pacity, etc.; vulnerability is a measure of the proportion of this value that might be lost as a result of an event, and hazard is used in the sense expressed above.

The generally quiescent and effusive style of basaltic eruptions in Hawaiʻi results in the characteristic shield volcanoes. The most obvious hazard associated with shield volcanoes is lava flows. Lava flows seldom cause deaths but can cause extensive damage by covering, burning, and crushing everything in their path. In Hawaiʻi, basaltic flows may reach distances of more than 50 km. Hawaiian volcanoes erupt two different morphological types of lava: pahoehoe and ʻaʻa. The differences between them are governed by the relationship between the increasing viscosity of lava as it cools and the rate of shear strain of the lava. Lava tubes, which can transport lava long distances with little cooling, commonly form in pahoehoe flows. The two longest flows of Mauna Loa were pahoehoe flows from the 1859 and the 1880-81 eruptions, 50 and 48 km long, respectively. ʻAʻa usually forms lava channels rather than lava tubes.

A number of factors may influence the hazard presented by lava flows. The most significant are distance from the vent and area covered. The distance reached by a lava flow is determined by several factors: eruption rate, eruption volume, viscosity, and topography over which the flow moves.

The dark, shaded region in Figure 2 presents the average downslope extent reached by historical lava flows. The data were compiled from 17 historical NERZ, SWRZ, and radial vent eruptions. Historical lava flows within the NER geographic subdivision attained an average length of 18.6 km (18 flows; from 7 different eruptions), the SWR flows averaged 17.2 km (24 flows; from 6 different eruptions) in length, and flows from radial vents, 24.6 km (8 flows; from 4 different eruptions). In order to determine the average downslope extent (dark, shaded region; Figure 2), the average flow length values were used; rift zones and summit regions of the volcano were assumed to be the source of the lava flows. Radial vents on the north, northwest and west flanks were not included. In order to account for RV flows and preclude the selection of particular RV as a source, the summit was arbitrarily selected as the source for RV flows on the north and west flanks.

Two geographic regions on the slopes of Mauna Loa have been protected from historical lava flows by topographic barriers. These are MKS and the area

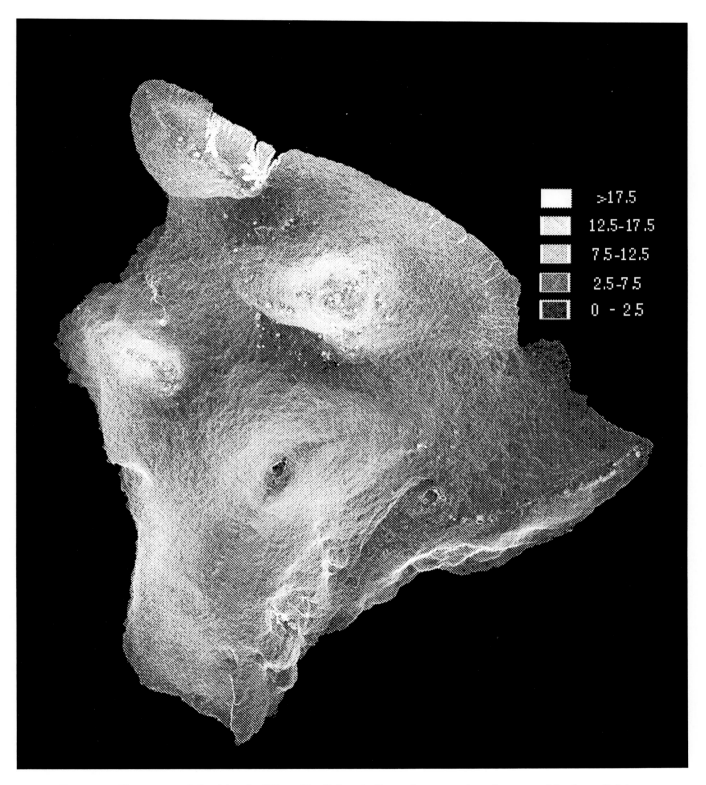

Figure 3. Slope map of the island of Hawaiʻi. Colors indicate 5o categories of topographic slope; lighter colors indicate steeper slopes. Modified from Moore and Mark; 1992.

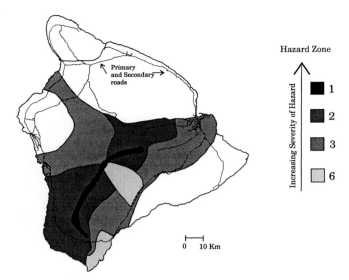

Figure 4. Hazards zone map of Hawaiʻi after Wright et al. (1992), modified to show only Mauna Loa. See Table 1 for criteria used to establish the hazard.

around Naʻalehu at the south coast of the Island. Slope and topography also affect flow advancement rates. Both aa and pahoehoe follow the topography; ʻaʻa is responsive to regional topographic gradients but not to local perturbations, whereas pahoehoe is more sensitive to topography at all scales, especially in areas where slopes are less than 5%. Pahoehoe can accumulate near and/or around an impediment or structure and eventually overtop it. Shallow slopes may hinder flow-front progress as much as steep slopes accelerate it, as demonstrated by the aa flow surges in Royal Gardens [Neal and Wolfe, 1987].

Figure 3 adapted from Moore, J.G., and R.K. Mark, 1992, presents the average slopes for the island of Hawaiʻi. All values of slope are reported in percentages: the NER has been broken down into two regions, with slopes ranging from greater than 10% in the higher elevations to 0-5% in the low elevations; the geographic subdivisions MKN and SWR also have two regions, with slopes ranging from 5% to greater than 10%. MKS has slopes exceeding 10% for the entire region (Figure 3).

The average effusion rates per day of historical eruptions within each of these geographic subdivisions are given in Figure 1 (NER produced 6 million m³/d, MKN 3 million m³/d, and SWR 12 million m³/d). Wherever there is a combination of steep slopes and high eruption rates, the lava flow hazard is increased (e.g. the southern and southwestern portion of the island). For

example, the Kaʻapuna flow from the 1950 SWRZ eruption advanced at an average rate of 9.6 km per hour [Finch and Macdonald, 1955] although the flows must have traveled much faster closer to the eruptive vents. This flow, which originated at 2,400-m elevation, reached the ocean in just over two hours from the beginning of the eruption [Finch and Macdonald, 1955].

HAZARD ZONE MAPS

Lava flow hazard maps for the Island of Hawaiʻi were prepared by Mullineaux and Peterson (1974) and have been revised successively by Mullineaux et al. (1987), Heliker (1990), and Wright et al. (1992). These maps are intended to educate the public, planners, and civil authorities about the degree and kind of hazard that prevail in each geographic region of the island. The Island of Hawaiʻi was divided into nine hazard zones by Wright et al., (1992), with zone 1 being the area of greatest hazard. The criteria used to establish these zones are (1) proximity to the summits and rift zones of the active volcanoes, (2) lava coverage during historical time (since 1800, the date speci-

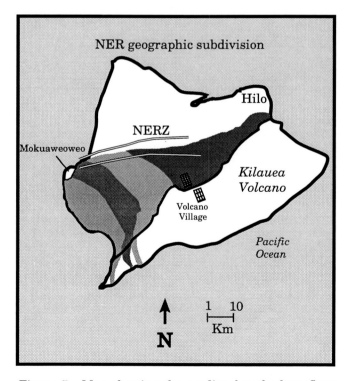

Figure 5. Map showing the predicted paths lava flows would take if erupted from the south side of the NERZ after following topography.

TABLE 1. Hazard Zones for Lava Flows

Zone	Percent of area covered by lava since 1843	Percent of area covered by lava in last 750 yrs	Explanation
1	greater than 25 pct	greater than 65 pct	Includes the summits and rift zones of Kilauea and Mauna Loa where vents have been repeatedly active in historic time.
2	20 pct	25-75 pct	Areas adjacent to and downslope of active rift zones.
3	1-3 pct	15-20 pct	Areas gradationally less hazardous than Zone 2 because of greater distance from recently active vents and/or because topography makes them less likely to be covered.
6	none	very little	Areas currently protected from lava flows by the topography of the volcano.

Modified from *Mullineaux and Peterson* (1972) and *Wright et al.* (1992).

fied by *Wright et al.*, 1992), (3) surface coverage within the last 750 years, and (4) topography. For Mauna Loa (Figure 4), Zone 1 includes the summit and rift zones, whereas Zone 2 includes the regions downslope and adjacent to the summit and the rift zones. Zone 3 includes regions less hazardous because of greater distance from the summit and rift zones or because of topography. (See Table 1 for further descriptions of the hazard zones.) Most of Mauna Loa is designated as hazard zones 1, 2, and 3, although part of MKS is in hazard zone 6, along with the region surrounding the town of Na'alehu. Both of these zone 6 regions are protected by topography from lava flows. The hazards map quantifies hazards based on existing data; it may be modified when new data, mapping, and analytical techniques improve our understanding of the history of the island.

New understanding of Mauna Loa's eruptive mechanisms may suggest changes in the location or style of future eruptive activity. Such understanding may alter our perceptions of future hazards and risk, but the actual hazards map must be based on the observed record of the past, and not on speculations about the future. As an example of future trends, *Lockwood* (1990) has suggested that the south flank of the NERZ may be more vulnerable to future lava flows than previously thought. Lockwood observed that loci of eruptive vents have generally migrated across the crest of the NERZ from north to south in historical time. If

this pattern of southerly dike emplacement continues, then the next series of volcanic vents should surface on the south side of the crest of the NERZ, resulting in lava flow coverage patterns depicted in Figure 5.

Geographical Information Systems (GIS) will greatly enhance our ability to quantify volcano hazards and risk (*Kauahikaua et al.*, this volume) through statistical analysis of the mapping data. As an example of GIS analysis, the NER geographic subdivision has 40%/1000 yrs coverage rate for the past one thousand years and 30%/1000 yrs before that (Figure 6). This trend implies that historical rates of coverage have been somewhat higher than in the prehistoric past.

VOLCANIC RISK

The early Hawaiians lived with ongoing volcanic activity comparable to the current level of activity of Mauna Loa [*Lockwood*, this volume]. These early colonizers of the islands lived in a manner more amenable to the changes wrought by nature. Land-use tenure was more flexible, and they could readily relocate to new sites if their villages were overrun by lava. In contrast, the present-day population is dependent on permanent structures, roads, and power lines. Private land ownership and investments made for infrastructure restricts easy relocation.

Prior to the time of first contact with Europeans, Hawai'i Island had an estimated population of 110,000

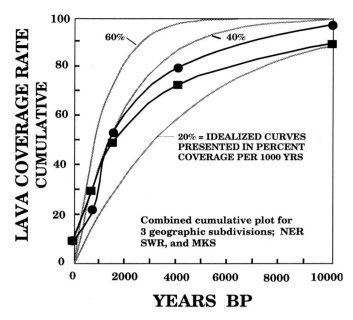

Figure 6. Coverage of Mauna Loa's surface as a function of time. Curves shown are for map compilation of Lockwood et al. (1990) and GIS and idealized coverage rates. Idealized curves indicate uniform coverage rate in percent per 1,000 years.

people (State of Hawai'i Data Book, 1993), but this number dropped dramatically after 1778 due to the Hawaiians' lack of immunity to imported diseases (Figure 7). The decline in population stopped in 1875 and then reversed. The population growth continues at present, with the Big Island now having the largest rate of increase state-wide (State of Hawai'i Data Book, 1993). The fastest-growing regions are the Puna and Kona districts (State of Hawai'i Data Book, 1993).

In the South Kona district, risks from volcanic activity are of special concern because population growth and new construction are proceeding rapidly in hazards zones 1 and 2. Hawaiian Ocean View Estates (HOVE), a subdivision established in September 1961 (County Planning Dept., oral comm., 1995), extends across the southwest rift zone. Thirteen percent lies within the SWRZ in hazard zone 1, and the remaining 87% in zone 2. HOVE currently has approximately 1,500 homes with a projected maximum of 26,000 homes on 13,000 residential lots and a projected population of 40,000 to 80,000. Another example of potential growth in a hazardous zone is a resort approved for development in South Kona in hazard zone 2. This proposed coastal development would be located sea-

ward of the southwest rift zone and includes an airport, a marina, three golf courses, a shopping complex, a hotel, 500 condominiums, 300 townhouses, 652 residential homes and a support community of 1,600 [*Reed*, 1987]. This area was partially covered by lava flows in 1887 and 1907.

Concurrent with the measured and projected population growth is increased capital investment. The aforementioned development in Ka'u has an estimated value of $1 billion. In the 1989-1994 period, 7,972 single-family building permits were issued for the slopes of Mauna Loa, representing a capital investment (assessed value) of approximately $1.4 billion (County of Hawai'i, 1994). Real estate investments (commercial, residential, and public) on the flanks of Mauna Loa since the last eruption in 1984 now exceed $2.3 billion, excluding the value of the land (Figure 8; County of Hawai'i, 1994). The above figures, reported as assessed real estate value, do not include pre-existing structures. In addition, these totals do not include infrastructure costs (such as roads, water lines, utility poles and lines, etc.). Risk will continue to grow with an expanding population and increasing capital investments.

MITIGATION THROUGH EDUCATION

The primary mission of the Hawaiian Volcano Observatory is to reduce risks due to volcanic activity.

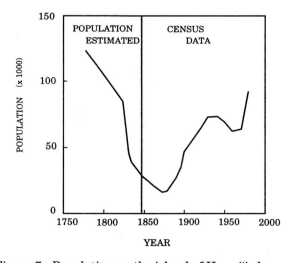

Figure 7. Population on the island of Hawai'i shown as a function of time since the late 18th century. The rapidly increasing population on the island will inevitably cause more people to occupy the higher hazard zones on the active volcanoes.

To this end, the observatory assesses volcanic hazards and educates the public and public officials about those hazards. The destruction of the town of St. Pierre by the devastating eruption of Mt. Pelee in 1902 led Thomas Jaggar to found the Hawaiian Volcano Observatory in 1912. His objective in establishing the observatory was humanitarian-"protecting life and property on the basis of sound scientific achievement." The first comprehensive volcanic hazards assessment of Hawai'i island was published in 1974 [*Mullineaux and Peterson*, 1974]. Since then, many articles have addressed volcanic hazards (e.g., Heliker, 1990). Currently, the observatory provides a weekly column to the local newspapers called "Volcano Watch," devoted to volcanic processes, hazards awareness and the monitoring work of the U.S. Geological Survey. In addition, scientists from the observatory serve as advisors to County and State Civil Defense and consult with, and advise, groups such as insurance representatives, realtors, planners, etc., who have an interest in the impact of volcanoes on the general public.

The observatory works hand-in-hand with the Hawaii County Civil Defense Agency to enhance communications and rapid response to future crises.

CONCLUSIONS

It is essential that civil authorities use volcanic hazards assessments to reduce the risk to people and property through education, land-use planning, and development of evacuation plans.

Land-use planning is a powerful, though little used, means to guide development and minimize risk in hazardous areas.

With significant capital investments in real estate being made on the Big Island, our responsibility in evaluating volcanic hazards becomes especially crucial. The well-documented historical period accounts for only the past 150 years of eruptive activity. Currently, we are extending this historical record by detailed geologic mapping of the prehistoric lava flows of Mauna Loa. By documenting Mauna Loa's long-term eruptive history, we can make more accurate volcanic hazards assessments. We will continue to inform the public about volcanic hazards so that our work will guide land-use planners to make decisions that reduce volcanic risks in the future.

Although this paper has focused only on the lava flow hazards, volcanic activity is indirectly associ-

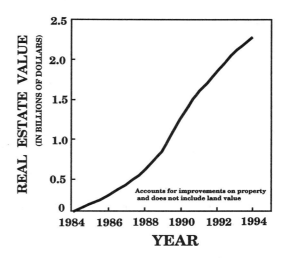

Figure 8. Real estate value on the flanks of Mauna Loa between 1984 and 1994. Investments in real estate improvements on the flanks of Mauna Loa reached $2.3 billion dollars by 1994. Inventory prior to 1984 is not included.

ated with other hazards such as earthquakes [*Swanson et al.*, 1976; *Buchanan-Banks*, 1987; *Wyss*, 1988; *Wyss and Koyanagi*, 1992; *Klein and Okubo*, 1993; *Klein*, 1994] and tsunami [*Pararas-Carayannis*, 1977, *Bolt et al.*, 1975; *Blong*, 1984] whose threat must be considered for any overall volcanic hazards planning.

Acknowledgments. This manuscript benefited greatly from reviews from C. Heliker, and S. Rowland. Critical and extensive reviews from D. Peterson, J. P. Lockwood and D. A. Clague greatly improved this paper. Special thanks to J. Takahashi and I. Tengan.

REFERENCES

Blong, R.J., *Volcanic Hazards*, Academic Press, Sydney, Australia, 424 pp., 1984.

Bolt, B.A., W.L. Horn, G.A. Macdonald, and R.F. Scott, *Hazards from Volcanoes*, Springer-Verlag, New York, 328 pp., 1975.

Buchanan-Banks, J.M., Structural damage and ground failures from the November 16, 1983, Kaoiki earthquake, island of Hawai'i, in Decker, R.W. , T.L. Wright, and P.H. Stauffer, eds., Volcanism in Hawai'i, *U.S. Geol. Surv. Prof. Pap. 1350*, 2, 1187-1220, 1987.

County of Hawai'i, Department of Public Works, *Building Permits Summary*, 1994.

County of Hawai'i, Planning Department, "First recorded

lot sale in HOVE was Sept. 1961," *Pers. comm.*, 1995.

Department of Business and Economic Development and Tourism, *State of Hawai'i Data Book*, Research and Economic Analysis Division, Honolulu, 618 pp. 1993.

Finch, R.H., and G.A. Macdonald, Hawaiian volcanoes during 1950, *U.S. Geol. Surv. Bull. 996-B* pp. 27-89, 1953.

Fournier d'Albe, E.M., Objectives of volcanic monitoring and prediction, *J. Geol. Soc. London*, 136, 321-326, 1979.

Garcia, M.O., T.P. Hulsebosch, and J.M. Rhodes, Glass and mineral chemistry of olivine-rich submarine basalts, Southwest Rift Zone, Mauna Loa Volcano: Implications for magmatic processes, AGU Monograph, this volume.

Heliker, C., Volcanic and seismic hazards on the island of Hawai'i, *U.S. Geol. Surv. Gen. Int. Pub.*, 48 pp., 1990.

Kauahikaua, J., S. Margriter, J.P. Lockwood, and F.A. Trusdell, Applications of GIS to the estimation of lava flow hazards on Mauna Loa Volcano, Hawai'i, AGU Monograph, this volume.

Klein, F.W., and P. Okubo, Hazards from earthquakes, submarine landslides and expected levels of peak ground acceleration on Kilauea and Mauna Loa Volcanoes, Hawai'i (abstr.), *Eos Trans. AGU supp.*, 74, 634-635, 1993.

Klein, F.W., Seismic hazards at Kilauea and Mauna Loa Volcanoes, Hawai'i: *U.S. Geol. Surv. Open-File Rept. 94-216*, var. pag, 1994.

Lipman, P.W., Growth of Mauna Loa Volcano during the last hundred thousand years: Rates of lava accumulation versus gravitational subsidence, AGU Monograph, this volume.

Lockwood, J.P., and P.W. Lipman, Holocene eruptive history of Mauna Loa Volcano, in Decker, R.W., T.L. Wright, and P.H. Stauffer, eds., Volcanism in Hawai'i, *U.S. Geol. Surv. Prof. Pap 1350*, 2, 509-536, 1987.

Lockwood, J.P., P.W. Lipman, Petersen, and F.R. Warshauer, Generalized ages of surface lava flows of Mauna Loa Volcano, Hawai'i, *U.S. Geol. Surv. Misc. Investig. Ser. Map I-1908*, scale 1:250,000, 1988.

Lockwood, J.P., Implications of historical eruptive-vent migration on the northeast rift zone of Mauna Loa Volcano, Hawai'i, *Geology*, 15, 611-613, 1990.

Lockwood, J.P., Mauna Loa eruptive history - the radiocarbon record, AGU Monograph, this volume.

Moore, J.G., D.J. Fornari, and D.A. Clague, Basalts from the 1877 submarine eruption of Mauna Loa, Hawai'i: new data on the variation of palagonitization rate with temperature, *U.S. Geol. Surv. Bull. 1663*, 11 pp., 1985.

Moore, J.G., and R.K. Mark, Morphology of the Island of Hawai'i, *GSA Today*, 2, 257-262, 1992.

Moore, J.G., W.R. Normark, and R.T. Holcomb, Giant Hawaiian landslides: *Annual Review of Earth and Planetary Sciences*, 22, 119-144, 1994.

Moore, R.B., P.T. Delaney, and J.P. Kauahikaua, Annotated bibliography: Volcanology and volcanic activity with a primary focus on potential hazard impacts for the Hawai'i Geothermal Project, *U.S. Geol. Surv. Open-File Rept. 93-512A*, 10 pp., 1993.

Mullineaux, D.R., and D.W. Peterson, Volcanic hazards on the island of Hawai'i, *U.S. Geol. Surv. Open-File Rept. 74-239*, 61 pp., 2 folded maps, 1974.

Mullineaux, D.R., D.W. Peterson, and D.R. Crandell. "Volcanic hazards in the Hawaiian Islands", in Decker, R.W., T.L. Wright, and P.H. Stauffer, eds., Volcanism in Hawai`i, *U.S. Geol. Surv Prof. Pap. 1350*, 1, 599-621, 1987.

Neal, C.A., and E.W. Wolfe, Observations of lava flows, Puu Oo eruption of Kilauea Volcano, Hawai'i (abstr.), *in International Union of Geodesy and Geophysics* a 19th General Assembly, Abstracts, 414 pp., 1987.

Okamura, A.T., J.J. Dvorak, R.Y. Koyanagi, and W.R. Tanigawa, Surface deformation during dike propagation, in Wolfe, E.W., ed., The Puu Oo eruption of Kilauea Volcano, Hawai'i: Episodes 1 through 20, Jaunary 3, 1983, through June 8, 1984, *U.S. Geol. Surv. Prof. Pap. 1463*, 165-182, 1988.

Pararas-Carayannis, G., Catalog of tsunamis in Hawai'i. *World Data Center A for Solid Earth Geophysics*, Boulder, SE-4, 78 pp., 1977.

Pukui, K, and J. Handy, *The Polynesian Family System in Ka'u, Hawai'i*, Bishop Museum Press, Honolulu, 300 pp., 1952.

Reed, C., Huge Ka'u resort seen, *Hawai'i Tribune -Herald*, pp. 1 and 10, Sept. 27, 1987.

Swanson, D.A., W.A. Duffield, and R.S. Fiske, Displacement of the south flank of Kilauea Volcano: The result of forceful intrusion of magma into the rift zones, *U.S. Geol. Surv. Prof. Pap. 963*, 39 pp., 1976.

United Nations Disaster Relief Organization, *Natural disasters and vulnerability analysis*, UNESCO, Paris, 97 pp., 1979.

Wright, T.L., J.Y. Chu, J. Esposo, C. Heliker, J. Hodge, J.P. Lockwood, and S.M. Vogt, Map showing lava-flow hazard zones, island of Hawai'i, *U.S. Geol. Surv. Misc. Field Stud. Map*, MF-2193, scale 1:250,000, 1992.

Wyss, M., A proposed source model for the great Kau, Hawai'i, earthquake of 1868, *Bull. Seis. Soc. Am.*, 78, 1450-1462, 1988.

Wyss, M., R.Y. Koyanagi, and D.C. Cox, The Lyman Hawaiian earthquake diary, 1833-1917, *U.S. Geol. Surv. Bull. 2072*, 34 pp., 1992.

Forecasting Eruptions of Mauna Loa Volcano, Hawaii

Robert W. Decker[1], Fred W. Klein[2], Arnold T. Okamura[3], and Paul G. Okubo[3]

U.S. Geological Survey

Past eruption patterns and various kinds of precursors are the two basic ingredients of eruption forecasts. The 39 historical eruptions of Mauna Loa from 1832 to 1984 have intervals as short as 104 days and as long as 9,165 days between the beginning of an eruption and the beginning of the next one. These recurrence times roughly fit a Poisson distribution pattern with a mean recurrence time of 1,459 days, yielding a probability of 22% (P=.22) for an eruption of Mauna Loa during any next year. The long recurrence times since 1950, however, suggest that the probability is not random, and that the current probability for an eruption during the next year may be as low as *6%*. Seismicity beneath Mauna Loa increased for about two years prior to the 1975 and 1984 eruptions. Inflation of the summit area took place between eruptions with the highest rates occurring for a year or two before and after the 1975 and 1984 eruptions. Volcanic tremor beneath Mauna Loa began 51 minutes prior to the 1975 eruption and 115 minutes prior to the 1984 eruption. Eruption forecasts were published in 1975, 1976, and 1983. The 1975 and 1983 forecasts, though vaguely worded, were qualitatively correct regarding the timing of the next eruption. The 1976 forecast was more quantitative; it was wrong on timing but accurate on forecasting the location of the 1984 eruption. This paper urges that future forecasts be specific so they can be evaluated quantitatively.

INTRODUCTION

Because it is probabilistic, forecasting volcanic eruptions is a dicey business. At best, a well-based forecast provides essential information for contingency planning in case of a volcanic crisis. At worst, an incorrect forecast can result in substantial social disruption and economic loss, or even human fatalities. Still, from a societal standpoint, it is important for scientists to try to make the best possible eruption forecasts with the available information.

Eruption forecasts are generally based on past eruption patterns and/or on various precursors [*Decker*, 1973, 1986; *Tilling*, 1995]. For both these basic approaches, Mauna Loa Volcano has been relatively well-studied

compared to most of the world's active or potentially active volcanoes.

Mauna Loa's historical eruption record consists of 39 events since 1832 (Table 1, Figure 1). The term "repose time" conventionally means the interval time between eruptions, the period during which the volcano is in repose. If eruptions are of short duration, the difference between repose time and the interval between the beginning of one eruption and the beginning of the next is negligible. However, in long-lived eruptions this difference can be significant. For this reason we use the term "recurrence time" in this paper to clearly indicate the time interval from the onset of one eruption to the onset of the next.

In addition to the historical record, Mauna Loa's hundreds of prehistoric lava flows have been mapped and dated in considerable detail [*Lockwood and Lipman*, 1987; *Lockwood*, this volume]. Almost 90% of the surface of Mauna Loa is covered by flows less than 4,000 years old. Monitoring the number and location of microearthquakes beneath Mauna Loa began in 1962, and deformation monitoring began in 1965. Increased seismic activity preceded the eruptions of 1975 and 1984 by about two years, and more rapid rates of inflation occurred for about

---

[1]Kawaihae Village 30, Kamuela, Hawaii.

[2]U.S. Geological Survey, Menlo Park, California.

[3]Hawaiian Volcano Observatory, Hawaii National Park, Hawaii.

Mauna Loa Revealed: Structure,
Composition, History, and Hazards
Geophysical Monograph 92

TABLE 1

Mauna Loa eruptions

Event	Starting date	Summit duration (days)	Flank duration (days)	Location[a]	Days since last eruption	Erupted volume (10^6 m^3)
1	6/20/1832	21	—	S	—	—
*2	1/9/1843	5	90	NF	3855	190
3	5/1849	15	—	S	2318	—
4	8/8/1851	21	—	S	815	70
5	2/17/1852	1	20	NER	193	110
6	8/11/1855	—	450	NER	1271	110
*7	1/23/1859	1	300	NF	1261	460
8	12/30/1865	120	—	S	2533	—
9	3/27/1868	1	15	SWR	818	145
10	1/1/1870	14	—	S	645	—
11	8/1/1871	30	—	S	577	—
12	8/10/1872	60	—	S	375	—
13	1/6/1873	2	—	S	149	—
14	4/20/1873	547	—	S	104	—
15	1/10/1875	30	—	S	630	—
16	8/11/1875	7	—	S	213	—
17	2/13/1876	1	—	S	186	—
18	2/14/1877	10	1	WF	367	—
19	5/1/1880	6	—	S	1172	—
*20	11/1/1880	—	280	NER	184	230
*21	1/16/1887	—	10	SWR	2267	230
22	11/30/1892	3	—	S	2145	—
23	4/21/1896	16	—	S	1238	—
24	7/4/1899	4	19	NER	1169	150
25	10/6/03	60	—	S	1554	—
26	1/9/07	1	15	SWR	1191	75
27	11/25/14	48	—	S	2877	—
28	5/19/16	—	14	SWR	541	60
*29	9/29/19	1	42	SWR	1228	270
30	4/10/26	1	14	SWR	2385	110
31	12/2/33	17	1	S	2793	75
32	11/21/35	1	42	NER	719	120
33	4/7/40	133	1	S	1813	75
34	4/26/42	2	13	NER	535	75
35	11/21/43	3	—	S	574	—
36	1/6/49	142	2	S	1873	60
*37	6/1/50	1	23	SWR	511	460
38	7/5/75	2	—	S	9165	3
*39	3/25/84	1	21	NER	3185	220

[a]NER = northeast rift zone; SWR = southwest rift zone; S = summit caldera; NF = north flank; WF = west flank. Data are complete through 1994.
*Largest eruptions: volume exceeds 160 × 10^6 m^3.

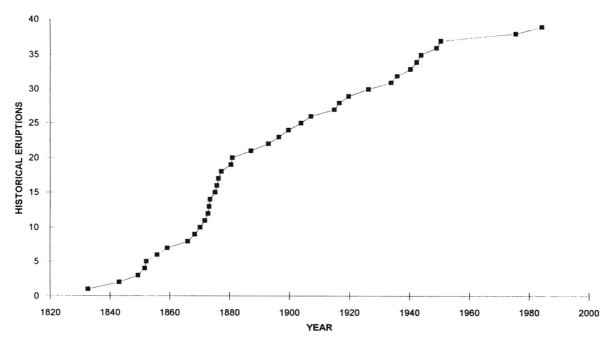

Fig. 1. Cumulative number of historical eruptions of Mauna Loa plotted against the onset dates of the 39 eruptions (Table 1). The frequent eruptions between 1870 and 1880, and the long recurrence times since 1950 suggest distinct clusters of non-random character.

a year or two both before and after the eruptions. The changing patterns of seismicity and surface deformation both indicated that magma was being injected and stored in a shallow reservoir beneath the summit. This accumulation of magma and eventual cracking of the reservoir margins are plausible physical precursors to eruption.

PAST ERUPTION PATTERNS

Thomas Jaggar, the founder of the Hawaiian Volcano Observatory, was fascinated by the patterns he saw in the historical record of Mauna Loa eruptions. Jaggar [1912, p.74] published the following forecast based on Mauna Loa's past behavior: "The last eruption of Mauna Loa came to an end about February 1, 1907. Applying the above average interval of repose the next eruption would be due four and three-quarter years thereafter, namely November 1, 1911. The shortest interval since 1868 has been three years and the longest eight years. Applying the maximum figure, we get February 1, 1915. It is fair to expect, therefore, at the time of this writing (September 14, 1912), that Mauna Loa will renew activity with a lava pool at the Mokuaweoweo center, and without flows at first, between now and February 1, 1915."

Jaggar continued his forecast saying that he expected a north flank eruption within 5 years after the summit erup-

tion at an elevation above the Dewey crater of 1899. These expectations were based on past alternation of summit and rift eruptions, past alternation of rift eruptions between the north and south rift zones, and progressively higher elevations of rift zone eruptions since 1868. Jaggar's forecasts were accurate with regard to timing but not entirely correct with regard to the locations of the eruptions. A summit eruption began November 25, 1914, and lasted 48 days; a flank eruption began on May 19, 1916, but on the south flank rather than the north. The main vents of the 1916 eruption were between 6,500 and 7,500 feet, well below the 11,300-foot elevation of the 1899 eruption at Dewey cone.

Jaggar was not shy about making eruption forecasts. Some were published in scientific papers, others were made in lectures and newspaper reports [Barnard, 1991]. He was often but not always right, and no quantitative evaluation of his forecasts is attempted here.

Wickman [1966] analyzed the historical time series of Mauna Loa eruptions and concluded that the time intervals between eruptions fit a Poisson distribution and were essentially random. His statistical analysis indicated that no matter how long it had been since the previous eruption, the probability that Mauna Loa would erupt during any next month was constant at $P = .022$. This is analogous to cutting a deck of cards to get an ace. No matter

TABLE 2. Probabilities of eruption of Mauna Loa tomorrow, next week, next month, next year, ..., next 19 years calculated for a random model with a mean recurrence time of 1,459 days (39 historical eruptions from 1832 through 1984). $P = 1 - e^{(-t/1459)}$ where t is the number of days of the specified time interval. P in this equation is derived by taking 1 minus the probability of no eruptions occurring during time t (*Klein*, 1982, equation 3). Probability in a random model does not change if the eruption does not occur during the forecast time period; it does change as future eruptions create new mean recurrence times.

Time interval	Probability
1 Day	.00069
1 Week	.0048
1 Month	.020
1 Year	.22
2 Years	.39
3 Years	.53
4 Years	.63
5 Years	.71
6 Years	.78
7 Years	.83
8 Years	.86
9 Years	.89
10 Years	.92
12 Years	.95
19 Years	.99

how many times a full deck is cut, the probability of getting an ace on the next cut is 1/13 ($P=.077$).

Klein [1982] made a more detailed statistical analysis of the historical record of Hawaiian volcanic eruptions. His conclusions regarding Mauna Loa were that the mean recurrence time was 1,412 days, and that the pattern of recurrence times fits a Poisson distribution and is random with the following exceptions: 1) Large volume eruptions increase the recurrence time to the next eruption. 2) The length of the longest recurrence time (1950 to 1975) is marginally unusual for a random process. 3) The sequence of 8 relatively short recurrence times from 1870 to 1877 differs from a random process.

The 1984 eruption began 3,185 days after the 1975 eruption (Table 1), extending the mean recurrence time between eruptions to 1,459 days. The probabilities in Table 2 are calculated for a random model using this updated mean recurrence time.

Ho [1991] applied a nonhomogeneous Poisson model (the Weibull distribution) to the repose times of several volcanoes, but not Mauna Loa. The Weibull distribution considers whether a time series of eruptions shows a constant mean recurrence rate or a smoothly increasing or decreasing mean recurrence rate. Applying Ho's test to the recurrence times of Mauna Loa's eruptions from 1832 to 1984 yields a slowly decreasing mean recurrence rate. The result, however, is not significantly different from the constant mean recurrence time of 1,459 days of a homogeneous Poisson distribution.

Figure 1 clearly shows that there is not an obvious smooth decrease in the recurrence times in the historical record of Mauna Loa; rather there is an abrupt change in slope in 1950. The question is whether this change is a statistical aberration of random chance or a real change in the eruptive behavior of Mauna Loa.

The probability that the change is simply a statistical aberration in a homogeneous Poisson distribution can be tested. Three of the four longest recurrence times in Mauna Loa's historical eruption record have occurred since 1950 -- 25 years (1950-1975), 9 years (1975-1984), and at least 11 years. The last eruption was in 1984; publication year of this paper is 1995. Since the 25 year recurrence time was marginally unusual for a random process [$P=.054$, *Klein*, 1982, page 19], the sequence of three consecutive long intervals is even less probable for an overall random process. There has been at least a 16,250 day period (6/1/1950 to late 1994) with only two eruptions. The probability of two or fewer random eruptions with a mean recurrence interval of 1,459 days occurring during this long interval is only $P=.00042$. Klein [1982, equations 1-3] derives the probability of no eruptions occurring during time t. Adding the probabilities for one and two eruptions occurring during time t yields

$$P = e^{(-t/a)} [1+(2/3)(t/a)+(1/6)(t/a)^2]$$

where $a=1,459$ days and $t=16,250$ days. Clearly this low probability ($P=.00042$) suggests a non-random process; Mauna Loa has apparently entered a mode of longer recurrence times. This cluster of long recurrence times suggests that at present the probability for an eruption of Mauna Loa during the next year is much lower than the $P=.22$ figure for the entire historical record. The probability may be as low as $P=.06$ per year if the long recurrence times since 1950 are typical of the future.

If the longer recurrence times between Mauna Loa eruptions since 1950 is not just a statistical fluke, what could be causing this apparent change? Perhaps the resupply rate of magma to Mauna Loa's summit reservoir diminished after 1950. Klein [1982] also demonstrated that the number of eruptions of Mauna Loa and Kilauea between

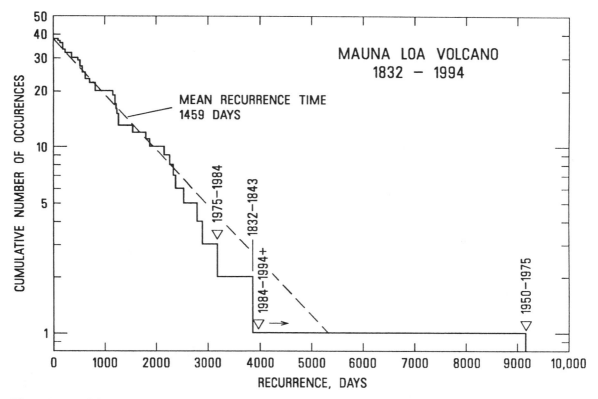

Fig. 2. Graph of the shortest (left) to longest (right) recurrence times of all historical Mauna Loa eruptions versus the logarithm of the cumulative number of eruptions. A random model will plot near the dashed line which has a slope equal to (-1/mean recurrence time). Derivation of this type of graph is explained in Wickman [1966] and Klein [1982]. All the historical recurrence times fit this random model except the 25-year-long recurrence time between the 1950 and 1975 eruptions. As explained in the text, it is not just the single 25-year-long recurrence time that invalidates this overall random model, but the cluster of three long recurrence times (25 years, 9 years, and at least 11 years following the 1984 eruption) since 1950. The three most recent recurrence times are marked with triangles.

1916 and 1980 are inversely correlated; that is, when Mauna Loa eruptions are numerous, Kilauea has fewer eruptions, and vice versa. With the long-lasting and voluminous eruption of Kilauea still in progress (1983 to present), this relationship becomes an important consideration in forecasting the next eruption of Mauna Loa.

King's [1989] analysis of the volume predictability of historical eruptions of Kilauea and Mauna Loa concludes that the post-1952 increase in eruption rate at Kilauea was accompanied by a decrease at Mauna Loa. In addition, his study indicates that in general the volume of an eruption episode of Mauna Loa is approximately proportional to the time elapsed since the previous eruption.

Lockwood and Lipman's [1987] mapping of Mauna Loa indicates that 13% of its surface is covered by flows erupted during the time interval from 1843 through 1984. The volume of these lava flows is about 4 cubic kilome-

ters, yielding an average eruption rate of 28 million cubic meters per year. From A.D.1200, about the time the present summit caldera formed, until 1843, areal coverage rates by lavas outside the caldera were lower than during the period 1843 through 1984. It is not known how much lava was ponded in the caldera during this 1200 to 1843 interval. During about 750 years prior to A.D.1200, eruptions were characterized by voluminous overflows of pahoehoe from a summit lava lake. More than 25% of Mauna Loa is covered by these overflow lavas. The record becomes more fragmentary with increasing age because over 50% of Mauna Loa has been covered by lavas in the past 1,500 years.

It is clear from both the historical and late prehistorical record that Mauna Loa has been -- and probably will continue to be -- a vigorously active volcano, albeit past its prime [Lipman, this volume].

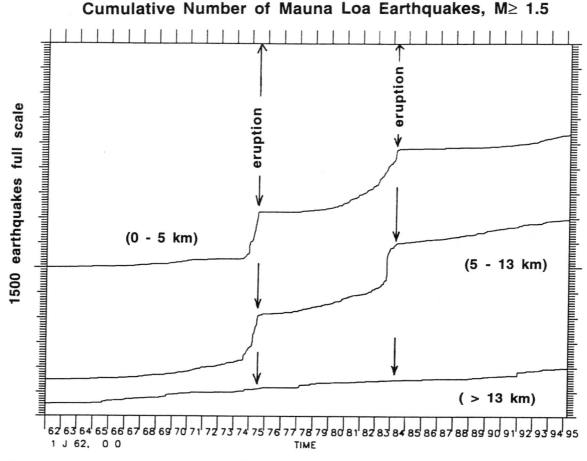

Fig. 3. Cumulative numbers of earthquakes of magnitude 1.5 and greater located beneath the summit region of Mauna Loa (the area in Figures 4 and 5) from 1962 through 1994. Three different depth ranges of earthquake foci are shown; notice that the increase in earthquakes prior to eruptions is limited to depths shallower than 13 km.

ERUPTION PRECURSORS

Wood [1915], and Finch [1943] published post-eruption accounts of the seismic preludes to the 1914 and 1942 eruptions of Mauna Loa. These were not forecasts; even so, it is clear that earthquake swarms were considered to be potentially important precursors to eruptions of Mauna Loa as far back as 80 years ago. However, it was not until the early 1960s that the Hawaiian Volcano Observatory's (HVO) network of seismographs was adequate to monitor the seismicity beneath Mauna Loa in a quantitative fashion. Earthquakes greater than magnitude 1.5 from the surface down to 13 km beneath Mauna Loa showed definite increases for one to two years before the 1975 and 1984 eruptions (Figures 3, 4 and 5). Earthquakes deeper than 13 kilometers did not show these increases, indicating that the increased seismicity was apparently related to

cracking in the rocks surrounding the summit reservoir complex.

Leveling and distance measurements near and across the summit caldera of Mauna Loa began in 1965 [Decker, 1968]. These measurements, although made only intermittently, indicated inflation of the summit of Mauna Loa prior to the 1975 eruption, inflation following the 1975 eruption, a small increase in rate of inflation prior to the 1984 eruption, major deflation during the 1984 eruption, and inflation following the 1984 eruption (Figures 6 and 7).

Strong, shallow volcanic tremor has been recorded on the Mauna Loa seismometers immediately before and during eruptions. Presumably the onset of tremor indicates the beginning of magma ascent towards the surface. Tremor began 51 minutes before the surface outbreak of lava on July 5, 1975 [Lockwood et al, 1987]. After 27

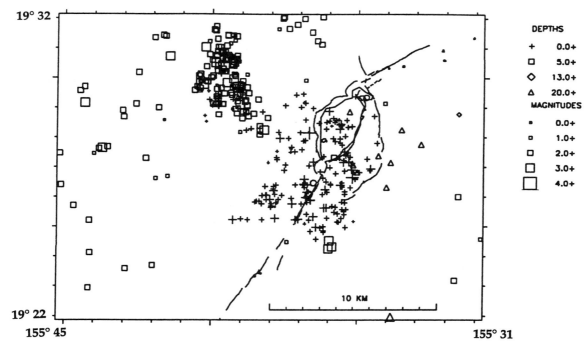

Fig. 4. Earthquake epicenters in the summit region of Mauna Loa during 16 months prior to the eruption of 1975 (January 1, 1974, to July 1, 1975). Outlines are shown of the summit caldera and the northeast and southwest rift zones.

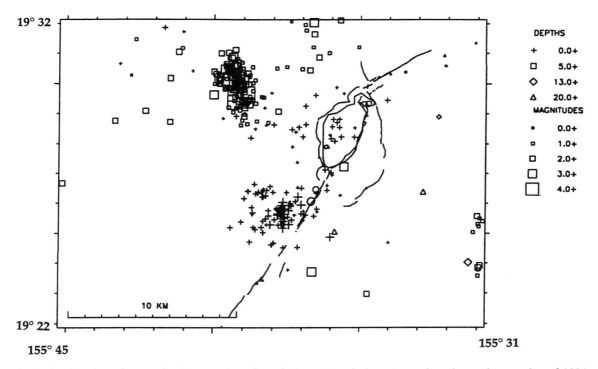

Fig. 5. Earthquake epicenters in the summit region of Mauna Loa during 16 months prior to the eruption of 1984 (September 24, 1982, to March 24, 1984). Notice the similarity to Figure 4.

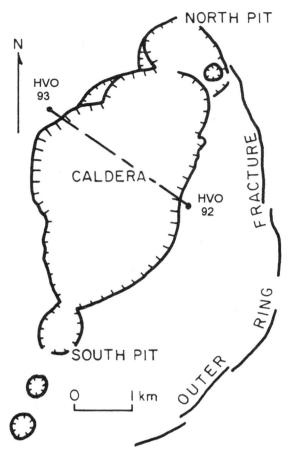

Fig. 6. Map of the summit caldera region of Mauna Loa showing the cross-caldera line that has been intermittently measured since 1965. The extensions and contractions of this line are plotted in Figure 7. Other distance-measurement lines, level lines, and tilt-measurement stations were established in this region after 1965.

minutes, this precursory tremor activated the telephone alarm system, and even though the eruption began just before midnight, HVO staff members were at the observatory 12 minutes before glow was seen above Mauna Loa.

Tremor began 115 minutes before the 1:25 AM eruption of March 25, 1984. Unfortunately, the tremor alarm had been turned off because high winds on Mauna Loa had been triggering the alarm during the previous day. Wind noise can be distinguished from volcanic tremor on the visible seismic recorders, but at that time the alarm trigger could not distinguish between these two types of signals. HVO staff did not reach the observatory until a half hour after the eruption began. They were alerted by staff from the astronomical observatories on neighboring Mauna Kea Volcano who saw the eruption glow above Mauna Loa.

Reports of other precursors to the 1984 eruption were received after the eruption began. Some hikers saw a "dull red glow in a crack" on the caldera floor on March 18, "steam clouds" rising above the caldera on March 23, and "rocks and steam being ejected from the 1975 fissures near the center of the caldera" on March 24 [*Lockwood et al*, 1987]. A remote recorder of hydrogen-gas-emission and temperature, installed in November, 1983 in a fumarole along the 1975 fissure, transmitted normal background levels one hour before it was destroyed by the onset of the 1984 eruption.

Changes in the amount of carbon dioxide presumably issuing from summit fumaroles are seen on the nearly continuous records of the atmospheric observatory at 11,000 feet on the north flank of Mauna Loa. Major increases in CO_2 emission occurred during and after the 1975 and 1984 eruptions. The 1975 eruption may have been preceded by a small increase in CO_2 emission, and a similar increase has occurred since early 1993 [*Ryan*, this volume].

THREE PREVIOUS FORECASTS OF MAUNA LOA ERUPTIONS

1975

Koyanagi, Endo, and Ebisu [1975] published a paper titled *Reawakening of Mauna Loa Volcano, Hawaii: A Preliminary Evaluation of Seismic Evidence*. In the abstract they state: "The significant increase in seismicity, combined with expansion measured at the summit, suggest that Mauna Loa may be reawakening after nearly 25 years of quiescence."

This paper was prepared in June, 1975, but was not published until September, 1975. The eruption occurred on July 5-6, 1975.

1976

Lockwood, Koyanagi, Tilling, Holcomb, and Peterson [1976] published a paper titled *Mauna Loa Threatening*. In this paper they state: "The northeast rift zone appears to be the most reasonable place to expect the next flank outbreak because of the seismic activity and dilation of the northeast rift zone near Puu Ulaula in the days immediately after the summit eruption of July 5-6 [1975]."

"Six eruptions have occurred on the northeast rift zone during written history ... Each was preceded by a summit outbreak and except for 1855 followed the initial summit eruption by 6-34 months."

"The historical record thus suggests that the summit eruption of July 1975 was the first phase of an eruptive sequence that will culminate with a major eruption on the flanks of Mauna Loa sometime before the summer of 1978. This flank eruption will likely be closely preceded by brief summit activity. If the Mauna Loa activity of July 6-10 is a reliable indicator, we can expect the flank vents to open between 2,800 and 3,000 meters on the northeast rift zone in the vicinity of Puu Ulaula."

1983

Decker, Koyanagi, Dvorak, Lockwood, Okamura, Yamashita, and Tanigawa [1983] published a paper titled *Seismicity and Surface Deformation of Mauna Loa Volcano, Hawaii*. Their concluding statement was: "The near-surface strain from the apparent intrusion of magma beneath the summit region of Mauna Loa has recently shown an accelerating trend on the basis of both seismic and ground-surface-deformation data. But since the present strength of Mauna Loa is not known, no precise forecast of the next eruption can be made. However, if the present rate of strain continues to increase (and we emphasize the "if"), the probability significantly increases for an eruption of Mauna Loa during the next 2 years."

The paper was published in September, 1983. The eruption began on March 25, 1984, and lasted for 21 days.

EVALUATION OF FORECASTS

With regard to estimating the onset time of an eruption of Mauna Loa, the 1975 and 1983 forecasts were qualitatively correct. However, the forecasts were so vaguely worded that no quantitative evaluation of their merit is possible. Not so the 1976 forecast; it was boldly and specifically stated, and can be evaluated by a simple betting scheme. For example, in the 1976 forecast, Lockwood et al stated that the next eruption will begin before the summer of 1978. Based on the historical record, the statistical odds that this would occur are 1.63/1 (Odds are determined by the probability of losing divided by the probability of winning; $P = .62$ that no eruption will occur divided by $P = .38$ that eruption will occur). Say they bet $1.00. They lose. The 1976 forecast also states that the next eruption will occur on the northeast rift zone. The odds on that bet are 5.33/1 (32/6; 32 historical eruptions did not occur on the northeast rift zone, 6 did). They

bet another dollar. They win $5.33 (that is, they get their dollar back plus $5.33). In addition, the 1976 forecast places the main vent of the next eruption at an elevation between 2800 and 3000 meters. The odds on that bet are 2/1 (4/2; 4 historical northeast rift eruptions had occurred outside that elevation interval; 2 had occurred within that interval). They bet another dollar. They win $2.00. Their net winnings are $6.33 (- 1.00, + 5.33, + 2.00), 71% of the maximum winnings ($8.96) had all their forecasts been correct.

A break-even or losing evaluation would indicate that the precursors on which the forecasts of 1976 were made were not valid. The specific precursors on which the location forecasts were made were the seismic swarm and surface deformation which immediately followed the brief summit eruption of July 5-6, 1975. These geophysical data indicated that a 25-km-long dike had been emplaced at a shallow level into the northeast rift zone from the summit to about the 2900-meter elevation near Puu Ulaula.

We are not trying to trivialize the serious business of forecasting eruptions by using a hypothetical betting scheme to evaluate forecasts after time has validated or invalidated their conclusions. However, we urge that this betting scheme, or some more sophisticated procedure be used to evaluate future forecasts so that they can be tested against one another and tested against the statistical record. Several similar evaluations over a long period of time are needed to see if the forecasting techniques are indeed valid or merely lucky (or unlucky). Keeping score is not possible with qualitative evaluations.

PRESENT FORECAST OF THE NEXT MAUNA LOA ERUPTION

At the time of this writing (March, 1995) no increased seismic activity is occurring beneath Mauna Loa. Deformation measurements indicate that about 50% of the summit deflation that occurred during the 1984 intrusion and eruption has been recovered. Kilauea is still erupting vigorously [*Mattox*, 1993]. None of these conditions provide much of a basis for a forecast based on precursors. Hence, our forecast is largely statistical:

The next eruption of Mauna Loa will occur before the end of 2007 ($P=.95$ using entire historical record; $P=.5(?)$ using the two recurrence times since 1950). That year, 2007, is chosen to give a probability of 50% or better that an eruption will occur before the end of that year. Note that this forecast calls for an eruption between now and the end of 2007, not necessarily during the year 2007. There are four reasons for estimating this long time interval: 1) The significant probability that the clustering of three long recurrence intervals since 1950 is not random; 2) The tendency for long recurrence intervals following large volume eruptions (the 1984 eruption -- 220 million cubic meters -- was a large eruption); 3) The continuing vigorous eruption of Kilauea and the inverse correlation between Kilauea and Mauna Loa eruptions; and 4) That the deformation measurements indicate only about 50% of the summit deflation during the 1984 eruption has been recovered.

When the next eruption occurs, it will begin at the summit of Mauna Loa, but will not necessarily be confined to the summit. Table 1 shows that 35 of the 39 historical eruptions began at the summit ($P=.90$). In addition, Klein [1982] demonstrates that there is a non-random tendency for flank and summit eruptions of Mauna Loa to alternate, and the 1984 eruption was predominantly a flank eruption.

Obviously, the advent of detectable precursors may call for a modification of this forecast.

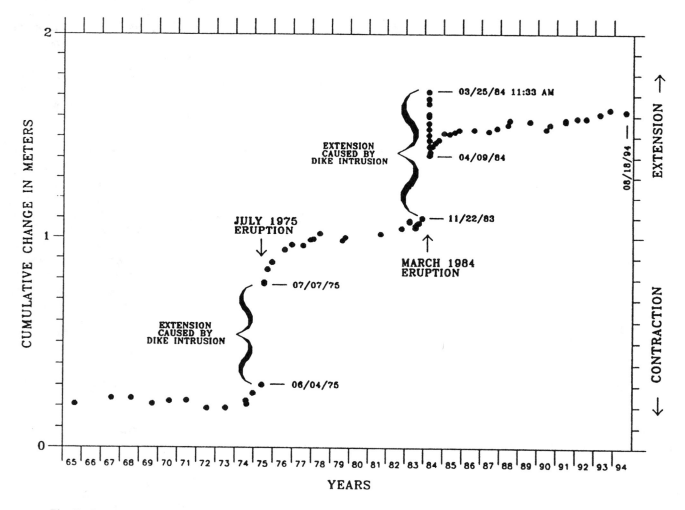

Fig. 7. Cumulative changes in the line length between bench marks HVO 93 and HVO 92 (see Figure 6) across the summit caldera of Mauna Loa. Significant extension (inflation) occurred for one to two years prior to the 1975 eruption. Major extension occurred during the emplacement of a SW-NE dike beneath and within the caldera at the start of the 1975 eruption. Extension (inflation) followed the 1975 eruption for about 2 years. Minor extension (inflation) preceded the 1984 eruption (note that the 11/22/83 measurement was made 4 months prior to the eruption). Major extension marked the dike intrusion that initiated the 1984 eruption, followed almost immediately by contraction (deflation) associated with magma withdrawal from the summit magma chamber during the eruption. Extension (inflation) again followed the 1984 eruption for about 2 years.

DISCUSSION

The fact that most of the recurrence intervals of historical eruptions of Mauna Loa fit a Poisson distribution suggests that the volcano is ready to erupt at any time. That is, it does not require any long reloading time. This conclusion is supported in part by the rapid reinflation of the summit during the first year following the 1975 and 1984 eruptions (Figure 7). However, the increase in shallow seismicity for about two years prior to these eruptions suggests that the reinflation must reach some point where it begins to crack the margins of the shallow magma reservoir. That point did not occur until about 22 years after the 1950 eruption, and about 6 years after the 1975 eruption. Unfortunately, there are no comparable seismic or deformation data for the intervals between eruptions prior to 1950.

The inverse correlation between eruption rates of Mauna Loa and Kilauea is also paradoxical. On one hand, this supports the concept that the hot spot starves Mauna Loa to feed Kilauea; on the other hand, it is well known that there is no direct hydraulic connection between the magma systems of Mauna Loa and Kilauea, and that there are significant differences in the geochemistry of their magmas.

What is clear is that the complex ways in which these volcanoes work will only be understood by the joint efforts of geologists, geochemists, and geophysicists working closely together.

SUMMARY AND RECOMMENDATIONS

The historical and prehistorical eruption patterns of Mauna Loa provide a basis for statistical forecasts of future eruptive behavior. In addition, seismic and deformation data have clearly provided important precursors to the past two eruptions with time scales of years and minutes. Unfortunately, no precursors with time scales of months or days preceding eruptions have yet been found.

Additional analyses of past eruption patterns should be investigated, for example a study similar to Shaw's [1987] dynamic attractor analysis of Kilauea eruptions. A statistician specializing in probability theory might be able to extract more conclusions from the existing record.

Improvements in the existing seismic and deformation data collection and analyses should be sought. For example, some technique to acquire more frequent or continuous deformation data at one or two locations would be a major step forward. Satellite radar interferometry would provide a much broader view of the surface deformation

picture. A tremor alarm that can distinguish between wind noise and volcanic tremor might avoid missing the onset of the next Mauna Loa eruption.

Other eruption precursors should be sought. These might include measuring changes in the temperature, volume, and composition of gases emitted from Mauna Loa fumaroles during repose periods; or measuring changes in volumetric strain with borehole strainmeters [*Linde and Sacks*, this volume]. Experimental studies and new instrumentation by investigators and institutions cooperating with HVO should be encouraged. Collaborative investigations have been fruitful in the past, and we forecast that they will be in the future.

The keys to successful eruption forecasting of Mauna Loa include not only its past behavior, and detection of various precursors, but also a clearer understanding of the physical and chemical processes involved in the evolution and dynamics of this great volcano.

Acknowledgements. The data on which this paper is based have been carefully collected and recorded by past and present staff of HVO. Our thanks to all of them. Thanks also for the gentle but thorough reviews by Robert Tilling, Allen Glazner, and Chi-Yu King.

REFERENCES

Barnard, W.M., *Mauna Loa - A source book*, vol. 2, 452 pp.,Dept. of Geosciences, State Univ. N.Y. at Fredonia, 1991.

Decker, R.W., Deformation measurements on Mauna Loa volcano Hawaii (abstract), *Bull. Volcanol.*, 32-2, 401, 1968.

Decker, R.W., State-of-the-art in volcano forecasting, *Bull. Volcanol.*, 37, 3, 372-393, 1973.

Decker, R.W., Forecasting volcanic eruptions, *Ann. Rev. Earth Planet. Sci.*, 14, 267-291, 1986.

Decker, R.W., R.Y. Koyanagi, J.J. Dvorak, J.P. Lockwood, A.T. Okamura, K.M. Yamashita, and W.R. Tanigawa, Seismicity and surface deformation of Mauna Loa Volcano, Hawaii, *Eos, Trans. AGU*, 64, 37, 545-547, 1983.

Finch, R.H., The seismic prelude to the 1942 eruption of Mauna Loa, *Bull. Seis. Soc. Amer.*, 33, 237-242, 1943.

Ho, C.-H., Nonhomogeneous Poisson Model for Volcanic Eruptions, *Math. Geol.*, 23, 2, 167-173, 1991.

Jaggar, T.A., Jr., *Report of the Hawaiian Volcano Observatory of the Massachusetts Institute of Technology and the Hawaiian Research Association, January-*

March 1912, 74 pp., M.I.T., Boston, 1912.

King, C.-Y., Volume predictability of historical eruptions at Kilauea and Mauna Loa volcanoes, *J. Volcanol. Geotherm. Res.*, 38, 281-285, 1989.

Klein, F.W., Patterns of historical eruptions at Hawaiian volcanoes, *J. Volcanol. Geother. Res.*, 12, 1-35, 1982.

Koyanagi, R.Y., E.T. Endo, J.S. Ebisu, Reawakening of Mauna Loa Volcano, Hawaii: A preliminary evaluation of seismic evidence, *Geophys. Res. Letters*, 2, 9, 405-408, 1975.

Lockwood, J.P., R.Y. Koyanagi, R.I. Tilling, R.T. Holcomb, and D.W. Peterson, Mauna Loa threatening, *Geotimes*, 21, 6, 12-15, 1976.

Lockwood, J.P., and P.W. Lipman, Holocene eruptive history of Mauna Loa volcano, *U.S. Geol. Sur., Prof. Paper 1350*, 1, 509-536, 1987.

Lockwood, J.P., J.J. Dvorak, T.T. English, R.Y. Koyanagi, A.T. Okamura, M.L. Summers, and W.R. Tanigawa, Mauna Loa 1974-1984: A decade of intrusive and extrusive activity, *U.S. Geol. Sur., Prof. Paper 1350*, 1, 537-585, 1987.

Mattox, T.N., Where Lava Meets the Sea: Kilauea Volcano, Hawaii, *Earthquakes and Volcanoes*, 24, 160-177, 1993.

Shaw, H.R., Uniqueness of volcanic systems, *U.S. Geol. Sur., Prof. Paper 1350*, 2, 1357-1394, 1987.

Tilling, R.I., The role of monitoring in forecasting volcanic events, in *Monitoring Active Volcanoes: Strategies, Procedures, and Techniques*, edited by McGuire, W.J., Kilburn, C.R.J., and Murray, J.B.: Ch. 14, pp. 369-401, University College London Press, 1995.

Wickman, F.E., Repose period patterns of volcanoes, *Ark. Mineral. Geol.*, 4, 291-367, 1966.

Wood, H.O., The seismic prelude to the 1914 eruption of Mauna Loa, *Bull. Seis. Soc. Amer.*, 5, 39-51, 1915.

R.W. Decker, Kawaihae Village 30, Kamuela, HI 96743

F.W. Klein, U.S. Geological Survey, 345 Middlefield Road, Menlo Park, CA 94025

A.T. Okamura and P.G. Okubo, Hawaiian Volcano Observatory, Hawaii National Park, HI 96718